T0328366

Skin Tissue Models

Skin Tissue Models

Edited by

Alexandra P. Marques

Rogério P. Pirraco

Mariana T. Cerqueira

Rui L. Reis

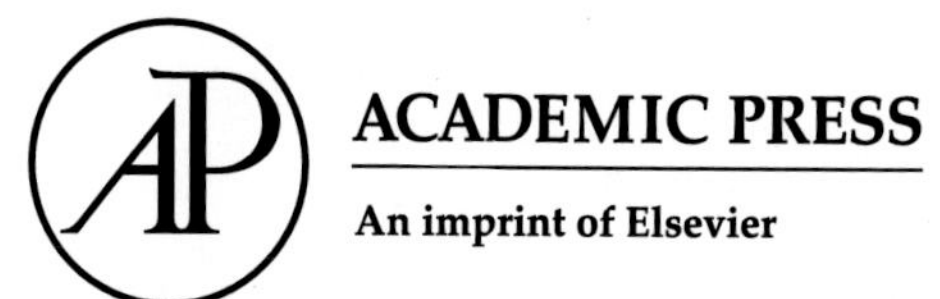

Academic Press is an imprint of Elsevier
125 London Wall, London EC2Y 5AS, United Kingdom
525 B Street, Suite 1800, San Diego, CA 92101-4495, United States
50 Hampshire Street, 5th Floor, Cambridge, MA 02139, United States
The Boulevard, Langford Lane, Kidlington, Oxford OX5 1GB, United Kingdom

Notices

Knowledge and best practice in this field are constantly changing. As new research and experience broaden our understanding, changes in research methods, professional practices, or medical treatment may become necessary.

Practitioners and researchers must always rely on their own experience and knowledge in evaluating and using any information, methods, compounds, or experiments described herein. In using such information or methods they should be mindful of their own safety and the safety of others, including parties for whom they have a professional responsibility.

To the fullest extent of the law, neither the Publisher nor the authors, contributors, or editors, assume any liability for any injury and/or damage to persons or property as a matter of products liability, negligence or otherwise, or from any use or operation of any methods, products, instructions, or ideas contained in the material herein.

Library of Congress Cataloging-in-Publication Data
A catalog record for this book is available from the Library of Congress

British Library Cataloguing-in-Publication Data
A catalogue record for this book is available from the British Library

ISBN: 978-0-12-810545-0

Publisher: Mica Haley
Acquisition Editor: Mica Haley
Editorial Project Manager: Samuel Young
Production Project Manager: Anusha Sambamoorthy
Cover Designer: Mark Rogers

Typeset by SPi Global, India

Contents

Contributors

Teresa Alonso-Rasgado Bioengineering Group, School of Materials, University of Manchester, Manchester, United Kingdom

Saeid Amini-Nik Sunnybrook Research Institute; Laboratory Medicine and Pathobiology; Department of Surgery, Division of Plastic Surgery, University of Toronto; Ross-Tilley Burn Centre, Sunnybrook Health Sciences Centre, Toronto, ON, Canada

Mohammed Ashrafi Plastic & Reconstructive Surgery Research, Centre for Dermatological Research, University of Manchester; University Hospital of South Manchester NHS Foundation Trust, Manchester; Bioengineering Group, School of Materials, University of Manchester, Manchester, United Kingdom

Mohamed Baguneid University Hospital of South Manchester NHS Foundation Trust, Manchester, United Kingdom

Jeremy Baskin Department of Diagnostic Sciences, School of Dental Medicine,Tufts University, Boston, MA, United States

Ardeshir Bayat Plastic & Reconstructive Surgery Research, Centre for Dermatological Research, University of Manchester; University Hospital of South Manchester NHS Foundation Trust; Bioengineering Group, School of Materials, University of Manchester, Manchester, United Kingdom

Audrey Bélanger Centre de recherche en Organogénèse expérimentale de l'Université Laval (LOEX), Sherbrooke; Université Laval, Quebec, QC, Canada

François-Xavier Bernard BIOalternatives, Gençay, France

Manuel Berning German Cancer Research Center, Heidelberg, Germany

Katia Boniface INSERM U1035 BMGIC Immuno-Dermatology, ATIP-AVENIR, University of Bordeaux, Bordeaux, France

Charbel Bouez L'Oréal Research and Innovation, Clark, NJ, United States

Petra Boukamp IUF Leibniz Research Institute for Environmental Medicine, Düsseldorf; German Cancer Research Center, Heidelberg, Germany

Nicholas Boulais Laboratory of Neurosciences of Brest (EA4685), University of Brest, Brest, France

Jennifer Bourland Centre de recherche en organogénèse expérimentale de l'Université Laval/LOEX; Division of Regenerative Medicine, CHU de Québec – Université Laval Research Center; Department of Surgery, Université Laval, Québec, QC, Canada

Virginie Buhé Laboratory of Neurosciences of Brest (EA4685), University of Brest, Brest, France

Muriel Cario-André INSERM U1035 BMGIC, University of Bordeaux, Bordeaux, France

Jean-Luc Carré Laboratory of Neurosciences of Brest (EA4685), University of Brest, Brest, France

Mariana T. Cerqueira 3B's Research Group—Biomaterials, Biodegradables and Biomimetics, Headquarters of the European Institute of Excellence on Tissue Engineering and Regenerative Medicine, University of Minho, Avepark, Barco; ICVS/3B's—PT Government Associate Laboratory, Braga/Guimarães, Portugal

Esteban Chacón-Solano Department of Bioengineering, Carlos III University of Madrid (UC3M); Health Research Institute-Jiménez Díaz Foundation, Madrid, Spain

David Y.S. Chau Research Centre in Topical Drug Delivery and Toxicology, Department of Pharmacy, Pharmacology and Postgraduate Medicine, School of Life and Medical Sciences, University of Hertfordshire, Hatfield, United Kingdom

Vivien Chen Department of Dermatology and Cutaneous Surgery, University of Miami Miller School of Medicine, Miami, FL,United States

Lin Chen Center for Wound Healing and Tissue Regeneration, College of Dentistry, University of Illinois at Chicago, Chicago, IL, United States

Jérémy Chéret Laboratory of Neurosciences of Brest (EA4685), University of Brest, Brest, France

José Cotovio L'Oréal Research and Innovation, Aulnay-sous-Bois, France

Stephen C. Davis Department of Dermatology and Cutaneous Surgery, University of Miami Miller School of Medicine, Miami, FL,United States

Marcela Del Río Department of Bioengineering, Carlos III University of Madrid (UC3M); Health Research Institute-Jiménez Díaz Foundation; Epithelial Biomedicine Division, CIEMAT-CIBERER, Madrid, Spain

Maria L. Dell'Anna San Gallicano Dermatology Institute, Rome, Italy

Marianne Deslauriers Centre de recherche en Organogénèse expérimentale de l'Université Laval (LOEX), Sherbrooke, QC, Canada

Luisa A. DiPietro Center for Wound Healing and Tissue Regeneration, College of Dentistry, University of Illinois at Chicago, Chicago, IL, United States

Stefan Drakulich Department of Dermatology and Cutaneous Surgery, University of Miami Miller School of Medicine, Miami, FL,United States

Alexandra Duque-Fernandez Centre de recherche en Organogénèse expérimentale de l'Université Laval (LOEX), Sherbrooke, QC, Canada

Julie Fradette Centre de recherche en organogénèse expérimentale de l'Université Laval/LOEX; Division of Regenerative Medicine, CHU de Québec – Université Laval Research Center; Department of Surgery, Université Laval, Québec, QC, Canada

Jonathan A. Garlick Department of Diagnostic Sciences, School of Dental Medicine,Tufts University, Boston, MA, United States

Behzad Gerami-Naini Department of Diagnostic Sciences, School of Dental Medicine,Tufts University, Boston, MA, United States

George D. Glinos Department of Dermatology and Cutaneous Surgery, University of Miami Miller School of Medicine, Miami, FL,United States

Olivier Gouin Laboratory of Neurosciences of Brest (EA4685), University of Brest, Brest, France

Sara Guerrero-Aspizua Department of Bioengineering, Carlos III University of Madrid (UC3M); Health Research Institute-Jiménez Díaz Foundation, Madrid, Spain

Adam Hague Plastic & Reconstructive Surgery Research, Centre for Dermatological Research, University of Manchester, Manchester, United Kingdom

Kyle Hewitt School of Stem Cell and Regenerative Medicine, University of Wisconsin Madison, Madison, WI, United States

Victoria Hutter Research Centre in Topical Drug Delivery and Toxicology, Department of Pharmacy, Pharmacology and Postgraduate Medicine, School of Life and Medical Sciences, University of Hertfordshire, Hatfield, United Kingdom

Marc G. Jeschke Institute of Medical Science, University of Toronto Sunnybrook Research Institute; Department of Surgery, Division of Plastic Surgery; Ross-Tilley Burn Centre, Sunnybrook Health Sciences Centre; Department of Immunology, University of Toronto, Toronto, ON, Canada

Olga Kashpur Department of Diagnostic Sciences, School of Dental Medicine,Tufts University, Boston, MA, United States

Stewart B. Kirton Research Centre in Topical Drug Delivery and Toxicology, Department of Pharmacy, Pharmacology and Postgraduate Medicine, School of Life and Medical Sciences, University of Hertfordshire, Hatfield, United Kingdom

Ana Korosec Skin & Endothelium Research Division, Department of Dermatology, Medical University of Vienna, Vienna, Austria

Dagmar Kulms Experimental Dermatology, Department of Dermatology; Center for Regenerative Therapies TU Dresden, TU-Dresden, Dresden, Germany

Manuela E.L. Lago 3B's Research Group—Biomaterials, Biodegradables and Biomimetics, Headquarters of the European Institute of Excellence on Tissue Engineering and Regenerative Medicine, University of Minho, Avepark, Barco; ICVS/3B's—PT Government Associate Laboratory, Braga/Guimarães, Portugal

Fernando Larcher Department of Bioengineering, Carlos III University of Madrid (UC3M); Health Research Institute-Jiménez Díaz Foundation; Epithelial Biomedicine Division, CIEMAT-CIBERER, Madrid, Spain

Christelle Le Gall-Ianotto Laboratory of Neurosciences of Brest (EA4685), University of Brest, Brest, France

Raphaële Le Garrec Laboratory of Neurosciences of Brest (EA4685), University of Brest, Brest, France

Nicolas Lebonvallet Laboratory of Neurosciences of Brest (EA4685), University of Brest, Brest, France

Raphaël Leschiera Laboratory of Neurosciences of Brest (EA4685), University of Brest, Brest, France

Killian L'Hérondelle Laboratory of Neurosciences of Brest (EA4685), University of Brest, Brest, France

Liang Liang Department of Dermatology and Cutaneous Surgery, University of Miami Miller School of Medicine, Miami, FL,United States

Beate M. Lichtenberger Skin & Endothelium Research Division, Department of Dermatology, Medical University of Vienna, Vienna, Austria

Isabelle Lorthois Centre de recherche en Organogénèse expérimentale de l'Université Laval (LOEX), Sherbrooke; Université Laval, Quebec, QC, Canada

Maxim Maheux Centre de recherche en Organogénèse expérimentale de l'Université Laval (LOEX), Sherbrooke; Université Laval, Quebec, QC, Canada

Alexandra P. Marques 3B's Research Group—Biomaterials, Biodegradables and Biomimetics, Headquarters of the European Institute of Excellence on Tissue Engineering and Regenerative Medicine, University of Minho, Avepark, Barco; ICVS/3B's—PT Government Associate Laboratory, Braga; The Discoveries Centre for Regenerative and Precision Medicine, Headquarters at University of Minho, Avepark, Barco, Guimarães, Portugal

Lucía Martínez-Santamaría Department of Bioengineering, Carlos III University of Madrid (UC3M); Health Research Institute-Jiménez Díaz Foundation, Madrid, Spain

Louis-Charles Masson Centre de recherche en Organogénèse expérimentale de l'Université Laval (LOEX), Sherbrooke, QC, Canada

Stephanie H. Mathes ICBT, Tissue Engineering, Zurich University of Applied Sciences, Waedenswil, Switzerland

Friedegund Meier Experimental Dermatology, Department of Dermatology; Center for Regenerative Therapies TU Dresden, TU-Dresden; National Center for Tumor Diseases (NCT), Dresden, Germany

Laurent Misery Laboratory of Neurosciences of Brest (EA4685), University of Brest, Brest, France

Alexandre Morin Centre de recherche en Organogénèse expérimentale de l'Université Laval (LOEX), Sherbrooke, QC, Canada

Nailia Mukhamedshina Department of Diagnostic Sciences, School of Dental Medicine,Tufts University, Boston, MA, United States

Deborah G. Nguyen Organovo Inc., San Diego, CA, United States

Christian N. Parker Novartis Institutes of Biomolecular Research, Basel, Switzerland

Irena Pastar Department of Dermatology and Cutaneous Surgery, University of Miami Miller School of Medicine, Miami, FL,United States

Elizabeth Pavez Loriè IUF Leibniz Research Institute for Environmental Medicine, Düsseldorf, Germany

Christian Pellevoisin EPISKIN Academy, Lyon, France

Ulysse Pereira Laboratory of Neurosciences of Brest (EA4685), University of Brest, Brest, France

Rogério P. Pirraco 3B's Research Group—Biomaterials, Biodegradables and Biomimetics, Headquarters of the European Institute of Excellence on Tissue Engineering and Regenerative Medicine, University of Minho, Avepark, Barco; ICVS/3B's—PT Government Associate Laboratory, Braga/Guimarães, Portugal

Roxane Pouliot Centre de recherche en Organogénèse expérimentale de l'Université Laval (LOEX), Sherbrooke; Université Laval, Quebec, QC, Canada

Rui L. Reis 3B's Research Group—Biomaterials, Biodegradables and Biomimetics, Headquarters of the European Institute of Excellence on Tissue Engineering and Regenerative Medicine, University of Minho, Avepark; ICVS/3B's—PT Government Associate Laboratory, Braga; The Discoveries Centre for Regenerative and Precision Medicine, Headquarters at University of Minho, Avepark, Barco, Guimarães, Portugal

Kelsey N. Retting Organovo Inc., San Diego, CA, United States

Geneviève Rioux Centre de recherche en Organogénèse expérimentale de l'Université Laval (LOEX), Sherbrooke, QC, Canada

Bryan Roy Centre de recherche en Organogénèse expérimentale de l'Université Laval (LOEX), Sherbrooke, QC, Canada

Mehdi Sakka Laboratory of Neurosciences of Brest (EA4685), University of Brest, Brest, France

Andrew P. Sawaya Department of Dermatology and Cutaneous Surgery, University of Miami Miller School of Medicine, Miami, FL,United States

Megan Schrementi Department of Biology, DePaul University, Chicago, IL, United States

Julien Seneschal Department of Dermatology, Hôpital Saint-André, CHU de Bordeaux and INSERM U1035 BMGIC Immuno-Dermatology, ATIP-AVENIR, University of Bordeaux, Bordeaux, France

Yulia Shamis Department of Diagnostic Sciences, School of Dental Medicine,Tufts University, Boston, MA, United States

Mélissa Simard Centre de recherche en Organogénèse expérimentale de l'Université Laval (LOEX), Sherbrooke; Université Laval, Quebec, QC, Canada

Philippe Simard Centre de recherche en Organogénèse expérimentale de l'Université Laval (LOEX), Sherbrooke, QC, Canada

Avi Smith Department of Diagnostic Sciences, School of Dental Medicine,Tufts University, Boston, MA, United States

Hans-Jürgen Stark German Cancer Research Center, Heidelberg, Germany

Olivera Stojadinovic Department of Dermatology and Cutaneous Surgery, University of Miami Miller School of Medicine, Miami, FL,United States

Alain Taieb Department of Dermatology, Hôpital Saint-André, CHU de Bordeaux and INSERM U1035 BMGIC, University of Bordeaux, Bordeaux, France

Matthieu Talagas Laboratory of Neurosciences of Brest (EA4685), University of Brest, Brest, France

Marjana Tomic-Canic Department of Dermatology and Cutaneous Surgery, University of Miami Miller School of Medicine, Miami, FL,United States

Tongyu Cao Wikramanayake Department of Dermatology and Cutaneous Surgery, University of Miami Miller School of Medicine, Miami, FL,United States

Yusef Yousuf Institute of Medical Science, University of Toronto; Sunnybrook Research Institute, Toronto, ON, Canada

Foreword

The skin is a complex organ comprising distinct cell populations with specific functions. The importance of this organ can truly be appreciated when one recognizes the great impact of skin loss in patients affected by large ulcers or extended wounds such as burns. This critical clinical need has been a major incentive for the development of the first therapies based on the expansion of epidermal cells in culture. Thanks to the progress in tissue engineering and regenerative medicine, various models of skin substitutes have now been developed through the years by several teams around the world. The power of cells to create living substitutes for wound healing and burn treatment holds not only great promises but also daunting challenges. In addition to the clinical goal of enabling wound repair, the human nature of these in vitro models provides excellent tools to study skin physiology, physiopathology, and pharmacotoxicology. Moreover, the well-characterized culture conditions of skin cells allow the amplification of human pathological cells in vitro and the production of three-dimensional (3D) models to investigate mechanisms of disease.

This book, "Skin Tissue Models," offers an overview of available cutaneous models and their applications. The eighteen chapters summarize the efforts of leading centers toward continued improvement of skin tissue models. The content of this volume is enhanced by the experience and opinions of experts with diverse backgrounds: burn surgeons, biologists, chemists, and engineers. The first section comprises two chapters on industrial and clinical applications: the cosmetic industry requirements (Chapter 1) and the clinical importance of skin models (Chapter 2). The second section focuses on 3D human models for skin diseases, which could favor the investigation of various pathophysiologies and therapies. Each of the five chapters of this section focuses on a specific disease: melanoma (Chapter 3), genodermatoses that comprise several rare inherited diseases (Chapter 4), psoriasis (Chapter 5), vitiligo (Chapter 6), and squamous cell carcinoma (Chapter 7). The third section encompasses four chapters addressing important parameters to consider for clinical applications: promoting angiogenesis and tissue vascularization (Chapter 8), minimizing scar formation (Chapter 9), promoting wound healing (Chapter 10), and targeting inflammation (Chapter 11). The fourth section regroups four chapters describing more complex models as platform to study skin biology: hair follicle generation (Chapter 12), skin innervation (Chapter 13), white adipose tissue impact on skin healing (Chapter 14), and response of immunocompetent cells (Chapter 15). The last section describes emerging technologies for the production of innovative 3D skin models: bioprinting (Chapter 16), induced pluripotent stem cells (Chapter 17), and high-throughput screening (Chapter 18).

The variety of skin models gathered in this book highlight all the progress made toward the development of human models reflecting the native normal or pathological skin. These substitutes and accompanying enabling technologies offer exciting possibilities for new discoveries in skin biology and improved testing of the safety and efficacy of drugs or cosmetic ingredients. Finally, these substitutes provide hope of new treatments for patients suffering from various types of skin wounds or diseases.

Professor Lucie Germain
Scientific Director, Centre de recherche en organogénèse expérimentale de l'Université Laval (LOEX), Canada

Section A

Therapeutic molecules and cosmetics testing

Cosmetic industry requirements regarding skin models for cosmetic testing

*Christian Pellevoisin**, *Charbel Bouez*[†], *José Cotovio*[‡]
[*]EPISKIN Academy, Lyon, France, [†]L'Oréal Research and Innovation, Clark, NJ, United States, [‡]L'Oréal Research and Innovation, Aulnay-sous-Bois, France

1. Introduction

Scientific models are a simpler representation of reality, which help to explain and predict the behavior of real complex systems. Models of tissues or organs such as the skin, reconstructed by tissue engineering, are of central importance in many scientific contexts and especially in cosmetics. In order to innovate, cosmetics research and development rely on reproductive, quantitative, and predictive methods, to assess the efficiency and safety of ingredients and cosmetic finished products at the earliest stages of development. In order to predict the response in human beings as accurately as possible, in vitro test systems rely more and more on cells of human origin, with a 3D organization that mimics the in vivo situation more closely. Since the 1980s, when the first reconstructed human epidermis (RHE) was produced, several models of the human epidermis and skin have been developed. Based on these models, numerous protocols are widely used in cosmetics research, making skin engineering a powerful and highly versatile technology used at all stages of development of a cosmetic product (Fig. 1).

2. Principles of skin engineering

Tissue and skin engineering have proved to be a very valuable tool in different fields such as research (biomedical research, skin pathology, skin grafting, etc.) and evaluation (safety and efficacy). Regardless of the application, successful tissue engineering projects always rely on the following key components:

- Access to cells (isolation, amplification, storage, and handling)
- Use of the proper culture media to control the balance between cell proliferation and differentiation
- Use of a scaffold that mimics the in vivo extracellular matrix (ECM) to support the growth of the model
- Designing and implementing the proper industrial setup to produce quality at all levels of production

Skin Tissue Models. https://doi.org/10.1016/B978-0-12-810545-0.00001-2

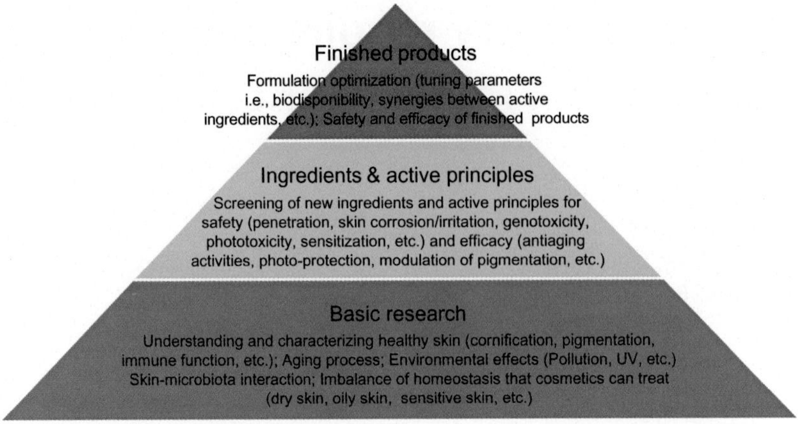

Finished products
Formulation optimization (tuning parameters i.e., biodisponibility, synergies between active ingredients, etc.); Safety and efficacy of finished products

Ingredients & active principles
Screening of new ingredients and active principles for safety (penetration, skin corrosion/irritation, genotoxicity, phototoxicity, sensitization, etc.) and efficacy (antiaging activities, photo-protection, modulation of pigmentation, etc.)

Basic research
Understanding and characterizing healthy skin (cornification, pigmentation, immune function, etc.); Aging process; Environmental effects (Pollution, UV, etc.) Skin-microbiota interaction; Imbalance of homeostasis that cosmetics can treat (dry skin, oily skin, sensitive skin, etc.)

Fig. 1 Application of reconstructed human epidermis and skin at different stages of the cosmetic product development process.

2.1 The reconstructed skin adventure: Key developments

The pioneers of tissue culture are the American biologist Ross Harrison and the French Nobel Prize winner Alexis Carrel [1,2]. In addition to medium and culture conditions, the limiting factor was access to a sufficient number of cells. Thanks to the findings of Theodore Puck, who demonstrated that a layer of lethally irradiated epithelial cells could be used as a source of mitogens allowing a second population of cells to proliferate [3], keratinocytes were first cultured at the laboratory level about 35 years ago. The reconstructed skin adventure began when James Rheinwald and Howard Green [4,5] at Harvard cultivated keratinocytes on top of a feeder layer of fibroblasts (irradiated Swiss 3T3 mouse fibroblasts) and succeeded in obtaining small sheets of epithelium two to three layers thick (cultured epithelial autographs) used to treat burns patients [6]. However, these first epidermal substitutes composed of a few layers of human keratinocytes did not present a functional horny layer and were far from resembling the in vivo tissue. The first reconstructed epidermis model topped with a horny layer was perfected by Michel Pruniéras in 1979 [7]. Mimicking the in vivo situation, the reconstructed epidermis consisted of seeded human keratinocytes (extracted from human skin biopsies) on the top of a cell-free dermal substitute for about 2 weeks at the air-liquid interface [8]. Simple dermal substitutes containing fibroblasts are also of importance for clinical use in the case of severe burns and wounds [9]. Today, skin equivalents can be divided into three categories: (i) epidermal substitutes, (ii) dermal substitutes, and (iii) full-thickness models containing both epidermal and dermal components.

2.2 Main steps in the reconstruction of the epidermis

Over the last few years, in vitro skin models have been perfected by adding different cell types, allowing various skin functions to be replicated. Epidermal reconstruction usually follows similar key steps: Primary keratinocytes are isolated from the epidermis of skin biopsies or from the external sheath of a hair follicle and further cultured as monolayers [8] in order to amplify the number of cells and to set up easy-to-use cell banks. In a second step, keratinocytes are "seeded" and allowed to proliferate onto the surface of a support, which can be either a deepidermized dermis (DED, a dermis whose cells have been eliminated by successive freezing and thawing) or a biomaterial often based on collagen. This second step in the culture process consists of obtaining a monolayer of keratinocytes by placing the cells and the dermal substitute immersed in the culture medium. During this step, other types of epidermal cells (e.g., melanocytes and Langerhans cell (LC) precursors) may be added to keratinocytes depending on the model to be produced. In a third step, keratinocytes are brought into contact with air (the air-exposed period) and then differentiated and stratified spontaneously to form all the layers of a fully differentiated epidermis, including the horny layer, which performs a barrier function for the reconstructed epidermis.

Furthermore, a dermal equivalent can be produced by mixing native collagen type I and living fibroblasts according to the conditions that lead to the formation of a dermal tissue. Thus, an important additional step is taken when this living dermal compartment containing fibroblasts is used during the epidermal reconstruction process, ending with a full-thickness skin equivalent incorporating the skin's two major cell types (keratinocytes and fibroblasts) [10].

Important developments in culture media composition and skin cell biology have allowed other important cells such as melanocytes and LC precursors to be incorporated, together with keratinocytes in the epidermal compartment generating more complex models and opening the way to new fields of research and applications [11,12]. Several specific substrates have been used to create the dermal structure in vitro, including DED [8,13,14] collagen lattices [15], collagen-glycosaminoglycan matrices [16,17], chitosan cross-linked collagen-glycosaminoglycan matrices [18], and biodegradable matrices [19] (Fig. 2).

In the future, new technologies could revolutionize the way we reconstruct skin equivalents. Bioprinting is a new area of research and engineering that involves printing devices and computer-aided design (CAD) technology to print 3D tissues or organs [20]. In this additive manufacturing approach, cells embedded in a gel are deposited onto a surface, layer by layer until the desired tissue has been reproduced. Bioprinting has the advantage of enabling a predefined arrangement of living cells, biomaterials, and growth factors to fabricate customizable constructs with a high degree of flexibility and repeatability. Applied to skin reconstruction, this technology could make it possible to reproduce the complex architecture and multicellular organization of the living skin [21]. The primary application would be medical, for wound management, but reconstructed skins could also be used in different applications such as predictive evaluation tools. Recently, an American team conducted a proof-of-concept study showing that this technology could be used to build a full-thickness skin with keratinocytes and fibroblasts [22].

2.3 Industrial standardized production of 3D skin models

The regular use of 3D skin models in anticipated and well-planned studies requires the availability of standardized reproducible tissues and easy access to them. Some biotechnology companies are devoted to developing and producing robust models aiming at

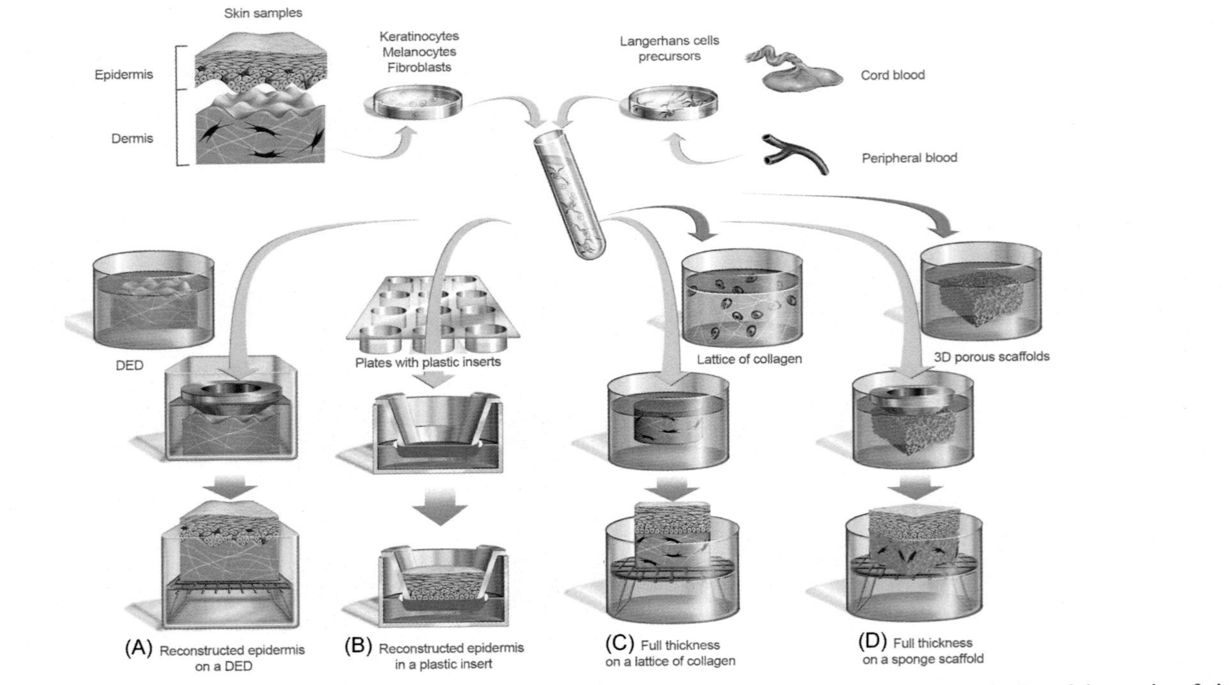

Fig. 2 Different types of reconstructed models depending on the support and the population of cells incorporated. Standard models consist of either an epidermis with keratinocytes or an epidermis and a living dermis with fibroblasts. However, it is possible to introduce other cells types such as melanocytes or Langerhans cells. Langerhans cells that reside in the skin, however, have a low survival and amplifying capacity in a culture, which is why it is more efficient to use blood circulating precursors from adult or cord blood. (A) Reconstructed epidermis on an acellular deepidermized dermis (DED). Keratinocytes are seeded directly onto a human dermis from which the resident cells have been eliminated. (B) Industrial models with keratinocytes seeded in plastic inserts on an artificial support (collagen sheets, polycarbonate, etc.). The inserts can be easily packed and shipped in 6, 12, 24, or even 96 well plates. (C) Full-thickness model or skin equivalent. A gel of collagen is reticulated by inserting living fibroblasts to form a solid lattice. Then, the keratinocytes are seeded onto this lattice to form a living epidermis on a living dermis. (D) The lattice of collagen can be replaced by an artificial porous scaffold made of different biomaterials such as collagen, chitosan, or synthetic polymer.

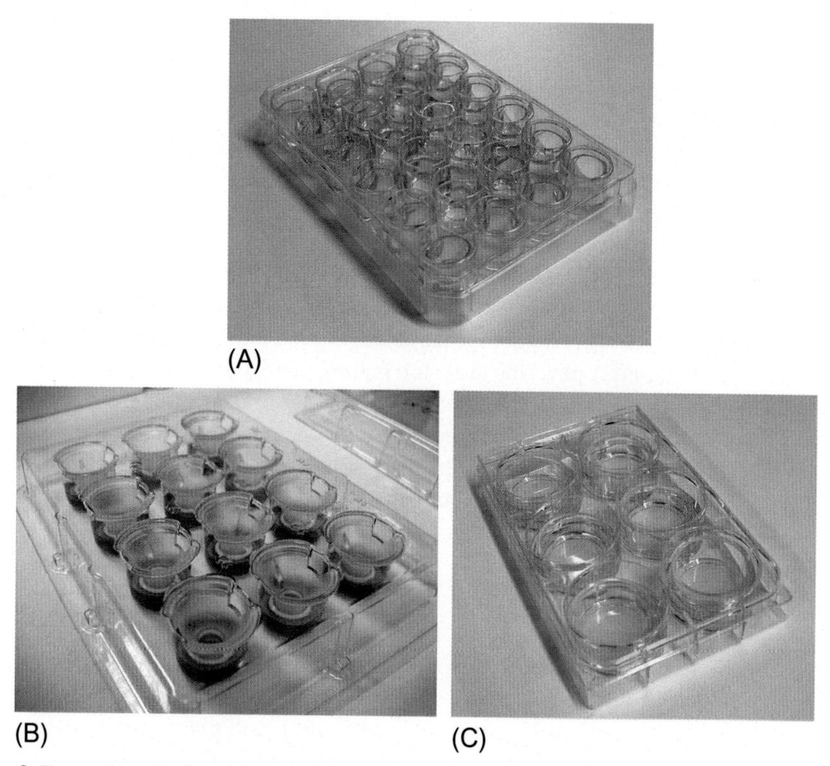

Fig. 3 Examples of industrial reconstructed 3D models in their container plates and specific inserts. (A) SkinEthic™ RHE small ($0.5\,cm^2$)×24 tissues, (B) EpiSkin™ large ($1\,cm^2$)×12 tissues, and (C) SkinEthic™ RHE large ($4\,cm^2$)×6 tissues.

providing tools for efficacy and safety screening of ingredients and/or finished cosmetic, pharmaceutical, and industrial products or medical devices. A specific cell culture process allows human tissues with well-characterized histology, functionality, and ultrastructure features to be produced on a large scale. The latter are mainly based on a RHE, using a biological or nonbiological support, possibly containing human melanocytes with different pigment levels (phototypes II–VI). Standardized large-scale manufacturing processes are certified by the ISO 9001 standard. Batch reproducibility and quality control are generally evaluated by histological scoring and response to a reference test chemical. Such models are well designed for industrial application. They are produced in different formats and 6, 12, 24, or 96 well plate (Fig. 3) sizes. They are robust and easy to handle and can be used for high-throughput screening (HTS) applications.

3. Skin models for research

In vitro skin models improve the understanding of the skin in several ways. In itself, the process of "building" the model generates knowledge by: (1) understanding the mechanism to being modeled; (2) identifying the cellular components; and

(3) finding the proper culture conditions to promote their proliferation and differentiation, etc. It is often a cross-disciplinary process that promotes exchanges and transfers of skills between different experts. In the same way, characterization of the model compared with the human skin increases the scientific knowledge that we have about this organ. Knowledge is still generated when the model does not respond as expected. Successfully incorporating melanocytes, LCs, or stem cells into the epidermis requires a high level of understanding of these cell types and their interactions. Verifying their presence and functionality leads to greater knowledge of the regulating mechanisms that govern their interactions and their survival and functionalities. Once constructed and characterized, the model is an exceptional tool for the researcher. The availability, reproducibility, and access to numerous parameters (proteomic and genomic analyses, assays, morphology, etc.) pave the way for innumerable fields of experimentation. The contributions of reconstructed skin models are significant in many areas of research, for example, UV sensitivity, specific ethnic characteristics, skin sensitization and allergy, chronological aging, and skin microbiome. The examples in the following paragraphs illustrate some of the contributions of reconstructed skin engineering to the general understanding of skin physiology.

3.1 Studying skin aging

Chronological aging is a complex biological phenomenon that affects all organs to a different extent. In the skin, it is accompanied by specific clinical signs, the most prominent of which are wrinkles, slackening of tissues, and pigmentation disorders. Studies have shown that aging induces significant changes in all skin layers: decline in epidermal turnover, thinning of the epidermal layers, and flattening and delamination of the dermal-epidermal junction (DEJ). The dermis is also affected by a decrease in ECM glycosaminoglycans (GAGs), in the number of fibroblasts and their activity, and a degradation of the ECM [23]. Actinic aging induced by exposure to ultraviolet (UV) increases and accelerates these changes and is accompanied by specific signs, such as solar elastosis (Fig. 4).

Reconstructed skins are tools of choice for understanding the biological mechanisms that accompany and induce the appearance of the clinical signs related to aging. Two regions are distinguished in the dermis: the papillary dermis, a region lying beneath the epidermis and the reticular dermis that lies below the papillary dermis. The influence of the origin, papillary or reticular, and the age of the fibroblasts (donors from 19 to 74 years of age) on the morphogenesis of the epidermis were evaluated in reconstructed skin models [24]. It was shown that dermal equivalents containing papillary fibroblasts were more potent in promoting morphogenesis of a correctly stratified and differentiated epidermis than those containing fibroblasts from the reticular dermis (Fig. 5).

These differences decline as donors increase in age. Matrix metalloproteinase (MMP)-1 and vascular endothelial growth factor (VEGF) secretion levels were increased in the presence of papillary fibroblasts, whereas MMP-3 and keratinocyte growth factor (KGF) levels were higher in the presence of reticular fibroblasts [25]. These models revealed that papillary and reticular fibroblasts exert distinct functions

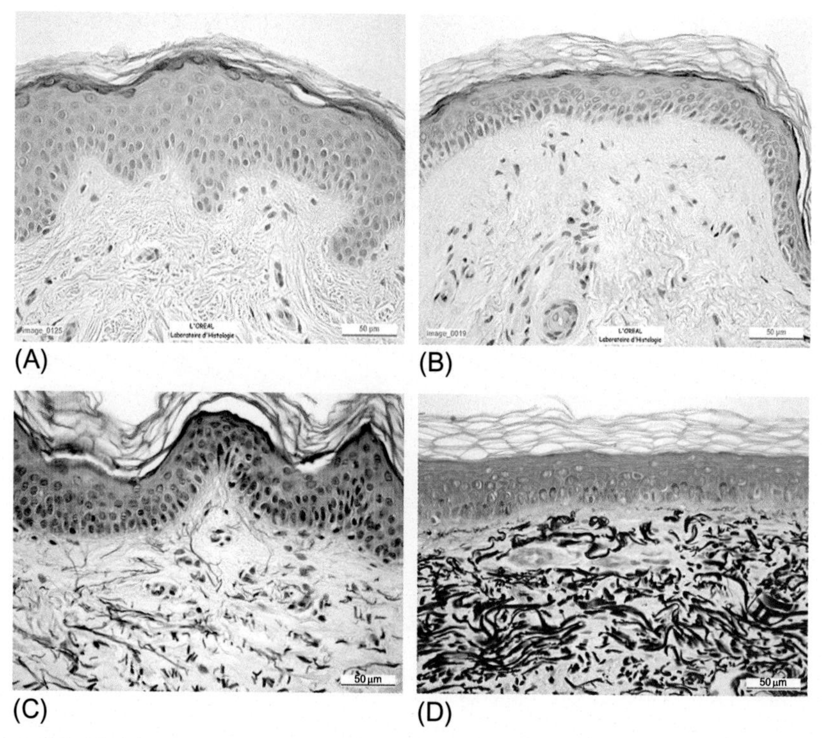

Fig. 4 Skin biopsies from a young (A) and a photoaged donor (B). Immunohistology shows epidermal alteration. Accumulation of elastin in the dermis of a photoaged skin (D) as compared with the skin of a younger donor (C).
From L'Oréal R&I.

and activities in the skin. These functional differences may have strong implications in wound healing and skin-aging processes.

Skin aging is also characterized in the dermal compartment by cross-linking of fibrillar proteins that occurs by glycation. Glycation (i.e., sugars reacting with the free amine moiety of lysine and arginine in proteins) is a nonenzymatic slow process with the formation of covalent cross-linking between collagen fibers, making the collagen network more rigid. A dermal equivalent made of collagen fibers preglycated by ribose was developed, providing new insights into the role played by advanced glycation end products (AGEs) in the skin. This work highlights the influence of collagen glycation not only at the dermal level but also on the quality of the epidermis through fibroblast regulation [26–29]. Treatment of human dermal fibroblasts by mitomycin C (MMC) induced an accelerated senescence through DNA damage and generation of reactive oxygen species (ROS), two main aspects of in vivo aging [30,31]. This property has been used to developed aged 3D skin models by coculturing skin fibroblasts exposed to MMC and keratinocytes on a collagen-glycosaminoglycan-chitosan scaffold. In this in vitro aged model, filaggrin expression is reduced in the epidermis, and the amount of elastin and collagen is lowered in the dermis [32].

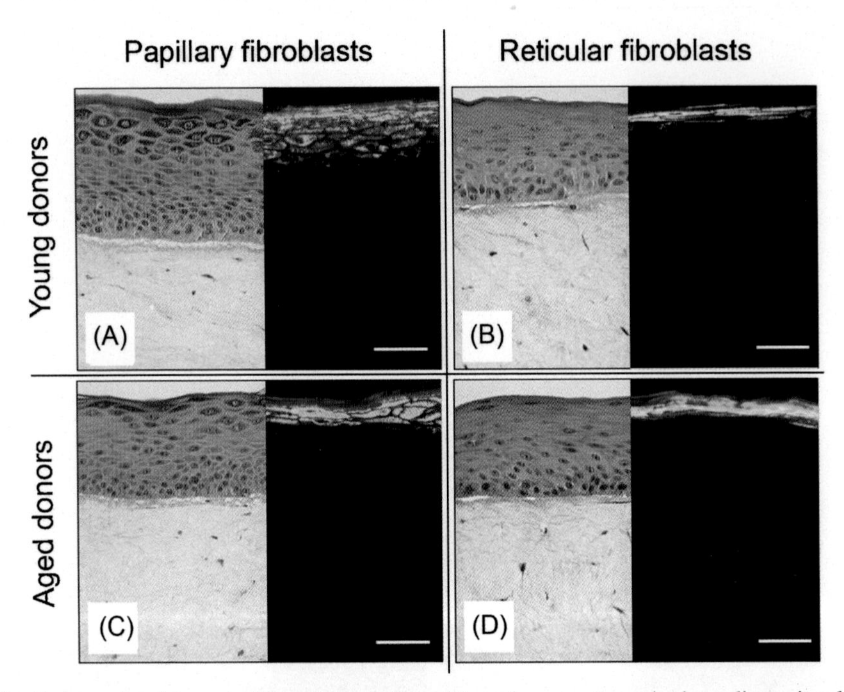

Fig. 5 Age-related impact of Fp and Fr on the epidermal compartment in three-dimensional reconstructed skin. Keratinocytes from the same batch were seeded onto dermal equivalents containing papillary fibroblasts (Fp) or reticular fibroblasts (Fr) from young or old donors. Typical histological sections (HES *blue coloration*) and filaggrin immunolabeling (*green labeling*) of reconstructed skin samples made with collagen constructed lattices containing (A) young papillary fibroblasts, (B) young reticular fibroblasts, (C) old papillary fibroblasts, and (D) old reticular fibroblasts. The histologies and the filaggrin immunolabeling show that only reconstructed skin with papillary fibroblasts from young donors (A) induces morphogenesis of a well-stratified and well-differentiated epidermis.
Data from Mine S, Fortunel NO, Pageon H, Asselineau D. Aging alters functionally human dermal papillary fibroblasts: a new view of skin morphogenesis and aging. PLoS One 2008;3(12):e4066.

Apart from the eyes, the human skin is the only organ directly exposed to solar (sun) UV rays, a well-recognized environmental damaging factor responsible for photoaging [33,34]. Acute overexposure of the human skin to UV rays causes sunburn, altered pigmentation, inflammation, immunosuppression, and damage to dermal connective tissue. Chronic UV exposure can induce a disruption in the normal architecture of the skin and ultimately cause premature photoaging. In worse cases, it may also cause skin cancer. Historically, UVB rays were thought to be the only UV rays detrimental to the skin because of their involvement in sunburn (apoptotic cells) and direct DNA damage in epidermal cells. Biologists have observed that UVA rays, which penetrate deeper into the skin, are also harmful as they directly alter the dermal compartment. Through an oxidative stress process (ROS generation), they damage the fibroblasts and promote the synthesis of proteins responsible for collagen breakdown. The effects of UV rays (UVB and UVA) have been investigated using models of reconstructed full-thickness skin. It was shown that keratinocytes indirectly participate in the photoaging process [35]. Under the influence of UVB, they release soluble factors, which cause dermal fibroblasts to synthesize MMP proteolytic enzymes that degrade collagen and other ECM proteins of the dermal connective tissue. The effects of cumulative exposure to each type of UV radiation and their possible synergy can be studied using full-thickness skin models and combining fine adjustments of simulated exposure and specific treatments [36,37]. Likewise, significant advances in the knowledge of chronological aging/photoaging are anticipated through the use of "aging"/UV-exposed skin models, thus providing new strategies in antiaging research.

Pigmentary disorders are also associated with photoaging. To test the hypothesis of a dermal fibroblasts modulation of skin pigmentation, Bernerd et al. developed and characterized a functional pigmented reconstructed skin model composed of a melanocyte-containing epidermis grown on a dermal equivalent with living fibroblasts [38]. Using the same pool of keratinocytes and melanocytes, different models have been reconstructed with and without fibroblasts of different origins: fetal, adult, and photoaged [39]. Compared with the model with fibroblasts, the model without fibroblasts shows a 10-fold higher melanin content assessed by quantifying Fontana-Masson staining (Fig. 6A), suggesting that fibroblasts down control melanocyte activity. To analyze the effect of the origin of fibroblasts on pigmentation, fibroblasts of fetal or adult origin have been used for reconstruction. An intense pigmentation appeared, macroscopically and on histological sections, in the epidermis reconstructed on the dermal equivalent with fetal fibroblasts. The hyperpigmentation of the samples prepared with fetal fibroblasts was quantified and found to be associated with a 6.6-fold increase in melanin content (Fig. 6B). To investigate the influence of chronically photoexposed fibroblasts on pigmentation, the epidermis reconstructed on dermal equivalents containing fibroblasts from a photoaged skin (three different strains, donors aged over 70 years of age) was compared with reconstructed skin models containing fibroblasts from a young, unexposed skin (16–21 years old, three strains). The data demonstrate that the presence within the dermal equivalent of natural photoaged fibroblasts, as compared with young unexposed fibroblasts, stimulates epidermal pigmentation as revealed by an increase in melanin content (Fig. 6C). These results clearly suggest that fibroblasts from a chronically sun-exposed skin may contribute to hyperpigmentation observed during photoaging.

The impressive variation in skin pigmentation at the macroscopic level, supported by quantitative measurement, demonstrates that the presence, origin, and, more importantly, the history and acquired characteristics of dermal fibroblasts are indeed modulators of the level of pigmentation. Although the precise action mechanisms have not been identified, the tools employed here allow further investigations on ECM proteins and soluble factors to being considered to unravel the underlying molecular mechanisms. This work underlines the importance and power of reconstructed human skin to model different biological systems for deciphering complex mechanisms and challenging different hypotheses.

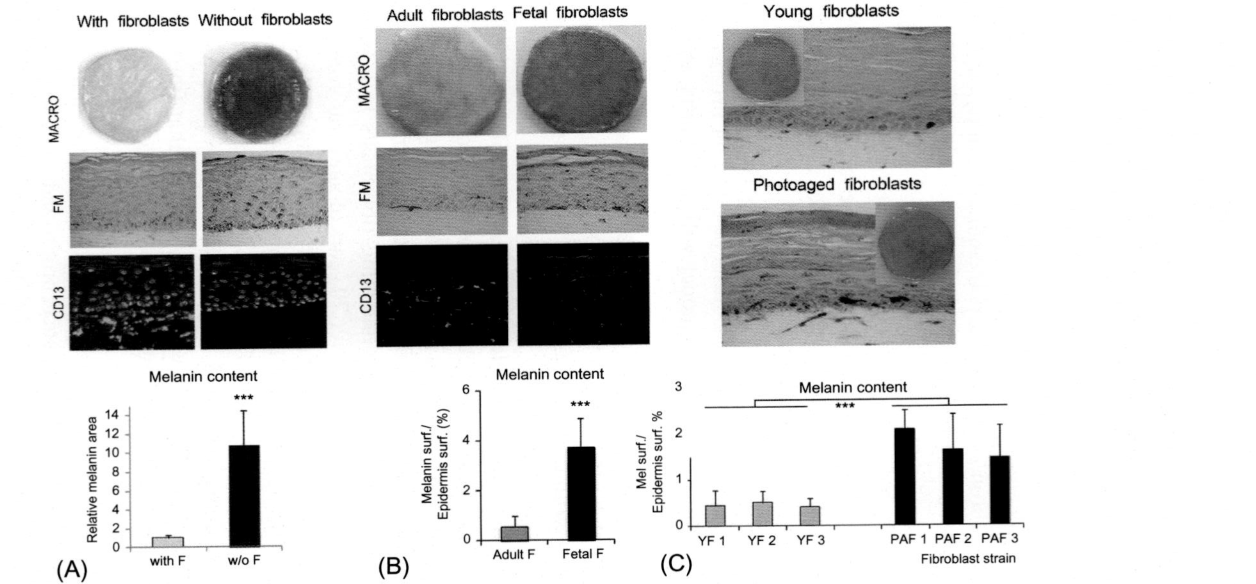

Fig. 6 Different reconstructed human skin with the same keratinocyte and melanocyte strains and fibroblasts from different origins. (A) Macroscopic pictures of reconstructed skin illustrate the drastic difference in pigmentation between the model with fibroblasts (from an adult 21-year-old donor) and the one without. This is confirmed by Fontana-Masson (FM) staining of melanin granules used and melanin content quantification. The CD13 staining (*green*) of cross sections confirmed the presence or absence of fibroblasts in the dermal equivalent. (B) Reconstructed skin models were reconstructed with either fetal fibroblasts (GM10) or adult (21-year-old donor) fibroblasts. A drastic increase in pigmentation of the model and activation of melanocytes in the presence of fetal fibroblasts compared with adult fibroblasts were noted macroscopically (macro) and on histological sections stained with Fontana-Masson (FM). (C) Pigmentation in the skin reconstructed with either natural photoaged fibroblasts or young fibroblasts was analyzed macroscopically and on histological sections stained with Fontana-Masson and melanin quantification. A significant increase in pigmentation was observed as shown by an increase in the concentration of melanin granules measured by image analysis on histological sections and the darkening of the skin samples. Values are expressed as a mean +/– SD calculated for three samples in three independent experiments.
Adapted from Duval C, Cohen C, Chagnoleau C, Flouret V, Bourreau E, Bernerd F. Key regulatory role of dermal fibroblasts in pigmentation as demonstrated using a reconstructed skin model: impact of photo-aging. PLoS One 2014;9(12):e114182. doi:10.1371/journal.pone.0114182.

3.2 Studying skin-microbiota crosstalks

Skin is not only a complex organ that performs several of the body's vital functions, but also with a surface area of nearly $2\,m^2$ providing humidity, nutrients (sweat, sebum, and stratum corneum), and controlled temperature, it is an ecosystem hosting microbial communities of up to 10^6 microorganisms per square centimeter. The diversity and complexity of this microbiota can begin to be unraveled using direct sequencing approaches for all microbial DNA (shotgun metagenomics), giving a detailed map of the biogeography of bacterial and fungal populations living in different areas of this skin ecosystem [40]. Characterizing the skin microbiome opens the way to a new understanding of the interactions between the human skin and the foreign organisms inhabiting its ecosystem. The classical pathogen versus commensal strains classification is challenged by a more complex system, where the cross talks between different actors modulate the homeostasis of the system. Currently, the skin microbiome is a major research focus in dermatology laboratories across the globe, and its role is well recognized as a major player in skin homeostasis, opening new opportunities in predictive evaluation of cosmetic products.

Disturbance of the equilibrium between microbiota and its host is associated with several disorders or diseases [41] such as acne, atopic dermatitis (AD) [42,43], dandruff, and psoriasis. The cross talk between the skin microbiota and the host's immune system is an important phenomenon in maintaining a healthy skin: it is clear that cutaneous innate and adaptive immune responses modulate the skin microbiota, while the microbiota also functions in educating the immune system [44]. RHE epidermis offers a unique experimental support for modeling skin-microbiota interactions, to help decipher the system's elementary mechanisms. Although simpler than the situation in vivo, reconstructed skin equivalents offer key advantages for building experimental systems that allow control of both the skin tissue and added microorganisms (bacteria and fungi) throughout the process of skin colonization and interactions.

The skin barrier is an important element of the innate immune function. Using RHE seeded with *Propionibacterium acnes*, *Staphylococcus aureus* (*S. aureus*), and *S. epidermis*, it has been shown that protection against bacteria disappears as soon as the barrier is impaired, leading to a strong inflammatory response from keratinocytes [45]. Some strains of *S. aureus* have developed resistance to antibiotics and are implicated in several nosocomial pathologies. Thanks to an experimental system using an RHE model, several markers of virulence of these strains have been identified, such as gene coding for surface adhesion molecules, which are important for host-pathogen interaction, and gene coding for bacterial surface proteins, which participate in biofilm formation [46]. *Acinetobacter* species are other opportunistic pathogens associated with nosocomial infections. Two different strains, *Acinetobacter baumannii* and *A. junii*, are able to colonize an RHE model, but only *A. baumannii* forms large biofilms on the stratum corneum [47]. The ability of *A. baumannii* to form a biofilm on the human skin could play an important role in its persistence on the skin. With the emergence of antibiotic resistance, such in vitro models may be advantageous for identifying and evaluating new targets for disinfection and antimicrobial strategies.

AD is a type of chronic inflammation of the skin characterized by erythema, pruritus, and severe impairment of the barrier function [48]. The lesioned skin in patients

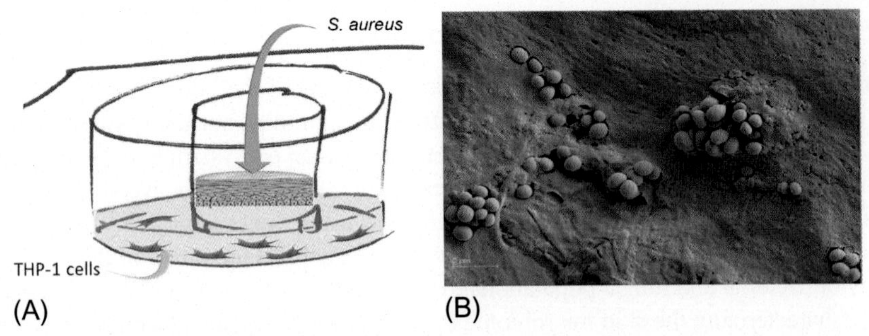

Fig. 7 (A) Experimental system comprising a reconstructed human epidermis (SkinEthic RHE™), THP-1 cells in the culture medium, and *S. aureus* seeded topically on the surface of the RHE model. (B) SEM image of *S. aureus* on the surface of the RHE model.
Data from De Servi B, Semenzato A, Baratto G, Meloni M. Modeling atopic dermatitis in reconstructed human epidermis. Paris: SFC-Journées Jean Paul Marty; 2015.

exhibits infiltration by lymphocytes T helper cells type 2 (Th2) and a predisposition to colonization by *S. aureus* [49,50]. Recently, a cocultured model of RHE (Vitroscreen, Italy) has been able to reproduce a specific keratinocyte response observed in the lesioned skin of AD patients [51]. The experimental system comprises an RHE topically colonized by *S. aureus* and, underneath in the culture medium, a population of THP-1 cells (Fig. 7). THP-1 is a human monocytic cell line derived from an acute monocytic leukemia patient. The response of the reconstructed epidermis evaluated by qRT-PCR reveals features close to the in vivo situation: overexpression of defensin beta 2, integrin alpha 6, integrin beta 1, and keratin 19. Other genes are downregulated, such as claudin 1, filaggrin, kallikrein 5, and TRLP2.

These results show that this experimental system, which mimics the features of AD lesion, takes into account the keratinocyte innate and inflammatory response and the adaptive immunomediated response (coculture with monocytes). After thorough characterization, this model was further used to assess the efficiency of two cosmetic formulations to restore normal homeostasis (barrier function) and to stimulate natural defenses (antimicrobial response and downregulation of adhesion molecules) of the skin.

The development of such in vitro models of skin-microbiota cross talk and coregulation will undoubtedly increase in the coming years and will allow innovative indirect strategies to be developed for addressing classical cosmetic end points (hygiene, antiaging, dandruff, sensitive skin, etc.).

4. Skin models and methods for hazard and risk assessment of cosmetics

The ability to reproduce several functions of the human skin in vitro broadens the scope for industrial applications. Thanks to in vitro skin models, it is now possible to predict, early in their development process, the effects, positive or negative, of new compounds, drugs, chemicals, cosmetics, medical devices, etc. The development of new cosmetics requires a precise assessment of their safety and activity. A number of in vitro methods have been developed and proposed for this purpose. Methods using reconstructed epidermis or full-thickness skin models have been introduced and are today commonly used to assess both the efficacy and safety of cosmetic ingredients and formulations. These 3D skin models offer many advantages:

- Their architecture is close to that of the native skin.
- The presence of a stratum corneum forms a barrier to exogenous substances, although it is less efficient than that of the skin in vivo.
- They can be used to measure a large variety of parameters related to the efficacy or safety of ingredients.
- The air-exposed surface allows ingredients/products with various physicochemical properties to be applied topically, thus mimicking in vivo use.
- Compared with traditional in vitro 2D cell cultures, 3D models reproduce the complexity and richness of in vivo cell-cell and cell-ECM interactions.

Reproducing this microenvironment is important, since cellular interactions regulate several mechanisms, such as viability, proliferation, differentiation, morphology, protein synthesis, and genome expression [52,53]. It has been shown that the gene expression profiles of keratinocytes from in vitro 3D reconstructed skin is much closer to the in vivo human skin than keratinocytes cultivated in 2D [54]. In 2D models, cells are immersed in an aqueous culture medium, which limits the testing to hydrosoluble substances. The use of in vitro skin models overcomes this limitation and makes it possible to apply ingredients or compounds, whatever their physicochemical forms (liquid, powder, or gel) directly onto the stratum corneum. The way the tested products are applied and the resulting exposure of the cells (barrier function and gradient of concentration) are closer to the in vivo conditions, as compared with 2D protocols. Over the last 20 years, substantial work and investments have been devoted to the development of both in vitro and ex vivo methods. As a result, several validation studies based on test methods using organotypic models, reconstructed human tissues, simple cell assays, physicochemical assays, or a combination of tests have been conducted. However, despite huge progress in the models used, numerous questions have been raised regarding action mechanisms, classification, the scope of applications and, finally, the need to identify testing strategies able to strengthen final predictions. Today, in vitro reconstructed skin models are widely used in both safety and efficacy prescreening tests [55,56]. Increased interest in them was followed in the last decade by several validations and acceptance for regulatory purposes.

4.1 New paradigm in toxicology

Any novel substance used in the composition of consumer products has to undergo a preliminary comprehensive toxicological evaluation process. The use of animal testing remains strongly questioned and rightly criticized for both ethical and scientific reasons, that is, methods with more or less subjective scoring, low interlaboratory reproducibility, and known differences in sensitivity with humans [57–59]. In Europe, the seventh amendment (2003/15/EC) [60] to the Cosmetics Directive (76/768/EEC) introduced the regulatory framework for the phasing out of animal testing for cosmetics purposes. Since 2009, this testing and marketing ban has covered "acute" toxicological end points, including those linked to adverse effects on the skin and eyes. In March 2013, it was extended to other tests requiring repeated applications, for example, repeat-dose toxicity, sensitization, reproductive toxicity, carcinogenicity, and toxicokinetics. This total ban on animal testing concerns not only products developed in Europe but also any cosmetic products intended to be marketed in Europe. In line with the movement initiated in Europe, we can see a strong trend worldwide toward a progressive shift to nonanimal methods for assessing the safety of ingredients and cosmetic products. In 2013, Israel banned imports of cosmetics tested on animals. This regulation supplemented a 2007 ban on animal testing for domestic cosmetics. In June 2013, India became the first country in South Asia to decide on a ban on animal testing for domestic manufacturing of cosmetics. In May 2014, the ban was extended to all imported cosmetics. For the ban to enter fully into force, the Bureau of Indian Standards (BIS) intends to set down standards to allow the Drugs Controller General of India (DCGI) to make relevant amendments to acts and laws. China's position on animal testing is also changing rapidly. A new policy passed in December 2013 authorizes certain categories of cosmetics (such as soap, shampoo, and some skin products) developed locally to be marketed without the need for animal testing from June 2014. In January 2014, the state of São Paulo in Brazil released a bill prohibiting cosmetics testing on animals. In October 2016, Taiwan approved a regulation that comes into effect in 2019 and will amend Taiwan's Control for Cosmetic Hygiene Act and ban cosmetic animal testing for both finished products and cosmetic ingredients. Cosmetics developed outside the country will not fall under the new regulation.

In light of the changes in the regulations, toxicology is undergoing a profound revolution, with a shift from toxicology based on the observation of effects in animals to mechanistic approaches based on in vitro and *in silico* tests to predict potential adverse effects in humans. In 2007, the US National Research Council proposed in its report "Toxicity testing in the 21st century" a road map for building a mechanistic understanding of biological action and exploiting an array of new technologies (i.e., high-throughput assays, –omics, bioinformatics, systems biology, and computational toxicology). The success of these mechanistic approaches relies on progress in fundamental knowledge of AOP to identify the cascade of events, from the molecular level up to the organism, which may ultimately produce an adverse effect. Advances in basic knowledge pave the way to the development of in silico, in chemico, and in vitro assays that can provide information about one or more key events of an AOP for a specific toxicological end point. This paradigm shift is accompanied by an astonishing scientific and technological revolution, in which the academic community and the cosmetics industry have been engaged for more than 30 years. The cosmetics industry is at the forefront of this new paradigm in toxicology, but it will undoubtedly and increasingly concern other industries, such as chemicals, medical devices, or drug development.

4.2 Validated versus valid methods

The diversity of existing in vitro models including reconstructed epidermis and skin tissues, together with their variety, represents a potential strong driver for the application of adapted and integrated in vitro methods. To date, several in vitro methods have been developed based on skin models to assess different toxicological end points (skin penetration [61,62], skin corrosion/irritation [63,64], phototoxicity [65–70], genotoxicity [71], and skin sensitization [72–74]). However, an in vitro alternative method must be validated before being recognized by the regulatory bodies concerned. Validation is the process whereby the reliability and relevance of a particular approach, method, procedure, or assessment is established for a defined purpose [75]. In Europe, the European Center for Validation of Alternative Methods (EURL-ECVAM) is devoted to the validation of alternative methods. EURL-ECVAM publishes a yearly report providing updates on the development, validation, and regulatory acceptance of alternative methods [76] and maintains the Tracking System on Alternative Methods (TSAR), which provides a view on the status of alternative methods from submitted scientific protocols up to their validation and acceptance within a regulatory context. Since the creation of ECVAM in 1991, four regional organizations have been created on the same principle in different countries (Table 1). Today, two alternative methods based on in vitro skin models have been validated as full replacement methods to animal testing. These methods have been integrated into the test guidelines of the Organization for Economic Cooperation and Development (OECD): TG 431 for in vitro skin corrosion and TG 439 for in vitro skin irritation potential of chemicals. Moreover, two other methods based on reconstructed human epidermis and full-thickness models have been submitted for validation in the field of sensitization and genotoxicity.

Table 1 List of validation centers of alternative methods worldwide

Country	Name	Acronym and website	Created
Europe	European Union Reference Laboratory for alternatives to animal testing	EURL-ECVAM (formerly ECVAM) http://ihcp.jrc.ec.europa.eu/our_labs/eurl-ecvam	1991
The United States	Interagency Coordinating Committee on the Validation of Alternative Methods	ICCVAM http://ntp.niehs.nih.gov/go/iccvam	1994
Japan	Japanese Center for the Validation of Alternative Methods	JacVAM http://www.jacvam.jp/	2005
Korea	Korean Center for the Validation of Alternative Methods	KocVAM http://www.nifds.go.kr/en/inter/kocvam.jsp	2009
Brazil	Brazilian Centre for the Validation of Alternative Methods	BraCVAM	2011

4.3 Skin penetration & metabolism

In order to better evaluate and understand the needs for safety evaluation and the role of the tissue compartments, studies using infinite doses versus finite doses have been conducted according to a validation-like format with several manufactured and "in-house" reconstructed human skin models [61]. A suitable in vivo/in vitro correlation has been demonstrated with regard to the test chemicals' ranking with EpiSkin™, EpiDerm™, and SkinEthic™ RHE, even though permeation of in vitro models still generally exceeds that of the human excised skin. Because of this higher penetration profile, 3D models are preferentially used as screening tools to study relative penetration of chemicals and mixtures [77] rather than as regulatory validated assay where absolute quantification is required. The extracellular lipids of the stratum corneum form a continuous separation between living layers and outside and are therefore important in the functionality of the skin barrier. All major classes of human skin lipids are expressed in the stratum corneum of RHE models [78], but deviation in some lipid profile composition could lead to differences in the spatial organization of the lamellar lipids that seems an important parameter for skin barrier function [79]. Different approaches have been tested to increase barrier function such as cytokines treatments [80], decrease of atmosphere humidity [81,82], or, more recently, by using specific dermal matrix [83]. These strategies showed mitigated success, and this subject is always of main concern to produce next generation of models with barrier function closer to in vivo situation.

Routine protocols for penetration studies have been developed to screen new chemicals and mixtures. Using the EpiSkin™ model, it has been demonstrated that percutaneous assays could be easily performed without a specific or complex device, such as Franz cell [62]. This system simplifies the development of screening tests for the evaluation of skin penetration of compounds with a high reliability and a high throughput. The permeation of essential oils was evaluated using SkinEthic™ RHE to establish the kinetics of release of active principles from cosmetic formulations. Diffusion and quantification of selected terpenes proved the method was sensitive, simple, and reproducible, thus providing a convenient model for safety/quality assessment of formulations [84]. Similarly, EpiSkin™ tissues were used to evaluate the absorption of vitamins C and E from topical microemulsions and to select the best formulation able to enhance penetration of the tested vitamins in the tissue [85]. Investigating the biotransformation of oxygen-sensitive aromatic diamines (ADA) (used in oxidative hair coloring) during penetration after topical application onto the EpiSkin™ model, it was shown, interestingly, that the reconstructed epidermis had a clear capacity to activate *N*-acetylation, glucuronidation, and sulfation enzymes, that is, to metabolize and detoxify these ADA derivatives. The data support the growing evidence that ADAs are transformed in the human skin and suggest that current practices for assessing the safety of ADAs should take these essential findings into account [86]. The latter results evidenced the role of the barrier function, together with the expression of metabolic capacities in 3D models and thus supported their use as simple and cost-effective tools suited to efficacy/safety studies, overcoming the availability problems usually encountered with human skin explants, but with care needed when extrapolating results to clinical levels [61]. Other approaches using noninvasive imaging technology, such as confocal Raman spectroscopy or two-photon imaging, provided new insights for studying the permeation process and development of local toxicity [87,88].

The skin plays an important role in the detoxification and metabolism of xenobiotics for purposeful (i.e., drugs and cosmetics) or environmental substances and pollutants in contact with the epidermis. The phase I metabolism of xenobiotics is catalyzed by cytochrome P450 and other oxidoreductases (esterases, epoxide hydrolases, alcohol, aldehyde dehydrogenases, etc.) that transform hydrophobic chemicals into more polar hydrophilic compounds. Phase 2 metabolism is catalyzed by transferases (UDP-glucuronosyl transferases, glutathione-S-transferases, sulfotransferases, etc.), which add conjugated groups to increase hydrophilicity and clearance, thus avoiding bioaccumulation and potential intracellular toxicity. The skin's role as a first-pass organ has been studied in detail through gene expression, protein level, or enzymatic activities and compared with reconstructed models. For example, gene expression profiles of 61 phases of I and II metabolizing enzymes in the reconstructed epidermis EpiSkin™ model and in a reconstructed full-thickness tissue showed strong similarity to those in the skin in vivo [89]. The study of protein expression profiles has confirmed the similarity observed at the mRNA level between reconstructed models and whole skin [90]. Recently, the enzymatic activities of xenobiotic enzymes have been measured and compared between different reconstructed epidermis models, reconstructed skin model, and whole skin [91,92]. Weak cytochrome P450-dependent monooxygenase (phase I enzymes) and high transferase (phase II enzymes) activities have confirmed that the normal human skin and reconstructed skin models are well provided with the main detoxifying drug-metabolizing enzymes and present similar basal levels of activity. This xenobiotic metabolism is characterized by a limited functionalization capacity and a much greater detoxification (hydrolytic and conjugating) capacity. These results demonstrate that skin models are similar to the in vivo skin, in terms of metabolic functionality toward the investigated xenobiotics [93].

4.4 Skin corrosion/irritation

Skin corrosion or dermal corrosion tests assess the potential of a substance to cause irreversible damage to the skin, namely, visible necrosis through the epidermis and into the dermis, following the application of a test substance for a period of between 3 min and 4 h. The historical Draize test, developed in the 1940s, causes severe discomfort and pain to the animal (rabbit) and was one of the animal tests banned by the 2013 deadline. Many attempts have been made to replace the in vivo Draize skin corrosion test. To date, six methods have been validated between 1999 and 2006 and officially quoted in EU statements from the ECVAM advisory committee (ESAC) (Table 2). Among these, two validated methods were not based on reconstructed tissues: the Corrositex (artificial biobarrier), a corrosion test peer-reviewed by ICCVAM (report 1999), quoted in the ESAC 2000 statement and in the OECD guideline TG435, and the rat skin transepidermal electrical resistance—TEER test—validated by ECVAM, quoted in the ESAC 1998 statement and in the OECD guideline TG430. Four validated test methods using in vitro reconstructed epidermis were validated by ECVAM and subsequently endorsed by ESAC: the EpiSkin™ model [94], the EpiDerm™

Table 2 **List of assays for safety evaluation of chemicals using reconstructed human epidermis**

End point	Method name	Endorsement of scientific validity		Regulatory acceptance	
Skin corrosion	In vitro skin corrosion: RHE test method: Episkin™ SCT, Epiderm™ SCT, SkinEthic™ RHE SCT, epiCS® SCT (formerly EST-1000); Vitrolife-Skin™	ESAC (1998, 2000, 2006, 2009)	ICCVAM 2002	OECD TG 431 (2004, updated 2015, updated 2016)13a	REACH regulation
Skin irritation	Reconstructed human epidermis test method: Episkin™ SIT, Epiderm™ SIT, SkinEthic™ RHE SIT, LabCyte EPI-MODEL24 SIT	ESAC (2007, 2008, 2009)	JaCVAM (2010, 2012)	OECD TG 439 (2010, updated 2015)13b	REACH regulation
Dermal absorption/ penetration	In vitro skin absorption methods	No		No	
Genotoxicity	Micronucleus test and comet assay in reconstructed skin Models for genotoxicity testing	Submitted at EURL-ECVAM for micronucleus		No	
Skin sensitization	SENS-IS method (Episkin™ model)	Submitted at EURL-ECVAM		Submitted	
Phototoxicity	EPISKIN phototoxicity assay (EPA) and Epiderm phototoxicity protocol	No		No	
Eye irritation	Reconstructed human cornea-like epithelium (RhCE) test for identifying chemicals not requiring classification and labeling for eye irritation or serious eye damage 22	EURL-ECVAM		OECD TG 492 (2015)	REACH regulation

model [94], the SkinEthic™ RHE model [95], and the EST-1000 model [64]. All these in vitro methods based on reconstructed epidermis were implemented in the EU testing methods for dangerous substances and included in the OECD test guideline 431 (dedicated to human in vitro skin model testing methods). Test Guideline 431 defines some general and functional rules and conditions that must be evaluated before using the models for routine purposes. The last revision of TG 431, released in July 2016, allows noncorrosive and corrosive substances and mixtures to be identified in accordance with the UN GHS. It further supports the subcategorization of corrosive substances and mixtures into optional subcategory 1A, versus 1B–1C versus noncorrosive [96,97]. A limitation of this test guideline is that it does not allow skin corrosive subcategory 1B to being discriminated from subcategory 1C due to the limited set of well-known in vivo corrosive subcategory 1C chemicals.

Skin irritation or dermal irritation is defined as reversible damage of the skin following the application of a test substance for up to 4h. Evaluation of the potential to irritate the human skin of any chemical or product used in the pharmaceutical or cosmetics industry is a requirement. The assessment of acute skin irritation was usually performed on a rabbit skin (Draize rabbit skin irritation test) [98] according to the test guideline 404 [99] and involved determining clinical signs, such as erythema and edema, after a single application to the skin. Animal data could be very questionable and are subject to regulatory issues [60] regarding potential misclassification [100] and ethical considerations. Consequently, alternatives to these animal testing procedures have been developed in different fields. EpiSkin™ and EpiDerm™ test methods were evaluated for their ability to predict skin irritation [101] throughout an ECVAM formal validation study. On the basis of this study, ESAC (2007) [63] concluded that the EpiSkin™ skin irritation test (SIT) showed evidence of being a reliable and relevant stand-alone test for predicting skin irritation as a replacement for the Draize skin irritation test. The test method is able to distinguish between skin irritant and no-label test substances. In 2008, ESAC endorsed two additional skin model test methods for skin irritation testing: the EpiDerm™ skin irritation test method and the SkinEthic™ RHE skin irritation test method. Both showed results similar to the validated reference method (VRM) EpiSkin™ test method and are considered as valid replacement methods. Validated models—EpiSkin™, EpiDerm™, SkinEthic™ RHE, and LabCyte EPI-MODEL24 SIT—are included in the last version of the test guideline TG439 (July 2015). The VRM (EpiSkin™) was based on a short treatment time (15min) followed by an extended 42h posttreatment incubation period. The prediction model mainly based on the formally validated viability measurement (MTT test) and appropriate cutoff values (prediction model) allowed two chemical classes to be drawn up in accordance with the EU's new classification labeling and packaging (CLP) regulation: irritants (=Globally Harmonized System class 2) and nonclassified substances. Applied to a set of 184 substances, including references and industrial chemicals covering diverse physical-chemical categories, the performance of the EpiSkin™ test method were well balanced, with 85% sensitivity, 86% specificity, and 86% accuracy [102]. Assessing the performance of test methods on a large number of chemicals and products relevant to industrial needs will help to concretely determine and adjust (industry-specific) areas of applicability while identifying the remaining gaps. These validated tools are part of the necessary and useful battery of alternative methods allowing industrial nonanimal testing strategies to be used. In 2014, the OECD released a document on integrated approaches to testing and assessment (IATA) for skin corrosion and irritation [103]. It proposes pragmatic, science-based approaches to chemical hazard characterization that rely on an integrated analysis of existing information, coupled with testing strategies based on validated methods using reconstructed skin models.

4.5 Phototoxicity

Phototoxicity is an adverse acute, inflammatory and nonimmunologically mediated skin reaction that occurs after a single contact with a photoreactive chemical, under sun exposure. Since 2000, the in vitro 3T3 NRU phototoxicity test has been the only in vitro test validated for evaluating the acute phototoxicity of a chemical. However, this monolayer culture of fibroblasts is a simple basic system lacking connections and interactions between different cell types and ECM [66]. To overcome these limitations, some methods using reconstructed skin models (Epiderm™ and EPISKIN™) have been developed. In 1999, ECVAM funded a prevalidation study on the EpiDerm™ phototoxicity test, with promising results. More recently, the EpiSkin™ phototoxicity assay's performance (on 17 reference chemicals) showed a very good overall accuracy of 94% [69]. Interestingly, the EPA takes into account the influence of the administration route (topical treatment vs a systemic-like administration) in the prediction of phototoxic potency. Similar protocols based on viability assessment were applied to different skin and epidermis reconstructed models [104,67]. The promising performance of EPA showed that the EpiSkin™ model is an interesting tool that is able to integrate decision-making processes to address the question of phototoxicity linked to the administration route. Using the industrial epidermis, SkinEthic™ RHE model showed that exposure to solar-simulated radiation (SSR) induced phototoxic features, such as the formation of sunburn cells and DNA damage (fragmentation) with the activation of caspase-3 protease in keratinocyte basal layers, which opens possibilities for evaluating the photoprotective capacity of cosmetic formulations [105].

4.6 Skin sensitization

An epidemiological study based on data collected on all age groups and all countries between 1966 and 2007 showed that prevalence of contact allergy to at least one allergen is around 20% [106]. This underlines the importance of this health end point in hazard and risk assessments of chemicals. The need for nonanimal data to assess the skin sensitization properties of chemicals, especially cosmetics ingredients, has driven the development of many in vitro methods [107]. It is now widely accepted that for this end point, no single alternative can provide a solution to replace animal methods. In 2012, the OECD described the first AOP (Fig. 8) for this adverse effect [108]. According to this document "an AOP is the sequence of events from the chemical structure of a target chemical or group of similar chemicals through the molecular initiating event to an in vivo outcome of interest." The first key event of the AOP for

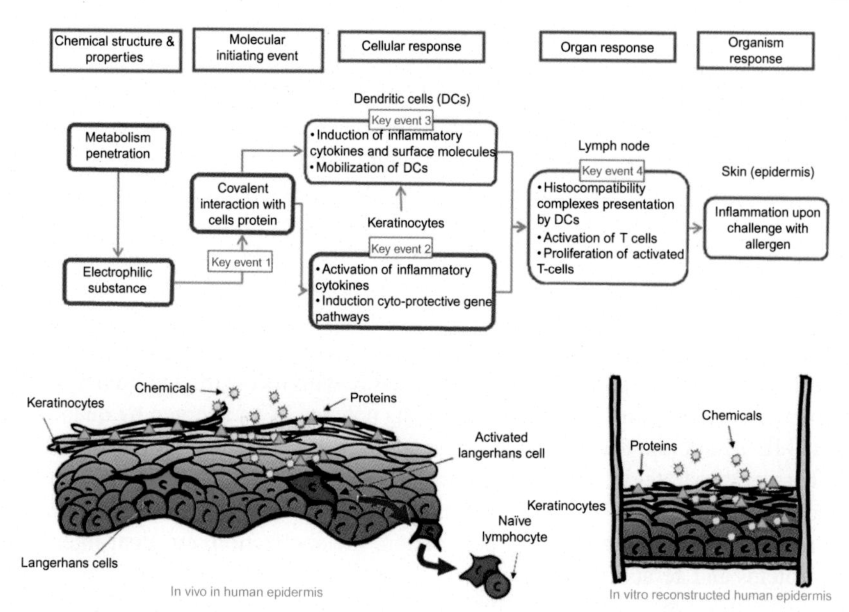

Fig. 8 Flow diagram of the pathways associated with skin sensitization adapted from "Adverse Outcome Pathway for Skin Sensitization Initiated by Covalent Binding to Proteins Part 1: Scientific Evidence," OECD document (2012). Representation of the key events in the human skin compared with reconstructed human model that mimics the bioavailability of the chemical, metabolism and penetration, and key events 1 and 2.

skin sensitization is the molecular interaction with skin proteins, the site of action. Specifically, the target chemical or a metabolite or abiotic transformation product of the target chemical covalently binds to cysteine and/or lysine residues. The second key event takes place in the keratinocyte. This includes inflammatory responses and gene expression associated with particular cell signaling pathways (e.g., antioxidant/electrophile response element-dependent pathways). The third key event is activation of dendritic cells, which is typically assessed by expression of specific cell surface markers, chemokines and cytokines. The final key intermediate event is T-cell proliferation.

This AOP provides a framework to develop, validate, and gain regulatory acceptance of various alternative approaches. For each key event, especially the first three, several alternative methods have been developed, and some are validated and OECD test guidelines released. Recent efforts have focused on the development of the integrated approach to testing and assessment (IATA) and on using the data generated with these methods for hazard and potency assessment [109-111], but the situation is much more complex than for skin corrosion/irritation.

Among the alternative methods developed and under evaluation, some are based on the use of RHE as test systems [107,76]. The SENS-IS assay is an in vitro method developed based on the EPISKIN™ model and the analyses of the expression of three groups of genes by the transcriptomics approach (qRT-PCR) measuring fold increases in the expression levels of genes [73]. Statistical analyses of data on 150 chemicals give an accuracy higher than 96% for a yes/no answer and 92% for potency prediction [74]. The results of a recent multicentric study on 19 blinded chemicals with this assay have been submitted to EURL-ECVAM and OECD for validation.

The epidermal equivalent (EE) potency assay is another assay that aims at classifying sensitizer potency using RHE. This test requires prior identification with another assay if the substance is a sensitizer. Then, the reconstructed epidermis is exposed to different concentrations of the chemical, and the potency of the allergen is calculated from the concentration inducing 50% of cell viability (EE-EC50) measured by MTT [112,72,113]. In response to allergens, keratinocytes are capable of producing a wide range of pro-inflammatory cytokines. The release of the inflammatory cytokine IL-18 has been shown to be specific to keratinocytes' exposure to contact allergens [114]. This property is also exploited to develop an assay with reconstructed human epidermis to predict the sensitization potential and its potency of an allergen by quantifying secreted IL-18 into the reconstructed tissue culture medium [115,116].

Whatever the strategy chosen, methods using RHE as a test system benefit from the capacity of the human epidermis to reproduce four steps of the skin sensitization AOP in vitro, including two key events: penetration, metabolization, covalent interaction with proteins and keratinocyte activation.

4.7 Genotoxicity

Genotoxicity refers to processes that alter the structure, information content, or segregation of DNA and that are not necessarily associated with mutagenicity. Thus, tests for genotoxicity include tests that assess induced damage to DNA (without direct evidence of mutation). Evaluation of genotoxicity and mutagenicity evaluations are important and early steps in the safety assessment of chemicals for industrial development and regulatory purposes. Although many in vitro tests are routinely used and accepted by regulatory agencies, their accuracy in predicting mutagenic/genotoxic potential in humans is frequently questioned [117]. In 2007, a workshop of experts underlined the fact that genotoxicity tests in 2D culture of mammalian cells in vitro produced a remarkably high and unacceptable occurrence of irrelevant positive results. The only way to confirm such results would imply an in vivo rodent carcinogenicity test. Among the proposals made to overcome this limitation, the experts concluded that the use of 3D in vitro skin models was a promising option. These do present clear advantages (organ structure, functionality, standardized production, availability, etc.) as in vitro screening tools for genotoxicity. Specific protocols to perform the comet assay (genotoxicity test) on EpiSkin™ were developed both with UV exposure and/or reference chemicals (lomefloxacin, 4-nitroquinoline-*N*-oxide) as stress agents. Pretreatment with the sunscreen Mexoryl SX® was able to clearly reduce the extent of UV-induced DNA damage assessed by the comet signal [118]. A second approach was developed, taking advantage of the barrier function and metabolic capacities of the EpiSkin™ model, to drive the response to a topically applied test chemical and its possible metabolites. A specific coculture system using a target lymphoma cell line (L5178Y) underneath the epidermis was carried out for evaluating micronucleus induction after topical application of the test substance. This approach aims to improve the relevance of exposure conditions for testing further products to be applied to the skin [118]. This could improve the predictive value of a genotoxicity assessment as compared with those afforded by existing in vitro tests and, therefore, could be used as follow-up tests in the event of positive results from the standard in vitro genotoxicity testing battery. Another approach, currently being validated, is to conduct the micronucleus test and the comet assay directly on 3D reconstructed human skin models [119]. Validation studies of the reconstructed skin micronucleus assay (RSMN) using the Epiderm™ model are coordinated and funded by the Cosmetics Europe Genotoxicity Task Force [120]. The 3D skin comet assay in full-thickness skin models, such as EpidermFT [71] or Phenion® FT, is being developed as a joint venture between the Cosmetics Europe Genotoxicity Task Force and a German consortium funded by the Federal Ministry of Education and Research (BMBF). The reproducibility and predictive capacity of this comet assay between laboratories is still being assessed.

These different projects on reconstructed epidermis and skin have helped improve the predictive capacity of in vitro assays. Optimization of the protocols has resulted in improved specificity of the micronucleus test, such that over 60% of irrelevant positive findings could be prevented by using the optimized methods. Both the micronucleus and the comet assay in reconstructed skin models will not be considered as stand-alone assays, but they will be ideal in genotoxicity animal-free testing strategies to address positive in vitro genotoxicity test results [76] as part of a second tier.

5. Skin models and methods for assessing the efficiency of new active ingredients and finished products

Objective evidence of the final efficacy of a cosmetic product should always be confirmed by clinical studies. At earlier stages of development, however, reconstructed human skin is an unbeatable tool for screening and assessing the efficacy of new active ingredients, deciphering their action mechanism, and finally, optimizing the composition of formulations to maximize in vivo benefits. Extrapolating the results obtained in vitro on reconstructed skins models to the in vivo situation in humans is an important line of research.

5.1 *In skin aging*

Skin aging induces dramatic changes in the extracellular dermal matrix (ECM) and in the DEJ. Reversing or slowing down these phenomena could reduce these visible signs of aging. It has been shown that topical application of a C-xylopyranoside derivative (C-beta-D-xylopyranoside-2-hydroxy-propane) onto a reconstructed full-thickness skin induced neosynthesis of some matrix proteins such as glycosaminoglycans (GAG) and heparan sulfate proteoglycans, as well as an improvement in DEJ morphogenesis together with increased collagen VII expression, thus suggesting beneficial effects on an aged skin [121]. GAGs are major components of dermal ECM and participate in tissue cohesiveness and hydration. Through their ability to bind to and modulate the activity of a number of growth factors, GAGs are also involved in cell adhesion and migration, as well as skin organogenesis and wound healing. The structure and integrity of GAGs is therefore essential for skin homeostasis and regeneration. Some research suggests that age-related alterations of the dermal connective tissue may involve a remodeling of GAG expression and structure [122,123]. Further studies with in vitro skin models have confirmed that C-xyloside strongly not only enhances synthesis of GAG chains but also induces significant changes in their size and structure. C-xyloside primed GAGs were exclusively chondroitin/dermatan sulfate with reduced chain size, increased O-sulfation, and changed in iduronate content and distribution [124]. This gives new insights into the effect of C-xyloside on GAG structure and activities, which opens up prospects and applications for such compounds in skin repair/regeneration.

The differentiation/proliferation balance in keratinocytes strongly influences skin homeostasis. Topical application of the antiaging gold standard retinol on an in vitro skin model has been shown to modulate differentiation. A combination of retinol and adenosine applied to the 3D skin model showed the ability of adenosine to potentiate retinol effects in accordance with results from a clinical application of a skin care product containing these active ingredients [125]. A derivative of jasmonic acid (LR2412), a signal molecule synthesized by plants under stress, was studied on EpiSkin™ models to evaluate its effect on their regenerative potential. Morphological studies, immunohistochemical measurements, and transcriptomic analyses were conducted. Preliminary results on thickness and proliferation markers (Ki67 markers) showed an increase in the regenerative potential of treated in vitro epidermis compared with a control (Fig. 9). Transcriptomics and immunohistochemistry studies showed that this activity was mediated by a stimulation of the hyaluronasome, a complex of proteins involved in the synthesis (hyaluronate synthetase HAS3 and HAS2), maturation (hyaluronidase) and fixation of hyaluronic acid on specific receptors (CD44). However, this increase of the epidermal hyaluronic content and of keratinocytic proliferation does not disrupt or alter the normal differentiation of the epidermis, unlike retinol, which tends to inhibit this process [126].

Further tests in a full-thickness model (Fig. 9) showed an increased synthesis of collagens IV and VII and laminin 5, essential components of the DEJ. The clinical studies that have followed confirmed the activity of this active and its antiaging potential in vivo [127].

5.2 In photoprotection

Despite the beneficial effects of sunlight, excessive or repeated exposure is known to induce oxidative stress resulting from ROS production in the skin, a major contributor to the harmful effects that follow UVA exposure. Among the mechanisms known to be involved, iron plays a central role [128] and can act as a catalyst of uncontrolled oxidation reactions with cell components [129]. Applying iron chelators to the EpiSkin™ model in vitro potentially offers an effective way to protect the skin against UVA damage, but it is usually difficult to achieve it without disturbing iron homeostasis. Recently, a novel compound (Sideroxyl), developed to avoid these potentially harmful side effects, has been evaluated in reconstructed human skin model. Designed to acquire its strong chelating capacity only during oxidative stress following an

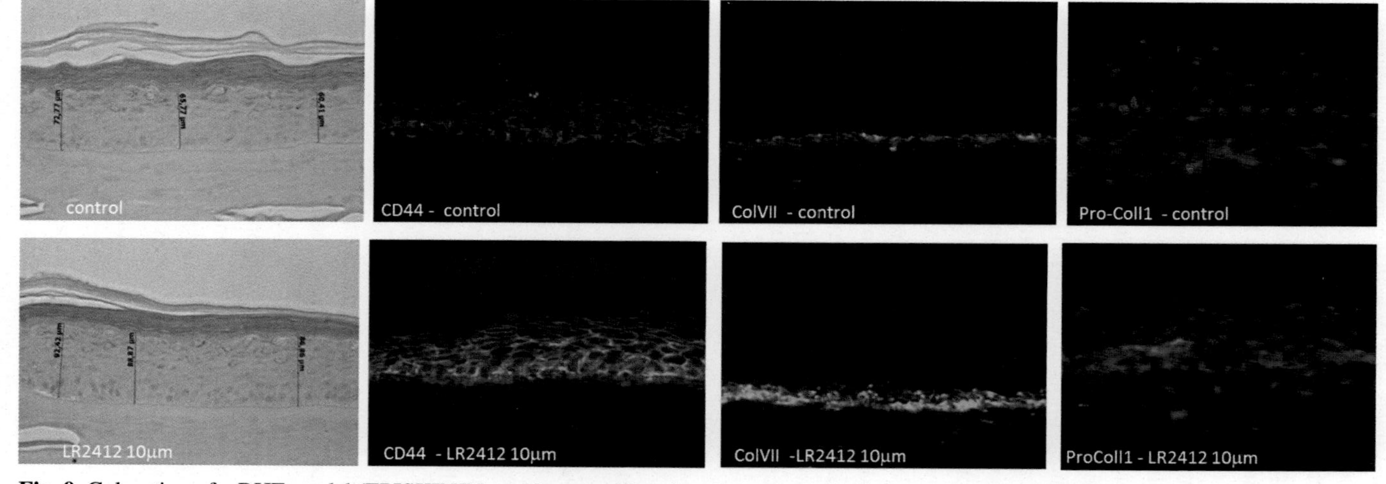

Fig. 9 Coloration of a RHE model (EPISKIN™) and immunohistochemistry of a full-thickness model (TSKIN™) after 5 days' exposure to 10 µmol/L *dehydrojasmonic acid*. HES coloration shows an increased thickness of the epidermis after treatment versus control. We observe also an increase in the labeling of the hyaluronic acid receptor CD44 in the treated epidermis. In the dermal-epidermal junction the treatment induces elevation of some markers such as Coll VII (and laminin 5, data not showed) and procollagen I in the upper dermis.
Adapted from Tran C, Michelet JF, Simonetti L, Fiat F, Garrigues A, Potter A, et al. In vitro and in vivo studies with tetra-hydro-jasmonic acid (LR2412) reveal its potential to correct signs of skin ageing. J Eur Acad Dermatol Venereol 2014;28(4): 415–23.

intramolecular hydroxylation process, this active ingredient demonstrated protective efficiency against the harmful effects of UVA at the molecular, cellular, and tissue levels. The use of in vitro reconstructed tissue allowed several end points to be measured including carbonyl increase, intracellular ROS, DNA damage (comet assay), and MMP-1 synthesis and release, altogether suggesting that such an active ingredient may prevent UVA-induced damage in the human skin alongside sunscreens, especially in the long-wavelength UVA range [130].

The structure of the 3D models and the presence of an air-exposed stratum corneum make them tools of choice for the topical application of finished products. Application of a product containing Mexoryl SX® sunscreen before solar-simulated sunlight exposure (conditions mimicking in vivo use) revealed the ability of the sunscreen to inhibit UV-induced pigmentation [131] almost completely. The in vitro evaluation of sunscreens after topical application allows their protective effects against UV to be investigated and proper selection of absorbers to be made and tested based on their absorption spectrum. Recent definition of a standard "daily UV radiation" (DUVR) allowed nonzenithal sun exposure conditions to be reproduced on in vitro models. Two commercial sunscreen products with a similar sun protection factor (SPF) but different absorption profiles over the UVA range were evaluated on reconstructed human skin. Using vimentin immunostaining for dermal fibroblasts and matrix metalloprotease (MMP)-1 secretion, it was shown that to ensure efficient daily photoprotection from DUVR, the filtration profile of the product should be well balanced over the UV range with a sufficient level of UVA absorption [132].

5.3 In the modulation of pigmentation

N-Acetyl glucosamine (NAG) is an amino sugar precursor in the human skin of the high molecular weight, water-binding polymer hyaluronic acid. It has been shown to be effective in reducing the appearance of hyperpigmented spots [133] through inhibiting the glycosylation of tyrosinase, a key enzyme in the production of melanin. To identify additional mechanisms by which NAG might affect melanin production, a transcriptomics study of the effect of NAG has been conducted on the SkinEthic™ model of reconstructed human pigmented epidermis (RHPE). The results showed several changes in gene expression (upregulation and downregulation) with some of them related to synthesis and regulation of intracellular cytoskeletal proteins that are involved in melanosome transport within the cell [134]. Thus, several mechanisms may be operative in the observed effects of NAG on pigmentation. Produced on an industrial scale, these in vitro pigmented skin models are able to reproduce different phototypes. They are powerful tools to screen the pro- or depigmentation efficacy of new ingredients or compounds [135,136]. A model of this kind has been also used to develop RNA silencing (siRNA) strategies to modulate skin pigmentation. After selection of a potent siRNA compound that targets the tyrosinase gene, its efficacy was assessed during reconstruction of an RHPE model. Melanocytes derived from the highly pigmented phototype VI skin were used for the reconstruction of a phenotypically silenced epidermis. A visible and a long-term pigmentation interference effect (lasting more than 30 days) were observed in the

SkinEthic RHPE model [137]. This knockdown of tyrosinase expression was transient since the reconstructed epidermis started recovering pigmentation at day 39 and thus did not irreversibly affect melanocyte differentiation. The use of a pigmented reconstructed epidermis is thus becoming a precious tool, allowing the action of different topical applications and vectors to be studied and new dermatological interventions, tools, and targets to be selected, evaluated, and designed.

References

[1] Alexis CarrelL MD, Montrose T, Burrows MD. Cultivation of adult tissues and organs outside the body. JAMA 1910;55(16):1379–81.
[2] Carrel A, Lindbergh CA. The culture of organs. Am J Med Sci 1938;196(5):732.
[3] Puck TT, Marcus PI. Clonal growth of mammalian cells in vitro; growth characteristics of colonies from single HeLa cells with and without a feeder layer. J Exp Med 1956;103:273–83.
[4] Rheinwald JG, Green H. Serial cultivation of strains of human epidermal keratinocytes: the formation of keratinizing colonies from single cells. Cell 1975;6:331–43.
[5] Green H, Kehinde O, Thomas J. Growth of cultured human epidermal cells into multiple epithelia suitable for grafting. Proc Natl Acad Sci U S A 1979;76:5665–8.
[6] O'Connor NE, Mulliken JB, Banks-Schlegel S, et al. Grafting of burns with cultured epithelium prepared from autologous epidermal cells. Lancet 1981;1:75–8.
[7] Pruniéras M. Epidermal cell cultures as models for living epidermis. J Invest Dermatol 1979;73:135–7.
[8] Pruniéras M, Régnier M, Woodley D. Methods for cultivation of keratinocytes with an air-liquid interface. J Invest Dermatol 1983;81:28–33.
[9] Boyce ST, Warden GD. Principles and practices for treatment of cutaneous wounds with cultured skin substitutes. Am J Surg 2002;183:445–56.
[10] Asselineau D, Bernard BA, Bailly C, et al. Epidermal morphogenesis and induction of 67kD keratin polypeptide by culture at the liquid-air interface. Exp Cell Res 1985;159:536–9.
[11] Haake AR, Scott GA. Physiologic distribution and differentiation of melanocytes in human fetal and neonatal skin equivalents. J Invest Dermatol 1991;96:71–7.
[12] Facy V, Flouret V, Regnier M, et al. Langerhans cells integrated into human reconstructed epidermis respond to known sensitizers and ultraviolet exposure. J Invest Dermatol 2004;122:552–3.
[13] Régnier M, Schweizer J, Michel S, et al. Expression of high molecular weight (67K) keratin in human keratinocytes cultured on dead de-epidermized dermis. Exp Cell Res 1986;165(1):63–72.
[14] Ponec M, Weerheim A, Kempenaar J, et al. Lipid composition of cultured keratinocytes in relation to their differentiation. J Lipid Res 1988;29:949–62.
[15] Bell E, Ehrlich HP, Sher S, et al. Development and use of a living skin equivalent. Plast Reconstr Surg 1981;67:386–92.
[16] Boyce ST, Christianson D, Hansbrough J. Structure of a collagen-GAG skin substitute optimized for cultured human epidermal keratinocytes. J Biomed Mater Res 1988;22:939–57.
[17] Black AF, Bouez C, Perrier E, et al. Optimization and characterization of an engineered human skin equivalent. Tissue Eng 2005;11(5–6):723–33.

[18] Shahabeddin L, Berthod F, Damour O, et al. Characterization of skin reconstructed on a chitosan-cross-linked collagen-glycosaminoglycan matrix. Skin Pharmacol 1990;3:107–14.
[19] El Ghalbzouri A, Lamme EN, van Blitterswijk C, et al. The use of PEGT/PBT as a dermal scaffold for skin tissue engineering. Biomaterials 2004;25:2987–96.
[20] Li J, Chen M, Fan X, Zhou H. Recent advances in bioprinting techniques: approaches, applications and future prospects. J Transl Med 2016;14:271. https://doi.org/10.1186/s12967-016-1028-0.
[21] Ng WL, Wang S, Yeong WY, Naing MW. Skin bioprinting: impending reality or fantasy? Trends Biotechnol 2016;34(9):689–99. https://doi.org/10.1016/j.tibtech.2016.04.006.
[22] Lee V, Singh G, Trasatti JP, Bjornsson C, Xu X, Tran TN, et al. Design and fabrication of human skin by three-dimensional bioprinting. Tissue Eng Part C Methods 2013;20:473–84. https://doi.org/10.1089/ten.tec.2013.0335.
[23] Lapierre CM. The aging dermis: the main cause for the appearance of "old" skin. Br J Dermatol 1990;122(35):5–11.
[24] Mine S, Fortunel NO, Pageon H, Asselineau D. Aging alters functionally human dermal papillary fibroblasts: a new view of skin morphogenesis and aging. PLoS One 2008;3(12):e4066.
[25] Pageon H, Zucchi H, Asselineau D. Distinct and complementary roles of papillary and reticular fibroblasts in skin morphogenesis and homeostasis. Eur J Dermatol 2012;22(3):324–32.
[26] Pageon H, Bakala H, Monnier VM, Asselineau D. Collagen glycation triggers the formation of aged skin in vitro. Eur J Dermatol 2007;17(1):12–20.
[27] Pageon H, Zucchi H, Rousset F, Monnier VM, Asselineau D. Skin aging by glycation: lessons from the reconstructed skin model. Clin Chem Lab Med 2014;52(1):169–74.
[28] Pageon H, Zucchi H, Dai Z, Sell DR, Strauch CM, Monnier VM, et al. Biological effects induced by specific advanced glycation end products in the reconstructed skin model of aging. Biores Open Access 2015;4(1):54–64. https://doi.org/10.1089/biores.2014.0053.
[29] Pennacchi PC, et al. Glycated reconstructed human skin as a platform to study the pathogenesis of skin aging. Tissue Eng Part A 2015;21(17–18):2417–25.
[30] Toussaint O, et al. Stress-induced premature senescence: from biomarkers to likeliness of in vivo occurrence. Biogerontology 2002;3(1):13–7.
[31] Ben-Porath I, Weinberg RA. The signals and pathways activating cellular senescence. Int J Biochem Cell Biol 2005;37(5):961–76. https://doi.org/10.1016/j.biocel.2004.10.013.
[32] Diekmann J, Alili L, Scholz O, Giesen M, Holtkötter O, Brenneisen P. A three-dimensional skin equivalent reflecting some aspects of in vivo aged skin. Exp Dermatol 2016;25(1):56–61. https://doi.org/10.1111/exd.12866.
[33] Leyden J. What is photoaged skin? Eur J Dermatol 2001;11(2):165–7.
[34] Helfrich YR, Sachs DL, Voorhees JJ. Overview of skin aging and photoaging. Dermatol Nurs 2008;20(3):177–83.
[35] Quan T, Qin Z, Xia W, et al. Matrix-degrading metalloproteinases in photoaging. J Investig Dermatol Symp Proc 2009;14:20–4.
[36] Bernerd F, Vioux C, Asselineau D. Evaluation of the protective effect of sunscreens on in vitro reconstructed human skin exposed to UVB or UVA irradiation. Photochem Photobiol 2000;71:314–20.
[37] Bernerd F, Asselineau D. An organotypic model of skin to study photodamage and photoprotection in vitro. J Am Acad Dermatol 2008;58(2):155–9.

[38] D C, C C, P F, S P, C E, B F. Human skin model containing melanocytes: essential role of keratinocyte growth factor for constitutive pigmentation—functional response to α-melanocyte stimulating hormone and forskolin. Tissue Eng Part C Methods 2012;18(12):947–57.

[39] Duval C, Cohen C, Chagnoleau C, Flouret V, Bourreau E, Bernerd F. Key regulatory role of dermal fibroblasts in pigmentation as demonstrated using a reconstructed skin model: impact of photo-aging. PLoS One 2014;9(12):e114182. https://doi.org/10.1371/journal.pone.0114182.

[40] Julia O, Byrd AL, Deming C, Sean C, NISC Comparative Sequencing Program, Kong HH, et al. Biogeography and individuality shape function in the human skin metagenome. Nature 2014;514(7520):59–64.

[41] Schommer NN, Gallo RL. Structure and function of the human skin microbiome. Trends Microbiol 2013;21(12):660–8. https://doi.org/10.1016/j.tim.2013.10.001.

[42] Williams MR, Gallo RL. The role of the skin microbiome in atopic dermatitis. Curr Allergy Asthma Rep 2015;15(11):65.

[43] Tott JEE, van der Feltz WT, Hennekam M, van Belkum A, van Zuuren EJ, Pasmans SGMA. Prevalence and odds of *Staphylococcus aureus* carriage in atopic dermatitis: a systematic review and meta-analysis. Br J Dermatol 2016;175(4):687–95. https://doi.org/10.1111/bjd.14566.

[44] Grice EA, Segre JA. The skin microbiome. Nat Rev Microbiol 2011;9:244–53.

[45] Duckney P, Wong HK, Serrano J, Yaradou D, Oddos T, Stamatas GN. The role of the skin barrier in modulating the effects of common skin microbial species on the inflammation, differentiation and proliferation status of epidermal keratinocytes. BMC Res Notes 2013;6:474. https://doi.org/10.1186/1756-0500-6-474.

[46] Paniagua-Contreras GL, Monroy-Pérez E, Vaca-Paniagua F, Rodríguez-Moctezuma JR, Negrete-Abascal E, Vaca S. Implementation of a novel in vitro model of infection of reconstituted human epithelium for expression of virulence genes in methicillin-resistant *Staphylococcus aureus* strains isolated from catheter-related infections in Mexico. Ann Clin Microbiol Antimicrob 2014;13:6. https://doi.org/10.1186/1476-0711-13-6.

[47] de Breij A, Haisma EM, Rietveld M, El Ghalbzouri A, van den Broek PJ, Dijkshoorn L, et al. Three-dimensional human skin equivalent as a tool to study *Acinetobacter baumannii* colonization. Antimicrob Agents Chemother 2012;56(5):2459–64. https://doi.org/10.1128/AAC.05975-11.

[48] Boguniewicz M, Leung DY. Atopic dermatitis: a disease of altered skin barrier and immune dysregulation. Immunol Rev 2011;242(1):233–46. https://doi.org/10.1111/j.1600-065X.2011.01027.x.

[49] Goh CL, Wong JS, Giam YC. Skin colonization of *Staphylococcus aureus* in atopic dermatitis patients seen at the National Skin Centre Singapore. Int J Dermatol 1997;36(9):653–7.

[50] Gong JQ, Lin L, Lin T, Hao F, Zeng FQ, Bi ZG, et al. Skin colonization by *Staphylococcus aureus* in patients with eczema and atopic dermatitis and relevant combined topical therapy: a double-blind multicentre randomized controlled trial. Br J Dermatol 2006;155(4):680–7.

[51] De Servi B, Semenzato A, Baratto G, Meloni M. Modeling atopic dermatitis in reconstructed human epidermis. Paris: SFC-Journées Jean Paul Marty; 2015.

[52] Bissell MJ, Hall HG, Parry G. How does the extracellular matrix direct gene expression? J Theor Biol 1982;99:31–68.

[53] Aplin AE, Howe A, Alahari SK, Juliano RL. Signal transduction and signal modulation by cell adhesion receptors: the role of integrins, cadherins, immunoglobulin-cell adhesion molecules, and selectins. Pharmacol Rev 1998;50(2):97–263.

[54] Bernard FX, Pedretti N, Rosdy M, Deguercy A. Comparison of gene expression profiles in human keratinocyte mono-layer cultures, reconstituted epidermis and normal human skin; transcriptional effects of retinoid treatments in reconstituted human epidermis. Exp Dermatol 2002;11(1):59–74.

[55] Roguet R. The use of standardized human skin models for cutaneous pharmacotoxicology studies. Skin Pharmacol Appl Ski Physiol 2002;15:1–3.

[56] Danilenko DM, Phillips GD, Diaz D. In vitro skin models and their predictability in defining normal and disease biology, pharmacology, and toxicity. Toxicol Pathol 2016;44(4):555–63.

[57] Olson H, Betton G, Robinson D, et al. Concordance of the toxicity of pharmaceuticals in humans and in animals. Regul Toxicol Pharmacol 2000;32:56–67.

[58] Andrew K. Systematic reviews of animal experiments demonstrate poor contributions toward human healthcare. Rev Recent Clin Trials 2008;3(2):89–96.

[59] Greaves P, Williams A, Eve M. First dose of potential new medicines to humans: how animals help. Nat Rev Drug Discov 2004;3(3):226–36.

[60] Directive 2003/15/EC of the European Parliament and of the Council of 27 February 2003 Amending Council Directive 76/768/EEC on the approximation of the laws of the Member States relating to cosmetic products (Text with EEA relevance). http://eur-lex.europa.eu/LexUriServ/LexUriServ.do?uri=CELEX:32003L0015:EN:HTML, 2003, Official Journal, Vol. 066, pp. 0026–0035. http://eur-lex.europa.eu/LexUriServ/LexUriServ.do?uri=CELEX:32003L0015:EN:HTML.

[61] Schäfer-Korting M, Bock U, Diembeck W, et al. The use of reconstructed human epidermis for skin absorption testing: results of the validation study. Altern Lab Anim 2008;36:161–87.

[62] Grégoire S, Patouillet C, Noé C, et al. Improvement of the experimental setup for skin absorption screening studies with reconstructed skin Episkin. Skin Pharmacol Physiol 2008;21(2):89–97.

[63] ESAC. Statement on the validity of in vitro tests for skin irritation, http://ecvam.jrc.it/; 2007.

[64] ESAC. Statement on the validity of in vitro tests for skin corrosion, http://ecvam.jrc.it/; 2009.

[65] Aardema MJ, Barnett BC, Khambatta Z, Reisinger K, Ouedraogo-Arras G, Faquet B, et al. International prevalidation studies of the EpiDerm™ 3D human reconstructed skin micronucleus (RSMN) assay: transferability and reproducibility. Mutat Res Genet Toxicol Environ Mutagen 2010;701:123–31.

[66] Augustin C, Collombel C, Damour O. Use of dermal equivalent and skin equivalent models for identifying phototoxic compounds in vitro. Photodermatol Photoimmunol Photomed 1997;13(1–2):27–36.

[67] Medina J, Elsaesser C, Picarles V, Grenet O, Kolopp M, Chibout SD, et al. Assessment of the phototoxic potential of compounds and finished topical products using a human reconstructed epidermis. In Vitr Mol Toxicol 2001;14(3):157–68.

[68] Jirová D, Kejlová K, Bendová H, Ditrichová D, Mezuláníková M. Phototoxicity of bituminous tars-correspondence between results of 3T3 NRU PT, 3D skin model and experimental human data. Toxicol in Vitro 2005;19(7):931–4.

[69] Lelievre D, Justine P, Christiaens F, et al. The EpiSkin phototoxicity assay (EPA): development of an in vitro tiered strategy using 17 reference chemicals to predict phototoxic potency. Toxicol in Vitro 2007;21:977–95.

[70] Rozmana B, Gosencaa M, Falsonb F, Gašperlin M. The influence of microemulsion structure on their skin irritation and phototoxicity potential. Int J Pharm 2016;499:228–35. https://doi.org/10.1016/j.ijpharm.2015.12.064.

[71] Reus AA, Reisinger K, Downs TR, Carr GJ, Zeller A, Corvi R, et al. Comet assay in reconstructed 3D human epidermal skin models—investigation of intra- and inter-laboratory reproducibility with coded chemicals. Mutagenesis 2013;28:709–20. http://mutage.oxfordjournals.org/content/28/6/709.long.
[72] Gibbs S, et al. An epidermal equivalent assay for identification and ranking potency of contact sensitizers. Toxicol Appl Pharmacol 2013;272:529–41.
[73] Cottrez F, Boitel E, Auriault C, Aeby P, Groux H. Genes specifically modulated in sensitized skins allow the detection of sensitizers in a reconstructed human skin model. Development of the SENS-IS assay. Toxicol in Vitro 2015;29(4):787–802. https://doi.org/10.1016/j.tiv.2015.02.012.
[74] Cottrez F, Boitel E, Ourlin JC, Peiffer JL, Fabre I, Henaoui IS, et al. SENS-IS, a 3D reconstituted epidermis based model for quantifying chemical sensitization potency: reproducibility and predictivity results from an inter-laboratory study. Toxicol in Vitro 2016;32:248–60.
[75] OECD. Guidance document on the validation and international acceptance of new. ENV/JM/MONO(2005)14:2005. http://search.oecd.org/officialdocuments/displaydocumentpdf/?cote=env/jm/mono(2005)14&doclanguage=en.
[76] Zuang V, Bifill DA, Barroso J, et al. EURL ECVAM status report on the development, validation and regulatory acceptance of alternative methods and approaches (2016), Publications Office of the European Union https://doi.org/10.2787/644905; 2016.
[77] Garrigues A, Gregoire S, Jean-Roch M, Patouillet C, Wargniez W, Patouillet C, et al. In: Promising alternative method for skin penetration evaluation using reconstructed epidermis model. Poster session presented at SOT Conference, Baltimore, March 15–19; 2009.
[78] Bouwstra JA, Gooris GS. The lipid organisation in human stratum Corneum and model systems. Open Dermatol J 2010;4:10–3.
[79] Ponec M, Boelsma E, Weerheim A, et al. Lipid and ultra structural characterization of human reconstructed skin models. Int J Pharm 2000;203:211–21.
[80] Danso MO, van Drongelen V, Mulder A, van Esch J, Scott H, van Smeden J, et al. TNF-α and Th2 cytokines induce atopic dermatitis-like features on epidermal differentiation proteins and stratum corneum lipids in human skin equivalents. J Invest Dermatol 2014;134(7):1941–50.
[81] Sun R, Celli A, Crumrine D, Hupe M, Adame LC, Pennypacker SD, et al. Lowered humidity produces human epidermal equivalents with enhanced barrier properties. Tissue Eng Part C Methods 2015;21(1):15–22.
[82] Cau L, Pendaries V, Lhuillier E, Thompson PR, Serre G, Takahara H, et al. Lowering relative humidity level increases epidermal protein deimination and drives human filaggrin breakdown. J Dermatol Sci 2017;86(2):106–13.
[83] Mieremet A, Rietveld M, Absalah S, van Smeden J, Bouwstra JA, El Ghalbzouri A. Improved epidermal barrier formation in human skin models by chitosan modulated dermal matrices. PLoS One 2017;12(3):e0174478https://doi.org/10.1371/journal.pone.0174478.
[84] Gabbanini S, Lucchi E, Carli M, et al. In vitro evaluation of the permeation through reconstructed human epidermis of essentials oils from cosmetic formulations. J Pharm Biomed Anal 2009;50:370–6.
[85] Rozman B, Gasperlin M, Tinois-Tessoneaud E, et al. Simultaneous absorption of vitamins C and E from topical microemulsions using reconstructed human epidermis as a skin mode. Eur J Pharm Biopharm 2009;72(1):69–75.
[86] Nohynek GJ, Duche D, Garrigues A, et al. Under the skin: biotransformation of para-aminophenol and para-phenylenediamine in reconstructed human epidermis and human hepatocytes. Toxicol Lett 2006;158:196–212.

[87] Pappinen S, Pryazhnikov E, Khiroug L, Ericson MB, Yliperttula M, Urtti A. Organotypic cell cultures and two-photon imaging: tools for in vitro and in vivo assessment of percutaneous drug delivery and skin toxicity. J Control Release 2012;161: 656–67.
[88] Fleischli FD, Mathes S, Adlhart C. Label free non-invasive imaging of topically applied actives in reconstructed human epidermis by confocal Raman spectroscopy. Vib Spectrosc 2013;68:29–33.
[89] Luu-The V, Duche D, Ferraris C, et al. Expression profiles of phase 1 and phase 2 metabolizing enzymes in human skin and the reconstructed skin models EpiSkin and full thickness model from EpiSkin. J Steroid Biochem Mol Biol 2009;116:178–86.
[90] van Eijl S, Zhu Z, Cupitt J, Gierula M, Götz C, Fritsche E. Elucidation of xenobiotic metabolism pathways in human skin and human skin models by proteomic profiling. PLoS One 2012;7(7):e41721.
[91] Wiegand C, Hewitt NJ, Merk HF, Reisinger K. Dermal xenobiotic metabolism: a comparison between native human skin, four in vitro skin test systems and a liver system. Skin Pharmacol Physiol 2014;27:263–75.
[92] Eilstein J, Léreaux G, Budimir N, Hussler G, Wilkinson S, Duché D. Comparison of xenobiotic metabolizing enzyme activities in ex vivo human skin and reconstructed human skin models from SkinEthic. Arch Toxicol 2014;88(9):1681–94.
[93] Hewitt NJ, Edwards RJ, Fritsche E, Goebel C, Aeby P, Scheel J, et al. Use of human in vitro skin models for accurate and ethical risk assessment: metabolic considerations. Toxicol Sci 2013;133(2):209–17.
[94] ESAC. Statement on the validity of in vitro tests for skin corrosion, http://ecvam.jrc.it; 1998.
[95] ESAC. Statement on the validity of in vitro tests for skin corrosion, http://ecvam.jrc.it/; 2006.
[96] Alépée N, Grandidier MH, Cotovio J. Sub-categorisation of skin corrosive chemicals by the EpiSkintm reconstructed epidermis skin corrosion test method according to UN GHS: revision of OECD test guideline 431. Toxicol in Vitro 2014;28:131–45.
[97] Desprez B, Barroso J, Griesinger C, Kandárová H, Alépée N, Fuchs H. Two novel prediction models improve predictions of skin corrosive sub-categories by test methods of OECD test guideline no. 431. Toxicol in Vitro 2015;29:2055–80. https://doi.org/10.1016/j.tiv.2015.08.015.
[98] Draize JH, Woodward G, Calvery HO. Methods for the study of irritation and toxicity of substances applied directly to the skin and mucous membranes. J Pharmacol Exp Ther 1944;377–390:377–90.
[99] OECD. OECD guidelines for the testing of chemicals no. 404: acute dermal irritation/corrosion, http://www.oecd-ilibrary.org/environment/test-no-404-acute-dermal-irritation-corrosion_9789264070622-en; 2002.
[100] York M, Griffiths HA, Whittle E, et al. Evaluation of human patch test for the identification and classification of skin irritation potential. Contact Dermatitis 1996;34:204–12.
[101] Spielmann H, Hoffmann S, Liebsch M, et al. The ECVAM international validation on in vitro tests for acute skin irritation: report on the validity of the EpiSkin and EpiDerm assays and on the skin integrity function test. Altern Lab Anim 2007;35:559–601.
[102] Cotovio J, Grandidier MH, Lelièvre D, et al. In: In vitro acute skin irritancy of chemicals using the validated EPISKIN model in a tiered strategy-results and performances with 184 cosmetic ingredients. AATEX Proc. 6th world congress on alternatives & animal use in the life sciences, Tokyo, Japan; 354. 2007. p. 351–8.
[103] OECD. New guidance document on an integrated approach on testing and assessment (IATA) for skin corrosion and irritation. Series on testing and assessment, No. 203. Paris: Organisation for Economic Co-operation and Development (OECD); 2014

[104] Bernard FX, Barrault C, Deguercy A, et al. Development of a highly sensitive in vitro phototoxicity assay using the SkinEthicTM reconstructed human epidermis. Cell Biol Toxicol 2000;6(16):391–400.

[105] Bacqueville D, Mavon A. Comparative analysis of solar radiation-induced damage between ex vivo porcine skin organ culture and in vitro reconstructed human epidermis. Int J Cosmet Sci 2009;31:293–302.

[106] Thyssen JP, Linneberg A, Menné T, Johansen JD. The epidemiology of contact allergy in the general population—prevalence and main findings. Contact Dermatitis 2007;57(5):287–99. https://doi.org/10.1111/j.1600-0536.2007.01220.x.

[107] Reisinger K, et al. Systematic evaluation of non-animal test methods for skin sensitisation safety assessment sensitisation safety assessment. Toxicol in Vitro 2015;29(1):259–70. https://doi.org/10.1016/j.tiv.2014.10.018.

[108] OECD. The adverse outcome pathway for skin sensitisation initiated by covalent binding to proteins. Part 1: scientific evidence. Series on testing and assessment, No. 168. Paris: Organisation for Economic Co-operation and Development (OECD); 2012

[109] Tollefsen KE, et al. Applying adverse outcome pathways (AOPs) to support integrated approaches to testing and assessment (IATA). Regul Toxicol Pharmacol 2014;70(3):629–40. https://doi.org/10.1016/j.yrtph.2014.09.009.

[110] Jaworska J. Integrated testing strategies for skin sensitization hazard and potency assessment—state of the art and challenges. Cosmetics 2016;3:16. https://doi.org/10.3390/cosmetics3020016.

[111] Ezendam J, Braakhuis HM, Vandebriel RJ. State of the art in non-animal approaches for skin sensitization testing: from individual test methods towards testing strategies. Arch Toxicol 2016;90:2861–83. https://doi.org/10.1007/s00204-016-1842-4.

[112] dos Santos GG, Spiekstra SW, Sampat-Sardjoepersad SC, Reinders J, Scheper RJ, Gibbs S. A potential in vitro epidermal equivalent assay to determine sensitiser potency. Toxicol in Vitro 2011;25:347–57.

[113] Teunis MA, Spiekstra SW, Smits M, Adriaens E, Eltze T, Galbiati V, et al. International ring trial of the epidermal equivalent sensitizer potency assay: reproducibility and predictive-capacity. ALTEX 2014;31(3):251–68. https://doi.org/10.14573/altex.1308021.

[114] Corsini E, Mitjans M, Galbiati V, Lucchi L, Galli CL, Marinovich M. Use of IL-18 production in a human keratinocyte cell line to discriminate contact sensitizers from irritants and low molecular weight respiratory allergens. Toxicol in Vitro 2009;23(5):789–96.

[115] Andres E, Barcham R, Barry M, Roggen EL, Dini C, Ferret PJ. Implementation of the rhe/il-18 sensitization assay on the skinethic rhe test system—prediction model optimization. Spain: Congress EUROTOX-Sevilla; 2016.

[116] Andres E, Barry M, Hundt A, Dini C, Corsini E, Gibbs S, et al. Preliminary performance data of the RHE/IL-18 assay performed on identification of contact sensitizers. Int J Cosmet Sci 2017;39(2):121–32.

[117] Kirkland D, Aardema M, Henderson L, et al. Evaluation of the ability of a battery of three in vitro genetoxicity tests to discriminate rodents carcinogens and non-carcinogens. Sensitivity, specificity and relative predictivity. Mutat Res 2005;584:1–256.

[118] Flamand N, Marrot L, Belaidi JP, et al. Development of genotoxicity test procedures with Episkin, a reconstructed human skin model: towards new tools for in vitro risk assessment of dermally applied compounds? Mutat Res 2006;606(1–2):39–51.

[119] Pfuhler S, Fellows M, van Benthem J, Corvi R, Curren R, Dearfield K, et al. In vitro genotoxicity test approaches with better predictivity: summary of an IWGT workshop. Mutat Res 2011;723(2):101–7.

[120] Roy S, Kulkarni R, Hewitt NJ, Aardema MJ. The EpiDerm™ 3D human reconstructed skin micronucleus (RSMN) assay: historical control data and proof of principle studies for mechanistic assay adaptations. Mutat Res Genet Toxicol Environ Mutagen 2016;805:25–37. https://doi.org/10.1016/j.mrgentox.2016.05.010.
[121] Sok J, Pineau N, Dalko-Csiba M, et al. Improvement of the dermal epidermal junction in human reconstructed skin by c-xylopyranoside derivative. Eur J Dermatol 2008;18(3):297–302.
[122] Vuillermoz B, Wegrowski Y, Contet-Audonneau JL, Danoux L, Pauly G, et al. Influence of aging on glycosaminoglycans and small leucine-rich proteoglycans production by skin fibroblasts. Mol Cell Biochem 2005;277:63–72.
[123] Shin JE, Oh JH, Kim YK, Jung JY, Chung JH. Transcriptional regulation of proteoglycans and glycosaminoglycan chain-synthesizing glycosyltransferases by UV irradiation in cultured human dermal fibroblasts. J Korean Med Sci 2011;26:417–24.
[124] Vassal-Stermann E, Duranton A, Black AF, Azadiguian G, Demaude J, Lortat-Jacob H, et al. A new C-xyloside induces modifications of GAG expression, structure and functional properties. PLoS One 2012;7(10):e47933. http://www.plosone.org/article/info%3Adoi%2F10.1371%2Fjournal.pone.0047933.
[125] Ruiz L, Benech F, Prunel A, et al. In: A new retinol against wrinkles. European Academy of Dermatology and Venerology Congress, p. Poster; 2009.
[126] Michelet JF, Olive C, Rieux E, Fagot D, Simonetti L, Galey JB, et al. The anti-ageing potential of a new jasmonic acid derivative (LR2412): in vitro evaluation using reconstructed epidermis Episkin™. Exp Dermatol 2012;21(5):398–400.
[127] Tran C, Michelet JF, Simonetti L, Fiat F, Garrigues A, Potter A, et al. In vitro and in vivo studies with tetra-hydro-jasmonic acid (LR2412) reveal its potential to correct signs of skin ageing. J Eur Acad Dermatol Venereol 2014;28(4):415–23.
[128] Pourzand C, Watkin RD, Brown JE, et al. Ultraviolet a radiation induces immediate release of iron in human primary skin fibroblasts: the role of ferritin. Proc Natl Acad Sci U S A 1999;96:6751–6.
[129] Reelfs O, Eggleston I, Pourzand C. Skin protection against UVA-induced iron damage by multiantioxidants and iron chelating drugs/prodrugs. Curr Drug Metab 2010;11(3):242–9.
[130] Pygmalion MJ, Ruiz L, Popovic E, et al. Skin cell protection against UVA by Sideroxyl, a new antioxidant complementary to sunscreens. Free Radic Biol Med 2010;49(11):1629–37.
[131] Duval C, Schmdt R, Regnier M, et al. The use of reconstructed human skin to evaluate UV-induced modifications and sunscreen efficacy. Exp Dermatol 2003;12(2):64–70.
[132] Lejeune F, Christiaens F, Bernerd F. Evaluation of sunscreen products using a reconstructed skin model exposed to simulated daily ultraviolet radiation: relevance of filtration profile and SPF value for daily photoprotection. Photodermatol Photoimmunol Photomed 2008;24(5):249–55.
[133] Bissett DL, Robinson LR, Raleigh PS, et al. Reduction in the appearance of facial hyperpigmentation by topical *N*-acetylglucosamine. J Cosmet Dermatol 2007;6:20–6.
[134] Bissett DL, Farmer T, McPhail S, Reichling T, Tiesman JP, Juhlin KD, et al. Genomic expression changes induced by topica N-acetyl glucosamine in skin equivalent cultures. J Cosmet Dermatol 2007;6:232–8.
[135] Hakozaki T, Laughlin T, Paradkar A, Zhao S. In: Undecylenoyl phenylalanine inhibits stem cell factor production in human keratinocytes. 22nd world congress, Seoul Korea, vol. P1073; 2011. http://pgbeautyscience.com/assets/files/Undecylenoyl%20Phenylalanine%20Inhibits%20Stem%20Cell%20Factor%20Production%20in%20Human%20Karetinocytes.pdf; .

[136] Dumont K, et al. In: Sepicalm® VG, a new skin lightening enable to modulate melanogenesis-related genes and to prevent UV-induced pigmentation thanks to its soothing properties. Asian Societies of Cosmetic Scientists meeting (ASCS); 2007.
[137] Collin-Djangoné C, Sahuc F, et al. In: Potential use of RNAi on human skin. RANi Europe meeting, vol. Poster; 2008.

Further reading

[1] EU. Directive 2003/15/EC of the European Parliament and the council of 27 February 2003 amending council Directive 76/768/EEC on the approximation of the laws of the member states relating to cosmetic products. Off J Eur Union 2003;L66:26–35.

Overall perspective on the clinical importance of skin models

2

Yusef Yousuf[*,†], *Saeid Amini-Nik*[†,‡,§,¶], *Marc G. Jeschke*[*,†,§,¶,||]
[*]Institute of Medical Science, University of Toronto, Toronto, ON, Canada, [†]Sunnybrook Research Institute, Toronto, ON, Canada, [‡]Laboratory Medicine and Pathobiology, University of Toronto, Toronto, ON, Canada, [§]Department of Surgery, Division of Plastic Surgery, University of Toronto, Toronto, ON, Canada, [¶]Ross-Tilley Burn Centre, Sunnybrook Health Sciences Centre, Toronto, ON, Canada, [||]Department of Immunology, University of Toronto, Toronto, ON, Canada

1. Introduction

The human skin is a complex organ in both anatomy and physiology. It is the largest organ by weight in the human body, playing important functions from the production of vitamin D to providing a protective barrier against the outside environment and pathogens [1]. Diseases affecting the skin are a common ailment, and their effective treatment requires a better understanding of their underlying pathophysiology.

Skin models are valuable to study skin development and its diseases. These models range from in vivo animal models to in vitro models created using primary cells or cell lines. Considering the evolutionary principle that all organisms share most of the genetic codes and somehow related due to common ancestry, well-designed animal models can recapitulate most of the disease characteristic. They are advantageous because studying skin disorders in the clinic is limited and in-depth investigation is only possible after biopsy of live tissue or postmortem analysis [2]. There is also limited access to normal and diseased human tissue, and therefore, there is a need for human tissue equivalents to be used in studying the underlying disease mechanisms in further detail. Various skin injuries can be mimicked in animal models that provide an easier way of studying the process of wound healing to find the main culprits of deficient healing versus excessive scar formation. Humanized mouse models are developed by growing human tissues transplanted into mice with severe combined immunodeficiency [3,4]. These models can be used to study diseased human cells or tissues as they continue to grow in the animal model, providing an exceptional opportunity to identify differential gene expression and response to drugs.

The use of animal models has been responsible for major advances in crucial in the understanding of human biology and disease in the last century. As noted, animal species share similar physiological characteristics with humans; therefore, it is possible to produce a condition artificially in animals that resembles human disease or injury. As a result, animal models have been employed in all fields of biomedical research including skin diseases. This has led to the development and testing of drugs, vaccines, and other biologicals that have improved and advanced human health. Historically, a

Skin Tissue Models. https://doi.org/10.1016/B978-0-12-810545-0.00002-4

variety of animal models has been utilized to develop treatments that revolutionize how physicians practice medicine. Most notably, in 1921, Drs. Banting and Best at the University of Toronto discovered the hormone insulin through the use of dogs. This discovery transformed the way physicians managed diabetic patients and provided relief for sufferers. Improving the quality of life became the main concern rather than how to prevent death in diabetic patients. The above example illustrates the substantive contributions of animal models in the investigation of human disease. The limitations of animal models should be acknowledged, however. Researchers should keep in mind that there is no "perfect" comprehensive animal model for every specific human disease. Rather, disease models need to focus on specific components of human illness to eventually develop a comprehensive understanding. For example, one model may be used to identify targets for compounds that treat acute tissue damage, whereas another may seek to find targets relevant to long-term recovery. The transferability and predictability of data obtained from animals is also a major concern [5]. This is because animals may not share all the properties necessary to completely recapitulate human conditions, particularly in the skin [6]. This requirement is not essential to creating an effective model, however, and animal models are required to discover new indications that may eventually be used in the clinic.

Interventional research studies play a fundamental role in understanding whether newly discovered drugs could be of clinical use. Long before these drugs are entered into clinical trials, the first step is to use surrogate models of the skin with closely matching physiology and anatomy. The relevant safety and efficacy of newly developed chemicals by drug discovery efforts can be tested extensively to ensure minimal risk and less number of failed attempts at finding the best remedy. Computational and mathematical models of skin bioprocesses are cost-effective ways of generating hypotheses that can later be tested in vitro or on animals.

Here, we will discuss the role of skin models in understanding the pathophysiology of skin diseases, drug discovery and treatment, and cell therapy. Various skin diseases ranging from skin cancer, trauma, and developmental disorders, and a wide variety of approaches with different levels of sophistication will be covered. We will address how these skin models are generated, their strengths and limitations, and finally their translation into the clinic.

2. Skin models for understanding the pathophysiology of diseases associated with skin

Skin models are useful tools for better understanding the pathophysiology of skin diseases. Understanding the process of skin development is essential for a better understanding of skin diseases and the process of wound healing. As such, models for skin development are important since they provide insight into the complex features observed during skin diseases. To develop better clinical treatments of skin diseases, there is a need to better understand the pathobiology of skin diseases through skin models. Several models have been developed to recapitulate the features of specific skin diseases, and such models have provided valuable insight into the pathobiology of these diseases. Moreover, the pathobiology differs greatly between skin diseases, and as such, understanding these differences is important to creating a useful skin model. These models need to encompass developmental and/or acquired skin diseases, skin cancer, and lastly wound healing in response to trauma such as burn injury.

3. Developmental and acquired skin diseases

An ideal model should recapitulate most of the characteristics of the organ or target disease. The search for the perfect animal model to study developmental skin diseases such as systematic sclerosis (SSc) continues to elude scientists. For instance, numerous in vitro and in vivo studies have shown that targeting tyrosine kinase inhibitors (TKI) such as nilotinib and imatinib can be used to target skin fibrosis [7–9]. In vitro data suggest that targeting TKI influences TGF-β and PDGF signaling leading to antifibrotic effects [7]. Moreover, in vivo studies utilizing bleomycin-treated mice (an experimental model of fibrosis) demonstrated the effect of nilotinib and imatinib in preventing fibrosis in several different organs, including the skin [9]. Despite these promising in vitro and in vivo studies, the use of these drugs in open-labeled and controlled clinical trials has led to poor results in SSc patients [10]. This illustrates the importance of properly developing a skin model that can be translated to the clinic. An ideal skin model should recapitulate most of the pathophysiological changes observed in a specific disease. To further complicate matters, the drug response depends heavily on the animal model used that has significant implications for clinical trials. Bleomycin-treated mice exhibit a drastic reduction in PDGFRB-positive cells, skin thickness, and myofibroblast activation in response to TKI treatment [11]. A newly developed animal model for SSc, called Fra2, however, has a completely different response. Like bleomycin-treated mice, Fra2 mice show decreased levels of PDGFRB in response to TKI treatment [11]. On the contrary, Fra2 mice do not show any phenotypic improvements as skin thickness was the same in treated and untreated mice [11]. It is clear that inconsistent results of the aforementioned clinical trials can be partially explained by the inaccuracies of these animal models in recapitulating SSc. This is especially important considering the heterogeneity observed in SSc and other skin diseases like skin cancer. Animal models, particularly mouse models, allow genetic manipulations of a gene throughout the body or in a specific cell type. For instance, labeling specific cells with a fluorescence reporter allows researchers to study the behavior of those specific cells in response to the disease [12] (Fig. 1). Animal models should be interpreted more cautiously when predicting treatment outcomes in humans. Another aspect that is often ignored when creating an animal model is whether the particular skin disease is observed in any other animal. As mentioned earlier, creating an accurate skin model for SSc has been difficult. SSc is a human-specific condition, and this key fact likely accounts for the difficulties scientists have encountered in creating a suitable model for this disease. Recently, in order to study skin diseases with an immunologic component such as allergic dermatitis, immunocompetent 3-D models including dendritic cells in addition to fibroblast and keratinocytes have been developed [13]. Overall, differences in the inflammatory response, immune system, or the metabolism of xenobiotics are only some of the issues limiting the applicability of animal models.

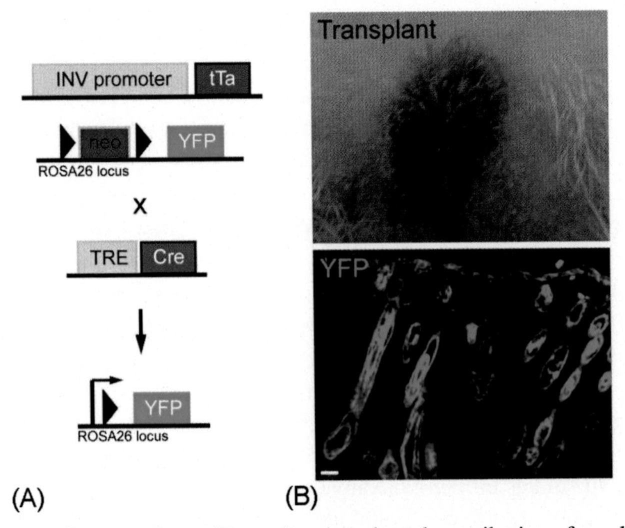

Fig. 1 Lineage-tracing experiment illustrating the role and contribution of myeloid cells in repairing the dermal compartment of the skin in mice.
Taken from Vorhagen S, Jackow J, Mohor SG, Tanghe G, Tanrikulu L, Skazik-Vogt C, Tellkamp F. Lineage tracing mediated by cre-recombinase activity. J Invest Dermatol 2015;135(1):1–4. https://doi.org/10.1038/jid.2014.472.

In vitro models that attempt to reconstruct the skin are important in studying skin diseases, particularly in increasing the speed and success rate of drug development. Monolayer (2-D)-based assays are not suitable for drug development as they do not mimic interactions of different cell types present in multilayer tissue [14]. 3-D constructs are needed to accurately recapitulate the human skin and mimic the biotransformation of drugs in such an environment. 3-D models are particularly important in diseases that arise from cell-cell interactions such as skin cancer. In 3-D skin models, a scaffold fosters cell adhesion to help guide tissue growth and development and supports the use of advanced coculture systems [14].

Several academic and industrial bodies have produced commercial in vitro human skin models composed of human keratinocytes cocultured with human fibroblasts (EpiSkin, EpiDerm, epiCS, and Labcyte) [15]. Interestingly, these skin models form multilayered epithelium and express markers of epidermal differentiation. The practicality of in vitro skin models lies in their utility in studying the toxicity, pharmacology, and transport ability of new drugs and their ability to be reproduced in large quantities [14]. While convenient, there are several limitations to these skin models. They are not exact replicas of the human skin in vivo; thus, assessing tissue response is problematic. Moreover, the biggest limitation is their weak barrier function that leads to false-positive results when assessing dermal penetration.

NativeSkin, an ex vivo skin model, attempts to address these limitations and improve upon currently available products (Fig. 2). To create NativeSkin, skin

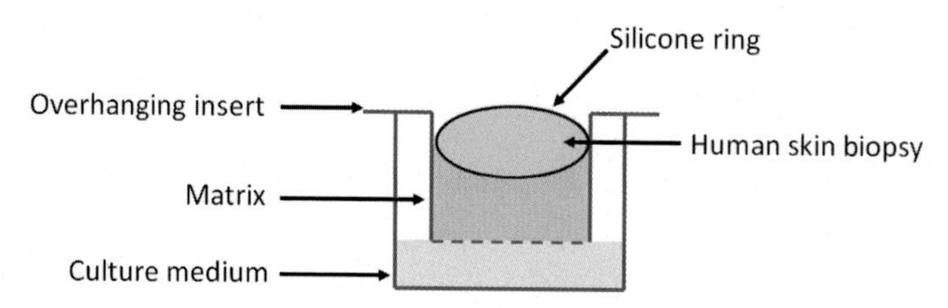

Fig. 2 NativeSkin, an ex vivo skin model created from human skin biopsies, can be used to test treatments prior to use in clinical trials.

biopsies are first obtained from patients undergoing surgery. The skin samples are then cultivated in multiwall transport plates surrounded by a proprietary matrix that nourishes the dermal of the skin using serum- and hydrocortisone-free culture medium. The tissue integrity, viability, and barrier properties are maintained for at least 7 days in culture allowing for studies to be conducted. NativeSkin has become a standardized ex vivo model of the human skin for final testing prior to clinical testing in humans due to its superior physiological barrier function and robust stratum corneum making it ideal for absorption studies. The major limitation of NativeSkin model is that one cannot replicate pathological conditions like SSc or skin cancer. Using a skin biopsy of these patients and developing an ex vivo skin model of the disease is another way to create disease models. However, the latter is deprived of the systemic microenvironment of the body. In summary, developing a skin model that recapitulates most of the hallmark features of the disease can be a useful tool to understand the accompanying pathophysiological features. For example, studying the signaling pathways that are activated (or inactivated) or cells that are recruited to the skin of a disease model will shed light on the pathophysiology of the disease.

4. Skin cancer

Several models have been developed for studying skin cancers. Developing skin models of nonmelanoma cancer is imperative as cutaneous squamous cell carcinoma (SCC) is one of the most common cancers in the Caucasian population causing significant morbidity and mortality [16]. Because most SCCs are caused by excessive exposure to UV radiation [17], some skin models have utilized immunocompetent SKH-1 mice that lack a dense, UV-impenetrable hair coat found in wild-type mice. UV exposure in this model has been shown to induce SCC in a time and dose and exposure time-dependent manner [18]. The use of this model has been questioned, however, as the hairless gene plays an important role in skin metabolism [19] and might serve as a major confounding factor. Therefore, its knockout may inadvertently influence tumor development. The classic and more accurate animal model of studying skin cancers is two-stage chemical carcinogenesis. First, there is a topical application of the carcinogen dimethylbenzanthracene (DMBA) that causes mutations in certain cells [20]. Next, there is repeated administration of the pro-inflammatory ester 12-O-tetradecanoylphorbol-1-acetate (TPA) that promotes the mutated cells into benign tumors that eventually develop into malignant ones [20]. The strength of this model is the ability to track the biological sequence of tumor formation, changes in signaling pathways, and the role of inflammation. The breadth of information acquired from the two-stage carcinogenesis model informs future approaches to treatment and chemoprevention in SCC. Skin models like this can enlighten further the signaling pathways and (ab)normal cells that have been recruited to the skin and therefore provide a valuable model for studying the pathophysiology of disease. Lastly, models of skin carcinogenesis caused by UVB rays have been created in vitro using human cells [21]. Even though in vitro models are deprived of the systemic and local microenvironment, these models are advantageous since human cells have been used, harboring the genetic and transcriptional machinery of human cells. One method of studying the pathophysiology of skin diseases is to culture patient-derived cells in vitro and studying disease mechanisms at the molecular and cellular levels. While animal models are useful for studying skin cancer, tumor cells can be readily isolated from patients allowing for in-depth experiments. It is also possible to model the genetic abnormalities in skin cancer by utilizing patient-derived cell lines isolated from tumor biopsies. Moreover, cell lines carrying a genetic defect of interest to model specific diseases can be accomplished [22]. This is advantageous because researchers can model a disease phenotype in undifferentiated or differentiated cells to elucidate disease etiology and develop novel therapies through drug screening.

5. Trauma and wound healing

The development of reconstructed skin to replace damaged or lost skin such as in burns is important. The engineering of the skin not only is useful for the treatment of the patients but also can be used for studying the process of skin healing in vivo and in vitro. Wound healing is a highly complex, well-orchestrated process that involves a series of coordinated and overlapping processes. Skin models provide a new platform to better understand these cellular and molecular mechanisms underlying skin regeneration in mammals, ultimately leading to better treatments in clinical settings. To study wound healing, in vitro assays are commonly used as they are inexpensive, fast, and excellent for examining the effect of certain agents on particular cell types (e.g., fibroblasts and keratinocytes). This allows for researchers to determine the effectiveness of various treatments, particularly healing-enhancing agents. For instance, 3-D skin models have been produced where keratinocytes are grown on a layer of fibroblasts within a collagen gel, mimicking basic skin layers [15]. This 3-D "skin equivalent" can be wounded, and interactions between these two different cell types during wound repair can be examined. Moreover, it is possible to evaluate the signaling pathways that are activated in each of these cell types after wounding and during different stages of healing. The strength of in vitro models of wound healing is the ability to easily control variables such as pH, salinity, and temperature. As previously mentioned, these models are incapable of recapitulating the in vivo wound environment nor capable of revealing the adverse effects of candidate therapeutic compounds since they are lacking the systemic and local microenvironment. Therefore, studying factors such as the inflammatory response or angiogenesis become difficult resulting in an inability to predict patient response to treatment.

Rodent and small mammal models of wound healing have become the model of choice for most researchers because they are inexpensive, easily obtainable, and undergo accelerated healing compared with humans [6]. For instance, rodent models are commonly used to study full-thickness wounds. Fig. 3 illustrates one such model in which a full-thickness scald burn is used to study skin healing after burn [6]. The main strength of small animal models is the ability to implement genetic manipulations. These genetic alterations can approximate human conditions such as diabetes and allows researchers to study the cellular and molecular intricacies involved in wound healing. As such, it is possible to map out the role of specific cells in skin healing that results in a better understanding of skin pathogenesis. The best in vivo animal model to study wound healing is the porcine model [6]. While most wound healing studies use the rodent system, the skin structure of pigs is most similar to humans as opposed to "loose-skinned" rodents. Both humans and pigs share similar epidermal thickness, similar patterns of hair follicles and blood vessels, and similar dermal collagen content. Pigs and humans also have similar molecular responses to various growth factors. Partial- or full-thickness surgical excisions can be performed on the dorsal trunk skin of the pig to accurately mimic a clinical situation tissue damage, and tissue loss occurs. Moreover, variations in the experimental design can be employed depending on which aspect of wound healing is being focused on. For instance, the method of anesthesia and pain control varies from study to study when porcine models are utilized. While the porcine model is an indispensable skin model in wound healing, the

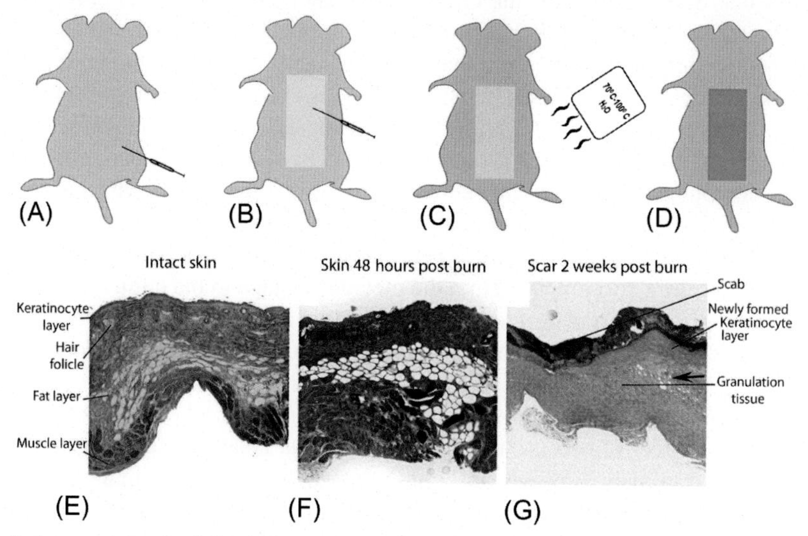

Fig. 3 An example of a full-thickness burn model used to study wound healing. Taken from Abdullahi A, Amini-Nik S, Jeschke MG. Animal models in burn research. Cell Mol Life Sci 2014;71(17):3241–55.

significant cost disadvantage and difficulty in handling make this a difficult model to carry out [6]. Overall, pig skin most closely resembles human skin, but due to the size of pigs, implementing a porcine model is quite challenging to execute and can pose a risk to the researcher despite.

The development of hypertrophic scarring and keloids (also known as fibrosis) due to the excess accumulation of extracellular matrix during wound healing is a major challenge in the clinic. Because of the complex nature of wound healing, in vivo models are best suited to study fibrosis. To evaluate potential antifibrotic therapeutics, bleomycin-induced skin fibrosis mouse model is commonly used. Repetitive subcutaneous injections of bleomycin cause dermal thickening and collagen accumulation that mimics fibrosis [23]. Histological analysis can subsequently be utilized to gain new insights into the mechanism of fibrosis. The complexity of the fibrotic process makes is quite difficult to mimic the process in vitro. Regardless, assays that evaluate key processes such as proliferation, migration, myofibroblast differentiation, and collagen production are well suited to evaluate the effectiveness of antifibrotic agents [24]. In burn clinics across the world, hypertrophic scarring is still a major challenge. There is a tremendous need to improve outcomes and quality of life of patients [25]. This can be accomplished by further elucidating the molecular mechanisms involved in fibrosis so that scar development can be attenuated or prevented by using new therapeutic strategies. Table 1 summarizes skin models used in understanding the pathophysiology of skin diseases and their clinical translation.

6. Skin models in drug discovery

Human diseases or injuries can be mimicked in animal models and then treated with experimental drugs to identify the ones with lowest complications and highest efficacy. Before drugs are tested on animals, high-throughput screening (HTS) can be performed on millions of drugs simultaneously using cells or small model organisms using robotic technology and computational analysis that show a high level of predictability for in vivo efficacy [26,27]. Many experiments can be conducted with more ease using in vitro model systems, such as assessment of repeated dosing effects [28]. In silico computational models are being used for drug discovery in skin cancers such as melanoma [29]. In silico models utilize computational models and/or computer simulation to screen and design drugs. Thus, these computational models are involved in all stages of drug development from the preclinical stage to the clinical trials themselves. Drug discovery and development is a lengthy and intensive venture; hence, in silico drug design is cost-effective in research and development of drugs. For example, in silico models have been used to study the mechanical stabilities of different cell types and keratinocyte migration during wound healing [30]. The use of skin models in HTS is cost-effective because it reduces the need for costly repeated trial and error experiments on animals. However, it should be noted that without animal experiments, results of HTS experiments do not initiate clinical trials. Furthermore, the translation of drug discovery to clinical trials is difficult. Many skin models used for drug discovery have yet to translate their discoveries in to the clinic (Table 2).

Table 1 **Skin models used in understanding the pathophysiology of skin diseases**

Skin disease model	In vitro	In vivo	Description	Clinical translation	References
Systemic sclerosis	✓ (Mouse)	✓	Antifibrotic effects by targeting PDGFRB and TGF-β1 via tyrosine kinase inhibitors (TKI)	TKI inhibitors had poor results in SSc patients	[7,10]
Systemic sclerosis		✓ (Mouse)	Fra2 knockout mice show reduced fibrotic response (PDFGBR)	N/A	[11]
Squamous cell carcinoma (SCC)		✓ (Mouse)	Two-stage chemical carcinogenesis in mice unravels pathophysiology and informs future treatments	N/A	[19]
Wound healing	✓ (Human and mouse)		3-D skin models where keratinocytes are grown on a layer of fibroblasts within a collagen gel, mimicking basic skin layers. Allows researchers to test the effectiveness of treatments	N/A	[14] [54,55]
Hypertrophic scarring		✓ (Mouse)	Bleomycin-induced fibrosis in rodents has led to the targeting of specific cytokines such as TGF-β3	Avotermin (recombinant, active, human TGF-β3) reduced total scar scores compared with placebo in phase I/II studies. Phase III trials failed, however	[22,56]

Table 2 Skin models used in drug discovery

Skin disease model	In vitro	In vivo	Description	Clinical translation	References
Melanoma	✓ (Human)		Three-dimensional full-thickness melanoma spheroids that respond to treatment. Cisplatin identified as potential treatment	N/A	[35]
Reconstructed human epidermis	✓ (Human)		Models such as EpiDerm, EpiSkin, and NativeSkin have been validated. Useful for testing drug toxicity, metabolization, and drug delivery	N/A	[57–59]
Explant	✓ (Human)		Explants of diseased skin used to study epithelial migration and compound screening	N/A	[60–63]

Since the discovery of induced pluripotent stem cells (iPSCs), are human pluripotent stem cells that are capable of self-renewal and have the potential to differentiate into almost any cell type. These cells can be used to help overcome the limitations of animal models for certain skin disorders leading to the establishment of new skin models that can be used for drug discovery. Recently, there has been a big boost in drug discovery using bioengineered human tissue such as the skin [31,32]. Skin models generated by iPSCs can be used to identify individual differences in treatment response for use in personalized medicine [33]. To illustrate this, organ-on-a-chip is a microfluidic device where cells can be cultured in chambers with continuous flow of media [34]. This is arranged on a microchip and allows for in vitro HTS of millions of drugs on live cells. Although most in vitro skin models are in a two-dimensional monolayer, 3-D models of the skin have been developed as better biomimetic to be used for HTS [35]. Better recapitulation of skin microenvironment provides a platform to study the response of skin cells to cosmetics and experimental drugs.

Three-dimensional full-thickness organotypic skin model of melanoma is being used to test for the most effective chemotherapy regimen [36]. Skin models can be used to evaluate the degree of topical drug penetration in vitro. By reducing skin barrier using chemicals, studies are able to model skin diseases with reduced barrier function to be tested for penetration of drugs [37]. Other proposed skin models for studying drug penetration include isolated flaps of the human skin maintained by continuous perfusion ex vivo [38]. They showed that these flaps can be maintained metabolically active for up to 6h and can provide a close-to-in vivo model of drug penetration in the human skin. The use of iPSCs to model specific skin disorders is limited by the ability to develop efficient and robust differentiation protocols [22]. Obtaining mature and functional cells for some cell types remains a challenge for some [22]. The greatest strength of iPSCs lies in the ability to screen new drugs, epically therapies relaying on nucleic acid-based and amino acid-based drugs [22]. iPSCs also become more useful when a skin disease lacks appropriate animal models. It is likely that many clinical trials will bypass animal models and begin due to new drugs being discovered in culture.

7. Skin models in cell therapy

Extensive skin injuries such as burn are associated with minimal viable cells including stem cells to repopulate the wound area. Tissue-engineered skin substitutes are a special type of 3-D skin model that are created in vitro using different cell types and biomaterials to replace the loss of skin with cells [39,40]. These models serve two purposes: (1) provide epidermal barrier to prevent further contamination of the wound and to prevent fluid and heat loss from the injured skin and (2) to introduce mature fibroblast and keratinocytes and/or stem cells to promote healing. These substitutes are in turn initially tested on animal models of skin injury to evaluate their safety and efficacy [41]. To date, decellularized human dermis has been the most effective scaffold clinically and is superior to the available artificially engineered cross-linked scaffolds [42]. These dermal matrices are usually isolated from a cadaveric skin and then processed with detergents to remove cells and antigenic material. Cells in a cellularized artificial skin substitute can be comprising of adult mesenchymal stem cells derived from multiple sources [43] or fully mature skin cells such as keratinocytes and fibroblasts. Bioprinting of engineered skin in 3-D is being developed as the next step in the advancement of skin substitutes and could potentially allow for a quick and efficient way of covering extensive skin loss in the operating room [44,45]. Another benefit would be their use in remote locations with limited availability of skilled plastic surgeons [45]. A crucial determinant of successful use of engineered skin substitutes is their engraftment and integration into the host skin. Vascularizing the scaffold can greatly improve its integration and viability at the graft site, and efforts are underway to create scaffolds that contain vascular networks derived from iPSCs [46].

Wounds caused by nonburn injuries are also a good candidate for the use of skin models. Skin healing post radiation therapy in mice has shown to be improved by bilayer tissue-engineered skin substitutes [47]. Any type of skin defect including developmental defects of the skin can be a good candidate for use of skin substitute. It has been used successfully in the treatment of abdominal wall defects such as giant exomphalos [48]. An artificial bilayer skin substitute derived from porcine tendon was clinically effective at covering wounds caused by necrotizing fasciitis in human with a low incidence of infection [49]. Table 3 summarizes skin models used in therapy and their clinical impact. Chronic nonhealing wounds are another major group of skin disorders that can benefit from skin substitutes. Venous leg ulcers are difficult to treat

Table 3 Skin models used in cell therapy

Skin disease model	In vitro	In vivo	Description	Clinical translation	References
Burn injury	✓ (Mouse)		Tissue-engineered skin substitutes are a 3-D skin model created from different cell types and biomaterials. Can introduce different cell types like fibroblasts and keratinocytes	Several clinical trials have shown that Integra dermal regenerative template is safe and effective in treating burn wounds	[38,39], [64–67]
Wound healing	✓ (Porcine)		Artificial bilayer skin substitute developed from the porcine tendon	Clinically effective at covering wound area in necrotizing fasciitis	[48]

clinically, and current treatment strategies are minimally effective. Randomized controlled trials show that bilayer tissue-engineered skin is better at healing these wounds than simple dressing [50]. In addition to cellularized versions, acellular biopolymers are also proving to be highly effective treatment strategies to treat skin wounds [51]. Another use of skin models for treatment discovery is for gene therapy [52]. In vitro and in vivo models of skin diseases can be subjected to gene-editing technologies such as CRISPR to remove known mutations in disease pathophysiology. Gene-edited iPSCs can potentially be used to generate gene-corrected tissue to be transplanted back into the patients [53].

8. Conclusion

The heterogeneity and complexity of human skin diseases make it so that no single skin model is capable of fully recapitulating each clinical scenario. The mechanisms of skin diseases have yet to be fully elucidated; however, the use of skin models is a promising avenue for clinical translation. The benefits and limitations of each skin model such as reproducibility, cost, and accuracy must be considered when choosing the optimal model for each study. Many of the models discussed thus far can be further improved to enable better represent the challenges facing the clinic. By optimizing skin models so that they better represent the clinical manifestations of skin diseases, we can use these models to more precisely understand the associated pathological mechanisms. Consequently, we can identify therapeutic targets and develop more appropriate treatments for patients in the clinic.

There has been a strong push recently in the scientific community to emphasize the importance of moving basic science into the clinic. Skin models provide a powerful tool for the discovery and development of new drugs and therapies for skin diseases. Clinical trials are expensive and time-consuming as they can require millions of dollars and many years to implement and are associated with ethical issues. Fully exploring potential therapeutics via skin models dramatically increases the chance of success in clinical trials. While the models described here are far from ideal, they are the best available presently and have provided new insights. The development of more robust models for evaluating new therapies for the treatment of skin diseases is of utmost importance. Despite imperfectness, skin models will continue to play an important role in the clinic.

References

[1] Proksch E, Brandner JM, Jensen JM. The skin: an indispensable barrier. Exp Dermatol 2008;17(12):1063–72.

[2] Kovarik CL, et al. Forensic dermatopathology and internal disease. J Forensic Sci 2005;50(1):154–8.

[3] de Oliveira VL, et al. Humanized mouse model of skin inflammation is characterized by disturbed keratinocyte differentiation and influx of IL-17A producing T cells. PLoS One 2012;7(10):e45509.

[4] Shimamura T, et al. Efficacy of BET bromodomain inhibition in Kras-mutant non-small cell lung cancer. Clin Cancer Res 2013;19(22):6183–92.
[5] Mestas J, Hughes CC. Of mice and not men: Differences between mouse and human immunology. J Immunol 2004;172(5):2731–8.
[6] Abdullahi A, Amini-Nik S, Jeschke MG. Animal models in burn research. Cell Mol Life Sci 2014;71(17):3241–55.
[7] Daniels CE, et al. Imatinib mesylate inhibits the profibrogenic activity of TGF-beta and prevents bleomycin-mediated lung fibrosis. J Clin Invest 2004;114(9):1308–16.
[8] Rosenbloom J, Jimenez SA. Molecular ablation of transforming growth factor beta signaling pathways by tyrosine kinase inhibition: the coming of a promising new era in the treatment of tissue fibrosis. Arthritis Rheum 2008;58(8):2219–24.
[9] Akhmetshina A, et al. Treatment with imatinib prevents fibrosis in different preclinical models of systemic sclerosis and induces regression of established fibrosis. Arthritis Rheum 2009;60(1):219–24.
[10] Daniels CE, et al. Imatinib treatment for idiopathic pulmonary fibrosis: randomized placebo-controlled trial results. Am J Respir Crit Care Med 2010;181(6):604–10.
[11] Maurer B, et al. Levels of target activation predict antifibrotic responses to tyrosine kinase inhibitors. Ann Rheum Dis 2013;72(12):2039–46.
[12] Amini-Nik S, et al. Beta-catenin-regulated myeloid cell adhesion and migration determine wound healing. J Clin Invest 2014;124(6):2599–610.
[13] Chau DY, et al. The development of a 3D immunocompetent model of human skin. Biofabrication 2013;5(3):035011.
[14] Alepee N, et al. State-of-the-art of 3D cultures (organs-on-a-chip) in safety testing and pathophysiology. ALTEX 2014;31(4):441–77.
[15] Bart De Wever SK, Descarg P. Human skin models for research applications in pharmacology and toxicology: introducing native skin, the “missing link” bridging cell culture and/or reconstructed skin models and human clinical testing. Appl In Vitro Toxicol 2015;**1**(1):6.
[16] Rogers HW, et al. Incidence estimate of nonmelanoma skin cancer in the United States, 2006. Arch Dermatol 2010;146(3):283–7.
[17] Koh HK, et al. Prevention and early detection strategies for melanoma and skin cancer. Current status. Arch Dermatol 1996;132(4):436–43.
[18] van Kranen HJ, et al. Frequent p 53 alterations but low incidence of ras mutations in UV-B-induced skin tumors of hairless mice. Carcinogenesis 1995;16(5):1141–7.
[19] Kumpf S, et al. Hairless promotes PPARgamma expression and is required for white adipogenesis. EMBO Rep 2012;13(11):1012–20.
[20] Huang PY, Balmain A. Modeling cutaneous squamous carcinoma development in the mouse. Cold Spring Harb Perspect Med 2014;4(9):a013623.
[21] Tyagi N, et al. Development and characterization of a novel in vitro progression model for UVB-induced skin carcinogenesis. Sci Rep 2015;5:13894.
[22] Avior Y, Sagi I, Benvenisty N. Pluripotent stem cells in disease modelling and drug discovery. Nat Rev Mol Cell Biol 2016;17(3):170–82.
[23] Avouac J. Mouse model of experimental dermal fibrosis: the bleomycin-induced dermal fibrosis. Methods Mol Biol 2014;1142:91–8.
[24] Benam KH, et al. Engineered in vitro disease models. Annu Rev Pathol 2015;10:195–262.
[25] Finnerty CC, et al. Hypertrophic scarring: The greatest unmet challenge after burn injury. Lancet 2016;388(10052):1427–36.
[26] Williams CH, Hong CC. Zebrafish small molecule screens: taking the phenotypic plunge. Comput Struct Biotechnol J 2016;14:350–6.

[27] Kenny HA, et al. Quantitative high throughput screening using a primary human three-dimensional organotypic culture predicts in vivo efficacy. Nat Commun 2015;6:6220.
[28] Spriggs S, et al. Effect of repeated daily dosing with 2,4-dinitrochlorobenzene on glutathione biosynthesis and Nrf 2 activation in reconstructed human epidermis. Toxicol Sci 2016;154(1):5–15.
[29] Pennisi M, et al. Computational modeling in melanoma for novel drug discovery. Expert Opin Drug Discov 2016;11(6):609–21.
[30] Safferling K, et al. Wound healing revised: a novel reepithelialization mechanism revealed by in vitro and in silico models. J Cell Biol 2013;203(4):691–709.
[31] Guo Z, et al. Building a microphysiological skin model from induced pluripotent stem cells. Stem Cell Res Ther 2013;4(Suppl. 1):S2.
[32] Mack DL, et al. Disease-in-a-dish: the contribution of patient-specific induced pluripotent stem cell technology to regenerative rehabilitation. Am J Phys Med Rehabil 2014;93(11 Suppl. 3):S155–68.
[33] Bilousova G, Roop DR. Induced pluripotent stem cells in dermatology: potentials, advances, and limitations. Cold Spring Harb Perspect Med 2014;4(11):a015164.
[34] Bhatia SN, Ingber DE. Microfluidic organs-on-chips. Nat Biotechnol 2014;32(8):760–72.
[35] Nam KH, et al. Biomimetic 3D tissue models for advanced high-throughput drug screening. J Lab Autom 2015;20(3):201–15.
[36] Vorsmann H, et al. Development of a human three-dimensional organotypic skin-melanoma spheroid model for in vitro drug testing. Cell Death Dis 2013;4:e719.
[37] Engesland A, Skalko-Basnet N, Flaten GE. In vitro models to estimate drug penetration through the compromised stratum corneum barrier. Drug Dev Ind Pharm 2016;42(11):1742–51.
[38] Ternullo S, et al. The isolated perfused human skin flap model: a missing link in skin penetration studies? Eur J Pharm Sci 2016;96:334–41.
[39] Nicholas MN, Jeschke MG, Amini-Nik S. Methodologies in creating skin substitutes. Cell Mol Life Sci 2016;73(18):3453–72.
[40] Chua AW, et al. Skin tissue engineering advances in severe burns: review and therapeutic applications. Burns Trauma 2016;4:3.
[41] Philandrianos C, et al. Comparison of five dermal substitutes in full-thickness skin wound healing in a porcine model. Burns 2012;38(6):820–9.
[42] Debels H, et al. Dermal matrices and bioengineered skin substitutes: a critical review of current options. Plast Reconstr Surg Glob Open 2015;3(1):e284.
[43] Leonardi D, et al. Mesenchymal stem cells combined with an artificial dermal substitute improve repair in full-thickness skin wounds. Burns 2012;38(8):1143–50.
[44] Patra S, Young V. A review of 3D printing techniques and the future in biofabrication of bioprinted tissue. Cell Biochem Biophys 2016;74(2):93–8.
[45] Bauermeister AJ, Zuriarrain A, Newman MI. Three-dimensional printing in plastic and reconstructive surgery: a systematic review. Ann Plast Surg 2016;77(5):569–76.
[46] Abaci HE, et al. Human skin constructs with spatially controlled vasculature using primary and iPSC-derived endothelial cells. Adv Healthc Mater 2016;5(14):1800–7.
[47] Busra MF, et al. Tissue-engineered skin substitute enhances wound healing after radiation therapy. Adv Skin Wound Care 2016;29(3):120–9.
[48] Almond SL, et al. Novel use of skin substitute as rescue therapy in complicated giant exomphalos. J Pediatr Surg 2006;41(3):e1–2.
[49] Akita S, Tanaka K, Hirano A. Lower extremity reconstruction after necrotising fasciitis and necrotic skin lesions using a porcine-derived skin substitute. J Plast Reconstr Aesthet Surg 2006;59(7):759–63.

[50] Jones JE, Nelson EA, Al-Hity A. Skin grafting for venous leg ulcers. Cochrane Database Syst Rev 2013;1:CD001737.
[51] Dickinson LE, Gerecht S. Engineered biopolymeric scaffolds for chronic wound healing. Front Physiol 2016;7:341.
[52] Levy A, Petit I, Aberdam D. Pluripotent stem cells as a cellular model for skin: relevance for physiopathology, cell/gene therapy and drug screening. Eur J Dermatol 2015;25(Suppl 1):12–7.
[53] Maeder ML, Gersbach CA. Genome-editing technologies for gene and cell therapy. Mol Ther 2016;24(3):430–46.
[54] Collawn SS, et al. Adipose-derived stromal cells accelerate wound healing in an organotypic raft culture model. Ann Plast Surg 2012;68(5):501–4.
[55] Nicholas MN, Jeschke MG, Amini-Nik S. Cellularized bilayer pullulan-gelatin hydrogel for skin regeneration. Tissue Eng Part A 2016;22(9–10):754–64.
[56] Ferguson MW, et al. Prophylactic administration of avotermin for improvement of skin scarring: three double-blind, placebo-controlled, phase I/II studies. Lancet 2009;373(9671):1264–74.
[57] Netzlaff F, et al. The human epidermis models epi skin, skin ethic and epi Derm: an evaluation of morphology and their suitability for testing phototoxicity, irritancy, corrosivity, and substance transport. Eur J Pharm Biopharm 2005;60(2):167–78.
[58] Hu T, et al. Xenobiotic metabolism gene expression in the epi Dermin vitro 3D human epidermis model compared to human skin. Toxicol In Vitro 2010;24(5):1450–63.
[59] Van Gele M, et al. Three-dimensional skin models as tools for transdermal drug delivery: challenges and limitations. Expert Opin Drug Deliv 2011;8(6):705–20.
[60] Peramo A, Marcelo CL. Visible effects of rapamycin (sirolimus) on human skin explants in vitro. Arch Dermatol Res 2013;305(2):163–71.
[61] Attia-Vigneau J, et al. Regeneration of human dermis by a multi-headed peptide. J Invest Dermatol 2014;134(1):58–67.
[62] Lu H, Rollman O. Fluorescence imaging of reepithelialization from skin explant cultures on acellular dermis. Wound Repair Regen 2004;12(5):575–86.
[63] Stenn KS, Dvoretzky I. Human serum and epithelial spread in tissue culture. Arch Dermatol Res 1979;264(1):3–15.
[64] Stiefel D, Schiestl C, Meuli M. Integra artificial skin for burn scar revision in adolescents and children. Burns 2010;36(1):114–20.
[65] Groos N, et al. Use of an artificial dermis (Integra) for the reconstruction of extensive burn scars in children. About 22 grafts. Eur J Pediatr Surg 2005;15(3):187–92.
[66] Danin A, et al. Assessment of burned hands reconstructed with Integra((R)) by ultrasonography and elastometry. Burns 2012;38(7):998–1004.
[67] Heimbach DM, et al. Multicenter postapproval clinical trial of Integra dermal regeneration template for burn treatment. J Burn Care Rehabil 2003;24(1):42–8.

Section B

Skin diseases: Clinical demands and diseased-skin in vitro models

In vitro models of melanoma

3

Dagmar Kulms[*,†], *Friedegund Meier*[*,†,‡]
[*]Experimental Dermatology, Department of Dermatology, TU-Dresden, Dresden, Germany, [†]Center for Regenerative Therapies TU Dresden, TU-Dresden, Dresden, Germany, [‡]National Center for Tumor Diseases (NCT), Dresden, Germany

1. Introduction

Malignant transformation of melanocytes, the pigment cells of the human skin, causes formation of malignant melanoma that is a highly aggressive cancer with increased metastatic potential. Despite remarkable efforts, metastatic melanoma still presents with significant mortality.

Recently, monochemotherapies are increasingly replenished by more cancer-specific combination therapies involving targeted kinase inhibitors and immunotherapy regimens. Still, metastatic melanoma remains a life-threatening disease because tumors exhibit primary resistance or develop resistance to novel therapies, thereby regaining tumorigenic capacity. In order to improve the therapeutic success of malignant melanoma, novel treatment options need to be explored. Therefore, determination of molecular mechanisms conferring resistance against conventional treatment approaches requires innovative cellular in vitro models. Since malignant melanoma presents with high heterogeneity, the need of predictive tools that specify dysregulation within cancer cells and focus on individual molecular targets is demanded to identify responders to selected combination therapies while sparing unnecessary treatment burden for nonresponders. It therefore appears that a more thorough and reliable preclinical evaluation of novel drugs and therapeutic combination will be essential to reduce attrition rates in clinical trials and improve the benefit from therapeutic interventions. The following chapter aims to introduce and discuss different 3D in vitro melanoma models that might be suitable to portray the in vivo architecture of malignant melanoma and may warrant new insights into intratumoral and tumor-host interactions.

2. Development and treatment of malignant melanoma

2.1 The role of melanocytes in human skin

The human skin is composed of two distinct layers that serve different functions in protecting the body from adverse environmental effects [1]. The lower dermal compartment consists of a fibroelastic connective tissue, being composed of loosely connected collagen and elastin fibers synthesized by fibroblasts, serving a mechanic barrier function. In addition, supplying nerves, vessels, and capillary and hair follicles and sweat glands are located in the dermis to further protect the body by regulating the temperature and itching [2]. The dermis is separated from the upper epidermal

Skin Tissue Models. https://doi.org/10.1016/B978-0-12-810545-0.00003-6

layer by basal lamina that is produced as an extracellular matrix due to a constant communication between both skin layers. In contrast to the dermis, the epidermis is a squamous epithelium that mainly consists of keratinocytes that can be differentiated into four sublayers. The *stratum basale* consists of undifferentiated basal keratinocytes that constantly derive from skin progenitor cells stratifying through the stages of the *stratum spinosum* and *stratum granulosum* into the *stratum corneum* to protect the body from dehydration and infection by microorganisms [3]. Immunologic protection is furthermore provided by epidermal Langerhans cells that as the first immunologic barrier recognize pathogens for elimination.

Melanocytes are aligned at the basal membrane and communicate through dendritic extensions with multiple keratinocytes to form an epidermal melanin unit. Melanocytes produce the pigment melanin which, when distributed to the surrounding keratinocytes, protects them from UV-induced DNA damage by quenching the radiation intensity. Accordingly, such melanin units serve the function to protect the body from the adverse effects of UV radiation like skin aging, immunosuppression, inflammation, and formation of nonmelanoma skin cancer (NMSC). UV-induced DNA damage concomitant with UV-induced immunosuppression are the key inducers of malignant transformation of epidermal keratinocytes into squamous cell carcinoma (SSC) and basal cell carcinoma (BCC); their contribution to transformation of melanocytes to malignant melanoma (MM), however, is still under debate [4].

2.2 Development of malignant melanoma

The source and/or risk factors for the development of malignant melanoma are multifactorial comprising environmental influences and genetic factors [5]. According to the traditional model postulated by Clark et al., melanoma development can be differentiated into different tumor progression stages, being characterized by certain genetic, morphological, and histological changes [6]. In this model, melanoma originates from either innate or acquired nevi due to a local increase of melanocyte proliferation causing benign neoplasia. This benign precursor lesion may convert into structurally modified dysplastic nevi containing atypic cells, which may continue to the first malignant stage, the radial growth phase (RGP). This early tumor progression phase is characterized by cells radially proliferating within the epidermis, showing few locally invasive cells within the papillary dermis. During the subsequent vertical growth phase (VGP), melanoma cells already show a metastatic and invasive phenotype by breaking through the basal lamina to infiltrate the deeper parts of the dermis and the subcutaneous tissue [7]. Finally, the metastatic melanoma (MM) represents the most aggressive progression stage with metastatic cells systemically spreading throughout the blood and lymph system to invade far organs like the liver, lung, and brain [8] (Fig. 1).

More recent melanoma models, however, could show that only 26% of all melanoma originate from nevi of which only 43% went through the stage of a dysplastic nevus [9]. Just like in any other neoplastic diseases, individual tumor progression steps may be omitted, or the tumor may develop from progenitor cells [10] or tumor-initiating stem cells [11].

While malignant melanoma represents only 5% of all skin cancers developed, it exhibits the highest mortality due to its very aggressive character, presenting with a

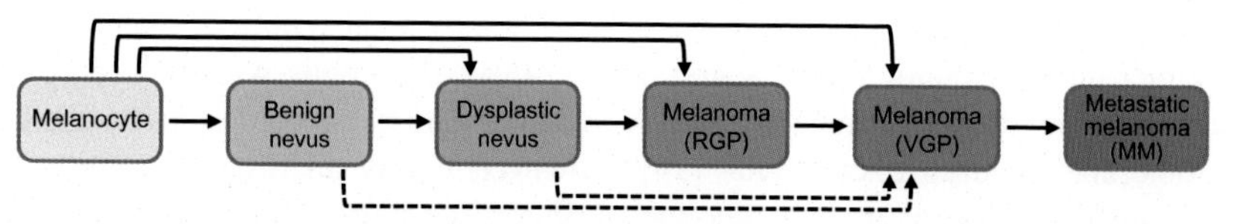

Fig. 1 Scheme of melanoma progression. Malignant transformation of primary melanocytes from benign or dysplastic nevi is a stepwise process involving the radial growth phase (RGP), followed by the vertical growth phase (VGP) finally leading to malignant/metastatic melanoma (MM).

sevenfold increase in incidences within the last four decades (Robert Koch Institute). Already since 2008, this type of cancer represents the fifth highest tumor entity for women and the eighth highest for men [12]. Worldwide, about 160,000 new incidences are being diagnosed per year (World Cancer Report).

2.3 Treatment options for malignant melanoma

2.3.1 Chemotherapy

To date, early diagnosis followed by surgery of malignant melanoma still remains the most effective therapy. While 90% of all melanoma are being diagnosed as primary tumors without metastatic indications, the 10-year survival rate adds up to 75%–85% [13]. The prognosis for patients with distant metastasis is particularly poor with 1-year survival rates ranging from 33% to 62% [14], with classical chemotherapy regimens found to confer little survival benefit. Until 2011, treatment with the alkylating agent dacarbazine (DTIC) was regarded as standard treatment for patients with inoperable metastatic melanoma. Up to that point, alternatives had been limited [15]. In several phase III studies, DTIC showed objective response rates between 5% and 14%, with a tolerable side effect profile but without a clinically significant impact on overall survival. The efficacy and toxicity of the alkylating agent temozolomide are comparable with those of DTIC [16]. Various combined chemotherapy regimens have failed to show any survival benefit compared to DTIC monochemotherapy [17]. Given their high toxicity and associated impairment of quality of life, they may be considered for patients with a high-tumor load or rapid progression after failure of other systemic therapies [15,17]. In view of the treatment options currently available, it is safe to assume that chemotherapy is to be considered a "last-line therapy." However, after decades of stagnation, recent advances in targeted therapies and immunotherapy have considerably improved the prognosis of stage IV melanoma.

2.3.2 Targeted therapy

Doubtlessly, dysregulation of two major mitogen-activated pathways, namely, the RAS-RAF-MEK-ERK and the PI3K-AKT-PTEN signaling pathways, presents key drivers of melanoma progression, especially when constitutively activating point mutations of the protooncogenes BRAFV600 and NRAS are present [18]. Accordingly, the invention of targeted kinase inhibitors promised therapeutic benefit for patients suffering from metastatic melanoma. A multitude of clinical trials conducted to this point has not achieved significant benefit for patients with metastatic melanoma. In this context, a phase III study showed that the BRAFV600E kinase inhibitor vemurafenib (Zelboraf) induces major tumor regression in 48% of patients with BRAFV600E-mutated metastatic melanoma compared with 5% of patients treated with the classical chemotherapeutic drug dacarbazine [19]. However, nearly all responses are partial with a subpopulation of patients showing primary resistance. Moreover, the acquisition of secondary resistance leading to relapse was observed in the majority of patients, with a median duration of response of ~7 months [20]. Administration of an alternative inhibitor of mutated BRAF, dabrafenib, showed comparable results in a phase III study [21].

Occurring in the majority of patients over the course of treatment, development of resistance poses a limitation to monotherapy with one of the aforementioned BRAF inhibitors. Most resistance mechanisms described to date result in reactivation of the MAP kinase signaling pathway [22]. By combining a BRAF and a MEK inhibitor, one may expect a delay in the development of resistance and a prolongation of the therapeutic response. Indeed, in a recent phase III study, the combined therapy with the BRAF inhibitor dabrafenib and the MEK inhibitor trametinib was superior to dabrafenib alone, with a response rate of 69% versus 53% and a median overall survival of 25.1 months versus 18.7 months [23]. Similar data were obtained with the BRAF inhibitor vemurafenib in combination with the MEK inhibitor cobimetinib [24].

2.3.3 Immunotherapy

Immunotherapy strategies for the treatment of melanoma have been used or investigated since a long time (e.g., interferon alpha and interleukin-2). They are based on the knowledge that, under certain conditions, the immune system is capable not only of controlling and combating tumor cells but also of "accepting" them [25]. Ipilimumab, a fully human IgG1 monoclonal antibody, blocks cytotoxic T-lymphocyte-associated antigen-4 (CTLA-4), a negative regulator of T-cells, thereby augmenting T-cell activation and proliferation. Ipilimumab is the first therapeutic agent that in phase III clinical trials has been shown to lead to a significantly longer survival time than in control groups when given as a monotherapy or in combination with dacarbazine [26,27]. However, the rate of best overall response was only 15.2% in the ipilimumab-dacarbazine group and 10.3% in the dacarbazine group. In the analysis of pooled survival data from phase II and III studies, a plateau in the survival curve was observed after 3 years, with a 3-year survival rate of at least 20%–26% [28]. Still, the plateau that started at 3 years continued through to 10 years.

Another relevant immune checkpoint involves the programmed death-1 (PD-1) signaling pathway. This PD-1 receptor inhibits T-cell activity by interaction with its ligands PD-L1 on tumor cells and antigen-presenting cells, resulting in an immunosuppressive tumor environment [29]. Therapeutic use of blocking antiPD-1 (nivolumab or pembrolizumab) antibodies interrupts this immunosuppression and increases the antitumor T-cell response.

In a phase III study, nivolumab versus chemotherapy achieved a response rate of 40% versus 13.9% and a 1-year survival rate of 72.9% versus 42.1% in patients with BRAF wild-type melanoma [27].

Combined inhibition of PD-1 and CTLA-4 allows inactivated tumor-specific T-cells to proliferate and exert their effector functions again. In a phase III trial investigating nivolumab and ipilimumab versus nivolumab or ipilimumab alone in patients with metastatic melanoma, both response rates and the median progression-free survival could significantly be improved in the combination therapy compared to the respective monotherapies [30]. The combination of nivolumab and ipilimumab is expected to achieve a long-lasting response; however, the benefit of this combination strategy comes at the expense of a high-level toxicity that results from overall increased immune activity [31,32], and a full cure of the disease is not yet tangible.

It becomes obvious that analysis of the mutation status alone is not sufficient to develop the most beneficial therapeutic strategy. In contrast, new fast and reliable diagnostic tools are demanded to systematically record and analyze responsiveness of individual cancer cells. The vast majority of currently available data on human melanoma have been obtained from homogeneous two-dimensional (2D) cultures of melanoma cell lines. Tumor cells, however, grow in a three-dimensional (3D) environment in which intercellular crosstalk exists between differentiated cancer cell subpopulations as well as between cancer and nontransformed neighboring host cells. It would therefore be desirable to reproduce the 3D context in which melanoma develops in a suitable preclinical model [33,34].

3. In vitro models of melanoma

3.1 Sphere formation assay

Sphere-forming assays have been widely used to retrospectively identify (tumor) stem cells and their progeny based on their reported capacity to evaluate self-renewal and differentiation at the single-cell level in vitro [35]. The assay essentially tests every single cell in the population for its ability to undergo unlimited division thereby taking advantage of the fact that tumorigenic cells grow anchorage independently in soft agar. A colony is defined to consist of at least 50 cells forming solid microspheres with close cell-cell junctions [36]. Traditionally, these types of sphere-forming assay have been used since many years to study therapeutic responses, like ionizing radiation or chemotherapy survival curves of melanoma cells [37,38]. For this purpose, melanoma cells can be treated in vitro or alternatively be isolated from human melanoma xenografts or patients, and the clonogenic survival and tumorigenic potential can be assessed by evaluating the number and size of colonies formed from an individual tumor cell in soft agar [37-40]. More recently, colony formation assays have also been used to study the effects of micro RNAs and expression of other tumor stem cell marker on melanoma stem cell potential in isolated tumor cells [41,42]. However, sphere formation assays may not exclusively monitor stem cells but may also evaluate the potential of a cell to behave as a stem cell when removed from its in vivo niche. Further limitations include medium composition and volume, surface area of the culture dish, duration in culture before quantification, and, most importantly, cell density—which may lead to fusion of individual spheres [35]. Above this, sphere formation assays per definition focus on (tumor) stem cell behavior and do not capture the holistic composition of tumor metastasis, being composed of a heterogeneous cell population in vivo. Consequently, more accurate cellular 3D models have been developed that better resemble the architecture and cellular differentiation of metastasis.

3.2 *In vitro homotypic melanoma spheroids*

In contrast to cells in 2D, culture cells organized in tissues acquire a specific 3D architecture by being embedded into an extracellular matrix (e.g., collagen), which serves as a mechanic support and conveys biochemical signals and intracellular communication [43]. This cellular organization influences the physiological properties of individual cells, like gene expression, adhesion, migration, invasion, and the therapeutic outcome [44,45]. Multicellular tumor spheroids (MCTS) represent spherical and compact cell aggregates, which mimic avascularized tumor nodes, micrometastasis, or intercapillary microregions of a solid tumor in vitro [46]. Just like in a native tumor, the architecture of tumor spheroids represents with cellular heterogeneity due to a differential distribution of oxygen, nutrients, and cellular waste within the organoid [45,47]. Hence, a sufficient supply with oxygen and nutrients results in formation of a proliferating cell population at the outer rim of the spheroid corresponding to peripheral cells within the tumor tissue being located close the vascularization. With increasing distance from the source of oxygen and nutrients and concomitant accumulation of cellular debris, a consecutive layer of quiescent cells is constituted, followed by a center being composed of necrotic and apoptotic cells, respectively (Fig. 2).

According to the pathophysiological gradient described, similar growth kinetics as in solid tumors occur in artificial tumor spheroids, starting with exponential cell proliferation in the periphery, followed by a layer of limited and linear proliferation, in which resting and dying cell are being generated to form the necrotic center [45,47]. Consequently, the size of the tumor spheroid critically determines the analogy to avascularized tumor tissue. While spheroids up to a diameter of 150–200 μm will be sufficiently supplied with nutrients and oxygen via diffusion, only spheroids of >500 μm in diameter will present with the diverse cellular layers and the typical necrotic center [45,48].

Conventionally, melanoma spheroids can be generated via different methods, while all of them require culture conditions under nonadhesive conditions. Different options include culturing under rotating conditions (e.g., spinner flasks) and static culturing on conic nonadhesive surfaces (e.g., on agarose-coated multiwell dishes) or within a "hanging drop" [46,48]. In all cases, gravitational forces cause aggregation of cells in regular culture medium, while the number of cells seeded plus the duration of culturing determines the size and complexity of individual melanoma spheroids in a very reproducible way [45]. Particularly for melanoma, an additional criterion determines the quality of spheroid formation, namely, the tumor progression stage melanoma cells were derived from. Accordingly, melanoma cells derived from the metastatic stage (MM) form more solid spheroids than melanoma cells derived from earlier radial growth phase, being indicative for genetic changes that must have occurred during tumor progression (Fig. 3).

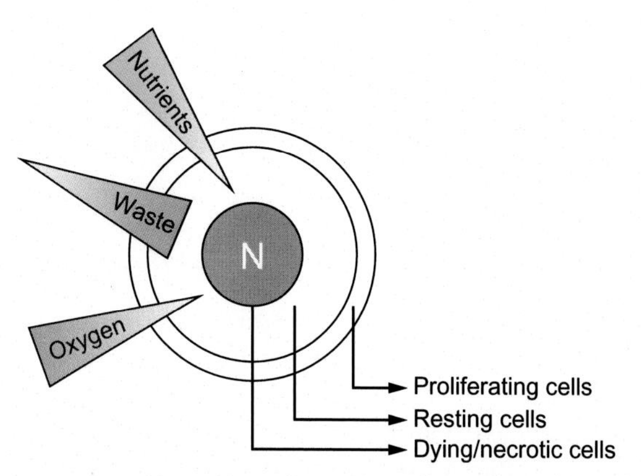

Fig. 2 Architecture of avascularized metastasis. Insufficient supply of nutrients and oxygen causes cell death in the center coinciding with an accumulation of cellular debris. As a consequence, only the cells of the periphery show high-proliferation rates, followed by a resting cell subpopulation and accordingly a necrotic/apoptotic center.

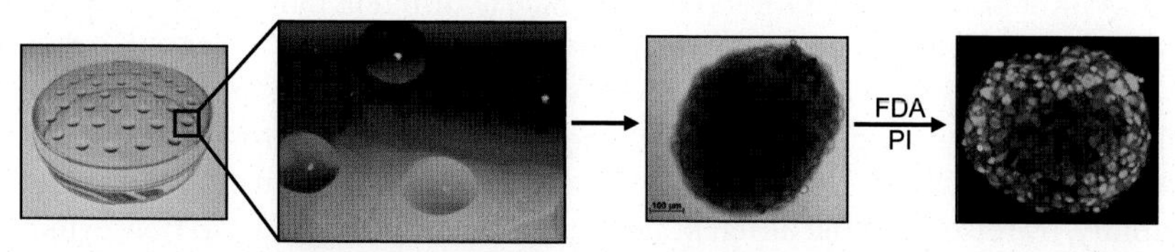

Fig. 3 Melanoma spheroids generated via the "hanging drop method." 250 cells per drop are being positioned on the lid of a cell culture dish supplied with PBS. After 10–14 days, melanoma spheroids of a diameter of about 500 μm can be harvested. Life/dead staining with fluorescein diacetate (FDA, *green*)/propidium iodide (PI, *red*) proofs that cells are well and alive within these spheroids.

The generation of melanoma spheroids allows a thorough analysis of therapy responsiveness in vitro. For this purpose, spheroids can be stained for cell death (fluorescein isothiocyanate (FITC, green)-coupled annexinV and propidium iodide (PI, red)) and analyzed by imaging or flow cytometry. Alternatively, individual cells can be recollected by intense trypsinization for further biochemical analysis. Certainly, melanoma spheroids can be generated from genetically manipulated cells to specifically study the influence of individual cellular components/molecules on spheroid formation, proliferation, and therapy susceptibility.

3.3 In vitro homotypic FUCCI melanoma spheroids

In order to study the specific impact of, for example, cytostatic drugs on melanoma proliferation or the influence of targeted drugs on the cell cycle, FUCCI-melanoma cells have been generated that can be analyzed individually or within a spheroid structure, as described above [49]. FUCCI-melanoma cells are based on the protocol invented for the epithelial cell line HeLa to monitor progression of cells through the different cell cycle phases [50]. Cells express an mCherry-fused DNA replication factor Cdt1 and mVenus-fused Geminin. These two chimeric proteins accumulate reciprocally during the cell cycle, labeling the nuclei of cells in G1 phase red and those of cells in S/G2/M phase green. Confocal live-cell imaging of FUCCI-melanoma spheroids can be utilized to document the relationship between the cell cycle and melanoma cell motility, invasion, and drug sensitivity [51].

3.4 In vitro heterotypic multi cellular melanoma spheroids

Since tumor nodes or micrometastasis in vivo do not exclusively exist of a homogeneous melanoma cell population but are surrounded or interspersed with stroma cells and endothelial cells [43], the development of heterotypic multicellular spheroids has become a particular challenge. On the one hand, a two-step process involves generation of melanoma microspheres first, which subsequently are exposed to fibroblasts or endothelial cells to surround or enclose the tumor (reviewed in [52]). On the other hand 2–3 cell types (melanoma cells, fibroblasts, and/or endothelial cells) might be incorporated at the same time into one single spheroid, probably reflecting best the heterogeneity of tumor metastasis in vivo [53]. Accordingly, melanoma cells, endothelial cells, and/or skin fibroblasts can be mixed at different ratios to generate heterotypic multicellular spheroids, for example, in a "hanging drop." Once the individual cell types included have been stained with different cell trackers before aggregation, multicolor heterotypic spheroids can be imaged by confocal microscopy (Fig. 4).

Above this, these heterotypic multicellular spheroids can be analyzed for drug susceptibility, by reisolating the different cell types and subjecting them to flow cytometry analysis after identification with selective antibodies. Consequently, heterotypic multicellular spheroids represent the smallest cellular unit to represent a heterogeneous

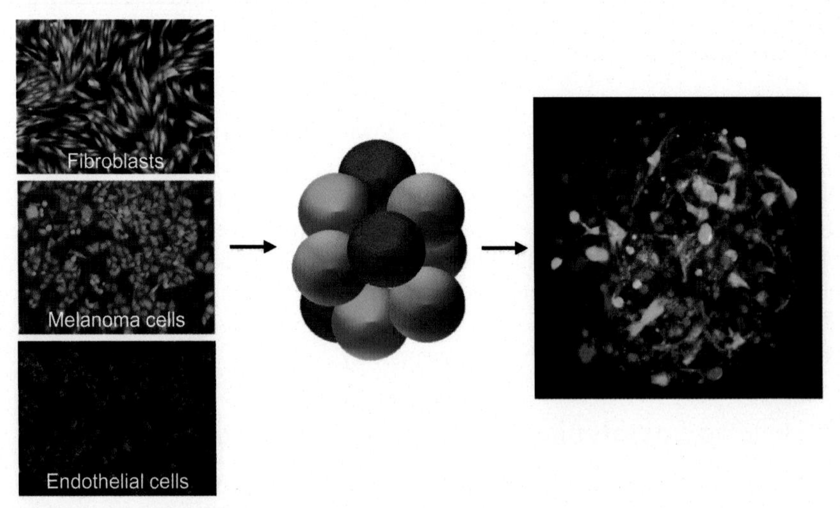

Fig. 4 Generation of multicellular spheroids. Fibroblasts (*green*), melanoma cells (*blue*), and endothelial cells (*red*) were stained with cell tracker and combined within a spheroid via the "hanging drop method." Three days later, spheroids were analyzed by confocal fluorescence microscopy.

tumor tissue including endothelial and stromal cells and may be supportive to understand how infiltration with stroma and vessels may impact on the therapeutic outcome. If applicable, the composition of heterotypic spheroids may be altered by substituting skin fibroblasts with lung fibroblasts, liver stellate cells or astrocytes to mimic different host tissues preferably targeted by melanoma metastasis. By these means, differences in growth behavior and drug susceptibility can be monitored dependent on the organ-specific tumor stroma. Still, these models have to deal with very small entities of cells, and an appropriate embedment of the tumor tissue into a respective microenvironment is not provided by these minimal models.

As an alternative, organotypic human skin models have been developed that harbor melanoma cells, melanoma nests, or melanoma spheroids to better mimic the in vivo setting of a primary tumor or primary metastasis embedded into the human skin.

3.5 Organotypic melanoma skin tissue models

Besides predictive tools and alternative treatment concepts, cellular models are mandatory that best resemble the in vivo situation of human melanoma. Although genetically engineered mouse models for metastatic melanoma have been developed [54], they provide only limited insight into mechanisms of development and therapy resistance of the human disease, because mice fundamentally do not develop melanoma. This is probably due to the fact that murine melanocytes are located in the hair follicle rather than in the epidermal basal cell layer [55] and the skin is tightly populated with hair, by both means protecting melanocytes from the adverse effects of UV radiation. Accordingly, in vivo animal studies on melanoma are carried out in transgenic mice or using xenograft models [56]. Above this, the architecture and the physiology of murine and human skin differs significantly, comprising only three layers of keratinocytes in the murine compared with 6–10 layers in the human epidermis [57,58]. Also for ethical reasons, the reputation of mice models decreases, while the impact of human organotypic skin reconstructs increases significantly.

Three-dimensional skin reconstructs (skin equivalents) have proved themselves to be very useful as test systems in the investigation of skin cancer, including melanoma. These skin equivalents can close the gap between investigations in different in vitro cell cultures and the in vivo testing in animal models by reproducing the three-dimensional arrangement of epidermal and mesenchymal cells. While solely dermal and epidermal skin reconstruction models exist, organotypic full skin equivalents best recapitulate a differentiation status homologous to normal human skin [59]. In these skin equivalents, primary human keratinocytes, seeded on a dermal layer of primary human fibroblasts embedded into a collagen scaffold, stratify from undifferentiated highly proliferative cells located at the base membrane to highly keratinized cells in the *stratum corneum*, thus forming distinct layers of the epidermis as seen in normal human skin. As shown in different studies, the close proximity of dermal fibroblasts and epidermal keratinocytes stimulates both cell types to form a basal membrane [60,61]. Due to cellular adhesion of molecules embedded in the basal membrane, the development of a mechanical resilient neodermis during the in vitro culture is possible [62]. These organotypic full skin models can host melanoma cells of different tumor progression stages and are therefore eminently suitable for investigating physiological and pathological processes of cutaneous melanoma under in vivo-mimicking conditions.

For this purpose, melanoma cells can be seeded into the epidermal or dermal part of the full skin reconstructs to mimic different stages of the disease, including migration, invasion, and tumor nest formation. Regarding the aggressiveness and the invasion potential, melanoma cells behave according to their tumor progression stage within this organotypic environment, thereby reflecting the clinical relevance of the model [58,63]. Cultured cells from RGP melanoma have characteristics of both malignant and nonmalignant cells: they are immortal but do not grow anchorage independently in soft agar nor are they tumorigenic in mice [64]. In a typical RGP lesion, melanoma cells predominantly reside in the epidermis, with little invasion into the dermis. RGP primary melanomas are considered metastasis-incompetent, implying that they do not invade lymphatics and capillaries [65]. Correspondingly, cells of the RGP reside in and distribute exclusively throughout the epidermal part of 3D organotypic skin reconstructs [45]. In contrast, melanoma cells from the biologically advanced VGP of primary lesions have an infinite lifespan [8] and an early metastatic potential. Accordingly, individual cells of the vertical growth phase penetrate through the basal lamina into the lower dermal part of the skin reconstruct. Finally, cells of the metastatic stage (MM) aggressively invade into the dermal part to spontaneously form melanoma nests [8] (Fig. 5).

By genetically manipulating melanoma cells prior to insertion, these types of melanoma skin reconstructs offer an opportunity to identify molecular determinants for the development of the disease and at the same time support the target identification for the treatment of malignant melanoma [66]. Both, dermal and full skin models containing dermal nests of metastatic melanoma cells have significantly prompted and improved studies on monotherapy and combination therapy for melanoma in a more physiological context [67,68]. However, melanoma nests form randomly showing huge variations in number and size, even if the same number of cells has been seeded into the skin reconstructs. Also it takes some days for melanoma nests to develop under these experimental conditions. Hence, due to the limited life span of skin reconstructs (~20 days), most studies conducted start with an early treatment thereby rather interfering with tumor melanoma nest formation or outgrowth than inducing regression of existing tumor nests.

Above this, these metastatic melanoma nests typically consist of only a limited number of cells and therefore do not adequately recapitulate the complexity of human melanoma metastases. Melanoma nests lack the intratumoral cell heterogeneity and the existence of a necrotic center, which both might have an impact on the therapeutic outcome.

In order to improve the increasing requests on new therapeutic alternatives, ethically unproblematic yet highly reliable and reproducible ex vivo human-based screening models are demanded that can be validated to translate the findings from basic cellular research into clinical applications. To narrow the gap between in vitro and in vivo studies, melanoma spheroids have been integrated into full skin reconstructs yielding an innovative organotypic skin-melanoma spheroid model that might be more suitable for translating new therapeutic approaches into the clinical setting [45].

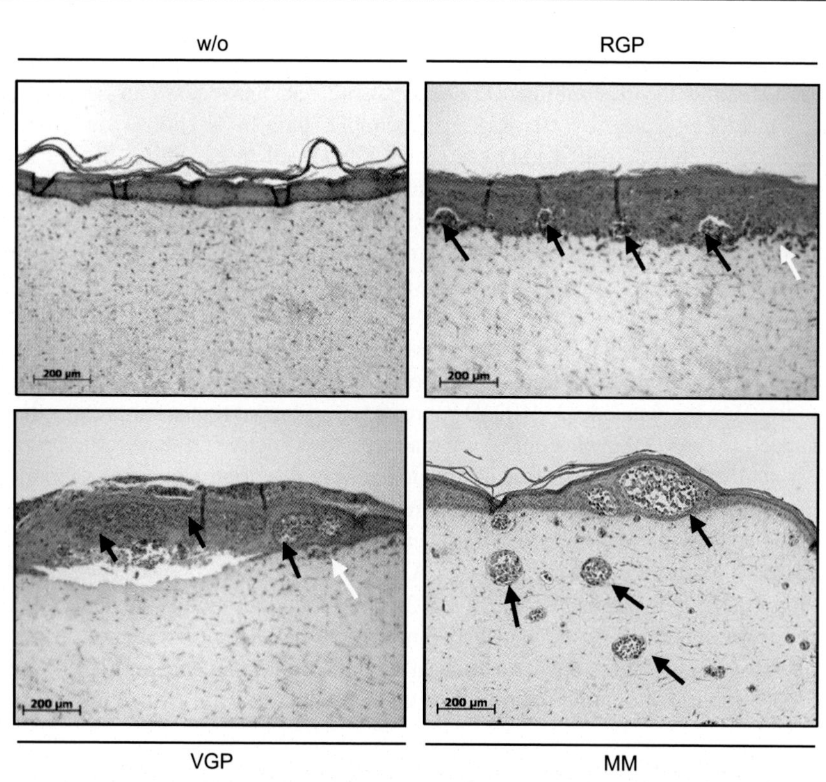

Fig. 5 Melanoma progression in an organotypic environment. Melanoma cells derived from different tumor progression stages were inserted into organotypic 3D skin reconstructs and their characteristic studied in hematoxylin-eosin stained paraffin sections. According to the tumor progression stage, cells from RGB remained within the epidermis, individual VGP cells invaded into the dermis, while cells from MM formed metastasis-like cell nests in the deeper part of the dermis.

3.6 Organotypic melanoma-spheroid skin tissue models

Organotypic skin-melanoma spheroid models harbor precise numbers of mature and well-differentiated melanoma spheroids of defined size, which have previously been raised by the "hanging drop" method. Thereby the number of cells applied per drop plus the duration of culturing can precisely be determined to gain spheroids of highly reproducible accuracy and differentiation. Subsequently, the number of spheroids inserted into the dermal part of the organotypic skin reconstructs can individually be altered depending on the experimental setting yielding a highly reproducible 3D model suitable for proper validation. This cell-based model recapitulates both the 3D organization and multicellular complexity of an organ/tumor in vivo but at the same time accommodates systematic experimental intervention. These organotypic skin-melanoma spheroid models—once terminally differentiated—still have a life span of about 2–3 weeks that allows extended experimental procedures and therapeutic interventions on existing metastasis-like melanoma spheroids. Embedded into organotypic 3D skin reconstructs, melanoma spheroids histologically show high similarity to primary cutaneous human melanoma metastases in vivo, including a similar degree of intratumoral melanoma cell differentiation and a necrotic center (Fig. 6).

Upon treatment with combination therapy, drug responsiveness can be evaluated immunohistochemically, by staining, for example, paraffin sections with antibodies against selected activated kinases, proteins involved in execution of cell death or TUNEL staining for necrotic and apoptotic cell death, respectively. Moreover, all cell types can be reisolated from the tissue models following collagenase digestion, separated via subsequent cell sorting, and might be subjected to further biochemical analysis or quantification. Making use of such techniques at least allows for a semi-quantitative prediction of therapeutic success.

In the context of melanoma therapy resistance, two molecules have moved into the focus of attention, because they had been shown to be upregulated in metastatic melanoma. On the one side, JARID1B represents a potential tumor suppressor that in normal cells exerts cell cycle control via maintenance of active retinoblastoma protein (Rb). Usually, JARID1B gets lost during melanoma development, but a subpopulation of tumor cells regains JARID1B expression resulting in downregulation of the cell cycle activity thereby escaping therapeutic effects [69,70]. On the other side, the membrane-bound drug efflux transporter ABCB5 was shown to be coexpressed with the tumor stem cell marker CD133 [71,72] on a subpopulation of melanoma cells derived from patients with advanced melanoma that had acquired therapy resistance [73]. Analogous to cutaneous metastasis in vivo, a certain percentage of cells within the proliferating cell subpopulation of melanoma spheroids are embedded into skin

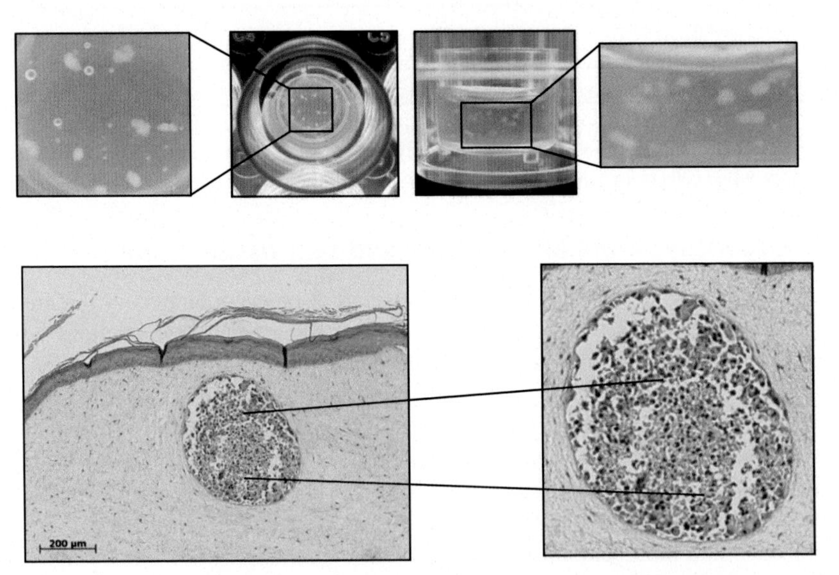

Fig. 6 Organization of melanoma spheroids embedded into organotypic 3D skin reconstructs. Melanoma spheroids were embedded into the dermal compartment (collagen matrix containing primary human fibroblasts) of full 3D skin reconstructs. H&E stained paraffin section reveals the architecture of melanoma spheroids that is homologous to avascularized metastasis in vivo, that is, proliferating peripheral population and necrotic center.

reconstructs, differentiated into JARID1B- and ABCB5-expressing cells [45]. Hence, melanoma spheroids may even represent the tumor stem cell subpopulation that is supposed to be present in metastasis in vivo and may hinder the therapeutic success. The degree of cellular differentiation observed is neither reflected in 2D cell culture nor sufficiently addressed in 3D models incorporating tumor nests or smaller spheroids. Moreover, melanoma spheroids generated from freshly isolated patient material can be generated, or small sections of tumor material can directly be inserted into the organotypic skin model for preclinical drug screening. Standardization of this model will allow to more reliably predict clinical effectiveness of novel therapeutic regimes to be taken to the clinical stage and may further allocate the basis for an individual preclinical drug screening tool to pave the way for personalized medicine.

4. Conclusion

Despite the recent development of novel effective targeted drugs and immune checkpoint inhibitors, metastatic melanoma remains a life-threatening disease because tumor cells show intrinsic resistance or acquire resistance to targeted therapy and immunotherapy, respectively, thereby regaining tumorigenic capacity. Yet, there is a high-unmet medical need for new treatment options for patients with metastatic BRAF wild-type and BRAF V600 mutant melanoma and for patients who developed resistance to BRAF inhibitors. In addition, alternative treatment options for patients suffering from NRAS-mutated melanoma need to be identified to offer alternative treatment options for patients that fail immunotherapeutic approaches.

This chapter summarizes numerous elegant 3D models that have been developed within the last years that are able to better reflect the human disease in vitro (Table 1). These cell-based models warrant new insights for a deeper understanding of intratumoral

Table 1 Summary of applications and limitations of different 3D melanoma models

Tissue model type	Application	Limitations
(3.1) Sphere formation assay	– Generated from individual cells – Tumor stem cell identification – Melanoma growth and survival – Studies on therapeutic responses – Studies on miRNA	– Not limited to tumor stem cells – Heterogeneous composition of tumor metastasis is not captured – No microenvironment – 50–100 cells only
(3.2) In vitro homotypic melanoma spheroids	– Generated from cell population – Mimic avascularized tumors – Architecture and physiology close to tumor nodes in vivo – >500 μm in diameter possible	– Vascularization and microenvironment missing – Size limited to <1 mm in diameter

Continued

Table 1 Continued

Tissue model type	Application	Limitations
(3.3) In vitro homotypic FUCCI-melanoma spheroids	– Cell cycle-dependent analysis of melanoma motility, invasion, and drug sensitivity	– Vascularization and microenvironment missing – Size limited to <1 mm in diameter
(3.4) In vitro heterotypic multicellular melanoma spheroids	– Contains melanoma cells, fibroblasts, and endothelial cells to better mimic the in vivo situation – Study therapy resistance depending on microenvironment – Separate analysis of drug susceptibility after cell sorting – >500 μm in diameter possible	– Very small entities of individual cell types – No embedment into microenvironmental scaffold (e.g., collagen)
(3.5) Organotypic melanoma skin tissue models	– Contains melanoma cells of different tumor progression stages within a full-thickness organotypic skin equivalent – Mimics melanoma cell behavior (migration, invasion, tumor nest formation) under physiological conditions within the skin – Drug testing/tumorselectivity	– Standardization as a screening model is problematic, since tumor nest numbers and sizes may vary significantly between different models – Tumor nest formation takes time; therefore, drug application rather inhibits tumor formation than causing regression of existing tumor nodes – Tumor nests do not reflect the complexity of metastasis in vivo – No immune cells present
(3.6) Organotypic melanomaspheroid skin tissue models	– Contains melanoma spheroids of defined number and size embedded within a full-thickness organotypic skin equivalent – Mimics avascularized metastasis under physiological conditions within the skin – Drug testing/tumorselectivity	– No immune cells present – Impossible to test drugs that interfere with immune checkpoints

and tumor-host interaction; however, they are not able to satisfactorily monitor immunotherapeutic approaches because insertion of immune cells faces problems with histocompatibility. Still, the different innovative 3D melanoma reconstruction models introduced here may provide better tools to study molecular mechanisms of tumor development and therapy resistance in the future. Each of them allows the in-depth investigation of genetically modified melanoma cells to pinpoint genes and proteins that may be responsible for treatment failure and drug resistance, respectively. In vitro characterization of transcriptomes, proteomes, and miRNomes comparing 2D and 3D settings of melanoma cells may help to understand more general changes from 2D melanoma cell culture to 3D in vivo-mimicking conditions, potentially helping to identify additional druggable targets. Moreover, utilizing melanoma cells that have previously been conditioned to specific drugs over months may be useful to investigate intratumoral changes under in vivo-mimicking conditions to study long-term survival of melanoma metastasis in vitro. While the development of novel therapeutic strategies has been beneficial for patients suffering from early stage systemic metastatic melanoma, patients with stage IV and V metastatic melanoma still have a poor overall survival. At these late tumor progression stages, metastasis usually no longer exists exclusively within the skin but has spread to more distant parts of the body including the liver, lung, and brain [8]. Thus, cross talk with the respective tumor microenvironment—different from the skin—may additionally alter the therapeutic outcome. Accordingly, development of dextran-based 3D matrices incorporating melanoma spheroids embedded into human astrocytes, lung fibroblasts, or liver stellate cells is currently in progress and may help to address the influence of different tumor microenvironments. For direct preclinical drug screening, embedment of whole melanoma pieces into full skin equivalents generated—if possible from primary keratinocytes and fibroblasts from the same patient's skin—would be most powerful and predictive. Conclusively, taking human cell-based models into the third dimension in the near future may foster the understanding of melanoma development, therapy resistance, tumor relapse, and metastatic outgrowth to pave the way to individualized medicine.

Abbreviations

2D two-dimensional
3D three-dimensional;
BCC basal cell carcinoma
CTLA-4 cytotoxic T-lymphocyte-associated antigen-4
DICT chemotherapeutic drug dacarbazine
FDA fluorescein diacetate
FUCCI fluorescence ubiquitin cell cycle indicator
MBM melanoma brain metastasis
MM malignant melanoma or metastatic melanoma
NMSC nonmelanoma skin cancer
PD-1 programmed death-1
PI propidium iodide
RGP radial growth phase
SCC squamous cell carcinoma
TUNEL terminal deoxynucleotidyl transferase dUTP nick end labeling
UV ultraviolet radiation
VGP vertical growth phase

References

[1] Boulais N, Misery L. The epidermis: a sensory tissue. Eur J Dermatol 2008;18(2):119–27.
[2] Nestle FO, Di MP, Qin JZ, Nickoloff BJ. Skin immune sentinels in health and disease. Nat Rev Immunol 2009;9(10):679–91.
[3] Blanpain C, Fuchs E. Epidermal homeostasis: a balancing act of stem cells in the skin. Nat Rev Mol Cell Biol 2009;10(3):207–17.
[4] Shain AH, Bastian BC. From melanocytes to melanomas. Nat Rev Cancer 2016;16(6):345–58.
[5] Leiter U, Eigentler T, Garbe C. Epidemiology of skin cancer. Adv Exp Med Biol 2014;810:120–40.
[6] Clark Jr WH. Human cutaneous malignant melanoma as a model for cancer. Cancer Metastasis Rev 1991;10(2):83–8.
[7] Hsu MY, Meier F, Herlyn M. Melanoma development and progression: a conspiracy between tumor and host. Differentiation 2002;70(9–10):522–36.
[8] Miller AJ, Mihm Jr MC. Melanoma. N Engl J Med 2006;355(1):51–65.
[9] Bhatia S, Tykodi SS, Thompson JA. Treatment of metastatic melanoma: an overview. Oncology (Williston Park) 2009;23(6):488–96.
[10] Meier F, Satyamoorthy K, Nesbit M, Hsu MY, Schittek B, Garbe C, et al. Molecular events in melanoma development and progression. Front Biosci 1998;3:D1005–10.
[11] Elder DE, Clark Jr WH, Elenitsas R, Guerry D, Halpern AC. The early and intermediate precursor lesions of tumor progression in the melanocytic system: common acquired nevi and atypical (dysplastic) nevi. Semin Diagn Pathol 1993;10(1):18–35.
[12] Erdmann F, Lortet-Tieulent J, Schuz J, Zeeb H, Greinert R, Breitbart EW, et al. International trends in the incidence of malignant melanoma 1953–2008—are recent generations at higher or lower risk? Int J Cancer 2013;132(2):385–400.
[13] Garbe C, Peris K, Hauschild A, Saiag P, Middleton M, Spatz A, et al. Diagnosis and treatment of melanoma: European consensus-based interdisciplinary guideline. Eur J Cancer 2010;46(2):270–83.
[14] Balch CM, Gershenwald JE, Soong SJ, Thompson JF, Atkins MB, Byrd DR, et al. Final version of 2009 AJCC melanoma staging and classification. J Clin Oncol 2009;27(36):6199–206.
[15] Kim T, Amaria RN, Spencer C, Reuben A, Cooper ZA, Wargo JA. Combining targeted therapy and immune checkpoint inhibitors in the treatment of metastatic melanoma. Cancer Biol Med 2014;11(4):237–46.
[16] Patel PM, Suciu S, Mortier L, Kruit WH, Robert C, Schadendorf D, et al. Extended schedule, escalated dose temozolomide versus dacarbazine in stage IV melanoma: final results of a randomised phase III study (EORTC 18032). Eur J Cancer 2011;47(10):1476–83.
[17] Eigentler TK, Caroli UM, Radny P, Garbe C. Palliative therapy of disseminated malignant melanoma: a systematic review of 41 randomised clinical trials. Lancet Oncol 2003;4(12):748–59.
[18] Lawrence MS, Stojanov P, Polak P, Kryukov GV, Cibulskis K, Sivachenko A, et al. Mutational heterogeneity in cancer and the search for new cancer-associated genes. Nature 2013;499(7457):214–8.
[19] Chapman PB, Hauschild A, Robert C, Haanen JB, Ascierto P, Larkin J, et al. Improved survival with vemurafenib in melanoma with BRAF V600E mutation. N Engl J Med 2011;364(26):2507–16.
[20] Flaherty KT, Puzanov I, Kim KB, Ribas A, McArthur GA, Sosman JA, et al. Inhibition of mutated, activated BRAF in metastatic melanoma. N Engl J Med 2010;363(9):809–19.

[21] Hauschild A, Grob JJ, Demidov LV, Jouary T, Gutzmer R, Millward M, et al. Dabrafenib in BRAF-mutated metastatic melanoma: a multicentre, open-label, phase 3 randomised controlled trial. Lancet 2012;380(9839):358–65.
[22] Solit DB, Rosen N. Resistance to BRAF inhibition in melanomas. N Engl J Med 2011;364(8):772–4.
[23] Long GV, Stroyakovskiy D, Gogas H, Levchenko E, De BF, Larkin J, et al. Combined BRAF and MEK inhibition versus BRAF inhibition alone in melanoma. N Engl J Med 2014;371(20):1877–88.
[24] Larkin J, Ascierto PA, Dreno B, Atkinson V, Liszkay G, Maio M, et al. Combined vemurafenib and cobimetinib in BRAF-mutated melanoma. N Engl J Med 2014;371(20):1867–76.
[25] Dunn GP, Old LJ, Schreiber RD. The three Es of cancer immunoediting. Annu Rev Immunol 2004;22:329–60.
[26] Hodi FS, Chesney J, Pavlick AC, Robert C, Grossmann KF, McDermott DF, et al. Combined nivolumab and ipilimumab versus ipilimumab alone in patients with advanced melanoma: 2-year overall survival outcomes in a multicentre, randomised, controlled, phase 2 trial. Lancet Oncol 2016;17(11):1558–68.
[27] Robert C, Long GV, Brady B, Dutriaux C, Maio M, Mortier L, et al. Nivolumab in previously untreated melanoma without BRAF mutation. N Engl J Med 2015;372(4):320–30.
[28] Schadendorf D, Hodi FS, Robert C, Weber JS, Margolin K, Hamid O, et al. Pooled analysis of long-term survival data from phase II and phase III trials of Ipilimumab in unresectable or metastatic melanoma. J Clin Oncol 2015;33(17):1889–94.
[29] Kyi C, Postow MA. Checkpoint blocking antibodies in cancer immunotherapy. FEBS Lett 2014;588(2):368–76.
[30] Larkin J, Chiarion-Sileni V, Gonzalez R, Grob JJ, Cowey CL, Lao CD, et al. Combined Nivolumab and Ipilimumab or monotherapy in untreated melanoma. N Engl J Med 2015;373(1):23–34.
[31] Heppt MV, Dietrich C, Graf SA, Ruzicka T, Tietze JK, Berking C. The systemic management of advanced melanoma in 2016. Oncol Res Treat 2016;39(10):635–42.
[32] Rauschenberg R, Garzarolli M, Dietrich U, Beissert S, Meier F. Systemic therapy of metastatic melanoma. J Dtsch Dermatol Ges 2015;13(12):1223–35.
[33] Kunz-Schughart LA, Freyer JP, Hofstaedter F, Ebner R. The use of 3-D cultures for high-throughput screening: the multicellular spheroid model. J Biomol Screen 2004;9(4):273–85.
[34] Pampaloni F, Reynaud EG, Stelzer EH. The third dimension bridges the gap between cell culture and live tissue. Nat Rev Mol Cell Biol 2007;8(10):839–45.
[35] Pastrana E, Silva-Vargas V, Doetsch F. Eyes wide open: a critical review of sphere-formation as an assay for stem cells. Cell Stem Cell 2011;8(5):486–98.
[36] Franken NA, Rodermond HM, Stap J, Haveman J, Van BC. Clonogenic assay of cells in vitro. Nat Protoc 2006;1(5):2315–9.
[37] Evans SM, Labs LM, Yuhas JM. Response of human neuroblastoma and melanoma multicellular tumor spheroids (MTS) to single dose irradiation. Int J Radiat Oncol Biol Phys 1986;12(6):969–73.
[38] Yohem KH, Bregman MD, Meyskens Jr FL. Effect of tumor colony definition on ionizing radiation survival curves of melanoma-colony forming cells. Int J Radiat Oncol Biol Phys 1987;13(11):1725–33.
[39] Courtenay VD, Selby PJ, Smith IE, Mills J, Peckham MJ. Growth of human tumour cell colonies from biopsies using two soft-agar techniques. Br J Cancer 1978;38(1):77–81.

[40] Selby PJ, Courtenay VD, McElwain TJ, Peckham MJ, Steel GG. Colony growth and clonogenic cell survival in human melanoma xenografts treated with chemotherapy. Br J Cancer 1980;42(3):438–47.
[41] Fomeshi MR, Ebrahimi M, Mowla SJ, Khosravani P, Firouzi J, Khayatzadeh H. Evaluation of the expressions pattern of miR-10b, 21, 200c, 373 and 520c to find the correlation between epithelial-to-mesenchymal transition and melanoma stem cell potential in isolated cancer stem cells. Cell Mol Biol Lett 2015;20(3):448–65.
[42] Roudi R, Ebrahimi M, Sabet MN, Najafi A, Nourani MR, Fomeshi MR, et al. Comparative gene-expression profiling of CD133(+) and CD133(−) D10 melanoma cells. Future Oncol 2015;11(17):2383–93.
[43] Flach EH, Rebecca VW, Herlyn M, Smalley KS, Anderson AR. Fibroblasts contribute to melanoma tumor growth and drug resistance. Mol Pharm 2011;8(6):2039–49.
[44] Chang TT, Hughes-Fulford M. Monolayer and spheroid culture of human liver hepatocellular carcinoma cell line cells demonstrate distinct global gene expression patterns and functional phenotypes. Tissue Eng Part A 2009;15(3):559–67.
[45] Vorsmann H, Groeber F, Walles H, Busch S, Beissert S, Walczak H, et al. Development of a human three-dimensional organotypic skin-melanoma spheroid model for in vitro drug testing. Cell Death Dis 2013;4:e719.
[46] Hirschhaeuser F, Menne H, Dittfeld C, West J, Mueller-Klieser W, Kunz-Schughart LA. Multicellular tumor spheroids: an underestimated tool is catching up again. J Biotechnol 2010;148(1):3–15.
[47] Sutherland RM. Cell and environment interactions in tumor microregions: the multicell spheroid model. Science 1988;240(4849):177–84.
[48] Lin RZ, Chang HY. Recent advances in three-dimensional multicellular spheroid culture for biomedical research. Biotechnol J 2008;3(9–10):1172–84.
[49] Beaumont KA, Anfosso A, Ahmed F, Weninger W, Haass NK. Imaging- and flow cytometry-based analysis of cell position and the cell cycle in 3D melanoma spheroids. J Vis Exp 2015;106:e53486.
[50] Sakaue-Sawano A, Kobayashi T, Ohtawa K, Miyawaki A. Drug-induced cell cycle modulation leading to cell-cycle arrest, nuclear mis-segregation, or endoreplication. BMC Cell Biol 2011;12:2. https://doi.org/10.1186/1471-2121-12-2.
[51] Beaumont KA, Hill DS, Daignault SM, Lui GY, Sharp DM, Gabrielli B, et al. Cell cycle phase-specific drug resistance as an escape mechanism of melanoma cells. J Invest Dermatol 2016;136(7):1479–89.
[52] Kulms D, Schwarz T. 20 years after—milestones in molecular photobiology. J Investig Dermatol Symp Proc 2002;7(1):46–50.
[53] Correa de SP, Auslaender D, Krubasik D, Failla AV, Skepper JN, Murphy G, et al. A heterogeneous in vitro three dimensional model of tumour-stroma interactions regulating sprouting angiogenesis. PLoS ONE 2012;7(2):e30753.
[54] Brohem CA, Cardeal LB, Tiago M, Soengas MS, Barros SB, Maria-Engler SS. Artificial skin in perspective: concepts and applications. Pigment Cell Melanoma Res 2011;24(1):35–50.
[55] El GA, Lamme E, Ponec M. Crucial role of fibroblasts in regulating epidermal morphogenesis. Cell Tissue Res 2002;310(2):189–99.
[56] Tuting T. T cell immunotherapy for melanoma from bedside to bench to barn and back: how conceptual advances in experimental mouse models can be translated into clinical benefit for patients. Pigment Cell Melanoma Res 2013;26(4):441–56.
[57] Pauwels M, Rogiers V. Safety evaluation of cosmetics in the EU. Reality and challenges for the toxicologist. Toxicol Lett 2004;151(1):7–17.

[58] Smalley KS, Lioni M, Herlyn M. Life isn't flat: taking cancer biology to the next dimension. In Vitro Cell Dev Biol Anim 2006;42(8–9):242–7.
[59] Breitkreutz D, Mirancea N, Nischt R. Basement membranes in skin: unique matrix structures with diverse functions? Histochem Cell Biol 2009;132(1):1–10.
[60] Andriani F, Margulis A, Lin N, Griffey S, Garlick JA. Analysis of microenvironmental factors contributing to basement membrane assembly and normalized epidermal phenotype. J Invest Dermatol 2003;120(6):923–31.
[61] Walles T, Weimer M, Linke K, Michaelis J, Mertsching H. The potential of bioartificial tissues in oncology research and treatment. Onkologie 2007;30(7):388–94.
[62] Schanz J, Pusch J, Hansmann J, Walles H. Vascularised human tissue models: a new approach for the refinement of biomedical research. J Biotechnol 2010;148(1):56–63.
[63] Bechetoille N, Haftek M, Staquet MJ, Cochran AJ, Schmitt D, Berthier-Vergnes O. Penetration of human metastatic melanoma cells through an authentic dermal-epidermal junction is associated with dissolution of native collagen types IV and VII. Melanoma Res 2000;10(5):427–34.
[64] Satyamoorthy K, Meier F, Hsu MY, Berking C, Herlyn M. Human xenografts, human skin and skin reconstructs for studies in melanoma development and progression. Cancer Metastasis Rev 1999;18(3):401–5.
[65] Meier F, Nesbit M, Hsu MY, Martin B, van Belle P, Elder DE, et al. Human melanoma progression in skin reconstructs: biological significance of bFGF. Am J Pathol 2000;156(1):193–200.
[66] Gaggioli C, Hooper S, Hidalgo-Carcedo C, Grosse R, Marshall JF, Harrington K, et al. Fibroblast-led collective invasion of carcinoma cells with differing roles for RhoGTPases in leading and following cells. Nat Cell Biol 2007;9(12):1392–400.
[67] Meier F, Schittek B, Busch S, Garbe C, Smalley K, Satyamoorthy K, et al. The RAS/RAF/MEK/ERK and PI3K/AKT signaling pathways present molecular targets for the effective treatment of advanced melanoma. Front Biosci 2005;10:2986–3001.
[68] Meier F, Busch S, Lasithiotakis K, Kulms D, Garbe C, Maczey E, et al. Combined targeting of MAPK and AKT signalling pathways is a promising strategy for melanoma treatment. Br J Dermatol 2007;156(6):1204–13.
[69] Roesch A, Mueller AM, Stempfl T, Moehle C, Landthaler M, Vogt T. RBP2-H1/JARID1B is a transcriptional regulator with a tumor suppressive potential in melanoma cells. Int J Cancer 2008;122(5):1047–57.
[70] Roesch A, Fukunaga-Kalabis M, Schmidt EC, Zabierowski SE, Brafford PA, Vultur A, et al. A temporarily distinct subpopulation of slow-cycling melanoma cells is required for continuous tumor growth. Cell 2010;141(4):583–94.
[71] Frank NY, Pendse SS, Lapchak PH, Margaryan A, Shlain D, Doeing C, et al. Regulation of progenitor cell fusion by ABCB5 P-glycoprotein, a novel human ATP-binding cassette transporter. J Biol Chem 2003;278(47):47156–65.
[72] Frank NY, Margaryan A, Huang Y, Schatton T, Waaga-Gasser AM, Gasser M, et al. ABCB5-mediated doxorubicin transport and chemoresistance in human malignant melanoma. Cancer Res 2005;65(10):4320–33.
[73] Kupas V, Weishaupt C, Siepmann D, Kaserer ML, Eickelmann M, Metze D, et al. RANK is expressed in metastatic melanoma and highly upregulated on melanoma-initiating cells. J Invest Dermatol 2011;131(4):944–55.

Organotypic and humanized animal models of genodermatoses

Esteban Chacón-Solano[*,†], *Sara Guerrero-Aspizua*[*,†], *Lucía Martínez-Santamaría*[*,†], *Marcela Del Río*[*,†,‡], *Fernando Larcher*[*,†,‡]
[*]Department of Bioengineering, Carlos III University of Madrid (UC3M), Madrid, Spain, [†]Health Research Institute-Jiménez Díaz Foundation, Madrid, Spain, [‡]Epithelial Biomedicine Division, CIEMAT-CIBERER, Madrid, Spain

1. Introduction

1.1 Genodermatoses

The skin that covers and protects us is renewed throughout our lives thanks to complex processes of proliferation and differentiation or specialization of its cells, particularly those of the epidermis, whose main representative is the keratinocyte. These processes can be affected in many ways giving rise to a vastness of dermatologic diseases of lower or greater severity and degree of extension. Skin disorders like cancer or psoriasis are familiar to the public, but other entities are much less known, but not less important. They are infrequent or rare hereditary diseases of the skin, also known as genodermatoses. They constitute, as a whole, a relevant part of the pathology and dermatologic clinic, accounting for about 8%–10% of the total number of rare diseases estimated in around 7000 [1]. Currently, the genetic basis of more than 500 genodermatoses, mostly monogenic diseases, has been elucidated. This knowledge enables precise disease classification, molecular diagnosis, and, to a certain extent, the understanding of the pathogenic mechanism [2]. However, despite this knowledge, in the vast majority of cases, there is no curative treatment but essentially palliative. The genodermatoses can be broadly grouped in (a) skin fragility disorders, (b) keratinization/cornification, (c) pigmentation disorders, (d) DNA repair, (e) ectodermal dysplasias, and (f) connective tissue disorders. Other more gene-oriented classification methods have been proposed [3]. Despite editorial-imposed text size limitations and the fact that not all genodermatoses have been modeled, this chapter intends to provide pertinent examples of studies using in vitro and in vivo models of the most relevant inherited skin disorders.

1.2 The importance of modeling genodermatoses

Ethical and practical constraints preclude in vivo skin studies on human beings, more so in patients affected by genodermatoses, some of them devastating diseases. Establishing reliable models to address mechanistic matters and to test the efficacy of novel therapeutic approaches thus remains as a major research challenge. The premise

Skin Tissue Models. https://doi.org/10.1016/B978-0-12-810545-0.00004-8

of replacing animals (mostly mice) by in vitro and other skin-humanized in vivo models of disease not only concerns to important ethical issues but also, in the case of human skin diseases, pertains to the major differences between organisms. Current knowledge about disease mechanisms is mainly due from the use of murine models, including transgenic and knockouts. However, based on the significant differences existing between human and murine skin architecture and physiology, the question remains as to how far the results can be extrapolated to the human scenario. Skin organotypic cultures (OTCs) are widely used to examine epidermal gene expression, epidermal and dermal interactions, wound repair, and pharmacotoxicological in vitro studies [4–6]. However, rare skin disease modeling lags behind other applications. In this chapter, we review specific advantages and limitations of current in vitro organotypic and humanized models of genodermatoses, with regard to the particular features of the diseases modeled (Fig. 1).

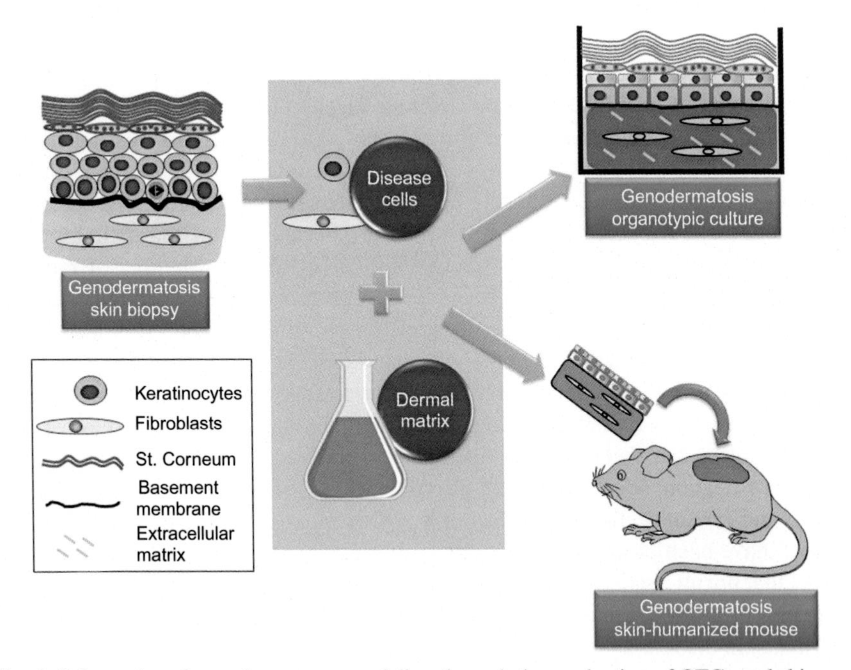

Fig. 1 Schematics of genodermatoses modeling through the production of OTCs and skin-humanized mice. The process starts with the dissociation of a genodermatoses patient skin biopsy to obtain keratinocytes and fibroblasts. Cells are combined with the appropriate matrix biomaterial to produce a bioengineered 3-D skin, which upon culture at the air-liquid interface will be able to induce epidermal stratification. The bioengineered skin without epidermal stratification is orthotopically transplanted in the immunodeficient mouse and engraft to generate the genodermatoses skin-humanized mouse. The cartoon could well represent the modeling of an ichthyosis since the stratum corneum in the biopsy and in the OTC is exaggerated.

2. Modeling genodermatoses in vitro

Reproducing the normal physiology and architecture of the human skin in durable 3-D systems remains challenging. This is the subject of other chapters in this book. The human skin is a stratified, multilayer epithelium possessing extensive capacity for self-renewal and repair. The original studies describing the ability to stratify and differentiate of human keratinocytes, when cultured in 3-D at the air-liquid interface, noticed that differentiation was highly dependent not only on the exposure to air but also on proper interactions of keratinocytes with mesenchymal fibroblasts, as well as the dermal matrix and culture media composition [7,8]. Thus, to authentically recapitulate the disease phenotype in three-dimensional organotypic cultures, the specific process affected (e.g., epidermal differentiation) should be fully recreated. In other words, if for instance a keratinization disorder (e.g., ichthyosis) is to be modeled, it is common sense that the control OTC, established with healthy keratinocytes, should be able to achieve terminal differentiation including a well-developed stratum corneum and barrier function. Unfortunately, this is not always the case as it will be discussed along the chapter. Another critical factor for the accurate modeling of skin diseases is the composition of the dermal equivalents, which normally is composed of deepidermalized dermis (DED), polycarbonate membranes, collagen hydrogels, fibrin gels, or commercial matrices. Efforts are now being put on the development of more authentic matrices using smart combinations of biodegradable biomaterials and embedded fibroblasts. These dermal cells produce natural extracellular matrix components and soluble factor that helps to maintain the epidermal homeostasis and the stem-cell niche for enduring cultures [9–12]. In Table 1, we summarize some features of the OTCs described in this chapter.

2.1 Skin fragility disorders

Adhesion of the epidermis to the underlying dermis is critical for skin function and depends on the proper assembly of basement membrane (BM) proteins to form specialized interconnected "gluing" structures. The keratinocytes of the basal layer of the epidermis are anchored to the dermis through hemidesmosomes that together, with the anchoring filaments of the basal lamina and the anchoring fibrils of the dermis, form the dermoepidermal junction zone [33]. The hemidesmosomes are multiprotein complexes that bind to the keratin filaments on their cytoplasmic face and to the basal lamina on their extracellular side through the anchoring filaments made of laminin 322 and type XVII collagen. The anchoring fibrils, composed almost exclusively of collagen type VII, anchor the basal lamina to type I collagen in the dermis. The different forms of inherited epidermolysis bullosa, that is, simplex, junctional, or dystrophic, are due to the faulty formation of these BM structures, in turn owing to a deficient or incorrect synthesis of their component proteins [33].

In the 3-D modeling of skin fragility disorders, the need to reconstruct a well-developed BM [34] with the proteins synthesized by the cellular components of the OTC equivalent is evident. Both epidermal and mesenchymal cells contribute in a

Table 1 Review of in vitro OTCs skin models of genodermatoses

Genodermatoses	OTC composition	Major achievements	Major limitations	Application	Reference
Epidermolysis bullosa simplex (EBS)	KRT5-deficient immortalized human keratinocytes grown on DED or polycarbonate membrane	Cells stratified and differentiated into a multilayered epidermis	Did not reproduce a cytolytic phenotype	- To test the therapeutic benefits of a chemical chaperone for EBS	[13]
	KRT5- and KRT14-deficient immortalized human keratinocytes grown on a collagen-treated membrane	Generation of a multilayered epidermis	Did not reproduce a full EBS phenotype	- To establish an in vitro model for EBS with expanded lifespan keratinocytes	[14]
Junctional epidermolysis bullosa (JEB)	LAMB3-transduced keratinocytes seeded on DED	Corrected cells showed normal assembly of hemidesmosomes	Short-term genetic correction evaluation	- To test a gene therapy approach for JEB	[15]
	Primary β4-integrin null keratinocytes grown on DED	Genetic and functional correction of β4-null keratinocytes	Short-term genetic correction evaluation	- To test a gene therapy strategy for JEB	[16]
Recessive dystrophic epidermolysis bullosa (RDEB)	RDEB keratinocytes seeded on a normal fibroblast-derived dermal matrix	Restoration of Col 7 at the DEJ by normal fibroblasts	Anchoring fibril formation in OTCs requires long time	- To study if C7 produced by fibroblasts could be transported to the DEJ	[17]
	COL7A1-transduced keratinocytes grown onto collagen-chitosan-glycosaminoglycan matrices with fibroblasts	Corrected cells generated a stratified epidermis with deposition of C7 at the DEJ	Limited functional restoration of the dermal-epidermal adhesion	- To test a gene therapy for RDEB	[18]
	RDEB and non-RDEB cSCC keratinocytes grown on fibroblast-dermal matrix, collagen-matrigel, or DED	Col7 deficiency promotes cSCC invasion and tumor progression	No major limitation for the raised objective	- To study the effect of C7 deficiency on tumorigenesis and CAFs phenotype	[19,20]

Ichthyosis vulgaris (IV)	FLG knockdown human keratinocytes grown on a collagen I matrix with wild-type fibroblasts	Barrier dysfunction and reduced keratohyalin granules	The use of healthy fibroblast may have attenuated the phenotype of the disease	- To study the impact of filaggrin deficiency on the skin barrier function	[21]
	FLG knockdown human keratinocytes grown on a polycarbonate membrane	Impaired keratinocytes differentiation and alterations on epidermal barrier function	The absence of a dermal matrix may precluded the fully recapitulation of the phenotype	- To study the impact of filaggrin deficiency on the keratinocytes	[22]
Autosomal recessive congenital ichthyosis (ARCI)	ALOX12B and TGM1 knockdown rat keratinocytes grown on DED	Partial recapitulation of the ARCI phenotype with identification of common gene expression in ARCI	Differences between murine and human skin may be limited in its applicability	- To identify common networks and mechanisms associated with hyperkeratosis - To test new therapy approaches	[23,24]
	ALOX12B-null mouse keratinocytes grown on a collagen I matrix with fibroblasts	Abnormal barrier function and increased vesicular structures in the granular layer	The diseased OTC seems quite similar to the healthy skin and its stability declines along the first 2 weeks	- To develop a murine OTC ARCI model	[25]
	ALOX12B, TGM1, and ALOXE3 knockdown and deficient human keratinocytes grown on a collagen I matrix with fibroblasts	Impaired barrier function and decrease in filaggrin expression	Did not recapitulate the thicker stratum corneum of ARCI	- To develop a human OTC ARCI model - To test some penetration assays with different nanocarriers	[26,27]
Netherton syndrome (NS)	SPINK5 knockdown human keratinocytes grown on collagen matrix with fibroblasts	Detachment of the stratum corneum and the loss of corneodesmosomes	The diseased OTC did not recapitulate all features of NS	- To test the therapeutic benefits of KLK5 and KLK7 inhibition in NS patients	[28]

Continued

Table 1 Continued

Genodermatoses	OTC composition	Major achievements	Major limitations	Application	Reference
Xeroderma pigmentosum (XP)	XPC keratinocytes grown on collagen matrix with XPC fibroblasts	Some proliferation and differentiation features consistent with XP	Abnormal expression of some epidermal differentiation markers	- To study invasive phenotype of the XP cells - To test a gene therapy approach	[72–74]
Fanconi's anemia(FA)	Immortalized keratinocytes with knockdown FA genes (FANCA or FANCD)	Recapitulation of DNA double-strand break sensitivity	Did not recapitulate the tumor progression phase	- To study oncogenesis in FA patients and its relation with HPV in cSCC	[29,30]
Ehlers-Danlos syndrome (EDS)	Primary human keratinocytes grown onto a 3-D decorin null mouse fibroblast matrix	OTC reveal new possible mechanisms involved in the wound healing of EDS	Abnormal and disorganized skin architecture	- To study the impact of the decorin absence in the dermis	[31]
	TNXB-deficient human keratinocytes and fibroblasts embedded in a collagen matrix	The deposition of tenascin-X and elastic fibers occurs in a similar pattern than in normal skin	No obvious phenotypic alterations were observed neither in vitro nor in vivo after grafting the OTC onto mice	- To study the potential role of tenascin-X in extracellular matrix biology	[32]

dynamic interactive way to the de novo synthesis of the BM protein components that need to localize and assemble properly. The majority of studies on the BM protein expression and assembly kinetics have been performed on organotypic cultures where type I collagen constitutes the dermal matrix (Table 1). In this setting, collagen IV and laminin-1 deposition occurred only in the presence of mesenchymal cells: patchy at day 4 and continuous after 1 week in after keratinocyte confluence. On the other hand, laminin 332 staining was first observed at day 4, in both mono- and cocultures. Although BM protein deposition was continuous at day 14, the ultrastructural organization was still fragmentary, eventually normalizing at 3 weeks [34]. These results indicate that the modeling of skin fragility disorders may not be considered valid with short-term OTCs. The presence of natural extracellular matrix components, properly synthesized by mesenchymal cells in the dermal equivalent, is guaranty of a healthy BM and dermoepidermal anchoring structures.

2.1.1 Epidermolysis bullosa simplex

Epidermolysis bullosa simplex (EBS), the first mechanobullous disease for which the underlying genetic lesion was characterized, is caused by dominant-negative mutations in the keratin 5 (*KRT5*) and keratin 14 (*KRT14*) genes [35]. Organotypic cultures were also used to study the behavior of immortalized dominant EBS [13,14]. Two immortalized cell lines from an EBS patient with a *KRT5* mutation were established using human papillomavirus (HPV) E6/E7 protein expression and were assessed for growth characteristics and keratin expression profiles. Their ability to differentiate in OTCs on acellular dermis or polycarbonate membrane was demonstrated, with better results on DED. Although the OTCs did not reproduce a clear cytolytic phenotype, the cytoskeletal abnormalities, namely, keratin aggregates, were evident particularly in response to heat stress [13]. Moreover, these organotypic systems allowed testing the beneficial effects of a chemical chaperone [13]. EBS keratinocytes with greatly extended life span have also been obtained by ectopic human telomerase reverse transcriptase (hTERT) expression, and their differentiation features were studied in OTCs. Upon air-liquid interface exposure, hTERT-transduced EBS cells are differentiated into a multilayered tissue. However, a blistering phenotype was not described for these cells in OTCs [14]. A recessive form of EBS (REBS) has been described in which the keratin intermediate filaments are absent, because of homozygous null mutation in the *KRT14* gene [36]. REBS is characterized by generalized cutaneous blistering in response to mechanical trauma, resulting from fragility of the basal keratinocytes. In addition, the program of epidermal differentiation is focally disturbed, resulting in exfoliative dyskeratotic plaques with histological appearance of epidermolytic hyperkeratosis (EHK) and suprabasal keratin clumping. El Ghalbzouri and coworkers developed a faithful OTC in vitro model for REBS, which reproduced basal cell vacuolization but not EHK. They focused on the role of fibroblasts to modulate the EBS phenotype using chimerical OTCs. To that end, fibroblasts from nondyskeratotic and dyskeratotic skin of a REBS patient were embedded into collagen matrices. Afterward, fresh biopsies from the nondyskeratotic and dyskeratotic skin of the patient were seeded on top and cultured at the air-liquid interface. In this case, the OTCs were instrumental to show that fibroblasts play a critical role in reproducing the REBS phenotype [37].

2.1.2 Junctional epidermolysis bullosa

Junctional epidermolysis bullosa (JEB) is caused by deficiency of proteins Col17, Lam332, or integrins α6β4, located either in the *lamina lucida* or in the connecting structures of the basement membrane [38]. The severity of the different JEB forms ranges from mild to lethal. The only current therapeutic approach of the disease relies on gene and cell therapy. In fact, the first skin gene therapy clinical trial was carried out for the correction of a severe, not lethal, form of JEB due to a deficiency of Lam332 using autologous grafts of genetically corrected cultured epidermal sheets [39]. The viability of the approach, however, had been tested and confirmed early on in vitro with corrected keratinocytes growing on different protein-coated plastic plates and, more significantly, with OTCs. In fact, the *LAMB3*-transduced keratinocytes seeded on DED showed normal assembly of the dermal-epidermal attachment structures (hemidesmosomes), missing in uncorrected JEB cells [15]. Analogously, organotypic cultures (also on DED) were useful to evaluate the adherence of primary integrin-β4 null keratinocytes, obtained from a newborn suffering from lethal JEB and stably transduced with a retroviral vector carrying a full-length β4 cDNA. The study demonstrated that integrin α6β4 expression, localization, and structure of hemidesmosome components with β4-corrected keratinocytes were indistinguishable from organotypic cultures generated with normal keratinocytes, suggesting full genetic and functional correction of β4-null keratinocytes [16]. However, in contrast to *LAMB3*-corrected cells, β4-corrected keratinocytes have not been yet tested in the clinics. More recently, collagen OTCs were also used to test a protein replacement strategy for the restoration of the laminin 332 assembly and reversion of the JEB phenotype. In that study, transfection of recombinant β3 chain in JEB keratinocytes (*LAMB3* mutant) was shown to restore the formation and functional activity of the trimeric laminin 332. In fact, when β3-treated JEB keratinocytes were used in OTCs, laminin 332 was formed, secreted, and deposited, although in a patchy manner, into the basement membrane zone (BMZ) of the reconstructed skin [40].

In addition to their use as models for JEB therapies, OTCs (on collagen gels) provided mechanistic insight regarding the assembly of hemidesmosomal adhesion complexes through inhibition of nidogen binding to laminin with a recombinant laminin γ1 fragment [41]. Thus, using a simplified human skin model, a functional link between compound structures of the extra- and intracellular space at the junctional zone was demonstrated [41].

2.1.3 Dystrophic epidermolysis bullosa

Dystrophic epidermolysis bullosa (DEB), particularly the recessive form (RDEB), is among the most severe epidermolysis bullosa forms and the one that has prompted the lead developments with regard to strategies for therapeutic intervention. RDEB is due to a deficiency of the anchoring fibril-forming type VII collagen (C7) as a result of mutations in the *COL7A1* gene [38]. Besides generalized mucocutaneous blistering and scarring, RDEB presents a highly elevated risk of early-onset aggressive cutaneous squamous cell carcinoma (cSCC). The majority of advanced therapies under scrutiny, which include protein replacement and several cellular and genetic strategies, aim at C7 replacement to restore the dermal-epidermal adhesion. Since the formation of ultrastructurally mature anchoring fibrils takes relatively long time, upon keratinocyte seeding on fibroblast-populated dermal matrices, standard OTCs may not be the best option to study full restoration of C7 functionality. With nonconventional OTCs in which the dermal matrix is produced by fibroblasts growing on a nylon mesh [42], it was, however, possible to show that normal fibroblasts are capable to provide functional C7 to the BMZ of RDEB keratinocytes [17]. These results were also confirmed in an in vivo model with corrected RDEB fibroblasts and keratinocytes [17,43]. Nonetheless, rather than showing *bona fide* established anchoring fibrils, skin OTCs have been useful to early detect C7 production and deposition at the BMZ by genetically corrected RDEB keratinocytes from different sources. In one of these studies, RDEB keratinocytes were transduced with a *COL7A1* retroviral vector and cultured on collagen-chitosan-glycosaminoglycan sponges [44] embedded with RDEB fibroblasts [18]. In both cases, genetically corrected cells generated cohesive epidermal sheets. More recently, OTCs with keratinocytes derived from normal donor or RDEB patient iPSC have been used. Preliminary work had proved the feasibility of in vitro skin reconstruction using iPSC-derived fibroblasts and keratinocytes [45,46]. Later on, OTCs-based approaches were essential to functionally demonstrate the genetic correction in keratinocytes derived from RDEB and JEB iPSC cells [47,48]. In addition to gene transfer strategies, OTCs have also been instrumental to assess the effectiveness of experimental pharmacological approaches enabling C7 reexpression in cells carrying particular *COL7A1* mutations [49]. In several cases, the therapeutic approaches mentioned in this section have already been translated to the in vivo preclinical and clinical stages. The robust and reliable results obtained with in vitro organotypic systems as testing platforms for therapeutic intervention, mainly gene therapy, have been useful in the launching of in vivo studies and predicting clinical outcomes as it has been the case for JEB [39] and RDEB [50].

In addition to C7 restoration, other emerging therapeutic alternatives focus on the amelioration of the fibrotic and pro-inflammatory extracellular matrix (ECM), due to the aberrant behavior of RDEB fibroblast associated to an increased bioavailability of TGFβ among other causes not yet fully elucidated [51]. In fact, several studies have shown dramatic changes in the dermal microenvironment including increased expression of pro-inflammatory cytokines and derangement in expression and post-translational modification of proteins secreted by RDEB fibroblasts [52,53]. It is thus tempting to speculate that OTCs, whose dermal equivalents are produced with RDEB fibroblasts secreting their own ECM, will soon become popular to test pharmacological and cellular approaches capable to modulate fibrosis and matrix stiffness. While the latter application of RDEB OTCs remains potential, their use is increasingly widespread to study carcinogenesis in RDEB. The trauma-exposed, fibrotic areas are prone to develop aggressive cSCC, the leading cause of premature death in RDEB [19,38]. Using different kinds of OTCs, it has been observed that C7-deficient keratinocytes became invasive, a behavior that unambiguously depends on the phenotypic changes in RDEB fibroblast and ECM [19,20,54].

2.2 Keratinization disorders

The genodermatoses classified as disorders of keratinization/cornification (DOC) encompass a large heterogeneous group of skin diseases that share an abnormal barrier function, leading to increased transepidermal water loss with concomitant compensatory hyperproliferation and epidermal acanthosis [55]. The general unifying term ichthyosis comes from the common feature of localized and/or generalized scaling of the diseases. Mutations in over 50 genes with very distinct functions have been reported to cause ichthyoses. However, despite the diversity in the pathogenesis pathways, all somehow converge in disrupted barrier function [56]. In brief, epidermal barrier function is maintained by the differentiation of keratinocytes as they transit, in their permanent renewal cycle, from the innermost stratum basal to the outermost stratum corneum where they are ultimately sloughed off as squames. This complex differentiation program is marked by site-specific expression of proteins such as cytoskeletal keratins and, in the suprabasal layers, by the expression of cross-linking enzymes like transglutaminase and others responsible for the synthesis of components necessary for the generation of the lipid barrier (phospholipids, cholesterol, sphingomyelin, and glucosylceramides). In 2009, the first ichthyosis consensus conference established an international consensus classification for the DOCs based on pathophysiology, clinical manifestations, and mode of inheritance [56]. Several in vitro skin models have been developed for keratinization disorders, including ichthyosis vulgaris [21], autosomal recessive congenital ichthyosis [23-27,57-59], epidermolytic ichthyosis [60], Netherton syndrome [28], and other forms [61,62]. Most of these models begin with the addition of fibroblasts to type I collagen matrix or DED, followed by seeding of keratinocytes and air-liquid interface differentiation. The difficult acquisition of biopsies from patients with rare diseases has led to the use of alternative ways to mimic the protein deficiency such as shRNA technology to knock down the expression of genes of interest in keratinocytes derived from human [21,26,28,57,62], mouse [25], and rat [23,24]. Other OTCs have used human cells from patients [27,61], and in some cases, keratinocytes have been immortalized to overcome the senescence of primary cells [60].

2.2.1 Ichthyosis vulgaris

The ichthyosis vulgaris (IV) is the most common form of DOC, related with loss-of-function mutations in *FLG* gene. The profilaggrin/filaggrin protein encoded by this gene is a filament-associated protein that binds to keratin fibers and plays an important role as natural moisturizing factor in the formation of the stratum corneum, affecting the barrier function and the pH balance of the skin [63,64].

A filaggrin-deficient human organotypic skin model mimics some part of the defects observed in IV patients, despite it does not recapitulate the thicker stratum corneum seen in patients. The model presented barrier dysfunction, reduced keratohyalin granules, and decreased amounts of urocanic acid. Also the model showed increased UV sensitivity, evidencing the role of filaggrin in the UVB protection [21]. Similar results were observed in other OTCs, produced with filaggrin knockdown human keratinocytes grown on a polycarbonate membrane. In contrast to the one described by

Mildner and coworkers [21], this OTC displayed additional abnormalities in the epidermal differentiation program [22]. The differences between these two models were attributed by the fact that one of the equivalents used normal fibroblast embedded in a collagen hydrogel as dermal matrix while the second did not include a dermal component. There is some evidence that the use of healthy fibroblast ameliorates the phenotype of diseased skin [17,65,66] and improves the regeneration and morphology of epidermal tissue in 3-D cultures [10,12]. This and additional evidence from other models underscore the importance of using diseased-like fibroblast to create an equivalent that resembles the histogenesis of the skin and the disease phenotype. It is important to keep in mind that a useful organotypic model needs to be maintained for a long period and that the presence of an authentic dermal matrix is considered essential for this.

2.2.2 Autosomal recessive congenital ichthyosis

Autosomal recessive congenital ichthyosis (ARCI) is a group of nonsyndromic forms of ichthyosis, related to hyperkeratosis and skin barrier dysfunction. These disorders include harlequin ichthyosis (HI), lamellar ichthyosis (LI), congenital ichthyosiform erythroderma (CIE), and other minor variants. At least six genes have been associated with ARCI: *ABCA12*, *TGM1*, *NIPAL4*, *ALOX12B*, *ALOXE3*, and *CYP4F22* [56]. The gene most commonly mutated in ARCI is *TGM1*, which encodes transglutaminase-1, an essential enzyme in the formation of the cornified envelope by cross-linking structural proteins such as involucrin, loricrin, and filaggrin [67]. Mutations in *ALOX12B* and *ALOXE3*, encoding the lipoxygenases, are the second most common cause of ARCI.

An in vitro rat OTC containing knockdown *TGM1* gene reproduced part of the disease and presented hyperkeratosis, hyperproliferation, and an increase in nonpolar lipids of the stratum corneum. Also, the model showed a defective cross-linking of loricrin and upregulation of interleukin-1 alpha (IL1A), as observed in LI patients. Treatment of these organotypic equivalents with an IL-receptor antagonist (IL1RA) prevented the hyperkeratosis showing improvement of the LI phenotype [23].

In addition, to identify common networks in ARCI disorders and to find new target drugs, another rat organotypic model was developed using shRNA knockdown for *ALOX12B*. The comparison of the common networks between altered genes of the *TGM1* and *ALOX12B* models demonstrated a new critical pathway in hyperkeratosis leading to the repositioning of Mdm2 inhibitors, such as nutlin-3, as a new therapy approach for treating ARCI [24].

Another OTC for ARCI was developed from keratinocytes isolated from *ALOX12B*-deficient newborn mice. The equivalent presented an epithelial architecture similar to that observed in vivo. The skin equivalents deficient in *ALOX12B* showed abnormal skin barrier function and increased vesicular structures in the granular layer. However, the phenotype of the diseased skin model rather resembled that of healthy skin and did not recapitulate hyperkeratosis, as the main feature of the disease. The epithelial architecture and stability of the OTCs declined when the cultures were prolonged for more than 2 weeks [25].

Human OTCs with keratinocytes were also developed for *ALOX12B*, *TGM1*, and *ALOXE3* using gene expression knockdown with specific shRNAs. These models did

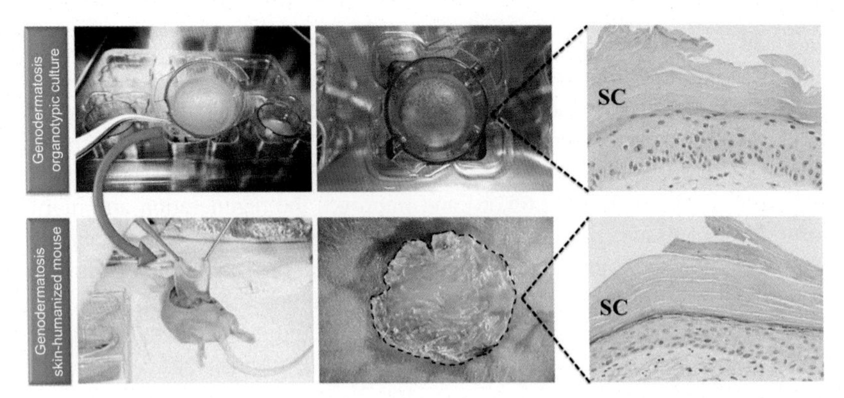

Fig. 2 Recapitulation of skin disease phenotype in vitro and in vivo. The figure shows the modeling of lamellar ichthyosis (LI), characterized by a thick stratum corneum (SC). The OTC is cultured at air-liquid interface to induce differentiation or is orthotopically transplanted onto the immunodeficient mouse. Both models present normal skin architecture, with a fully stratified epidermis, and faithfully reproduce the hyperkeratotic disease phenotype.

not recapitulate the thicker stratum corneum of ARCI, but showed an impaired barrier function and decrease in filaggrin expression [26]. These models were subsequently used to perform penetration and permeation assays of different nanocarriers, as a potential approach for the topical treatment of ARCI. In this study, some OTCs were also developed with patient-derived primary cells [27]. We have recently developed a robust skin OTC for LI, containing *TGM1*-deficient human keratinocytes and fibroblasts. This model, which uses keratinocytes seeded onto a fibroblast-containing fibrin dermal matrix, fairly recapitulates the major features of this ARCI, including a packed and faulty stratum corneum (Fig. 2) (Larcher and Chacón-Solano, unpublished results).

2.2.3 Netherton syndrome

Netherton syndrome (NS) is a syndromic form of inherited ichthyosis caused by loss-of-function mutations in *SPINK5* encoding lympho-epithelial Kazal-type inhibitor (LEKTI), a serine protease inhibitor expressed in the epidermis and other stratified epithelia. Clinically, this genodermatosis presents severe skin inflammation and scaling, specific hair shaft defect, and constant allergic manifestations, although the symptoms and severity of NS can vary from one patient to another. Lethal variants of the disease have also been described [68]. LEKTI deficiency results in exacerbated activity of proteases such as kallikreins 5, 7, and 14 (KLK5, KLK7, and KLK14), which consequently produce increased corneodesmosomal degradation and desquamation [69].

A *SPINK5* knockdown organotypic human model recapitulated part of the NS phenotype, especially by the detachment of the parakeratotic stratum corneum and the loss of corneodesmosomes. Furthermore, the simultaneous silencing of *KLK5* and *KLK7* in this LEKTI-deficient model led to a partial improvement of the skin architecture. This OTC was therefore useful to confirm the potential therapeutic benefits of the inhibition of serine proteases in NS patients [28].

2.3 Disorders of pigmentation

The major inherited disorder characterized by the complete or partial absence of pigment in the skin, hair, and eyes is albinism. In the majority of cases, this genodermatosis is due to a defect or an absence of tyrosinase, a critical enzyme expressed in melanocytes involved in the production of the human skin pigment melanin. The severe deficit of skin pigmentation in albinism makes individuals to be more susceptibility to sunburn and skin cancers. Although a careful review of the literature has not retrieved studies involving skin OTCs constructed with tyrosinase-deficient cells, there are several studies where OTCs are produced not only with keratinocytes but also with normal epidermal melanocytes in order to modulate the pigmentation and the concomitant responses to UV light [70].

2.4 DNA repair disorders

2.4.1 Xeroderma pigmentosum

The paradigm of genodermatoses involving of DNA repair disorders is xeroderma pigmentosum (XP), a rare autosomal recessive genetic disorder characterized by a deficient ability to repair DNA damage caused by UV light. Consequently, XP patients have an extremely high susceptibility to develop skin cancer. Skin malignancies occur at a young age, being metastatic malignant melanoma and cSCC, the two most common causes of death in XP patients. The absorption of the UV light leads to DNA damage in the form of cyclobutane pyrimidine dimers and pyrimidine(6–4)pyrimidone photoproducts. In a healthy, normal human being, DNA damage is repaired by a process known as nucleotide excision repair (NER). In XP, mutation in one of the repair enzyme genes (*XPA-XPG* and *XPV*) leads to a deficit of NER. If left unchecked, DNA damage caused by UV light causes mutations in cancer-relevant genes. Some forms of XP are associated with neurological abnormalities whose cause remains poorly understood, but are not connected with exposure to UV light [71]. Bernerd and coworkers developed an OTC model of the disease through the culture of several strains of primary XP keratinocytes and XP fibroblasts embedded in a collagen matrix. In this setting, some proliferation and differentiation features consistent with XP disease were found [72]. This model system was also used to study the invasive phenotype of the XP cells related to their cancer predisposition [73] and also to assess the efficacy of a gene therapy strategy for xeroderma pigmentosum C (XPC) [74].

2.4.2 Fanconi anemia

Fanconi's anemia (FA) is an inherited disease mainly characterized by congenital abnormalities, progressive bone marrow failure, and an increased predisposition to cancer. In FA patients, hematologic problems typically appear between ages of 5 and 10 years with an actuarial risk of bone marrow failure above 90% by 40 years of age. FA patients also have a high risk of developing solid tumors, principally cSCCs. FA is due to a deficiency of enzymes involved in the recombinational repair of DNA double-strand damage. So far, 21 different genes have been implicated in FA [75]. While studies of hematopoietic cells derived from FA patients have provided insight into bone marrow failure and leukemogenesis, oncogenic events in FA-deficient keratinocytes, the cell type of origin of cSCC, are much less understood. Using skin OTCs containing HPV E6/E7-immortalized keratinocytes with knockdown FA genes *FANCA* or *FANCD*, Hoskins and coworkers showed the predicted but not previously described DNA repair defects and apoptosis in FA keratinocytes upon mitomycin C treatment [29]. Also, hyperplasia due to increased basal and suprabasal cell proliferation was shown, although not observed in 2-D culture. The hyperplastic phenotype shown in the FA-reconstructed epithelium of the OTC model might thus represent a clinically relevant precursor of cSCC. Studies using a similar FA OTC setting showed that FA pathway downregulation increased HPV genome amplification in the epithelia, indicating that the HPV life cycle is attenuated by the intact FA pathway [30], a relevant finding with regard to HPV-related cSCC development in FA patients.

2.5 Ectodermal dysplasias

The ectodermal dysplasias (ED) comprise a heterogeneous group of genodermatoses that affect the development of the ectoderm, the embryonic layer that forms the skin and its attachments (hair, nails, teeth, and sweat glands). The most frequent ED is hypohidrotic ectodermal dysplasia (HED), characterized by hypotrichosis, hypohidrosis, and hypodontia. Clinical manifestations in ED will depend on the involvement of different ectodermal derivatives. The most frequent HED is the X-linked recessive form due to mutations in the *EDA* gene, which encodes ectodysplasin, a member of the tumor necrosis factor (TNF) family. Mutations in the *EDAR* gene, encoding the EDA receptor, or in the *EDARADD* gene encoding the domain associated with *EDAR,* cause autosomal dominant and recessive forms of HED. Recently, HED patients with mutations in the *WNT10A*, *TRAF6*, and other genes have also been identified [76]. Another rare autosomal dominant ED is ectrodactyly-ectodermal dysplasia-cleft syndrome (EEC) characterized by the triad of ectrodactyly, ectodermal dysplasia, and facial clefts. EEC is caused by heterozygous mutations in the *TP63* gene coding for transcription factor p63, a master gene for the commitment of ectodermal-derived tissues [76].

OTCs were used to study different aspects of EEC. By introducing various EEC-causing *TP63* mutations in human keratinocytes with retroviral vectors, Zarnegar and coworkers were able to disrupt epidermal differentiation and characterized a series of alterations in critical transcriptional activators of epidermal differentiation such as *HOPX*, *GRHL3*, *KLF4*, *PRDM1*, and *ZNF750*, shedding light on the mechanisms of these complex genodermatoses [77]. In another therapy-oriented study, naturally revertant oral mucosal epithelial stem cells from a patient with EEC, carrying a heterozygous R311K-p63 mutation, were isolated, and their capacity to differentiate was tested in OTC. The study showed that revertant EEC stem cells were able to generate a well-organized and stratified epithelium, lending support to a cell therapy strategy to counteract and resolve the progressive corneal degeneration due to a premature aging of limbal epithelial stem cells, the characteristic of this genodermatoses [78].

2.6 Connective tissue disorders

Connective tissue disorders comprise a large family of genodermatoses with genetic defects affecting any of the constitutive molecules of the connective tissue, including collagen, elastin, fibrillin, fibulin, and proteins that modify these component elements (e.g., metalloproteinases and lysyl hydroxylase 3). Collagen, the major connective tissue component, is not a single protein but a family of related proteins comprising at least 19 members identified to date, being dermal collagen type I the most abundant. The constituent strands of the known collagens are encoded by 30 different genes, dispersed in the human genome. Alterations in either specific collagen genes or enzymes involved in their synthesis give rise to various genetic diseases of collagen [79].

2.6.1 Ehlers-Danlos syndrome

Under this syndrome, various hereditary diseases exist, including some of autosomal dominant or autosomal recessive inheritance, and others linked to the X chromosome, whose common pathogenic base is an alteration in collagen biosynthesis and whose clinical characteristics consist in hyperextensibility. Once divided into 11 different types defined by different clinical manifestations, inheritance patterns, and genetic *loci*, Ehlers-Danlos syndrome (EDS) is now classified into 6 types, with 3 types receiving the most attention in both research and clinical practice: classical EDS, hypermobile EDS, and vascular EDS [80]. The genes associated with EDS include collagen genes like *COL1A1*, *COL1A2*, *COL3A1*, *COL5A1*, and other genes including *ADAMTS2*, *PLOD1*, and *TNXB*, which code for proteins that process or interact with collagen [81]. There are also mutations described in genes coding for enzymes involved in the posttranslational modifications of proteoglycans. For example, mutation in the gene *B4GALT7*, an enzyme involved in the synthesis of the linkage region of glycosaminoglycan (GAGs), leads to an EDS with skin fragility and wound healing deficiency. Notably, the fibroblasts from the affected patients display defective biosynthesis of decorin that in turn leads to abnormal fibrillar collagen [82,83].

Using an OTC system of keratinocytes grown on the top of fibroblasts embedded in their own matrix in the presence of abnormal GAGs, it has been shown that a decorin-deficient matrix caused a delay in epidermal differentiation. This provides an explanation for the established defects in wound healing observed in some EDS patients [31]. The tenascin X-deficient form of EDS, due to *TNXB* mutation, was also investigated using OTC and in vivo after OTC grafting to immunodeficient mice. In this case, when the OTC dermal equivalent was made with tenascin-deficient fibroblasts embedded in a collagen matrix, no obvious phenotypic abnormalities were found in vitro or in vivo, although the system showed potential to study the role of tenascin-X in dermal matrix assembly and stability [32]. No studies have been reported, however, with OTCs populated by fibroblasts from patients with the most common forms of EDS.

3. Modeling genodermatoses in vivo

Grafting of whole human skin to immunodeficient mice provides the ultimate and closer model to the patient/donor cutaneous situation. However, availability of the human skin is very limited, particularly from rare skin diseases. Therefore, engraftment of readily available bioengineered skin, which includes epidermal and dermal cells and authentic dermal matrices, represents an excellent alternative to skin biopsy grafting. We review here, with some exceptions, the studies carried out by the authors.

3.1 The skin-humanized mouse model

Even though skin OTCs have been growing in complexity through the introduction of various cutaneous cell types, soluble mediators, and different matrix/scaffold materials, the in vitro approaches are frequently insufficient to recreate complex dermatologic conditions. One of the major drawbacks of OTCs, related to the pathological scenarios, is their inability to recapitulate regenerative processes involving repeated cycles of epidermal proliferation and differentiation. In other words, OTCs are not particularly suitable to assess the epidermal stem-cell performance during tissue renewal. In fact, the majority of current skin OTCs do not even match a single native skin epidermal turnover cycle lasting at least 4 weeks. Other important shortcomings of OTCs relate to their isolation from environmental cues and the lack of interaction with the other tissues. Hence, modeling skin diseases such as genodermatoses, especially with the purpose of testing long-lasting therapeutic approaches, such as gene therapy, require accurate in vivo systems. As mentioned at the beginning of the chapter, murine disease models often fail in the trustful recapitulation of the human condition. Thus, bioengineering-based skin-humanized mouse models seem to mirror the pathophysiological characteristics of human dermatologic disorders. These platform models are based on the engraftment of human skin OTCs in immunodeficient mice. Our group has been working on a fibroblast-containing fibrin-based dermal matrix, where human keratinocytes are seeded [84]. In this system, patient fibroblasts are embedded in a fibrinogen solution that is allowed to clot by the addition of thrombin or calcium ions, depending if fibrinogen is provided as a raw material or contained in plasma. The resulting fibrin hydrogel containing live fibroblasts is seeded with human keratinocytes that have been previously grown using the Rheinwald and green culture method [85]. When keratinocytes on top of the fibrin dermal equivalent reach confluence, the OTC is manually detached from the culture flask or plate and is ready for grafting (Fig. 2).

We have demonstrated that this fibrin-based dermal equivalent provides excellent human keratinocyte growth support without the need of feeder layer cells [86]. Upon engraftment, a mature skin with human architecture was shown to persist for a large number of epidermal turnover cycles, as demonstrated during long-term follow-up periods in immunodeficient mice or in burn patients (autologous OTC) [84]. Experiments with single genetically marked epidermal stem-cell clones, as the keratinocyte component of the grafted fibrin-based OTCs, demonstrated enduring regenerative properties of epidermal stem cells in vivo [87,88] including those from RDEB patients [89].

One of the main advantages of this method is the possibility to generate a large number of engrafted mice containing a significant area of homogeneous single donor-derived human skin, available in a relatively short period of time. Using skin cells isolated from patient biopsies, we and others have been able to reconstruct skin disorders including several rare human monogenic skin diseases (Table 2).

3.1.1 Epidermolysis bullosa

We and others have used graftable fibrin-based OTCs to generate RDEB skin-humanized mice, where it was possible to demonstrate the efficacy of an ex vivo gene therapy strategy that corrected RDEB keratinocytes [18,89] or both RDEB keratinocytes and fibroblasts [92]. The same kind of engrafted OTCs has been recently used to demonstrate the value of a *COL7A1* exon skipping approach in vivo [93]. Using a JEB skin-humanized model, we demonstrated the accurate basal cell expression achieved with a lentiviral vector encoding *LAMB3* cDNA under the control of the *KRT14* gene regulatory sequences [90] for gene therapy. More recently, we could show that, upon engraftment, skin OTCs populated with *COL17A1* revertant JEB human keratinocytes could regenerate a clinically normal skin, paving the way to a clinic application of natural ex vivo gene therapy strategies [91].

3.1.2 Keratinization disorders

Our skin-humanized mouse platform has been widely used for the modeling of genodermatoses with altered keratinization. Using sole keratinocytes from a patient with pachyonychia congenita (PC), a dominant inherited disease due to mutations in either *KRT6*, *KRT16*, or *KRT17* coding genes, we were able to recapitulate in vivo the keratoderma and suprabasal blistering of this disease [94]. We could also show that the inclusion of PC fibroblasts in the graftable OTCs enhanced the disease phenotype upon engraftment (Larcher, unpublished results). Similarly, the engraftment of OTCs prepared with NS keratinocytes allowed a faithful recreation of the disease and asserted the efficacy of a *SPINK5* lentiviral gene therapy approach [95], now in the clinics [96]. Also, we established an ARCI (LI) skin-humanized model that mirrored the disease in humans with regard to defective barrier function and hyperkeratosis (Fig. 2), allowing a better characterization of the alterations in the expression of epidermal differentiation markers through a proteomic approach [97]. Using this LI model, we were able to preclinically assess a novel transglutaminase replacement approach that efficaciously restored barrier function and skin architecture [98].

3.1.3 DNA repair disorders: Xeroderma pigmentosum

In studies aimed to investigate the response to UV light under low and high sensitivity skin settings, we developed a skin-humanized model for the cancer-prone disease XPC that fairly recapitulated the DNA-damage-induced skin responses seen in XPC patients [99]. The model was subsequently used to prove the value of a gene therapy approach based on the transfer of the *XPC* gene to human keratinocytes [100].

Table 2 **Review of different skin-humanized mouse models of genodermatoses**

Genodermatoses	Characteristics	Applications	Reference
Junctional epidermolysis bullosa (JEB, LAMB3)	Unable to detect blistered regenerated JEB skin due to slough off of epidermis. Gene (LAMB3)-corrected keratinocytes were capable of normal skin regeneration	- To test a lentiviral gene therapy approach that directs the expression of laminin β-3 to the basal cells of the epidermis	[90]
Junctional epidermolysis bullosa (JEB, COL17A1)	Revertant CO17A1 cells showed regeneration of a normal epidermis	- To test a natural gene therapy approach using revertant COL17A1 human keratinocytes	[91]
Recessive dystrophic epidermolysis bullosa (RDEB)	Recapitulates general blistering caused by COL7A1 deficiency	- To test different gene therapy strategies and establish translational research toward long-term treatments of RDEB	[18,89,92,93]
Pachyonychia congenita (PC)	Full recapitulation of the PC phenotype: marked acanthosis and suprabasal epidermal blistering after minor trauma	- To test novel pharmacological or gene-based therapies for PC	[94]
Netherton syndrome (NS)	Recapitulates NS features: marked hyperplasia, hypogranulation, and the loss of stratum corneum	- To test different long-term gene therapy approaches for NS	[95,96]
Lamellar ichthyosis (LI)	Recapitulates full LI phenotype: hyperkeratosis and severely decreased barrier function	- To test novel protein (transglutaminase I) replacement strategies	[97,98]
Xeroderma pigmentosum (XPC)	Recapitulates XPC features: severe deficiency in DNA repair mechanisms. Maintenance of sunburn cells and pyrimidine dimers after UVB irradiation	- To provide a UV hypersensitive skin model - To test gene therapy strategies	[99,100]
Cutaneous scleroderma (cSC)	Marked fibrosis of the dermis induced by patient anti-PDGFR autoantibodies	- To study pathogenic mechanisms - To test the effect of cSC patient IgGs against PDGFR	[101]

All the models summarized were generated using the methodology developed by the authors.

3.1.4 Scleroderma

Although not cataloged as a connective tissue genodermatoses given its frequency and cause, cutaneous scleroderma is a rare disorder that results in hardening of the skin due to increased synthesis of collagen. Using a skin-humanized mouse model, we could demonstrate that anti-PDGF receptor (PDGFR) autoantibodies present in patients were able to induce a scleroderma-like phenotype in the human skin engrafted in the mouse, providing an explanation for the pathogenesis of this serious disease [101].

4. Conclusions

Three-dimensional human OTCs and in vivo humanized skin mouse systems represent robust biological tools to mimic a wide range of cutaneous conditions, including many inherited skin disorders. OTCs provide versatility with regard to cellular, matrix, and media components. Their value also lies in the fact that large quantities can be made suitable for multiassay settings without the ethical issues of animal experimentation. Major improvements have being achieved concerning duration, stability, and faithful recreation of certain skin pathology features with OTCs. The shortcomings of the in vitro 3-D cultures are evident, when the experimental approach (e.g., gene therapy) challenges epidermal stem-cell regenerative capability. Matrix and basement membrane remodeling are also long-term processes difficult to recreate using an in vitro setting. Therefore, enduring processes need to be addressed in vivo. The skin-humanized mouse models thus represent a fantastic platform to cope with that challenge, in the intense search for curative treatments for genodermatoses. We greatly rely on these models as trustworthy predictors of clinical responses. In fact, the results obtained with these models have led to current gene therapy trials on patients as in RDEB, JEB, and NS. With the advent of novel gene-editing strategies suitable for the treatment of many monogenic skin diseases, it is expected that the OTCs and skin-humanized mice herein introduced and many new ones will be pivotal to assess the efficacy and safety of these coming therapeutic approaches. Important advances may come from the new developments in the field of biomaterials and from the understanding of morphogenic interactions between epidermal and mesenchymal cells. It is foreseeable that, by including additional cell types, perfected matrix components, and coadjuvant factors, OTCs will be ready to match current in vivo models.

Acknowledgments

FL was supported by grant PI14/00931 from Spanish Instituto de Salud Carlos III. SGA, LMS, and MDR were supported by grant SAF2013-43475-R from Spanish Ministerio de Economía y Competitividad. ECS was supported by grant from Costa Rica Ministerio de Ciencia, Tecnología y Telecomunicaciones (PINN-MICITT).

Abbreviations

ARCI autosomal recessive congenital ichthyosis
BM basement membrane
cSCC cutaneous squamous cell carcinoma
C7 collagen VII
DEB dystrophic epidermolysis bullosa
DED deepidermalized dermis
DOC disorders of keratinization/cornification
EBS epidermolysis bullosa simplex
ECM extracellular matrix
ED ectodermal dysplasia
EDS Ehlers-Danlos syndrome
EEC ectrodactyly-ectodermal dysplasia-cleft syndrome
EHK epidermolytic hyperkeratosis
FA Fanconi's anemia
HED hypohidrotic ectodermal dysplasia
HPV human papillomavirus
iPSC induced pluripotent stem cell
IV ichthyosis vulgaris
JEB junctional epidermolysis bullosa
LI lamellar ichthyosis
NER nucleotide excision repair
NS Netherton syndrome
OTCs organotypic cultures
XP xeroderma pigmentosum
2-D two-dimensional
3-D three-dimensional

References

[1] Orphanet. http://www.orpha.net/consor/cgi-bin/index.php?lng=EN [accessed 09.12.16].
[2] Leech SN, Moss C. A current and online genodermatosis database. Br J Dermatol 2007;156(6):1115–48.
[3] Moss C. A new way to classify genetic skin disease. J Invest Dermatol 2009;129(11):2543–5.
[4] Parenteau NL, Nolte CM, Bilbo P, Rosenberg M, Wilkins LM, Johnson EW, et al. Epidermis generated in vitro: practical considerations and applications. J Cell Biochem 1991;45(3):245–51.
[5] Harding KG, Morris HL, Patel GK. Science, medicine and the future: healing chronic wounds. BMJ 2002;324(7330):160–3.
[6] Ali N, Hosseini M, Vainio S, Taïeb A, Cario-André M, Rezvani HR. Skin equivalents: skin from reconstructions as models to study skin development and diseases. Br J Dermatol 2015;173(2):391–403.
[7] Pruniéras M, Régnier M, Woodley D. Methods for cultivation of keratinocytes with an air-liquid interface. J Invest Dermatol 1983;81(1 Suppl):28s–33s.
[8] Smola H, Thiekotter G, Fusenig NE. Mutual induction of growth factor gene expression by epidermal-dermal cell interaction. J Cell Biol 1993;122(2):417–29.

[9] Stark HJ, Willhauck MJ, Mirancea N, Boehnke K, Nord I, Breitkreutz D, et al. Authentic fibroblast matrix in dermal equivalents normalises epidermal histogenesis and dermoepidermal junction in organotypic co-culture. Eur J Cell Biol 2004;83(11–12):631–45.

[10] Stark HJ, Boehnke K, Mirancea N, Willhauck MJ, Pavesio A, Fusenig NE, et al. Epidermal homeostasis in long-term scaffold-enforced skin equivalents. J Investig Dermatol Symp Proc 2006;11(1):93–105.

[11] Muffler S, Stark H-J, Amoros M, Falkowska-Hansen B, Boehnke K, Bühring H-J, et al. A stable niche supports long-term maintenance of human epidermal stem cells in organotypic cultures. Stem Cells 2008;26(10):2506–15.

[12] Boehnke K, Mirancea N, Pavesio A, Fusenig NE, Boukamp P, Stark HJ. Effects of fibroblasts and microenvironment on epidermal regeneration and tissue function in long-term skin equivalents. Eur J Cell Biol 2007;86(11–12):731–46.

[13] Chamcheu JC, Lorie E, Akgul B, Bannbers E, Virtanen M, Gammon L, et al. Characterization of immortalized human epidermolysis bullosa simplex (KRT5) cell lines: trimethylamine *N*-oxide protects the keratin cytoskeleton against disruptive stress condition. J Dermatol Sci 2009;53:198–206.

[14] Jensen TG, Sørensen CB, Jensen UB, Bolund L. Epidermolysis bullosa simplex keratinocytes with extended lifespan established by ectopic expression of telomerase. Exp Dermatol 2003;12(1):71–7.

[15] Vailly J, Gagnoux-Palacios L, Dell'Ambra E, Romero C, Pinola M, Zambruno G, et al. Corrective gene transfer of keratinocytes from patients with junctional epidermolysis bullosa restores assembly of hemidesmosomes in reconstructed epithelia. Gene Ther 1998;5(10):1322–32.

[16] Dellambra E, Prislei S, Salvati AL, Madeddu ML, Golisano O, Siviero E, et al. Gene correction of integrin beta4-dependent pyloric atresia-junctional epidermolysis bullosa keratinocytes establishes a role for beta4 tyrosines 1422 and 1440 in hemidesmosome assembly. J Biol Chem 2001;276(44):41336–42.

[17] Woodley DT, Krueger GG, Jorgensen CM, Fairley JA, Atha T, Huang Y, et al. Normal and gene-corrected dystrophic epidermolysis bullosa fibroblasts alone can produce type VII collagen at the basement membrane zone. J Invest Dermatol 2003;121(5):1021–8.

[18] Gache Y, Baldeschi C, Del Rio M, Gagnoux-Palacios L, Larcher F, Lacour JP, et al. Construction of skin equivalents for gene therapy of recessive dystrophic epidermolysis bullosa. Hum Gene Ther 2004;15(10):921–33.

[19] Ng YZ, Pourreyron C, Salas-Alanis JC, Dayal JH, Cepeda-Valdes R, Yan W, et al. Fibroblast-derived dermal matrix drives development of aggressive cutaneous squamous cell carcinoma in patients with recessive dystrophic epidermolysis bullosa. Cancer Res 2012;72(14):3522–34.

[20] Martins VL, Vyas JJ, Chen M, Purdie K, Mein CA, South AP, et al. Increased invasive behaviour in cutaneous squamous cell carcinoma with loss of basement-membrane type VII collagen. J Cell Sci 2009;122(Pt11):1788–99.

[21] Mildner M, Jin J, Eckhart L, Kezic S, Gruber F, Barresi C, et al. Knockdown of filaggrin impairs diffusion barrier function and increases UV sensitivity in a human skin model. J Invest Dermatol 2010;130(9):2286–94.

[22] Pendaries V, Malaisse J, Pellerin L, Le Lamer M, Nachat R, Kezic S, et al. Knockdown of filaggrin in a three-dimensional reconstructed human epidermis impairs keratinocyte differentiation. J Invest Dermatol 2014;134(10):2938–46.

[23] RFL O'S, Choudhary I, Harper JI. Interleukin-1 alpha blockade prevents hyperkeratosis in an in vitro model of lamellar ichthyosis. Hum Mol Genet 2010;19(13):2594–605.

[24] Youssef G, Ono M, Brown SJ, Kinsler VA, Sebire NJ, Harper JI, et al. Identifying a hyperkeratosis signature in autosomal recessive congenital ichthyosis: Mdm2 inhibition prevents hyperkeratosis in a rat ARCI model. J Invest Dermatol 2014;134(3):858–61.
[25] Rosenberger S, Dick A, Latzko S, Hausser I, Stark HJ, Rauh M, et al. A mouse organotypic tissue culture model for autosomal recessive congenital ichthyosis. Br J Dermatol 2014;171(6):1347–57.
[26] Eckl K-M, Alef T, Torres S, Hennies HC. Full-thickness human skin models for congenital ichthyosis and related keratinization disorders. J Invest Dermatol 2011;131(9):1938–42.
[27] Eckl KM, Weindl G, Ackermann K, Kuchler S, Casper R, Radowski MR, et al. Increased cutaneous absorption reflects impaired barrier function of reconstructed skin models mimicking keratinisation disorders. Exp Dermatol 2014;23(4):286–8.
[28] Wang S, Olt S, Schoefmann N, Stuetz A, Winiski A, Wolff-Winiski B. SPINK5 knockdown in organotypic human skin culture as a model system for Netherton syndrome: effect of genetic inhibition of serine proteases kallikrein 5 and kallikrein 7. Exp Dermatol 2014;23(7):524–6.
[29] Hoskins EE, Morris TA, Higginbotham JM, Spardy N, Cha E, Kelly P, et al. Fanconi anemia deficiency stimulates HPV-associated hyperplastic growth in organotypic epithelial raft culture. Oncogene 2009;28(5):674–85.
[30] Hoskins EE, Morreale RJ, Werner SP, Higginbotham JM, Laimins LA, Lambert PF, et al. The fanconi anemia pathway limits human papillomavirus replication. J Virol 2012;86(15):8131–8.
[31] Nikolovska K, Renke JK, Jungmann O, Grobe K, Iozzo RV, Zamfir AD, et al. A decorin-deficient matrix affects skin chondroitin/dermatan sulfate levels and keratinocyte function. Matrix Biol 2014;35:91–102.
[32] Zweers MC, Schalkwijk J, van Kuppevelt TH, van Vlijmen-Willems IM, Bergers M, Lethias C, et al. Transplantation of reconstructed human skin on nude mice: a model system to study expression of human tenascin-X and elastic fiber components. Cell Tissue Res 2005;319(2):279–87.
[33] Has C, Nyström A. Epidermal basement membrane in health and disease. Curr Top Membr 2015;76:117–70.
[34] Smola H, Stark HJ, Thiekötter G, Mirancea N, Krieg T, Fusenig NE. Dynamics of basement membrane formation by keratinocyte-fibroblast interactions in organotypic skin culture. Exp Cell Res 1998;239(2):399–410.
[35] Lane EB, McLean WH. Keratins and skin disorders. J Pathol 2004;204(4):355–66.
[36] Chan Y, Anton-Lamprecht I, Yu QC, Jäckel A, Zabel B, Ernst JP, et al. A human keratin 14 "knockout": the absence of K14 leads to severe epidermolysis bullosa simplex and a function for an intermediate filament protein. Genes Dev 1994;8(21):2574–87.
[37] El Ghalbzouri A, Jonkman M, Kempenaar J, Ponec M. Recessive epidermolysis bullosa simplex phenotype reproduced in vitro: ablation of keratin 14 is partially compensated by keratin 17. Am J Pathol 2003;163(5):1771–9.
[38] Fine JD, Bruckner-Tuderman L, Eady RA, Bauer EA, Bauer JW, Has C, et al. Inherited epidermolysis bullosa: updated recommendations on diagnosis and classification. J Am Acad Dermatol 2014;70(6):1103–26.
[39] Mavilio F, Pellegrini G, Ferrari S, Di Nunzio F, Di Iorio E, Recchia A, et al. Correction of junctional epidermolysis bullosa by transplantation of genetically modified epidermal stem cells. Nat Med 2006;12(12):1397–402.
[40] Igoucheva O, Kelly A, Uitto J, Alexeev V. Protein therapeutics for junctional epidermolysis bullosa: incorporation of recombinant beta3 chain into laminin 332 in beta3−/− keratinocytes in vitro. J Invest Dermatol 2008;128(6):1476–86.

[41] Breitkreutz D, Mirancea N, Schmidt C, Beck R, Werner U, Stark HJ, et al. Inhibition of basement membrane formation by a nidogen-binding laminin gamma1-chain fragment in human skin-organotypic cocultures. J Cell Sci 2004;117(Pt 12):2611–22.
[42] Contard P, Bartel RL, Jacobs L, Perlish JS, MacDonald ED, Handler L, et al. Culturing keratinocytes and fibroblasts in a three-dimensional mesh results in epidermal differentiation and formation of a basal lamina-anchoring zone. J Invest Dermatol 1993;100(1):35–9.
[43] Chen M, Kasahara N, Keene DR, Chan L, Hoeffler WK, Finlay D, et al. Restoration of type VII collagen expression and function in dystrophic epidermolysis bullosa. Nat Genet 2002;32(4):670–5.
[44] Berthod F, Germain L, Li H, Xu W, Damour O, Auger FA. Collagen fibril network and elastic system remodeling in a reconstructed skin transplanted on nude mice. Matrix Biol 2001;20(7):463–73.
[45] Itoh M, Kiuru M, Cairo MS, Christiano AM. Generation of keratinocytes from normal and recessive dystrophic epidermolysis bullosa-induced pluripotent stem cells. Proc Natl Acad Sci U S A 2011;108(21):8797–802.
[46] Itoh M, Umegaki-Arao N, Guo Z, Liu L, Higgins CA, Christiano AM. Generation of 3D skin equivalents fully reconstituted from human induced pluripotent stem cells (iPSCs). PLoS One 2013;8(10):e77673.
[47] Sebastiano V, Zhen HH, Haddad B, Bashkirova E, Melo SP, Wang P, et al. Human COL7A1-corrected induced pluripotent stem cells for the treatment of recessive dystrophic epidermolysis bullosa. Sci Transl Med 2014;6(264). 264ra163.
[48] Umegaki-Arao N, Pasmooij AM, Itoh M, Cerise JE, Guo Z, Levy B, et al. Induced pluripotent stem cells from human revertant keratinocytes for the treatment of epidermolysis bullosa. Sci Transl Med 2014;6(264). 264ra164.
[49] Cogan J, Weinstein J, Wang X, Hou Y, Martin S, South AP, et al. Aminoglycosides restore full-length type VII collagen by overcoming premature termination codons: therapeutic implications for dystrophic epidermolysis bullosa. Mol Ther 2014;22(10):1741–52.
[50] Siprashvili Z, Nguyen NT, Gorell ES, Loutit K, Khuu P, Furukawa LK, et al. Safety and wound outcomes following genetically corrected autologous epidermal grafts in patients with recessive dystrophic epidermolysis bullosa. JAMA 2016;316(17):1808–17.
[51] Nyström A, Thriene K, Mittapalli V, Kern JS, Kiritsi D, Dengjel J, et al. Losartan ameliorates dystrophic epidermolysis bullosa and uncovers new disease mechanisms. EMBO Mol Med 2015;7(9):1211–28.
[52] Küttner V, Mack C, Rigbolt KT, Kern JS, Schilling O, Busch H, et al. Global remodelling of cellular microenvironment due to loss of collagen VII. Mol Syst Biol 2013;9:657.
[53] Odorisio T, Di Salvio M, Orecchia A, Di Zenzo G, Piccinni E, Cianfarani F, et al. Monozygotic twins discordant for recessive dystrophic epidermolysis bullosa phenotype highlight the role of TGF-β signalling in modifying disease severity. Hum Mol Genet 2014;23(15):3907–22.
[54] Mittapalli VR, Madl J, Löffek S, Kiritsi D, Kern JS, Römer W, et al. Injury-driven stiffening of the dermis expedites skin carcinoma progression. Cancer Res 2016;76(4):940–51.
[55] Marukian NV, Choate KA. Recent advances in understanding ichthyosis pathogenesis. F1000Res 2016;5. pii: F1000 Faculty Rev-1497.
[56] Oji V, Tadini G, Akiyama M, Blanchet Bardon C, Bodemer C, Bourrat E, et al. Revised nomenclature and classification of inherited ichthyoses: results of the first ichthyosis consensus conference in Sorèze 2009. J Am Acad Dermatol 2010;63(4):607–41.

[57] Li H, Vahlquist A, Törmä H. Interactions between FATP4 and ichthyin in epidermal lipid processing may provide clues to the pathogenesis of autosomal recessive congenital ichthyosis. J Dermatol Sci 2013;69(3):195–201.
[58] Thomas AC, Tattersall D, Norgett EE, O'Toole EA, Kelsell DP. Premature terminal differentiation and a reduction in specific proteases associated with loss of ABCA12 in Harlequin ichthyosis. Am J Pathol 2009;174(3):970–8.
[59] Amsellem C, Haftek M, Hoyo E, Thivolet J, Schmitt D. Evidence of increased keratinocyte proliferation in air-liquid interface cultures of non-bullous congenital ichthyosiform erythroderma. Acta Derm Venereol 1993;73(4):262–9.
[60] Chamcheu JC, Pihl-Lundin I, Mouyobo CE, Gester T, Virtanen M, Moustakas A, et al. Immortalized keratinocytes derived from patients with epidermolytic ichthyosis reproduce the disease phenotype: a useful in vitro model for testing new treatments. Br J Dermatol 2011;164(2):263–72.
[61] Oji V, Eckl KM, Aufenvenne K, Nätebus M, Tarinski T, Ackermann K, et al. Loss of corneodesmosin leads to severe skin barrier defect, pruritus, and atopy: unraveling the peeling skin disease. Am J Hum Genet 2010;87(2):274–81.
[62] Dahlqvist J, Törmä H, Badhai J, Dahl N. SiRNA silencing of proteasome maturation protein (POMP) activates the unfolded protein response and constitutes a model for KLICK genodermatosis. PLoS One 2012;7(1):e29471.
[63] Smith FJD, Irvine AD, Terron-Kwiatkowski A, Sandilands A, Campbell LE, Zhao Y, et al. Loss-of-function mutations in the gene encoding filaggrin cause ichthyosis vulgaris. Nat Genet 2006;38(3):337–42.
[64] Sandilands A, Sutherland C, Irvine AD, McLean WHI. Filaggrin in the frontline: role in skin barrier function and disease. J Cell Sci 2009;122(Pt 9):1285–94.
[65] Berroth A, Kühnl J, Kurschat N, Schwarz A, Stäb F, Schwarz T, et al. Role of fibroblasts in the pathogenesis of atopic dermatitis. J Allergy Clin Immunol 2013;131(6):1547–54.
[66] Kern JS, Loeckermann S, Fritsch A, Hausser I, Roth W, Magin TM, et al. Mechanisms of fibroblast cell therapy for dystrophic epidermolysis bullosa: high stability of collagen VII favors long-term skin integrity. Mol Ther 2009;17(9):1605–15.
[67] Iismaa SE, Mearns BM, Lorand L, Graham RM. Transglutaminases and disease: lessons from genetically engineered mouse models and inherited disorders. Physiol Rev 2009;89(3):991–1023.
[68] Diociaiuti A, Castiglia D, Fortugno P, Bartuli A, Pascucci M, Zambruno G, et al. Lethal Netherton syndrome due to homozygous p.Arg371X mutation in SPINK5. Pediatr Dermatol 2013;30(4):e65–7.
[69] Furio L. Hovnanian a Netherton syndrome: defective kallikrein inhibition in the skin leads to skin inflammation and allergy. Biol Chem 2014 Sep;395(9):945–58.
[70] Duval C, Chagnoleau C, Pouradier F, Sextius P, Condom E, Bernerd F. Human skin model containing melanocytes: essential role of keratinocyte growth factor for constitutive pigmentation-functional response to α-melanocyte stimulating hormone and forskolin. Tissue Eng Part C Methods 2012;18(12):947–57.
[71] Dupuy A, Sarasin A. DNA damage and gene therapy of xeroderma pigmentosum, a human DNA repair-deficient disease. Mutat Res 2015;776:2–8.
[72] Bernerd F, Asselineau D, Frechet M, Sarasin A, Magnaldo T. Reconstruction of DNA repair-deficient xeroderma pigmentosum skin in vitro: a model to study hypersensitivity to UV light. Photochem Photobiol 2005;81(1):19–24.
[73] Bernerd F, Asselineau D, Vioux C, Chevallier-Lagente O, Bouadjar B, Sarasin A, et al. Clues to epidermal cancer proneness revealed by reconstruction of DNA repair-deficient xeroderma pigmentosum skin in vitro. Proc Natl Acad Sci U S A 2001;98(14):7817–22.

[74] Arnaudeau-Bégard C, Brellier F, Chevallier-Lagente O, Hoeijmakers J, Bernerd F, Sarasin A, et al. Genetic correction of DNA repair-deficient/cancer-prone xeroderma pigmentosum group C keratinocytes. Hum Gene Ther 2003;14(10):983–96.
[75] Mamrak NE, Shimamura A, Howlett NG. Recent discoveries in the molecular pathogenesis of the inherited bone marrow failure syndrome Fanconi anemia. Blood Rev 2016;31(3):93–9.
[76] DiGiovanna JJ, Priolo M, Itin P. Approach towards a new classification for ectodermal dysplasias: integration of the clinical and molecular knowledge. Am J Med Genet A 2009;149A(9):2068–70.
[77] Zarnegar BJ, Webster DE, Lopez-Pajares V, BVS H, Qu K, Yan KJ, et al. Genomic profiling of a human organotypic model of AEC syndrome reveals ZNF750 as an essential downstream target of mutant TP63. Am J Hum Genet 2012;91(3):435–43.
[78] Barbaro V, Nasti AA, Raffa P, Migliorati A, Nespeca P, Ferrari S, et al. Personalized stem cell therapy to correct corneal defects due to a unique homozygous-heterozygous mosaicism of ectrodactyly-ectodermal dysplasia-clefting syndrome. Stem Cells Transl Med 2016;5(8):1098–105.
[79] Murphy-Ryan M, Psychogios A, Lindor NM. Hereditary disorders of connective tissue: a guide to the emerging differential diagnosis. Genet Med 2010;12(6):344–54.
[80] De Paepe A, Malfait F. The Ehlers-Danlos syndrome, a disorder with many faces. Clin Genet 2012;82(1):1–11.
[81] Byers PH, Murray ML. Ehlers-Danlos syndrome: a showcase of conditions that lead to understanding matrix biology. Matrix Biol 2014;33:5–10.
[82] Miyake N, Kosho T, Mizumoto S, Furuichi T, Hatamochi A, Nagashima Y, et al. Loss-of-function mutations of CHST14 in a new type of Ehlers-Danlos syndrome. Hum Mutat 2010;31(8):966–74.
[83] Shimizu N, Okamoto N, Miyake K, Taira Y, Sato K, Matsuda N, et al. Delineation of dermatan 4-O-sulfotransferase 1 deficient Ehlers-Danlos syndrome: observation of two additional patients and comprehensive review of 20 reported patients. Am J Med Genet 2011;155A:1949–58.
[84] Llames SG, Del Rio M, Larcher F, García E, García M, Escamez MJ, et al. Human plasma as a dermal scaffold for the generation of a completely autologous bioengineered skin. Transplantation 2004;77(3):350–5.
[85] Rheinwald JG, Green H. Serial cultivation of strains of human epidermal keratinocytes: the formation of keratinizing colonies from single cells. Cell 1975;6(3):331–43.
[86] Meana A, Iglesias J, Del Rio M, Larcher F, Madrigal C, Fresno MF, et al. Large surface of cultured human epithelium obtained on a matrix based on live fibroblast-containing fibrin gels. Burns 1998;24(7):621–30.
[87] Larcher F, Dellambra E, Rico L, Bondanza S, Murillas R, Cattoglio C, et al. Long-term engraftment of single genetically modified human epidermal holoclones enables safety pre-assessment of cutaneous gene therapy. Mol Ther 2007;15(9):1670–6.
[88] Duarte B, Miselli F, Murillas R, Espinosa-Hevia L, Cigudosa JC, Recchia A, et al. Long-term skin regeneration from a gene-targeted human epidermal stem cell clone. Mol Ther 2014;22(11):1878–80.
[89] Droz-Georget Lathion S, Rochat A, Knott G, Recchia A, Martinet D, Benmohammed S, et al. A single epidermal stem cell strategy for safe ex vivo gene therapy. EMBO Mol Med 2015;7(4):380–93.
[90] Di Nunzio F, Maruggi G, Ferrari S, Di Iorio E, Poletti V, Garcia M, et al. Correction of laminin-5 deficiency in human epidermal stem cells by transcriptionally targeted lentiviral vectors. Mol Ther 2008;16(12):1977–85.

[91] Gostynski A, Llames S, García M, Escamez MJ, Martinez-Santamaria L, Nijenhuis M, et al. Long-term survival of type XVII collagen revertant cells in an animal model of revertant cell therapy. J Invest Dermatol 2014;134(2):571–4.

[92] Titeux M, Pendaries V, Zanta-Boussif MA, Décha A, Pironon N, Tonasso L, et al. SIN retroviral vectors expressing COL7A1 under human promoters for ex vivo gene therapy of recessive dystrophic epidermolysis bullosa. Mol Ther 2010;18(8):1509–18.

[93] Turczynski S, Titeux M, Tonasso L, Décha A, Ishida-Yamamoto A, Hovnanian A. Targeted exon skipping restores type VII collagen expression and anchoring fibril formation in an in vivo RDEB model. J Invest Dermatol 2016;136(12):2387–95.

[94] García M, Larcher F, Hickerson RP, Baselga E, Leachman SA, Kaspar RL, et al. Development of skin-humanized mouse models of pachyonychia congenita. J Invest Dermatol 2011;131(5):1053–60.

[95] Di WL, Larcher F, Semenova E, Talbot GE, Harper JI, Del Rio M, et al. Ex-vivo gene therapy restores LEKTI activity and corrects the architecture of Netherton syndrome-derived skin grafts. Mol Ther 2011;19(2):408–16.

[96] Di WL, Mellerio JE, Bernadis C, Harper J, Abdul-Wahab A, Ghani S, et al. Phase I study protocol for ex vivo lentiviral gene therapy for the inherited skin disease, Netherton syndrome. Hum Gene Ther Clin Dev 2013;24(4):182–90.

[97] Aufenvenne K, Rice RH, Hausser I, Oji V, Hennies HC, Rio MD, et al. Long-term faithful recapitulation of transglutaminase 1-deficient lamellar ichthyosis in a skin-humanized mouse model, and insights from proteomic studies. J Invest Dermatol 2012;132(7):1918–21.

[98] Aufenvenne K, Larcher F, Hausser I, Duarte B, Oji V, Nikolenko H, et al. Topical enzyme-replacement therapy restores transglutaminase 1 activity and corrects architecture of transglutaminase-1-deficient skin grafts. Am J Hum Genet 2013;93(4):620–30.

[99] García M, Llames S, García E, Meana A, Cuadrado N, Recasens M, et al. In vivo assessment of acute UVB responses in normal and Xeroderma Pigmentosum (XP-C) skin-humanized mouse models. Am J Pathol 2010;177(2):865–72.

[100] Warrick E, Garcia M, Chagnoleau C, Chevallier O, Bergoglio V, Sartori D, et al. Preclinical corrective gene transfer in xeroderma pigmentosum human skin stem cells. Mol Ther 2012;20(4):798–807.

[101] Luchetti MM, Moroncini G, Escamez MJ, Svegliati Baroni S, Spadoni T, Grieco A, et al. Induction of scleroderma fibrosis in skin-humanized mice by administration of anti-platelet-derived growth factor receptor agonistic autoantibodies. Arthritis Rheumatol 2016;68(9):2263–73.

In vitro models of psoriasis

5

Bryan Roy, Mélissa Simard*,†, Isabelle Lorthois*,†, Audrey Bélanger*,†, Maxim Maheux*,†, Alexandra Duque-Fernandez*, Geneviève Rioux*, Philippe Simard*, Marianne Deslauriers*, Louis-Charles Masson*, Alexandre Morin*, Roxane Pouliot*,†*

*Centre de recherche en Organogénèse expérimentale de l'Université Laval (LOEX), Sherbrooke, QC, Canada, †Université Laval, Quebec, QC, Canada

This chapter presents an overview of psoriasis, the latest developments in the multidisciplinary field of tissue engineering, and psoriatic in vitro models that have been developed. An emphasis is made on psoriatic substitutes produced according to a self-assembly approach, which maintained many characteristics of the disease including the presence of a disorganized and thicker epidermis compared with normal skin substitutes. This self-assembly approach allows the understanding of pathological skin complexity through the possibility of [1] dissecting step by step the mechanisms of skin pathologies according to which kind of cells is present in the model at this time and/or [2] using various cell combinations. Unlike animal models, alternative in vitro models offer the opportunity to add, at the same time or successively, one or more components to assess their role in the disease. Section 4 of this chapter addresses the role of immune and endothelial cells in psoriasis. Moreover, lipidomic studies and transcriptomic characterization are also discussed. Finally, sections on the pharmaceutical screening of new antipsoriatic drugs and the metabolic capabilities of the model are presented. Reconstructed artificial skin substitutes mimic in the best way the native physiological context and represent sophisticated important tools for basic and applied skin research.

1. Psoriasis

Psoriasis is an immune-mediated, genetic skin, and joint disorder affecting both men and women likewise [1,2]. An estimated 2%–3% of the world population suffers from this dermatosis, and of those, about 20 million are affected by the moderate or severe forms [3]. Therefore, psoriasis is probably the most prevalent immune-mediated skin disease in adults [4]. The specific pathogenesis of psoriasis is not yet fully understood, but the underlying mechanism seems to involve an interplay between genetic, immunologic, and environmental factors, all playing important roles in the onset of psoriasis [5,6]. This pathology is characterized by the formation of sharply demarcated red raised plaques that may occasionally show a silvery white appearance and are further covered with thick scales [7,8]. Certain body regions are more likely to be

Skin Tissue Models. https://doi.org/10.1016/B978-0-12-810545-0.00005-X

affected, as is the case for the elbows, knees, and scalp [9]. In fact, plaque formation occurs mainly at sites undergoing intensive mechanical stress [2]. Increased keratinocyte proliferation leading to the thickening of the epidermis, an extensively altered differentiation process, the development of epidermal rete ridges forming long and thin downward projections into the dermis, and a greatly reduced or absent granular layer are all characteristics of psoriasis. All of these factors combine to yield a highly perturbed *stratum corneum* and, thus, impaired permeability function [4,10–12]. This break in the permeability barrier is also due to the psoriatic corneocytes' failure to secrete extracellular lipids allowing these cells to adhere to one another [4,13]. Previous studies have also demonstrated alterations in lipid metabolism of psoriatic *stratum corneum* [14,15]. Other defining features of psoriasis include the presence of neutrophils within the *stratum corneum*, leukocytic infiltration into both the dermis and epidermis, along with marked angiogenesis and dilation of blood vessels causing the reddish aspect of psoriatic lesions [4]. This disorder may be found in several subtypes, such as guttate, pustular, inverse, erythrodermic, and plaque, the latter being the most common, accounting for about 90% of cases [2,16]. Psoriasis is currently incurable, although some treatments have shown worthwhile disease remission [17–22]. In fact, some treatments can revert back to symptomless or visibly healthy skin, with little or no trace of preexisting disease activity [23]. On the other hand, a large number of patients must live with consequently induced aftereffects, such as immunosuppression, since immunosuppressants still are widely prescribed for severe psoriasis cases, and secondly, some patients are simply unresponsive to current therapies [24].

2. Tissue engineering

Tissue engineering is an interdisciplinary field that brings together cell knowledge and engineering coupled with biochemical cues toward the development of biological substitutes in order to repair, replace, or regenerate tissues and their function [25]. Cells and various artificial or natural matrices are combined in such a manner as to obtain tissue mimics for transplants up to pharmaceutical models [26]. It's in the 1980s that the interest for tissue engineering and biomaterials really took off, thereby resolutely opening a new chapter in regenerative biomedicine [26]. First great efforts in this field have been made in skin research, and this was motivated by the acute need of solutions to help impaired burn patients in dire need of grafts. Indeed, this technique represents a pragmatic alternative where extensive wounds can neither pull through by themselves nor be satisfyingly cured or subdued using conventional techniques such as autografts or cadaveric transplantation. Assuredly, these approaches have saved and improved countless lives, but they remain imperfect solutions [25]. In addition, tissue engineering happens to be a thriving area in virtually every human tissue research for surgery-related application just as much as fundamental experiments or pharmaceutical assessments [27,28]. In this regard, tissue-engineered skin provides insights into the behavior of healthy and lesional skin cells [29]. It's an inestimable addition for pharmaceutical research purposes, providing models where research is hindered by the lack of naturally occurring disorders in laboratory animals that mimic the complex phenotype and pathogenesis of human diseases, as is the case for psoriasis [2]. Seizing this possibility to reconstruct human living tissues displaying distinctive lesional phenotypes, which could not have been envisioned not so long ago, has been achieved by several laboratories. They have contributed step by step to the understanding of psoriasis by putting together in vitro psoriatic models. In order to do so, they used various approaches, which will be discussed just below, thus opening tremendous perspective to enhance the knowledge into the causes and treatments of psoriasis.

3. In vitro models of psoriasis

The nonexistence of validated in vivo and in vitro models, or even the knowledge of a discriminating psoriasis biomarker, impede future findings and surely constitute one of the major challenges for antipsoriatic drug development. Better models are needed in order to dissect the interactions of many complex molecular pathways or networks in the skin [4]. Indeed, psoriasis consists of a complex disease involving the alteration of over 1300 genes and the interaction of 30 or more upregulated cytokines and chemokines [11,30].

Firstly, no naturally occurring affected animal models mirroring both phenotype and immunopathogenesis of psoriasis are currently known, which has resulted in severely hampered research efforts [31]. Consequently, several in vivo models have been developed to counterbalance this acute lack in psoriasis research and for preclinical drug development purposes. This category includes spontaneous mutations, xenotransplantation, knockout, and transgenic animal models [32]. Overall, rodent models are satisfyingly representative of the disease, but no one can fully mimic its particular genomic, histological, and inflammatory signature [31]. Indeed, dissimilar architecture and metabolism between the mouse and human skin limit this approach [33]. Accordingly, we must not jump to early conclusions that results obtained from murine models are always predictive of the human condition and its response to treatments. Lastly, another issue is that in vivo models are exposed to increasing ethical concerns surrounding the use of laboratory animals.

On the other hand, following tissue-engineering development, in vitro psoriatic models have also been widely used as test systems in order to better understand its physiopathology and for pharmaceutical industry assessments. They represent suitable alternatives to in vivo models for numerous technical, practical, and ethical reasons. Besides being consequently more faithful models due to the direct use of human cells, they allow broad control over cellular composition and production as well as being assessment friendly [34]. However, we must again avoid making hasty conclusions on the fact that in vitro models are better than in vivo in all cases. Indeed, the lack of a surrounding living system and hence of certain cell types may lead us to believe that a number of results gathered from some psoriatic models are of questionable value. In this vein, considering that in vitro models have been developed in order to acutely reproduce key structural components and functional aspects of the natural skin, the relevance of these models has undergone growing criticism in regard to their lack of an important component: immune cells [4,11]. However, it's worth mentioning that a group has recently been able to populate their psoriatic equivalents with $CD4^+$ T cells [35]. Seeing this opportunity, a focus of research manages to further improve the model we're using by adding an immune component. Moreover, several teams, including our own, have previously tried to partially get around this immune cell addition through an induction of the equivalents with various cytokines in order to alter their gene expression patterns in a similar manner to what would be found in native psoriasis [36–40]. Subsequent sections will aim to introduce the different psoriatic in vitro models and provide detailed explanations necessary for their understanding.

3.1 Monolayer

Monolayer models enable researchers to consider only one cell type at a time. Either keratinocytes or fibroblasts can be separately cultured to test different conditions or to observe psoriatic skin features, such as keratinocyte hyperproliferation or abnormal differentiation [32]. Thereby, this technique allows for the isolation of normal or pathological cell types, thus enabling the study of specifically involved mechanisms. Indeed, human skin cells grow within an organized three-dimensional environment, and monolayer cultures cannot capture this relevant complexity [41]. Yet, even if this monolayer culture does not allow this direct interaction, we can find in that same weakness all of its capability: through this application, it was possible to demonstrate the importance of this interaction between fibroblasts and keratinocytes and secondly that functional interactions between keratinocytes and T lymphocytes require direct cellular contact [42]. It has also been shown that psoriatic keratinocytes alter the function of normal T lymphocytes, which conversely modulates these keratinocytes, in a negative feedback loop [42]. Indeed, these models have led to a better understanding of the pathology [43]. For instance, Detmar et al. demonstrated that psoriasis involves a TGF-α regulation of vascular endothelial growth factor (VEGF) expression via an autocrine mechanism, which leads to vascular hyperpermeability and angiogenesis [44]. Also, monolayer cultures are still widely used to conduct toxicity assessments on each cell type independently [45]. Nonetheless, monoculture keratinocytes only generate thin epidermal layers, which anyhow, without any mesenchymal feeder layer, undergo fast apoptosis [46]. Furthermore, these kinds of artificial skin aren't suitable for percutaneous absorption tests, which cannot be properly assessed without essential characteristics provided by three-dimensional organization [33].

As previously discussed, complex interactions occurring between skin cells and their microenvironment further call for three-dimensional skin models, which can capture in a more relevant way the complexity of in vivo situations. It's undeniable that monolayers have played and still play an important role in the comprehension of psoriasis, but laboratories cannot be limited only to this type of model in order to fully understand the disease. Therefore, they have developed in vitro three-dimensional models in order to overcome its inherent limitations. These psoriatic equivalents are intended to be introduced in hereafter sections.

3.2 De-epidermized dermis

This approach has been first described by Pruniéras et al., as an attempt to achieve more physiological characteristics in the tissue-engineered skin [47]. By culturing epidermal cells on dead de-epidermized dermis (DED), an acellular structure, all morphological markers of differentiation were seen except for keratin patterns [47]. Using this same technique, Ponec further demonstrated that air-liquid interface cultures reproduce to a high extent the lipid composition of the native epidermis, except for higher triglyceride and lower glycosphingolipid content, as well as very low linoleic acid content [48]. Afterward, specialized research groups have introduced psoriatic skin equivalent models using various patterns arising from this approach. Some of them used explant cultures [49], while others turned out seeding directly keratinocytes belonging to psoriasis patients [50,51] or instead simply using healthy cells paired with an induction of a psoriatic phenotype by inhibition of transglutaminase [52]. Finally, others introduce psoriatic skin equivalents through the addition of lymphocytes or cytokines [53,54].

3.2.1 Explant culture on de-epidermized dermis

From psoriatic skin biopsies cultured on deepidermized dermis, Mils et al. have been able to develop a reconstructed epidermal psoriatic model in an attempt to accurately reproduce its phenotype [49]. They began by depositing a 4mm punch biopsy on the epidermal side of mortified de-epidermized human dermis, making contact with the remnant basement membrane components [55]. The DED was maintained at the air-liquid interface using a metallic support. Within 15days, they were able to obtain a well-differentiated epidermis entirely covering the DED. Their results did not show any differences whatever using normal or psoriatic biopsy, thus yielding a unique in vitro phenotype, which is in dissimilarity with physiopathologic differences found between the normal and psoriatic skin in vivo. In fact, both conditions shared similar morphology with slight hyperacanthosis and orthokeratotic hyperkeratosis [49]. Those results seem to suggest a need for extracutaneous stimuli to maintain the psoriatic phenotype in vitro.

3.2.2 Keratinocytes on de-epidermized dermis

Using the same basic technique, another in vitro epidermal model has been developed, but rather using normal keratinocytes seeded directly on a DED [53]. DED were generated from surgery-discarded abdominal skin, where the epidermis has been dislodged from the dermis. Healthy keratinocytes are then seeded within a hollow metal ring placed on DED punches. Once keratinocytes were crammed on the dermis, they removed the ring and submerged the whole assembly in culture medium. To obtain a well-differentiated epidermis, they also underwent an air-liquid phase using a metal grid as a shim [53]. Since they're using nonpathological keratinocytes, they induced psoriasis-associated features and gene expression by adding, at the last 4days of the air-liquid interface culture, various combinations of pro-inflammatory cytokines, such as TNF-α, IL-lα, IL-6, and IL-22. The addition of this cytokine blend results in a strongly induced expression of hBD-2, SKALP/elafin, CK16, IL-8, and TNF-α, as well as keratinocyte hyperproliferation. Tjabringa et al. showed that retinoic acid inhibits cytokine-induced gene expression at both the mRNA and protein levels, as found in the native form of psoriasis; different results were obtained using cyclosporin A. The model also demonstrated reduced expression of keratin 10, a differentiation marker [53]. In conclusion, this approach gathers major psoriatic model criteria and thereby attests promising potential in further pathology research efforts and pharmacological applications.

As introduced earlier, van den Bogaard et al. have been able, through this same technique, to produce psoriatic skin equivalents containing an immune component: $CD4^+$ T cells [35]. They showed that the migration of those lymphocytes into the dermal equivalent initiated keratinocyte activation within 2 days, with hallmarks of a psoriasiform inflammation after 4 days. They also observed upregulated expression of genes, which code for epidermal psoriasis markers along with various pro-inflammatory cytokines and chemokines. Furthermore, disturbed epidermal differentiation was established by downregulated filaggrin and involucrin expression in the spinous layer [35].

3.3 Collagen gels

The earliest dermal matrix used as tissue-engineered skin incorporated fibroblasts into exogenous collagen hydrated gels or lattices [56–58]. This path followed by tissue-engineering pioneers was sustained by the fact that collagen is the main matrix component of connective tissues and thus provides support for the whole cutaneous structure [59,60]. Once added, fibroblasts reorganized collagen fibers to form the dermis, and afterward, those fibroblast-populated matrices are overlayed with a suspension of keratinocytes or a full punch biopsy to recreate the epidermis [61]. Drawbacks of this method lie behind the use of exogenous material and weak mechanical resistance along with the fact that this matrix undergoes severe contraction early in the culture process [62]. Nevertheless, this skin culture approach benefits from relatively fast production time and is well suited for chemical toxicity assessments [62]. Since then, collagen gels are used in many laboratories to produce healthy and lesional skin substitutes. As it will be presented below, slight technical particularities may be observed between different research groups in the production of psoriatic skin equivalents.

3.3.1 Explant culture

Saiag and coworkers used a technique that involved laying full-thickness punch biopsies on fibroblast-populated collagen gels in order to measure the effects of various combinations of keratinocytes and fibroblasts from the psoriatic skin, normal skin from psoriatic patients, and normal skin from nonpsoriatic controls on skin equivalent keratinocytes proliferation kinetics [63]. Results have convincingly showed that the basic defect in psoriasis appears to be related with both uninvolved and involved psoriatic fibroblasts and that psoriatic keratinocytes' hyperproliferative growth was maintained even with the use of normal fibroblasts [43,63,64]. From these results, psoriatic fibroblasts prove to be key regulators of abnormal proliferation in psoriatic skin. Even so, it is still unproved whether they act autonomously or, more likely, under the direction of some local or distant stimuli that dictates its altered regulatory behavior [64].

3.3.2 Models using multiple cellular types

Collagen gels have also been used as scaffolds in models composed from separately extracted healthy and pathological fibroblasts along with keratinocytes. Autologous fibroblasts are obtained through a biopsy, where they are removed from the dermis and expanded, as is the same for keratinocytes, which are, however, extracted from the epidermis and cultured on a 3 T3 fibroblasts feeder layer. As described above, fibroblasts are first embedded in the collagen matrix, and then, keratinocytes are seeded on top of the dermal equivalent [65]. Konstantinova et al. were first to introduce this technique, borrowed from field pioneers Bell et al., in order to produce psoriatic skin substitutes [58,60]. Their skin equivalents were thicker, lost filaggrin, showed more intense IL-8 staining when performing immunohistochemistry assays, and developed invaginations, all of which are factors found in psoriasis in vivo. These results led to the hypothesis that an IL-8 paracrine loop between fibroblasts and keratinocytes may play a key role in epidermal regeneration and, thus, in hyperproliferation [65]. Still, in order to be able to characterize the psoriasis in vitro, Barker et al. developed a psoriatic collagen gel skin model derived from normal involved psoriatic and uninvolved psoriatic cells. These psoriatic skin substitutes were well representative of the pathology [51]. Results showed that classical keratinocyte differentiation markers exhibited similar patterns of distribution in psoriatic models to those derived from normal cells and generally reflected in vivo observations [51]. Moreover, their model retains many other psoriatic characteristics such as keratinocyte hyperproliferation and abnormal differentiation, as well as enhancement of IL-6 and IL-8 concentrations [51]. Models derived from uninvolved and involved psoriatic skin share the same gene expression profile and pathological characteristics as psoriatic human skin [43,51].

3.4 Commercially available model

MatTek corporation has developed an in vitro model of psoriasis in order to enable the screening of therapeutic candidates for safety and efficacy. The model is produced using collagen gels, exploiting, more specifically, the multiple cellular types approach described just above. They use normal human epidermal keratinocytes and psoriatic fibroblasts derived from psoriatic explants. Their skin substitutes exhibit and maintain a psoriatic phenotype as evidenced by increased basal cell proliferation; expression of psoriasis-specific markers; and enhanced release of psoriasis-specific cytokines, such as IL-6, IL-8, GM-CSF, and IP-10. Likewise, the model morphologically emulates lesional psoriatic human tissues. Finally, this in vitro psoriatic model is responsive to some marketed antipsoriatic therapeutics, thus representing a useful tool for pharmaceutical assessments [66–69].

3.5 The self-assembly method

The self-assembly method, developed by the LOEX group, allows the production of complete bilayered skin substitutes devoid of exogenous collagen or any other synthetic materials [50]. The fundamental concept behind this method is to reconstruct an organ in a fashion closely emulating its formation in vivo. Accordingly, this tissue-engineering approach is based on the capacity of mesenchymal cells, such as fibroblasts, to create their own extracellular matrix in vitro. Thus, cells are grown in such a way that they literally embed themselves in their very own extracellular matrix [26].

Following the path of tissue-engineering pioneers, the development of this method was firstly motivated by the need of solutions to help impaired burn patients in dire need of grafts. Nowadays, the method is used for wound healing not only in the case of extensive burns but also in the case of chronic wounds (the lack of cadaver skin to protect burn patients and plasters for diabetic ulcers), not to mention cornea reconstruction and various other research purposes [70,71]. Research upon these skin equivalents has been motivated by the fact that they closely emulate many characteristics found in the normal human skin and avoid the possible bias caused by the contribution of exogenous materials [72,73].

From a dermopharmaceutical point of view, there is a need for skin models to test new formulations developed in vitro. The self-assembly method represents an alternative method to replace animal testing and, secondly, provides models where research efforts are impeded by the absence of naturally occurring disorders in laboratory animals or validated in vitro models, as is the case for psoriasis [2]. Thus, facing the absence of exogenous material-free models, our group developed a pathological skin model to study psoriasis in vitro, using this unique approach [72]. Indeed, the method can be adaptable to produce pathological skin substitutes from affected patients' cells [43].

Foremost, it was shown that self-assembled equivalents partially maintained psoriasis-like features such as epidermis thickening and hyperproliferation, as well as abnormal keratinocyte differentiation [50]. We also demonstrated that our model positively reacts to treatments with an antipsoriatic molecule such as observed in psoriatic skin in vivo, thus providing a solid proof of principle [74]. It is clear that self-assembled skin substitutes emulate the normal human skin and psoriatic human skin much better than artificial membranes and thereby could become an effective and innovative dermopharmaceutical tool for the screening of new treatments (Fig. 1) [43,75].

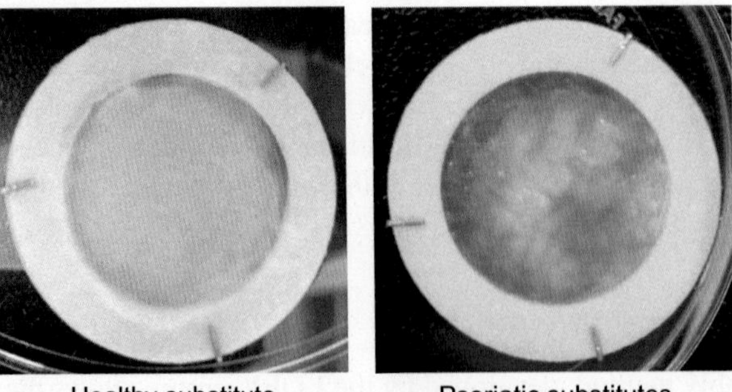

Fig. 1 Macroscopic aspect of substitutes produced according to the self-assembly method. (A) Healthy substitute and (B) psoriatic substitute.

3.5.1 Skin cells source

Dermal fibroblasts and epidermal keratinocytes were obtained from healthy skin specimens originating from surgical procedures, thereby serving as controls, and from 6 mm punch biopsies belonging to patients with plaque psoriasis who were recruited from affiliated hospitals [76,77]. Psoriatic punches were taken either from uninvolved or involved skin portions depending on experimental purposes [43].

Keratinocytes were initially isolated with a thermolysin digestion, followed by a mechanical separation from the dermis with forceps [77,78]. Afterward, keratinocytes were dissociated from the epidermis using a trypsin-EDTA solution [78]. Cells were then amplified on a lethally irradiated 3 T3 feeder layer and finally banked. On the other hand, human fibroblasts were dissociated using collagenase and, alike, subsequently banked for ulterior use [77].

3.5.2 Cell culture media

Keratinocytes were cultured in complete DME-HAM, a combination of Dulbecco-Vogt modification of Eagle's medium (DME) with Ham's F12 in a 3:1 proportion. The medium is supplemented with insulin, hydrocortisone, cholera toxin (which enhances growth factor uptake), 5% serum, epidermal growth factor, and antibiotics [79]. Fibroblasts, for their part, were cultured in DME supplemented with 10% serum and antibiotics [72]. In addition, fibroblasts were cultured in the presence of ascorbic acid. This additive increases procollagen mRNA synthesis in fibroblasts, thus allowing the posttranslational hydroxylation of lysine and proline, which leads to efficient collagen helix formation and increases secretion of collagen in cultured cells [80–83]. In short, we are able to obtain manipulable fibroblast sheets through the addition of this latter compound. This represents the cornerstone of the self-assembly approach, allowing the production of skin equivalents devoid of any synthetic components or other exogenous materials.

3.5.3 Production of the self-assembled substitutes

In order to obtain psoriatic skin equivalents, normal and pathological fibroblasts are firstly thawed and cultured with ascorbic acid over a period of 4 weeks. During this period, dermal sheets are produced, and cohesive dermal tissues are obtained by stacking two of these fibroblasts sheets together upon which normal or pathological keratinocytes are seeded 7 days later. After another 7 days of culture, the substitutes are raised to the air-liquid interface to promote cell differentiation and stratification. Finally, maturation ends after 21 days of culture at the air-liquid interface, and samples are analyzed using histological, immunohistochemical, physicochemical, or permeability assays, as required [50].

One must admit that this approach, at the present time, is labor-intensive and time-consuming. Therefore, the development of new protocols and techniques and optimizing cell culture conditions in order to reduce production and maturation delays remain relevant. This incessant development is likely worth it, considering, for now, that those skin equivalents are the purest form of in vitro models for various physiological, pathophysiological, pharmaceutical, and toxicological studies [26].

4. Results from ongoing research projects and future insights

4.1 Understanding the role of immune cells in psoriasis through tissue engineering

Psoriasis is an organ-specific T-cell-driven autoimmune pathology that is engaged following T cells' activity along with an array of other stimuli and subsequently maintained by the secretory activity of dendritic cells, macrophages, and again T cells [84,85]. Indeed, clinical and fundamental research observations have indicated that psoriasis involves an interplay between the innate and adaptive immune systems, including complex feedback loops from antigen-presenting cells, neutrophilic granulocytes, keratinocytes, vascular endothelial cells, and cutaneous nervous system [2,84]. These cells secrete cytokines such as IL-17, TNF-α, and IFN-γ, the first two being responsible for keratinocyte hyperproliferation, T-cell activation, and proliferation, as well as immune cell migration within both the dermis and epidermis [84,86].

Originally, there was a debate about whether the dominant process in psoriasis involved hyperplasic keratinocytes or the immune system [87]. However, the remarkable efficacy of cyclosporine A, an immunosuppressive agent that inhibits T-cell proliferation and cytokine production, essentially ended the controversy [84]. Nowadays, numerous other immunosuppressive therapies have been introduced for the therapeutic armamentarium of psoriasis, thus confirming the immune system's key role in its pathogenesis [88]. Even so, the further connection between immunologic, environmental, and genetic factors, all leading to this pathology, still remains somewhat puzzling.

Taking into account all the evidence showing the importance of underlying immune components in psoriasis, it becomes of prime importance to possess relevant models to investigate all of the involved mechanisms. Certainly, in vivo models could be used, but owing to human and animal immunologic differences, as well as price and ethical considerations surrounding their use, they constitute a less appealing avenue [89]. Thus, these concerns made it necessary to develop new human immune-competent in vitro models in order to study the crucial and yet enigmatic interactions between skin and immune cells [41]. Moreover, besides being suitable for the understanding of this complex pathology, immune-competent in vitro equivalents have the potential to serve as preclinical screening tools, where in vivo models often poorly predict the human immune response, leading to potential new drugs not reaching clinical trials or clinical trial failures [35,89]. Accordingly, laboratories have investigated two types of psoriatic in vitro models, cocultures and three-dimensional models, as presented below.

Cocultures are very useful models to study psoriasis, even if they may seem elementary. Fibroblasts or keratinocytes are seeded and expanded on a plastic supportive material before adding a second cellular component to the culture. This way, as it will be presented in the following paragraph, many findings that all contribute to a greater understanding of the pathology have been possible. In counterpart, although they are and will remain invaluable tools, cocultures are bound to be transitional solutions inasmuch as three-dimensional models containing immune components, which are currently rare due to technical challenges related to their production, are more comprehensive models that should be preferentially used when they'll become further available.

First, in the study of Schirmer et al., dendritic cells were cocultured with healthy fibroblasts, which allowed them to discover that dendritic cells secrete a greater amount of IL-23 and IL-6 when activated by fibroblast-derived TNF-α [90]. IL-23 involvement in psoriasis occurs by stimulating T cells to secrete IL-17, a compound that acts as a pro-inflammatory cytokine, which further inducts TNF-α secretion from fibroblasts [90]. In another study, Wittman et al. added T cells to a culture of healthy keratinocytes [91]. This way, they observed that preactivated keratinocytes with IFN-γ secrete more CXCL10, which is responsible for the attraction of T cells to the lesion site [91]. Moreover, in Prinz et al. study, psoriatic keratinocytes were separately cultured from T cells, and their supernatants were subsequently transferred to the keratinocyte cultures in order to study their interaction with T-cell-secreted cytokines [92]. They showed that keratinocytes' proliferation rate was higher in the presence of T-cell supernatants [92].

Our coculture models are unlike any other, this being due in part to the use of psoriatic epithelial cells. With this model, it appears that psoriatic keratinocytes lowered T cells' apoptotic rate. Furthermore, this study demonstrated that inflammatory cytokines secreted by psoriatic keratinocytes, such as MCP-1, IL-8, and TNF-α, were significantly increased compared with cocultures using healthy keratinocytes [93]. Consecutive studies on the interaction between T cells and psoriatic keratinocytes showed that cell-to-cell contact increased cellular cytokine secretion [42].

On the other hand, studies on three-dimensional psoriatic in vitro skin equivalents, despite inherent difficulties in their production, are becoming more prevalent. It may be first mentioned the van den Bogaard et al. model previously introduced, where preactivated and polarized T cells were inserted under an acellular dermis, over which are seeded healthy keratinocytes [35]. They demonstrated that preactivated T cells migrate in the acellular dermis and were able to activate keratinocytes and to induce their proliferation. After 2 days of migration, cytokines like IFN-γ, IL-17, IL-22, and TNF-α concentrations in the medium were at their peak. As priorly mentioned, this three-dimensional model is physiologically very accurate in regard to psoriatic lesions, because many proteins usually found in psoriatic skin are as well expressed in their equivalents. Furthermore, many genes normally expressed in psoriasis are alike expressed, and these include DEFB4, PI3, IL8, and S100A7 just to name a few [35]. In another study, Saalbach et al. introduced monocyte-derived dendritic cells to a collagen 1 lattice over which were seeded healthy human fibroblasts [94]. They demonstrated that monocyte-derived dendritic cells preactivated with LPS migrate at higher rates in the presence of dermal fibroblasts than inactivated dendritic cells. These results suggest a possible interaction between these two cell types. Their model proved to be genetically representative of the pathology; indeed, several genes such as MMP-1, MMP-3, and cox-2, all involved in psoriasis inflammatory responses, were expressed by skin equivalents' fibroblasts, as is observed in psoriatic lesions [94]. Conclusively, both models surely represent great achievements on a biological and technical basis and further allowed, in an unparalleled way, to gather precious evidence that points us toward the right direction in the understanding of this pathology.

4.2 Pharmaceutical screening of new antipsoriatic drugs

At the present moment, as was stated earlier, none of the available treatments has the potential to fully cure psoriasis; they only manage resulting symptoms. Despite their effectiveness in resolving psoriatic lesions, their use is limited because many of them have significant adverse effects on patients when used over extended periods of time or in large doses [95]. To address this problem, new therapeutic compounds are constantly sought in order to, if curing is not feasible at the moment, be able to better and more efficiently control cutaneous lesions, in an attempt to improve patients' quality of life. Developing such treatments for skin pathologies as psoriasis requires extensive and fastidious assessments to evaluate the therapeutic potential of the imposing amount of substances. These tests are mainly conducted on in vitro models, which closely emulate normal skin reactions. Monolayers are well suited for primary tests but are superseded by three-dimensional models later in the characterization process. Thus, in pilot pharmacological screenings, tissue-engineered skin constructs, such as our own self-assembled, become opportunistic tools to measure cytotoxic drawbacks, allowing thereafter to select effective and harmless compounds.

To this end, our three-dimensional model is a valid and efficient experimental tool to develop potential treatments [50]. Several aspects of this model have been closely correlated to the native human skin, particularly in terms of pharmacological response, where equivalents positively react to retinoic acid and methotrexate, both commonly used systemic treatments [74]. Hence, the laboratory focuses on the screening of chemical compounds that may possess beneficial outcomes in psoriasis, in hope of resolving toxic aftermaths caused by current therapies. This interest is especially oriented toward natural source molecules such as polyphenols, due to their antioxidant, antiproliferative, and anti-inflammatory potential [96]. This class of molecules could be of great interest for generating symptomatic treatments, which would be as effective as currently commercially available treatments, without related noxious effects, thus allowing extended use. The multiple phenol structural units, which consist of polyphenols, are responsible for its antioxidant potential by interrupting oxidation cascades, thus protecting cells against oxidative stress [97]. Increasing oxidative stress or decreased antioxidant enzymatic activity leads to cellular damage and could partially explain the outbreak of various pathologies such as psoriasis [98]. As aforementioned, polyphenols also detain anti-inflammatory properties, which may be explained by the inhibition of two noteworthy cytokines involved in the complex inflammatory process in psoriasis: TNF-α and IL-6 [99,100].

Results gathered from studies, which were realized in this respect, further proved the toxicological and antiproliferative properties of polyphenolic extracts, more specifically those originating from Canadian bark species, and especially crude extracts from black spruce (*Picea mariana*) [101,102]. Purification of this promising extract yielded a fraction of enriched polyphenols, which allow for the identification and characterization of their effect on the NF-kB signaling pathway [103]. Results showed that this enriched fraction has the ability to inhibit cytokines, chemokines, nitric oxide, and prostaglandins produced by TNF-α-activated psoriatic keratinocytes following downregulation of the NF-kB pathway, thus potentially enabling their use as anti-TNF-α [103]. Moreover, pharmacological screenings of diverse other polyphenolic molecules, extracted and purified from plant biomass, have been also carried out. These last analyses demonstrated that particular classes of phenols reduced the proliferation rate of lesional keratinocytes and improved their differentiation pattern, which is confirmed by a reduction of the living epidermis thickness, and the restoration of various protein expression levels in the epidermis. Results also suggest that these polyphenolic compounds would be as efficient as methotrexate, but at concentrations about a hundredfold lower, demonstrating the effectiveness and potency of this kind of plant source molecules.

4.3 Lipodomics studies and epidermis lipids characterization

The uppermost layer of the epidermis, the *stratum corneum*, consists of nearly 20 layers of corneocytes, completely keratinized dead cells, embedded in an extracellular lipid matrix [104]. The structure of this paper-thin layer has often been compared to a "brick-and-mortar" system [105]. Its main purpose consists of forming a semipermeable barrier, which prevents foreign matter from entering and protects underlying tissues from excessive water loss. Most of the lipids involved in its composition are ceramides, cholesterol, and various free fatty acids [106]. These lipids originate from a differentiation process occurring stepwise through maturation in the different epidermal layers. One of these lipids is of particular interest: ceramide 1. This lipid binds to involucrin, a corneocyte membrane protein, thus creating a solid scaffold for the subsequent lipid matrix organization. Therefore, increased fatty acid and ceramide content, following the differentiation process occurring in epidermal layers, plays an important role in the skin barrier function onset [13,107]. Thus, knowing that psoriasis is marked by an abnormal epidermal lipid composition, it makes sense that psoriatic skin has proved to be more permeable than the normal human skin [2].

When producing skin equivalents, it's essential to work on keeping the permeability levels as close as what is found in healthy or lesional human skin in order for the model to be representative. This standardization is a prerequisite for the use of such models in pharmacological assessments. Thereby, in our specific case concerning psoriasis, there is an urgent need for reproducible alternative systems mimicking accurately the barrier properties of human skin if one can hope to effectively find a way to cure psoriasis [108]. Although the self-assembly model is one of those that reproduces as faithfully the skin barrier function, absorption studies have shown that both healthy and psoriatic skin substitutes are more permeable than the normal human skin (Fig. 2) [72,109]. It has been shown that this increase could be correlated with an alteration of the skin substitutes' lipid profile; as introduced earlier, lipids have a major role to play in the formation and maintenance of the skin barrier function [48]. Therefore, several research groups, including our own, focus on restoring the lipid profile of tissue-engineered equivalents via supplementation in order to ameliorate their permeability, and this is usually achieved by directly adding lipids to the culture medium. Linoleic acid, given its direct role in the synthesis of the ceramide 1 and thus in staging of the lipid matrix, is one of the most commonly used fatty acid supplements, along with α-linolenic acid and palmitic acid [110–112].

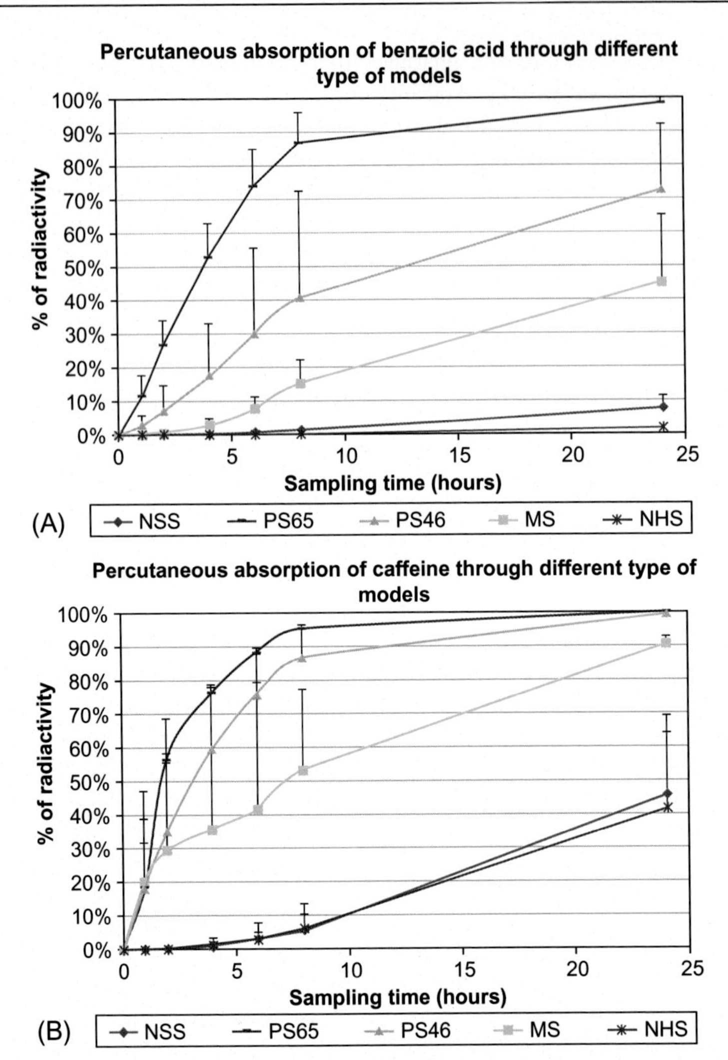

Fig. 2 Percutaneous absorption of benzoic acid (A) and caffeine (B) through the normal skin substitute (NSS), psoriatic substitute 65 (PS65) and 46 (PS46), mouse skin (MS), and normal human skin (NHS) over 24h. The number of each PS represents the age of the donor ($N=2$, $n=3$).

4.4 *In vitro* cutaneous vascularisation in psoriatic self-assembly skin substitutes

Angiogenesis is an important biological process through which new blood vessels are formed by sprouting or splitting from preexisting vasculature [113]. It involves the activation and expansion of endothelial cells, which are the constituents of the blood lining. It's of utmost importance in development, reproduction, and wound repair throughout our whole life [114]. Under these circumstances, angiogenesis is tightly regulated by a balance of repressive and inductive mediators [115]. However, it occurs not only under physiological conditions, where induced only transiently, but also in a variety of pathological conditions, including psoriasis. Once again, the establishment of new vessels represents a critical need for the disease development [115]. Indeed, capillaries grow and regress in healthy tissues according to functional demands, whereas increased metabolic needs and hypoxia of inflammatory and resident cells in psoriasis require greater and ceaseless blood flow in order to meet the extensive needs of these cells [113]. This vascular alteration is alike required for the extracellular matrix modifications supporting the leukocyte extravasation [116–119]. In psoriasis, a key vascular alteration consists of the outgrowth of capillaries within the papillary dermis in early stages of the disease, which are further elongated, dilated, and abnormally orientated [116,120]. Moreover, disappearance of those previous changes is being observed with disease clearance [115]. These outcomes suggest that angiogenesis plays an important role in the onset and complex pathology of psoriasis. This brief introduction leads us to our problematic concerning this angiogenic process and production of in vitro psoriatic skin models.

The issue lies in the fact that some aspects of tissue engineering have lagged behind following the common endeavor to improve existing models, namely, vascularization and innervation [26]. These critical features have not received all the attention they deserve maybe due to the fact that there are occurrences, such as regenerative medicine, where clinically applicable skin grafts do not absolutely need preexisting vascularization [26,121]. Nevertheless, these critical aspects must be addressed more efficiently in order to achieve sizeable skin substitutes [121].

This deficiency is particularly adverse in the case of psoriasis, where angiogenesis, as stated earlier, is one of its important histological hallmarks [2]. Indeed, the elucidation of biochemical pathways leading to angioproliferation may reveal promising novel therapeutic targets, but in vitro models including vascular components, which could serve as vehicles to get to that end, are rare or simply nonexistent [122]. With a view to thwart this deficiency, an in vitro vascularized psoriatic equivalent that could be used for fundamental research and pharmaceutical purposes has been reconstructed [116,123].

Foremost, our team tested whether it was feasible to produce psoriatic skin equivalents using endothelial cells combined with the self-assembly method [116]. Histological analysis and immunofluorescence staining targeting specific endothelial cells and basal membrane component markers demonstrated that endothelial cells can penetrate and migrate within the matrix and then divide, lengthen, and break up this matrix via enzyme secretion. Thus, endothelial cells spontaneously organize capillary-like structures [116]. Moreover, the addition of these cells did not alter the macroscopic and histological aspects of the psoriatic equivalents. Among others, it was possible to observe a well-fused dermis, the presence of differentiation markers, such as filaggrin, loricrin, and keratin 10, whose expressions were farther decreased in psoriatic substitutes than proliferation marker Ki67 [116]. In the same way, the granular layer of psoriatic substitutes was greatly reduced and sometimes completely absent, in comparison with healthy skin, which is in accordance with the current knowledge of this dermatosis [4,116]. Three-dimensional whole-mount immunofluorescence staining showed that the quantity, complexity, density, and disorganization of capillary-like structures were increased in psoriatic substitutes [116]. This chaotic vasculature organization in psoriatic equivalents, which includes various angiogenic provocation features and differential expression of cellular markers, provides a good proof of principle concerning the self-assembled model, which was taken a step further by introducing these endothelial cells.

Subsequently, now that the relevance of this vascularized psoriatic model has been confirmed, the new focus is on investigating the behavior of endothelial cells within this in vitro environment [123]. Rochon et al. and Liu et al. previously studied the effect of epithelial cells on capillary-like structures and observed that keratinocytes were able to regulate their size and morphology [124,125]. Based on these results, it was important to establish whether such a phenomenon could be observed in the self-assembled model and looked at the consecutive steps of angiogenesis occurring in this kind of substitutes. In order to preliminarily determine if keratinocytes could enhance the formation of capillary-like structures, tests on endothelial monolayer cultures were conducted using conditioned media previously collected from keratinocytes or fibroblast cultures. Results showed that keratinocyte-conditioned medium impacts the physiology of endothelial cells by inducing their reorganization in a circular pattern [123]. In the same course of events, endothelial cell proliferation ceased, and a sharp decline in viable cell count was observed. Fibroblast-conditioned medium have, however, shown to induce cell proliferation, and a mix of both conditioned media produced only few circular structures, which means branch formation was delayed. Additionally, specific endothelial cell medium EBM-2 alone did not induce the cells to form branches even with the addition of angiogenic growth factors and other supplements [123]. In order to further prove that the epidermis increased in vitro angiogenesis, the expression of angiogenic mediators by the endothelial cells was analyzed. ELISA and IHC staining showed increased expression of two distinct angiogenic inducers, VEGF and angiopoietin-1 (Ang-1). However, the coculture with an epidermis did not affect the interleukin-8 (IL-8) release. This chemoattractant cytokine being identified as having a pro-inflammatory function, it apparently means that the increasing capillary-like structure formation in psoriatic bilayered models is not due to inflammation process [123]. Lastly, matrix metalloproteinase activity and matrix degradation are important processes in angiogenesis, allowing the establishment of new blood vessels by degrading and remodeling the extracellular matrix in order to allow endothelial cells to migrate and invade surrounding tissue [126]. Hence, it was not surprising to observe an increase MMP activity in this bilayered model [123]. Conclusively, our results showed that in vitro microvascular network formation in tissue-engineered psoriatic skin models is predominantly managed by the epidermal components of substitutes [123].

4.5 Transcriptomic characterization of self-assembled psoriatic cutaneous model

The etiology and specific role of genetics in psoriasis, albeit its involvement is extensively accepted, remains uncertain. Indeed, it has been shown that genetic background greatly increases the risk of being affected, although segregation does not occur in accordance with Mendelian inheritance rules [125,127]. Genetic predisposition, on the other hand, cannot alone explain the dawn and sustaining of this pathology, since certain patients present no genetic influence [125]. Thus, these observations suggest a certain synergistic and imperative involvement of environmental, immunologic, and of course genetic factors. Among these factors, more than 30 putative susceptible loci have been identified as related to psoriasis, and these are named ##PSORS, an abbreviation of psoriasis susceptibility [128]. Genes contained within these regions are involved in inflammatory metabolic pathways, epidermal cell proliferation, and skin barrier function, all of which being impacted mechanisms in psoriasis [129]. Yet, the first and only consistently associated psoriasis susceptibility locus is PSORS1, a segment that spans a series of genes encoding for proteins such as class I MHC and further linked to early onset of chronic plaque, aberrant keratinocyte turnover rate, and corneocyte cohesion [4,11,130,131]. Genome-wide linkage scans also permitted to localize polymorphisms within or in the vicinity of cytokine-encoding genes, their receptors, or signal transduction pathway components, further implicating cytokines in the psoriasis pathomechanism [132]. Indeed, cytokines have been already linked to this pathology through their role in its initiation orchestration and afterward exacerbation of the cutaneous inflammatory response [4,133]. Among others, IL-17 and TNF, which promote the influx of inflammatory cells and onset of detrimental self-amplifying immunopathologic loops, were also linked to gene regulation and expression alterations in psoriasis [4]. Thus, a deeper knowledge of the multiple key roles of cytokines could suggest potential therapeutic targets or allow the development of more relevant and highly predictive tissue-engineered constructs.

As outlined earlier, human psoriatic skin being a rare commodity for laboratories, in vitro models are well-suited surrogates in order to study cytokines and their interwoven effects on genetic, angiogenesis, and cutaneous inflammatory response. Hence, our laboratory carried out a study consisting of two facets, either to investigate the effect of cytokines on skin cell transcriptomics or, secondly, if the addition of a cytokine cocktail could consequently compensate the current lack of immune cells in our model [36]. The cocktail was composed of IL-1α, IL-6, IL-17A, and TNF-α and was added at the same moment that keratinocytes were overlayed on the dermal equivalent. Microarray and scatter plot analysis revealed important gene pattern variations in subjected psoriatic equivalents. For instance, various late cornified envelope (LCE)-related genes were downregulated, as well as *KYNU*, *IL8*, *S100A12*, and *DEFB4A* genes showing considerable increases in transcription levels, all of which being consistent with the expression pattern found in psoriasis in vivo [36]. Along, many other strongly associated genes to psoriasis that further played important functions in angiogenesis, cell proliferation, and chemotaxis together with both immune cell activation and keratinocyte differentiation, such as *CEACAM*, *CCL27*, *SERPINA12*, and *CX3CL1*, were identified among deregulated genes [36].

Thus, our model is well representative of the pathology, as it appropriately reproduces the triggered repercussions owing to cytokines on the involved genetics in psoriasis. Cytokine cocktail addition also turned out to be a promising solution where in vitro skin models are not endowed with the capacity to retain gene expression levels near those found in native psoriasis, that being most likely due to the lack of immune cells within the equivalents. Besides being meaningful models for transcriptomic analysis and everything that might ensue, future cytokine responsive in vitro models could promote studies focusing on the identification of autoantigen-triggering T cells and understanding, in minute detail, the potential role of $CD4^+$ and $CD8^+$ T cells in lesion development and how they maintain psoriasis [36].

4.6 Determining the ability of In Vitro psoriatic models to metabolize antipsoritatic drugs

The psoriatic self-assembled in vitro model has been greatly characterized and consequently improved in recent years. However, drug metabolism assays were not yet attempted. Such studies could allow us to validate that this model is even more representative of native psoriatic skin and, therefore, is a valuable pharmaceutical tool in psoriasis investigation.

Treatments administered to patients are biotransformed during their travel within skin to yield metabolites, which are often the pharmacologically active agents. Thereby, in order to ensure that a treatment has the desired therapeutic effects, it must be properly metabolized. This metabolism process is influenced by several skin factors including, among others, its permeability, moisture, structure, temperature, and the presence or absence of specific enzymes. In this sense, it is known that the organization of psoriatic skin is impaired, which consequently leads to abnormal permeability barrier function, and furthermore that in vitro models are already more permeable than the native skin [72,109,134]. Thus, it would be interesting to observe if compounds are metabolized in the same manner even in these differential conditions, which is a prerequisite for the use of any model in pharmaceutical assessments.

Results have shown that self-assembled psoriatic skin equivalents possess a functional metabolic capacity to what could be observed in native lesional skin. These experiments were conducted on the metabolism of tazarotene, an acetylenic retinoid, which is effective for topical treatment of plaque psoriasis [135,136]. Thereby, besides being a relevant model for metabolism assays and consequently a promising tool for the pharmaceutical industry, these results could pave the way to multiple other perspectives. Indeed, due to its autologous character, the self-assembly approach could allow for personalized treatments based on the patient's very own cells' response. It could be possible to carry out reliable and personalized therapeutic screening in a greatly reduced amount of time compared with similar in vivo analyses, where tests cannot be carried out simultaneously. As such, access to adequate medicine could be sped up, which would further improve patients' quality of life. Moreover, one can see the potential to study the metabolism of antipsoriatic compounds in order to achieve effective formulations and concentration for the psoriasis treatment. It might also be interesting to avail the pairing of high-performance chromatography and percutaneous absorption in order to study the course of therapeutic compounds over an extended period of time, which is normally limited to 24 h. Thus, this merge could allow researchers to retrieve the whole administered dose and, afterward, precisely determine where it was accumulated, used, or lost.

One can note that there are only very few studies concerning the metabolism among in vitro skin models. If other groups follow suit in this promising direction and if already actively participating laboratories optimized their techniques, it might be possible in upcoming years to screen a phenomenal amount of auspicious compounds on a broad range of in vitro psoriatic models. Indeed, the metabolic capacity of in vitro models is a promising discovery for pharmaceutical development.

5. Conclusion

Psoriasis has many yet unexplained mechanisms and even if there's a spectrum of local and systemic therapeutic alternatives, none of them can do more than merely manage its repercussions. The absence of validated models or even the knowledge of a specific profile that defines the psoriatic phenotype surely constitutes one of the major challenges for antipsoriatic drug development and hinders research in this direction. Hence, both research and pharmaceutical laboratories resolutely need relevant and reliable psoriatic in vitro models to use as testing platforms in order to investigate the complex interplay between keratinocytes, fibroblasts, vascular endothelium, genetics, and the innate and adaptive immune systems. Indeed, in vitro tissue-engineered constructs represent suitable alternatives to in vivo models, which ever so often failed to faithfully mimic the human pathology and face, in addition, an increasingly strengthened ethical control. Existing psoriatic in vitro models use just as different as innovative techniques for their production, of those, we can mention the monolayers, the DED and collagen gel-based methods and the self-assembly approach. While none of these previous models is foolproof, they allowed us to gather valuable information about the pathogenic interplay underlying psoriasis. Thus, the development of more relevant psoriasis models could pave the way for new and less harmful treatments and for personalized medicine, but most importantly, it would be an important step toward a cure for this disorder, affecting patients far beyond the skin.

References

[1] Wippel-Slupetzky K, Stingl G. Future perspectives in the treatment of psoriasis. Curr Probl Dermatol 2009;38:172–89.
[2] Boehncke WH, Schon MP. Psoriasis. Lancet 2015;386(9997):983–94.
[3] Baker BS, Owles AV, Fry L. A possible role for vaccination in the treatment of psoriasis? Giornale italiano di dermatologia e venereologia: organo ufficiale. Soc Ital Dermatol Sifilogr 2008;143(2):105–17.
[4] Lowes MA, Bowcock AM, Krueger JG. Pathogenesis and therapy of psoriasis. Nature 2007;445(7130):866–73.
[5] Anandarajah AP, Ritchlin CT. Pathogenesis of psoriatic arthritis. Curr Opin Rheumatol 2004;16(4):338–43.
[6] Grossman RM, Krueger J, Yourish D, Granelli-Piperno A, Murphy DP, May LT. Interleukin 6 is expressed in high levels in psoriatic skin and stimulates proliferation of cultured human keratinocytes. Proc Natl Acad Sci U S A 1989;86(16):6367–71.
[7] Tzu J, Krulig E, Cardenas V, Kerdel FA. Biological agents in the treatment of psoriasis. G Ital Dermatol Venereol 2008;143(5):315–27.

[8] Luger T, Seite S, Humbert P, Krutmann J, Triller R, Dreno B. Recommendations for adjunctive basic skin care in patients with psoriasis. Eur J Dermatol 2014;24(2):194–200.

[9] Ovalle WK, Nahirney PC, Netter FH. Netter's essential histology. Philadelphia, PA: Elsevier/Saunders; 2013.

[10] Bernard G, Auger M, Soucy J, Pouliot R. Physical characterization of the stratum corneum of an in vitro psoriatic skin model by ATR-FTIR and raman spectroscopies. Biochim Biophys Acta 2007;1770(9):1317–23.

[11] Bowcock AM, Krueger JG. Getting under the skin: The immunogenetics of psoriasis. Nat Rev Immunol 2005;5(9):699–711.

[12] Danilenko DM. Review paper: preclinical models of psoriasis. Vet Pathol 2008;45(4):563–75.

[13] Elias PM, Choi EH. Interactions among stratum corneum defensive functions. Exp Dermatol 2005;14(10):719–26.

[14] Motta S, Sesana S, Monti M, Giuliani A, Caputo R. Interlamellar lipid differences between normal and psoriatic stratum corneum. Acta Derm Venereol Suppl 1994;186:131–2.

[15] Holtje M, Forster T, Brandt B, Engels T, von Rybinski W, Holtje HD. Molecular dynamics simulations of stratum corneum lipid models: fatty acids and cholesterol. Biochim Biophys Acta 2001;1511(1):156–67.

[16] Camisa C. Handbook of psoriasis. New Jersey: Wiley-Blackwell; 2008.

[17] Mackay IR, Rose NR. The Autoimmune Diseases. Amsterdam: Elsevier Science; 2013.

[18] Gordon KB. Phase 3 trials of Ixekizumab in moderate-to-severe plaque psoriasis. N Engl J Med 2016;375(21):2101–2.

[19] Papp KA, Reid C, Foley P, Sinclair R, Salinger DH, Williams G. Anti-IL-17 receptor antibody AMG 827 leads to rapid clinical response in subjects with moderate to severe psoriasis: results from a phase I, randomized, placebo-controlled trial. J Invest Dermatol 2012;132(10):2466.

[20] Canavan TN, Elmets CA, Cantrell WL, Evans JM, Elewski BE. Anti-IL-17 medications used in the treatment of plaque psoriasis and psoriatic arthritis: a comprehensive review. Am J Clin Dermatol 2016;17(1):33–47.

[21] Karle A, Spindeldreher S, Kolbinger F, editors. Secukinumab, a novel anti-IL-17A antibody, shows low immunogenicity potential in human in vitro assays comparable to other marketed biotherapeutics with low clinical immunogenicity. MAbs. UK: Taylor & Francis; 2016.

[22] Smith SH, Peredo CE, Takeda Y, Bui T, Neil J, Rickard D. Development of a topical treatment for psoriasis targeting RORγ: from bench to skin. PLoS One 2016;11(2):e0147979.

[23] Nickoloff BJ, Nestle FO. Recent insights into the immunopathogenesis of psoriasis provide new therapeutic opportunities. J Clin Invest 2004;113(12):1664–75.

[24] Rashmi R, Rao KS, Basavaraj KH. A comprehensive review of biomarkers in psoriasis. Clin Exp Dermatol 2009;34(6):658–63.

[25] Langer R, Vacanti JP. Tissue engineering. Science (New York, NY) 1993;260(5110):920–6.

[26] Auger FA. The LOEX perspective on the role of tissue engineering in regenerative medicine. Biomed Mater Eng 2006;16(4 Suppl):S19–25.

[27] Arosarena O. Tissue engineering. Curr Opin Otolaryngol Head Neck Surg 2005;13(4):233–41.

[28] Auger FA, Berthod F, Moulin V, Pouliot R, Germain L. Tissue-engineered skin substitutes: from in vitro constructs to in vivo applications. Biotechnol Appl Biochem 2004;39(Pt 3):263–75.

[29] Mac NS. Progress and opportunities for tissue-engineered skin. Nature 2007;445(7130):874–80.

[30] Guttman-Yassky E, Krueger JG. Psoriasis: Evolution of pathogenic concepts and new therapies through phases of translational research. Br J Dermatol 2007;157(6):1103–15.
[31] Schon MP. Animal models of psoriasis—what can we learn from them? J Invest Dermatol 1999;112(4):405–10.
[32] Jean J, Garcia-Pérez ME, Pouliot R. Psoriatic skin models: a need for the pharmaceutical industry. Rijeka, Croatia: INTECH Open Access Publisher; 2012.
[33] Brohem CA, Cardeal LB, Tiago M, Soengas MS, Barros SB, Maria-Engler SS. Artificial skin in perspective: concepts and applications. Pigment Cell Melanoma Res 2011;24(1):35–50.
[34] Groeber F, Holeiter M, Hampel M, Hinderer S, Schenke-Layland K. Skin tissue engineering—in vivo and in vitro applications. Clin Plast Surg 2012;39(1):33–58.
[35] van den Bogaard EH, Tjabringa GS, Joosten I, Vonk-Bergers M, van Rijssen E, Tijssen HJ. Crosstalk between keratinocytes and T cells in a 3D microenvironment: a model to study inflammatory skin diseases. J Invest Dermatol 2014;134(3):719–27.
[36] Pouliot-Bérubé C, Zaniolo K, Guérin SL, Pouliot R. Tissue-engineered human psoriatic skin supplemented with cytokines as an in vitro model to study plaque psoriasis. Regen Med 2016;11(6):545–57.
[37] Bergers LI, Reijnders CM, van den Broek LJ, Spiekstra SW, de Gruijl TD, Weijers EM. Immune-competent human skin disease models. Drug Discov Today 2016;21(9):1479–88.
[38] Carrier Y, Ma H-L, Ramon HE, Napierata L, Small C, O'toole M. Inter-regulation of Th17 cytokines and the IL-36 cytokines in vitro and in vivo: implications in psoriasis pathogenesis. J Invest Dermatol 2011;131(12):2428–37.
[39] Chiricozzi A, Romanelli M, Panduri S, Donetti E, Prignano F. Relevance of in vitro 3-D skin models in dissecting cytokine contribution to psoriasis pathogenesis. Histol Histopathol 2017;32(9):893–8.
[40] Teng X, Hu Z, Wei X, Wang Z, Guan T, Liu N. IL-37 ameliorates the inflammatory process in psoriasis by suppressing proinflammatory cytokine production. J Immunol 2014;192(4):1815–23.
[41] Mazzoleni G, Di Lorenzo D, Steimberg N. Modelling tissues in 3D: The next future of pharmaco-toxicology and food research? Genes Nutr 2009;4(1):13–22.
[42] Martin G, Guerard S, Fortin MM, Rusu D, Soucy J, Poubelle PE. Pathological crosstalk in vitro between T lymphocytes and lesional keratinocytes in psoriasis: necessity of direct cell-to-cell contact. Lab Invest—J Tech Methods Pathol 2012;92(7):1058–70.
[43] Jean J. Applications dermopharmaceutiques: Développement d'un modèle de substituts cutanés psoriasiques par génie tissulaire. Thèse de Doctorat Faculté des études supérieures de l'Université Laval; 2010.
[44] Detmar M, Brown LF, Claffey KP, Yeo KT, Kocher O, Jackman RW. Overexpression of vascular permeability factor/vascular endothelial growth factor and its receptors in psoriasis. J Exp Med 1994;180(3):1141–6.
[45] Vichai V, Kirtikara K. Sulforhodamine B colorimetric assay for cytotoxicity screening. Nat Protoc 2006;1(3):1112–6.
[46] Wong T, McGrath JA, Navsaria H. The role of fibroblasts in tissue engineering and regeneration. Br J Dermatol 2007;156(6):1149–55.
[47] Prunieras M, Regnier M, Woodley D. Methods for cultivation of keratinocytes with an air-liquid interface. The Journal of investigative dermatology 1983;81(1 Suppl):28s–33s.
[48] Ponec M. Reconstruction of human epidermis on de-epidermized dermis: expression of differentiation-specific protein markers and lipid composition. Toxicol In Vitro 1991;5(5–6):597–606.

[49] Mils V, Basset-Seguin N, Moles JP, Tesniere A, Leigh I, Guilhou JJ. Comparative analysis of normal and psoriatic skin both in vivo and in vitro. Differentiation 1994;58(1):77–86.
[50] Jean J, Lapointe M, Soucy J, Pouliot R. Development of an in vitro psoriatic skin model by tissue engineering. J Dermatol Sci 2009;53(1):19–25.
[51] Barker CL, McHale MT, Gillies AK, Waller J, Pearce DM, Osborne J. The development and characterization of an in vitro model of psoriasis. J Invest Dermatol 2004;123(5):892–901.
[52] Harrison CA, Layton CM, Hau Z, Bullock AJ, Johnson TS, Mac Neil S. Transglutaminase inhibitors induce hyperproliferation and parakeratosis in tissue-engineered skin. Br J Dermatol 2007;156(2):247–57.
[53] Tjabringa G, Bergers M, van Rens D, de Boer R, Lamme E, Schalkwijk J. Development and validation of human psoriatic skin equivalents. Am J Pathol 2008;173(3):815–23.
[54] Engelhart K, El Hindi T, Biesalski HK, Pfitzner I. In vitro reproduction of clinical hallmarks of eczematous dermatitis in organotypic skin models. Arch Dermatol Res 2005;297(1):1–9.
[55] Basset-Seguin N, Culard JF, Kerai C, Bernard F, Watrin A, Demaille J. Reconstituted skin in culture: a simple method with optimal differentiation. Differentiation 1990;44(3):232–8.
[56] Lopez Valle CA, Auger FA, Rompre P, Bouvard V, Germain L. Peripheral anchorage of dermal equivalents. Br J Dermatol 1992;127(4):365–71.
[57] Bouvard V, Germain L, Rompre P, Roy B, Auger FA. Influence of dermal equivalent maturation on the development of a cultured skin equivalent. Biochem Cell Biol—Biochim Biol Cell 1992;70(1):34–42.
[58] Bell E, Ehrlich HP, Buttle DJ, Nakatsuji T. Living tissue formed in vitro and accepted as skin-equivalent tissue of full thickness. Science (New York, NY) 1981;211(4486):1052–4.
[59] Berthod F, Germain L, Guignard R, Lethias C, Garrone R, Damour O. Differential expression of collagens XII and XIV in human skin and in reconstructed skin. J Invest Dermatol 1997;108(5):737–42.
[60] Bell E, Ivarsson B, Merrill C. Production of a tissue-like structure by contraction of collagen lattices by human fibroblasts of different proliferative potential in vitro. Proc Natl Acad Sci U S A 1979;76(3):1274–8.
[61] Ponec M, Gibbs S, Weerheim A, Kempenaar J, Mulder A, Mommaas AM. Epidermal growth factor and temperature regulate keratinocyte differentiation. Arch Dermatol Res 1997;289(6):317–26.
[62] Auger FA, Rouabhia M, Goulet F, Berthod F, Moulin V, Germain L. Tissue-engineered human skin substitutes developed from collagen-populated hydrated gels: clinical and fundamental applications. Med Biol Eng Comput 1998;36(6):801–12.
[63] Saiag P, Coulomb B, Lebreton C, Bell E, Dubertret L. Psoriatic fibroblasts induce hyperproliferation of normal keratinocytes in a skin equivalent model in vitro. Science 1985;230(4726):669–72. (New York, NY).
[64] Fine RM. The fine page. Psoriatic fibroblasts induce hyperproliferation of normal keratinocytes in skin equivalent model in vitro. Int J Dermatol 1986;25(4):232–3.
[65] Konstantinova NV, Duong DM, Remenyik E, Hazarika P, Chuang A, Duvic M. Interleukin-8 is induced in skin equivalents and is highest in those derived from psoriatic fibroblasts. J Invest Dermatol 1996;107(4):615–21.
[66] Zhao JF, Zhang YJ, Kubilus J, Jin XH, Santella R, Athar M. Reconstituted 3-dimensional human skin as a Novelin vitro model for studies of carcinogenesis. Biochem Biophys Res Commun 1999;254(1):49–53.

[67] Roguet R. The use of standardized human skin models for cutaneous pharmacotoxicology studies. Skin Pharmacol Physiol 2002;15(Suppl. 1):1–3.
[68] Monteiro-Riviere NA, Inman AO, Snider TH, Blank JA, Hobson DW. Comparison of an in vitro skin model to normal human skin for dermatological research. Microsc Res Tech 1997;37(3):172–9.
[69] Sharma M, Levenson C, Clements I, Castella P, Gebauer K, Cox ME. East Indian sandalwood oil (EISO) alleviates inflammatory and proliferative pathologies of psoriasis. Front Pharmacol 2017;8:.
[70] Germain L, Auger FA, Grandbois E, Guignard R, Giasson M, Boisjoly H. Reconstructed human cornea produced in vitro by tissue engineering. Pathobiol: J Immunopathol Mol Cell Biol 1999;67(3):140–7.
[71] Larouche D, Paquet C, Fradette J, Carrier P, Auger FA, Germain L. Regeneration of skin and cornea by tissue engineering. In: Audet J, Stanford WL, editors. Stem cells in regenerative medicine. Totowa, NJ: Humana Press; 2009. p. 233–56.
[72] Michel M, L'Heureux N, Pouliot R, Xu W, Auger FA, Germain L. Characterization of a new tissue-engineered human skin equivalent with hair. In Vitro Cell Dev Biol Anim 1999;35(6):318–26.
[73] Pouliot R, Germain L, Auger FA, Tremblay N, Juhasz J. Physical characterization of the stratum corneum of an in vitro human skin equivalent produced by tissue engineering and its comparison with normal human skin by ATR-FTIR spectroscopy and thermal analysis (DSC). Biochim Biophys Acta 1999;1439(3):341–52.
[74] Jean J, Soucy J, Pouliot R. Effects of retinoic acid on keratinocyte proliferation and differentiation in a psoriatic skin model. Tissue Eng Part A 2011;17(13–14):1859–68.
[75] Leroy M, Labbé J-F, Ouellet M, Jean J, Lefèvre T, Laroche G. A comparative study between human skin substitutes and normal human skin using Raman microspectroscopy. Acta Biomater 2014;10(6):2703–11.
[76] Michel M, Auger FA, Germain L. Anchored skin equivalent cultured in vitro: a new tool for percutaneous absorption studies. In Vitro Cell Dev Biol Anim 1993;29A(11):834–7.
[77] Germain L, Auger FA. Tissue engineered biomaterials: biological and mechanical characteristics. In: Encyclopedic handbook of biomaterials and bioengineering; 1995.
[78] Germain L, Rouabhia M, Guignard R, Carrier L, Bouvard V, Auger FA. Improvement of human keratinocyte isolation and culture using thermolysin. Burns 1993;19(2):99–104.
[79] Okada N, Kitano Y, Ichihara K. Effects of cholera toxin on proliferation of cultured human keratinocytes in relation to intracellular cyclic AMP levels. J Invest Dermatol 1982;79(1):42–7.
[80] Geesin JC, Darr D, Kaufman R, Murad S, Pinnell SR. Ascorbic acid specifically increases type I and type III procollagen messenger RNA levels in human skin fibroblast. J Invest Dermatol 1988;90(4):420–4.
[81] Hata R, Senoo H. L-ascorbic acid 2-phosphate stimulates collagen accumulation, cell proliferation, and formation of a three-dimensional tissuelike substance by skin fibroblasts. J Cell Physiol 1989;138(1):8–16.
[82] Murad S, Tajima S, Johnson GR, Sivarajah S, Pinnell SR. Collagen synthesis in cultured human skin fibroblasts: effect of ascorbic acid and its analogs. J Invest Dermatol 1983;81(2):158–62.
[83] L'Heureux N, Paquet S, Labbe R, Germain L, Auger FA. A completely biological tissue-engineered human blood vessel. FASEB J: Off Publ Feder Am Soci Exp Biol 1998;12(1):47–56.

[84] Cai Y, Fleming C, Yan J. New insights of T cells in the pathogenesis of psoriasis. Cell Mol Immunol 2012;9(4):302–9.
[85] Sweeney CM, Tobin AM, Kirby B. Innate immunity in the pathogenesis of psoriasis. Arch Dermatol Res 2011;303(10):691–705.
[86] Coimbra S, Figueiredo A, Castro E, Rocha-Pereira P, Santos-Silva A. The roles of cells and cytokines in the pathogenesis of psoriasis. Int J Dermatol 2012;51(4):389–95 [quiz 95-8].
[87] Kim J, Krueger JG. The immunopathogenesis of psoriasis. Dermatol Clin 2015;33(1):13–23.
[88] Dubertret L. Retinoids, methotrexate and cyclosporine. Curr Probl Dermatol 2009;38:79–94.
[89] Bergers LI, Reijnders CM, van den Broek LJ, Spiekstra SW, de Gruijl TD, Weijers EM. Immune-competent human skin disease models. Drug Discov Today 2016;.
[90] Schirmer C, Klein C, von Bergen M, Simon JC, Saalbach A. Human fibroblasts support the expansion of IL-17-producing T cells via up-regulation of IL-23 production by dendritic cells. Blood 2010;116(10):1715–25.
[91] Wittmann M, Purwar R, Hartmann C, Gutzmer R, Werfel T. Human keratinocytes respond to interleukin-18: Implication for the course of chronic inflammatory skin diseases. J Invest Dermatol 2005;124(6):1225–33.
[92] Prinz JC, Gross B, Vollmer S, Trommler P, Strobel I, Meurer M. T cell clones from psoriasis skin lesions can promote keratinocyte proliferation in vitro via secreted products. Eur J Immunol 1994;24(3):593–8.
[93] Rosa-Fortin M-M, Poubelle P-E, Soucy J, Pouliot R. Cellular interactions in vitro: psoriatic keratinocytes enhance lymphocyte survival. Psoriasis Forum 2010;16(1):12–5.
[94] Saalbach A, Janik T, Busch M, Herbert D, Anderegg U, Simon JC. Fibroblasts support migration of monocyte-derived dendritic cells by secretion of PGE2 and MMP-1. Exp Dermatol 2015;24(8):598–604.
[95] Lebwohl M, Ting PT, Koo JY. Psoriasis treatment: traditional therapy. Ann Rheum Dis 2005;64(Suppl 2):83–6.
[96] Grimm T, Chovanova Z, Muchova J, Sumegova K, Liptakova A, Durackova Z. Inhibition of NF-kappa B activation and MMP-9 secretion by plasma of human volunteers after ingestion of maritime pine bark extract (Pycnogenol). J Inflamm (Lond) 2006;3:1.
[97] Pandey KB, Rizvi SI. Plant polyphenols as dietary antioxidants in human health and disease. Oxid Med Cell Longev 2009;2(5):270–8.
[98] Gabr SA, Al-Ghadir AH. Role of cellular oxidative stress and cytochrome c in the pathogenesis of psoriasis. Arch Dermatol Res 2012;304(6):451–7.
[99] Denis MC, Furtos A, Dudonne S, Montoudis A, Garofalo C, Desjardins Y. Apple peel polyphenols and their beneficial actions on oxidative stress and inflammation. PLoS One 2013;8(1):e53725.
[100] Nickoloff BJ, Qin JZ, Nestle FO. Immunopathogenesis of psoriasis. Clin Rev Allergy Immunol 2007;33(1–2):45–56.
[101] Garcia-Perez ME, Royer M, Duque-Fernandez A, Diouf PN, Stevanovic T, Pouliot R. Antioxidant, toxicological and antiproliferative properties of Canadian polyphenolic extracts on normal and psoriatic keratinocytes. J Ethnopharmacol 2010;132(1):251–8.
[102] Garcia-Perez ME, Royer M, Herbette G, Desjardins Y, Pouliot R, Stevanovic T. Picea mariana bark: a new source of trans-resveratrol and other bioactive polyphenols. Food Chem 2012;135(3):1173–82.
[103] Garcia-Perez ME, Allaeys I, Rusu D, Pouliot R, Janezic TS, Poubelle PE. Picea mariana polyphenolic extract inhibits phlogogenic mediators produced by TNF-alpha-activated

psoriatic keratinocytes: impact on NF-kappa B pathway. J Ethnopharmacol 2014;151(1):265–78.

[104] El Maghraby GM, Barry BW, Williams AC. Liposomes and skin: from drug delivery to model membranes. Eur J Pharm Sci 2008;34(4–5):203–22.

[105] Elias PM. Epidermal lipids, barrier function, and desquamation. J Invest Dermatol 1983;80(Suppl):44s–9s.

[106] Breiden B, Sandhoff K. The role of sphingolipid metabolism in cutaneous permeability barrier formation. Biochim Biophys Acta 2014;1841(3):441–52.

[107] Candi E, Schmidt R, Melino G. The cornified envelope: a model of cell death in the skin. Nat Rev Mol Cell Biol 2005;6(4):328–40.

[108] Basse LH, Groen D, Bouwstra JA. Permeability and lipid organization of a novel psoriasis stratum corneum substitute. Int J Pharm 2013;457(1):275–82.

[109] Roy SD, Fujiki J, Fleitman JS. Permeabilities of alkyl p-aminobenzoates through living skin equivalent and cadaver skin. J Pharm Sci 1993;82(12):1266–8.

[110] Boyce ST, Williams ML. Lipid supplemented medium induces lamellar bodies and precursors of barrier lipids in cultured analogues of human skin. J Invest Dermatol 1993;101(2):180–4.

[111] Ponec M, Kempenaar J, Weerheim A, de Lannoy L, Kalkman I, Jansen H. Triglyceride metabolism in human keratinocytes cultured at the air-liquid interface. Arch Dermatol Res 1995;287(8):723–30.

[112] Marcelo CL, Rhodes LM, Dunham WR. Normalization of essential-fatty-acid-deficient keratinocytes requires palmitic acid. J Invest Dermatol 1994;103(4):564–8.

[113] Adair TH, Montani JP. Angiogenesis. San Rafael, CA: Morgan and Claypool Life Sciences; 2010.

[114] Otrock ZK, Mahfouz RA, Makarem JA, Shamseddine AI. Understanding the biology of angiogenesis: review of the most important molecular mechanisms. Blood Cells Mol Dis 2007;39(2):212–20.

[115] Heidenreich R, Rocken M, Ghoreschi K. Angiogenesis drives psoriasis pathogenesis. Int J Exp Pathol 2009;90(3):232–48.

[116] Ayata RE, Bouhout S, Auger M, Pouliot R. Study of in vitro capillary-like structures in psoriatic skin substitutes. Biores Open Access 2014;3(5):197–205.

[117] Hendriks AG, Steenbergen W, Hondebrink E, van Hespen JC, van de Kerkhof PC, Seyger MM. Whole field laser Doppler imaging of the microcirculation in psoriasis and clinically unaffected skin. J Dermatolog Treat 2014;25(1):18–21.

[118] Costa C, Incio J, Soares R. Angiogenesis and chronic inflammation: cause or consequence? Angiogenesis 2007;10(3):149–66.

[119] Pober JS, Sessa WC. Evolving functions of endothelial cells in inflammation. Nat Rev Immunol 2007;7(10):803–15.

[120] Pinkus H, Mehregan AH. The primary histologic lesion of seborrheic dermatitis and psoriasis. J Invest Dermatol 1966;46(1):109–16.

[121] Auger FA, Gibot L, Lacroix D. The pivotal role of vascularization in tissue engineering. Annu Rev Biomed Eng 2013;15:177–200.

[122] Creamer D, Sullivan D, Bicknell R, Barker J. Angiogenesis in psoriasis. Angiogenesis 2002;5(4):231–6.

[123] Ayata RE, Chabaud S, Auger M, Pouliot R. Behaviour of endothelial cells in a tridimensional in vitro environment. BioMed Res Int 2015;2015:630461.

[124] Rochon MH, Fradette J, Fortin V, Tomasetig F, Roberge CJ, Baker K. Normal human epithelial cells regulate the size and morphology of tissue-engineered capillaries. Tissue Eng Part A 2010;16(5):1457–68.

[125] Liu Y, Luo H, Wang X, Takemura A, Fang YR, Jin Y. In vitro construction of scaffold-free bilayered tissue-engineered skin containing capillary networks. BioMed Res Int 2013;2013:561410.

[126] Rundhaug JE. Matrix metalloproteinases and angiogenesis. J Cell Mol Med 2005;9(2):267–85.

[127] Deng Y, Chang C, Lu Q. The inflammatory response in psoriasis: a comprehensive review. Clin Rev Allergy Immunol 2016;50(3):377–89.

[128] Tsoi LC, Spain SL, Knight J, Ellinghaus E, Stuart PE, Capon F. Identification of 15 new psoriasis susceptibility loci highlights the role of innate immunity. Nat Genet 2012;44(12):1341–8.

[129] Stern RS. Psoriasis. Lancet 1997;350(9074):349–53.

[130] Monteleone G, Pallone F, Mac Donald TT, Chimenti S, Costanzo A. Psoriasis: from pathogenesis to novel therapeutic approaches. Clin Sci (Lond) 2011;120(1):1–11.

[131] Tiala I, Wakkinen J, Suomela S, Puolakkainen P, Tammi R, Forsberg S. The PSORS1 locus gene CCHCR1 affects keratinocyte proliferation in transgenic mice. Hum Mol Genet 2008;17(7):1043–51.

[132] Baliwag J, Barnes DH, Johnston A. Cytokines in psoriasis. Cytokine 2015;73(2):342–50.

[133] Nedoszytko B, Sokolowska-Wojdylo M, Ruckemann-Dziurdzinska K, Roszkiewicz J, Nowicki RJ. Chemokines and cytokines network in the pathogenesis of the inflammatory skin diseases: atopic dermatitis, psoriasis and skin mastocytosis. Postepy Dermatol Alergo 2014;31(2):84–91.

[134] Ghadially R, Reed JT, Elias PM. Stratum corneum structure and function correlates with phenotype in psoriasis. J Invest Dermatol 1996;107(4):558–64.

[135] Morin A. Détermination de la capacité des substituts cutanés à métaboliser le tazarotène, une molécule antipsoriasiques [Mémoire (M. Sc.)]: Université Laval.

[136] Tang-Liu DD, Matsumoto RM, Usansky JI. Clinical pharmacokinetics and drug metabolism of tazarotene: a novel topical treatment for acne and psoriasis. Clin Pharmacokinet 1999;37(4):273–87.

In vitro models of vitiligo

6

Muriel Cario-André, Katia Boniface†, François-Xavier Bernard‡, Alain Taieb§, Maria L. Dell'Anna¶, Julien Seneschal‖*
*INSERM U1035 BMGIC, University of Bordeaux, Bordeaux, France, †INSERM U1035 BMGIC Immuno-Dermatology, ATIP-AVENIR, University of Bordeaux, Bordeaux, France, ‡BIOalternatives, Gençay, France, §Department of Dermatology, Hôpital Saint-André, CHU de Bordeaux and INSERM U1035 BMGIC, University of Bordeaux, Bordeaux, France, ¶San Gallicano Dermatology Institute, Rome, Italy, ‖Department of Dermatology, Hôpital Saint-André, CHU de Bordeaux and INSERM U1035 BMGIC Immuno-Dermatology, ATIP-AVENIR, University of Bordeaux, Bordeaux, France

1. Introduction

Vitiligo is an acquired chronic depigmenting disorder of the skin resulting from a selective loss of melanocytes. The prevalence of vitiligo is often referred to as 0.5%–1% of the world's population [1,2]. Various theories have been suggested for the cause of melanocyte loss in vitiligo. Vitiligo can be considered as a multifactorial disease, with both genetic and environmental factors implicated in the initiation of the disease. Over the past years, several mechanisms of melanocyte loss in vitiligo have been presented. Researchers have shown the presence of melanocyte-intrinsic abnormalities in vitiligo leading to impaired melanocyte degeneration and/or proliferation, supporting the hypothesis that the disease could be due to a primary defect of melanocytes [3]. However, recent observations strongly support the role of the autoimmune system, particularly in chronic and progressive conditions [4–6]. Indeed, vitiligo is often associated with autoimmune diseases such as autoimmune thyroiditis, atopic dermatitis, and rheumatoid arthritis; genome-wide association studies and functional pathway analyses have shown that most vitiligo susceptibility loci encode components of the immune system [7–12]; and innate and adaptive immune cells are found in the perilesional margin of the actively depigmenting skin of vitiligo patients [13,14]. Since none of these processes alone are sufficient to fully explain the pathomechanisms of the disease, there are to date ample evidences supporting an interplay between all hypotheses [15]. Vitiligo can result from a combination of biochemical, environmental, and immunologic factors, in genetically predisposed patients. Therefore, only a multidisciplinary approach may unveil the pathogenic puzzle of the disease. Actually, in vitro strategies to study vitiligo range from the analysis of different skin cell subsets individually to the whole skin, using different techniques that will be described in this manuscript: primary cell cultures (melanocytes, keratinocytes, or fibroblasts), isolation of blood or skin-resident inflammatory cells to identify inflammatory cells

* This chapter has no funding source.

Skin Tissue Models. https://doi.org/10.1016/B978-0-12-810545-0.00006-1

involved in disease development, and three-dimensional skin models to reproduce the in vivo network. These models allow the characterization of the phenotype and function of all cell subsets involved in vitiligo pathogenesis.

2. Primary cell culture of skin cell subsets

Melanocytes and keratinocytes can be cultured from both the skin and hair follicle. Indeed, epidermal stem cells are present in the basal layer of the epidermis and within hair follicles; keratinocyte stem cells are situated in the bulge area, whereas those for melanocytes are found in the subbulge area [16].

2.1 Isolation and culture of skin melanocytes and keratinocytes

Vitiligo or healthy skin cells are obtained from the perilesional or nonlesional skin of vitiligo patients or healthy skin samples (e.g., the foreskin), respectively. The method for the isolation of primary melanocytes or keratinocytes is similar for both vitiligo and normal skin samples. Split-thickness skin samples are cut in small pieces and trypsinized. Trypsin disrupts the epidermis above the basal layer and is neutralized with fetal calf serum or trypsin-soybean inhibitor. The epidermis is removed, and the basal layer is scraped to dissociate melanocytes and basal keratinocytes. Cells are seeded at a density of 200,000/cm^2 for melanocytes and 100,000/cm^2 for keratinocytes culture [17]. Usually, the culture media for melanocytes are M2 medium (PromoCell), M254 (Gibco), or MCDB153 (Sigma), MGM4 BulletKit (Lonza), supplemented with specific growth factors. Several different handmade growth factors' cocktails for culture are used (Table 1). The specific composition of the growth factor cocktail will affect the proliferative or differentiation signature and growth rate and the delay between isolation and standard growth [18]. Moreover, the presence of α-melanocyte-stimulating hormone (MSH) analogue or precursor could influence the effect of some in vitro treatments [19]. The culture media used for keratinocytes are CellnTechn-07 (Chemicon), MCDB153 (Sigma), M154 (Gibco), KGM-2 (Promocell), or KSFM (Invitrogen), with the appropriate growth factors. Cells at passage 2–3 can be used to perform functional studies, reconstructed epidermis, or to start cocultures for studies investigating cellular network.

2.2 Isolation and culture of skin fibroblasts

After obtaining keratinocyte and melanocyte suspensions, the scraped dermis is cut in small pieces and incubated with collagenase, allowing the fibroblasts to exit from the extracellular matrix. Fibroblasts are cultured in DMEM supplemented with 10% fetal calf serum (FCS).

2.3 Genetic modification of cells

Since it is difficult to get enough cells from vitiligo patients to test all hypotheses, we can, according to genetic data, modulate susceptible genes using overexpressing or

Table 1 Types of cell culture media used to culture (A) vitiligo melanocytes and (B) vitiligo keratinocytes

A

Medium	bFGF (ng/mL)	PMA (nM)	Transferrin (μg/mL)	a-Toc (μg/mL)	Insulin (μg/mL)	FCS (%)	hCS	BPE (μg/mL)
M154	3	16	5		5	0.5	0.18 μg/mL	2
MCDB153	0.3		5	1	5	5	0.5 μg/mL	30
MCDB153	0.6	8		1	5	4		13
MCDB153	0.6	8	5	1	5	5	0.5 μg/mL	30*
MCDB153					20	3	1.75 μM	140

B

Medium	EGF (ng/mL)	Glutamine	Adenine	Cholera toxin	Thyronine	Insulin (μg/mL)	FCS	Hydrocortisone	BPE (μg/mL)
Celln Tech-07									
MCDB 153	10					5		1.4 μM	70
DMEM: HamF 12 (2:1)	10	4 mmol/L	0.18 mmol/L	0.1 nM	2 nM	5	10%	0.4 μg/mL	70

In all culture media, penicillin and streptomycin are added, 10,000 U/mL and 10,000 ng/mL, respectively. *a-Toc*, tocopherol; *hCS*, hydrocortisone.
Ham F12 contains 1% ultracer, 2 mM Glu, 10 ng/mL PMA No, and 0.1 nM IBMX.
* Also contains 20 μg/mL catalase.

silencing vectors such as nephroblastoma-overexpressed protein (NOV) also known as CCN3 (homeostasis of melanocyte), discoidin domain receptor 1 (DDR1) (adhesion of melanocyte to basal membrane), and cadherins (cell adhesion) [20,21]. Two types of vectors can be used in cells according to the time of the experiment or to the laboratory agreement. For short time, studies on transient modification (transfection) are made using plasmid encoding the gene of interest or siRNA [20]. However, for long-term studies, viral infection (transduction) using viral particles is necessary. Different types of virus coding the gene of interest or shRNA can be used. Viruses (retrovirus, lentivirus, adenovirus, or adeno-associated virus (AAV)) differ mostly by their capacity to infect proliferative or quiescent cells and to integrate or not host genome. To circumvent the fact that shRNA does not induce 100% inhibition, the CRISPR technology has been developed. Another key point especially for long-term studies is to decide if the gene has to be expressed or repressed permanently or at a precise time. For the last case, inducible promoter or Tet-on/Tet-off technologies are used.

After the first or the second passage, cells are incubated with viral particles at multiplicity of infection (MOI) 10 in a small volume of serum-free medium. After 6–16h, fresh FCS-containing medium is added to complete the volume to usual one (e.g., in 25 cm^2 flasks, transduction is made in 0.5 mL serum-free medium, and then, 3 mL FCS-containing medium is added). Medium is changed after 48h, and percentage of transduction using a fluorescent reporter gene (e.g., RFP, GFP, and YFP) present in the vector is estimated 72h after transduction to avoid overestimation due to pseudotransduction. To obtain 100% transduction, an antibiotic resistance gene such as puromycin resistance gene can be added in the vector construct. Otherwise, cells can be sorted by flow cytometry to obtain around 95% transduction.

3. Isolation of immune cells

Besides the role of epidermal cells (melanocytes-keratinocytes) and/or dermal cells such as fibroblasts, vitiligo is often associated with an exaggerated response of both the innate and the adaptive immune system [4]. Indeed, infiltration of T cells is consistently found in perilesional margin of the actively depigmenting skin of vitiligo patients and plays a crucial role in the loss of melanocytes. T cells expressing CXCR3 infiltrate the vitiligo skin in response to chemokine ligands CXCL9/CXCL10 produced by keratinocytes [22,23]. Isolating these cells to analyze their precise phenotype, function, and their capacity to respond against melanocytes is critical in order to decipher the link between the response of the immune system and defect of melanocytes in vitiligo.

3.1 Isolation of peripheral blood mononuclear cells

First, the isolation of vitiligo peripheral blood mononuclear cells (PBMC) is performed through the stratification onto a Ficoll density gradient (1.077), which allows the separation of mononuclear from polynuclear and red cells. PBMC localize at the interface between plasma and Ficoll, whereas polynuclear cells, after centrifugation, go to the bottom of the tube. After recovery, PBMC are washed twice with a saline solution and used as planned. The procedure should be carefully performed (short time, <30 min, between blood withdrawal and PBMC isolation; gentle manipulation) in order to avoid any physical stress able to affect vitiligo PBMC independently of in vitro tests. Theoretically, vitiligo PBMC may be used even after freezing in DMSO/serum mixture. Then, T cells can be isolated through positive or negative selection (MicroBeads) or using cell-sorting technology.

3.2 Isolation of skin antigen presenting cells (skin dendritic cells and Langerhans cells)

Antigen-presenting cells such as Langerhans cells (LC) or dermal dendritic cells (DC) play an important role in vitiligo by activating skin T cells. These cell subsets can be purified from the normal human skin [24]. Subcutaneous fat is excised, and the remaining tissue is washed with phosphate-buffered saline (PBS). The dermal layer is heavily scored with a scalpel and incubated with 1 mg/mL type 1 collagenase (Invitrogen), 1 mg/mL dispase (Invitrogen), and penicillin-streptomycin solution overnight at 37°C. Epidermal and dermal sheets are separated and placed in RPMI 1640 supplemented with 10% pooled human serum (Cellgro Mediatech), 0.1% gentamicin reagent solution (Invitrogen), and 1% 1 M HEPES buffer (Sigma-Aldrich). After 48 h at 37°C, cells are collected and filtered with 40 μm cell strainers (BD Biosciences) and then enriched with a Ficoll-diatrizoate gradient (GE Healthcare Bio-Sciences). LC are further purified by positive selection using anti-CD1a MACS (Miltenyi Biotec). Dermal DC are purified by positive selection using $BDCA1^+$ isolation kit (Miltenyi Biotec).

3.3 Isolation of skin T cells

Various techniques have been described to extract skin T cells. First, full-thickness human skin samples are extensively minced and incubated for 2 h at 37°C in RPMI 1640 containing 0.2% type I collagenase (Life Technologies) and deoxyribonuclease I (30 Kunitz units/mL, Sigma-Aldrich). Thereafter, cells are collected by filtering the collagenase-treated tissue through a 40 mm cell strainer. Another method to extract T cells out of the skin uses three-dimensional matrices. A 4 mm-diameter punch biopsy taken from vitiligo patients or the normal skin obtained from plastic surgery can be used to extract skin T cells. First, subcutaneous fat is removed, and the tissue is minced into explants of approximately 2×2×2 mm in size. Skin explants are placed on the surface of Cellfoam matrices (grids, Cytomatrix Pty Ltd., Melbourne, Australia) coated with type I collagen and allowed to briefly air-dry to maintain adherence of the skin to the matrix. The matrices and skin are then placed in wells of a 12 mm-diameter, 0.4 mm-pore-size polyester membrane transwells (Corning, Corning, NY), and the culture is maintained submerged in 2 mL of Iscove's modified medium (Mediatech, Herndon, VA) containing 20% heat-inactivated fetal bovine serum (Sigma, St Louis, MO), penicillin and streptomycin, and 3.5 mL/L β-mercaptoethanol with feeding three times per week. For feeding, 1 mL (of total 2 mL) is aspirated and replaced with fresh medium. The use of transwells was optional; T cells are then isolated by aspiration of medium in the wells and by thorough flushing of the matrices. To expand T cells without affecting their phenotype, interleukin (IL)-2 and/or IL-15 is added from the initiation of culture until collection of T cells after up to 21 days. Human recombinant IL-2 is included at 100 U/mL. Human recombinant IL-15 is added at 20 ng/mL [25].

4. Next generation cultures

Isolating primary cell subsets do not completely reflect the complexity of the skin microenvironment. Growing estimation is recently gained by the extracellular environment role on cellular behavior. The network between mechanical, chemical, and structural aspects of the environment heavily affects the cell function. Considering that physiologically, the cells are not seeded on plastic or glass but grow in a 3-D space, biomaterials, including hydrogels, have been developed to investigate the complex world of live cells. Most of functional data on cultured cells arise from 2-D support, such as flat stiff materials (polystyrene and glass), determining per se flattened shape, aberrant polarization, the loss of differentiation, and altered response to drugs. Hydrogels, that are water-swollen networks of polymers, have been developed to overcome this problem by mirroring, more than standard stiff supports, the physiological extracellular matrix. Currently, several different types of hydrogels are commercially available, both natural and synthetic (or mixed), characterized by specific physical and chemical properties. The companies BD Biosciences, Baxter, Johnson & Johnson, Sigma, Pronova, and BioTime, Inc. are the major vendors of the different hydrogels for both 2-D and 3-D cultures. According to their different physical features, each hydrogel is more or less suitable, based on epitope accessibility or stability and protein/RNA recovery, for microscopy, flow cytometry, or molecular studies [26].

5. Three-dimensional models to study pigmentation

To better understand cell subsets behavior in their proper microenvironment and their relation between each other, three-dimensional models can be developed. As early as in 1979, Pruniéras et al. have demonstrated that it is possible to obtain a fully differentiated epidermis in vitro by simply raising keratinocytes up to the air-liquid interface [27]. Apparently, the interface with air stimulates synthesis of profilaggrin by keratinocytes and thus the appearance of the granular phenotype when keratohyalin granules develop [28]. The epidermis can be reconstructed using various supports: dead deepidermized dermis (DDD) [27], gel of collagen (EpiSkin), lattices including fibroblasts and collagen [29], lattices including fibroblasts and collagen-glycosaminoglycan-chitosan [30], human fibrous sheet [31], and basal inert substrates such as porous filters [32]. Reconstructed epidermis models have been perfected by adding melanocytes [33]. These reconstructed epidermis containing both keratinocytes and melanocytes on DDD [34] and polycarbonate filters (BIO*alternatives* and SkinEthic RHPE) reproduce the epidermal melanin unit (EMU) and are suitable to study pigmentation. In addition, reconstructed models containing also immune cells, as LC [35] or T cells [36], or endothelial cells [37] have been developed; nonetheless, they have not been tested in vitiligo experiments.

5.1 In vitro reconstructed pigmented epidermis using polycarbonate filters

Melanocytes and keratinocytes (from healthy skin or nonlesional vitiligo skin) at passage 2 or 3 are seeded at 5×10^5 cells/insert (melanocyte/keratinocyte ratio of 1:10) in a 12mm-diameter polycarbonate Millicell inserts (Millipore) placed in six well plates, in complete EpiLife medium (Cascade Biologics) supplemented by 5µg/mL insulin and 1.5mM $CaCl_2$. Twenty-four hours after seeding, the medium is renewed and completed with 50µg/mL vitamin C and 3ng/mL KGF, and the inserts are shifted to the air/liquid interface for 10days before functional studies (Fig. 1).

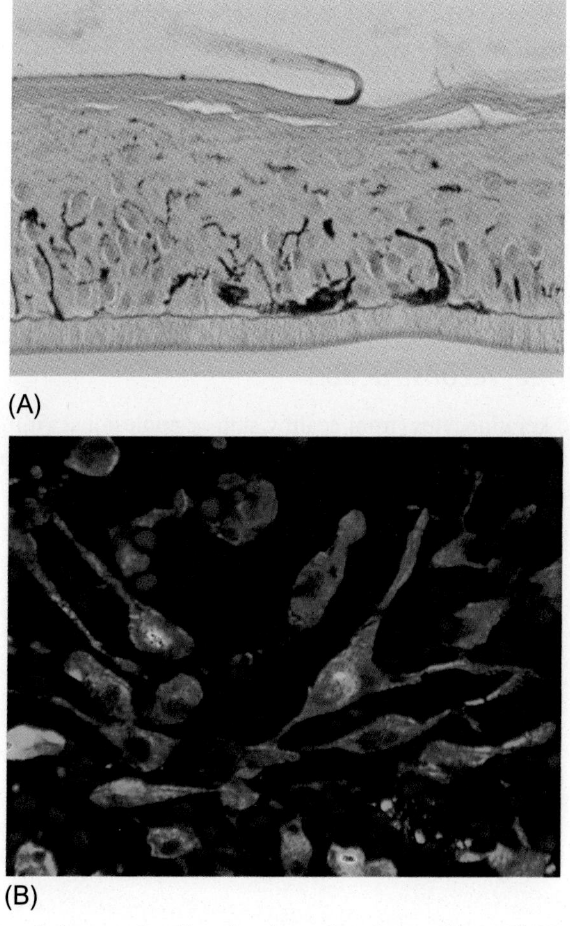

Fig. 1 Reconstructed pigmented epidermis using polycarbonate filter: (A) Hematoxylin and eosin staining associated with ES staining associated with Warthin-Starry staining to visualize melanocytes at the basal layer of the epidermis. (B) Immunofluorescence using anti-TRP-1 antibody (*green*, melanocytes); cell nuclei are labeled with DAPI *(red)*; observation of the basal side by confocal microscopy.

5.2 In vitro reconstructed pigmented epidermis containing fibroblasts

This model uses the same method and media as described above, but epidermal cells (keratinocytes and melanocytes 10:1 ratio) are uploaded on a previously prepared collagen gel containing fibroblasts (dermis equivalent). The dermis equivalent is prepared by mixing a rat tail collagen solution to fibroblast suspension (1.3 mg/mL and 0.5×10^6 cells/mL, final concentration/content, in a volume of 400 μL/insert). After 2 h, polymerization/solidification, the epidermal cell suspension is added for 2 days. Inserts are then placed at the interface air medium on a sterile gauze with a medium change every 2 days (Fig. 2).

5.3 In vitro reconstructed pigmented epidermis using dead deepidermized dermis

5.3.1 Preparation of dead deepidermized dermis

Human dermis is obtained from plastic surgery specimen from normal adults, mostly breast reduction specimen. Skin samples are thinned, cut into very small pieces, and incubated at 37°C in Hank's balanced salt solution until epidermis can be removed without excessive scraping. After removal, the dermis is rinsed in 70° ethanol and submitted to two cycles of freezing and thawing and stored in Hank's balanced salt solution at −20°C until use.

5.3.2 Epidermal reconstruction

Melanocytes and keratinocytes (from healthy skin or nonlesional vitiligo skin) at passage 2 or 3 are seeded in an incubation chamber placed on the epidermal side of

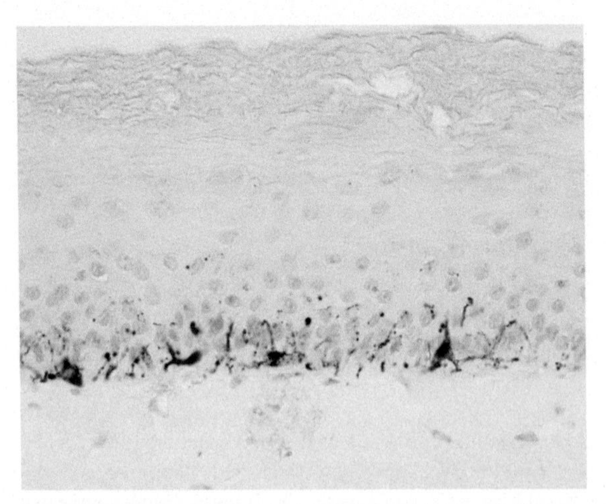

Fig. 2 Reconstructed pigmented epidermis containing fibroblasts. Melanocytes were stained for the PMEL17 protein (HMB45).

DDD at 4×10^5 cells/cm^2 at a melanocyte/keratinocyte ratio of 1:20 (5%) for normal melanocytes [34] and 1:20 or 1:10 (5%–10%) for vitiligo melanocytes since vitiligo melanocytes have a defective adhesion [38,39]. Twenty-four hours after seeding, the incubation chamber is removed, and the DDD is immersed for 3 days. Epidermal reconstruction is shifted to the air/liquid interface for 8 days before functional studies. The model can be improved by seeding fibroblasts in an incubation chamber placed on the dermal side of DDD 72 h before seeding keratinocytes and melanocytes (Figs. 3–5).

5.4 Inflammatory reconstructed skin model

To better understand the interplay between the immune system and melanocyte loss in vitiligo, T cells can be incorporated to the reconstructed skin. A human 3-D skin equivalent model containing keratinocytes and populated with CD4 T cells has been developed [36]. Blood or skin isolated T cells (0.25×10^6 to 2.5×10^6 T cells) could be introduced in the reconstructed skin to analyze the consequences of the presence of immune cells on keratinocytes. After 4 days of culture, T cells can be observed in the dermis or in the epidermis of the reconstructed skin. A similar model with reconstructed pigmented epidermis (containing keratinocytes and melanocytes) where vitiligo T cells would be added is of interest to study the impact of T cells in melanocyte loss. The possibility to include antigen-presenting cells into the equivalent skin model,

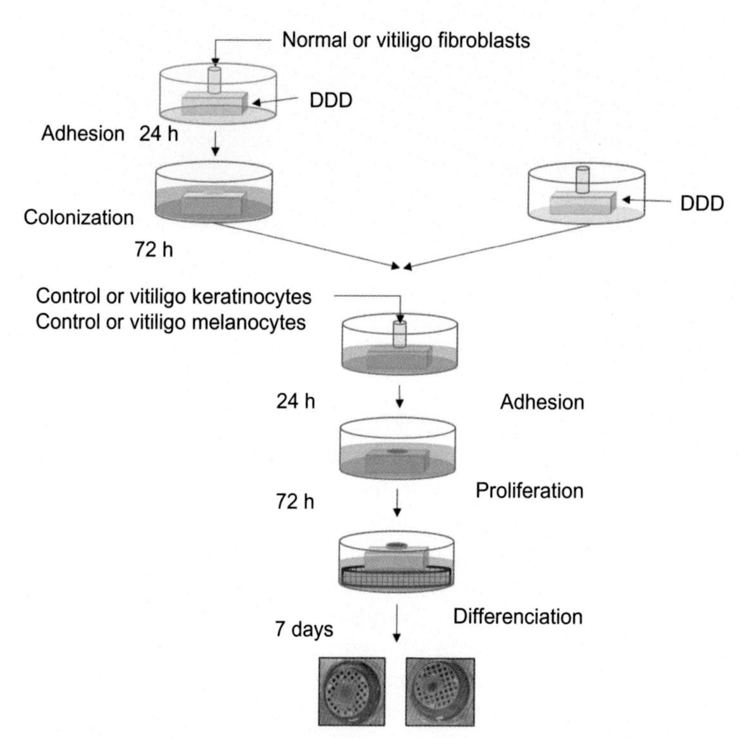

Fig. 3 Model of reconstructed pigmented epidermis using dead deepidermized dermis (DDD).

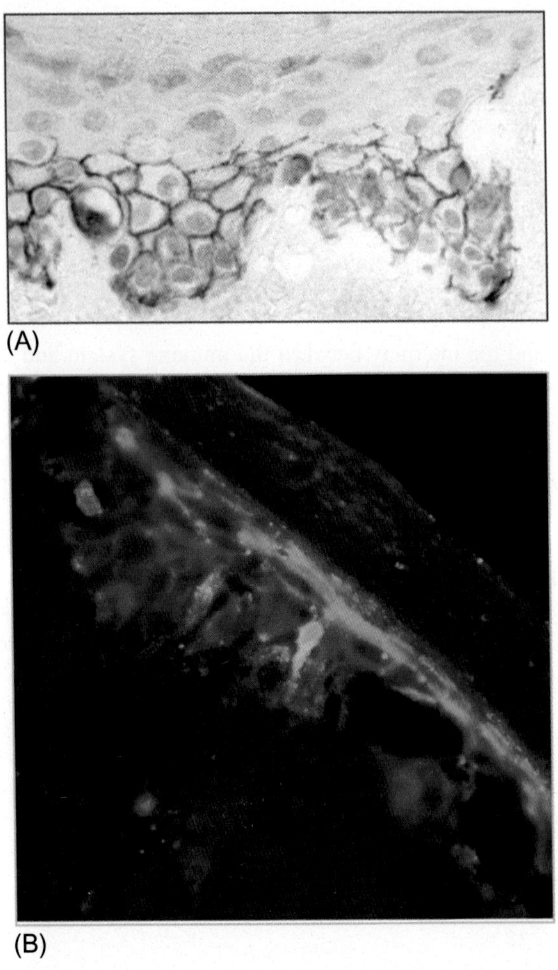

Fig. 4 Reconstructed pigmented epidermis using DDD: double staining melanA *(red)*-E-cadherin *(green)* on reconstructed epidermis using normal keratinocytes and normal melanocytes (A and B).

as LC, has also been demonstrated [35]. Yet, models with either T cells or LC have not been tested in vitiligo experiments so far. Moreover, due to the difficulty to obtain from the same patient sufficient skin samples to isolate both epidermal and immune cells, an autologous model is often difficult to perform, introducing an important bias to consider in the analysis of the results obtained.

Pigmented epidermis can therefore be reconstructed on dead deepidermized dermis (DDD) or polycarbonate filters. The main difference between these two types of support is the formation of epidermis with rete ridges only possible on DDD. Thereby, this model allows testing molecules that modulate the dermoepidermal interaction [40]. However, due to biological origin of DDD, quality of pigmented reconstructs is variable, and standardization of reconstructs is not possible. On filter, all the processes

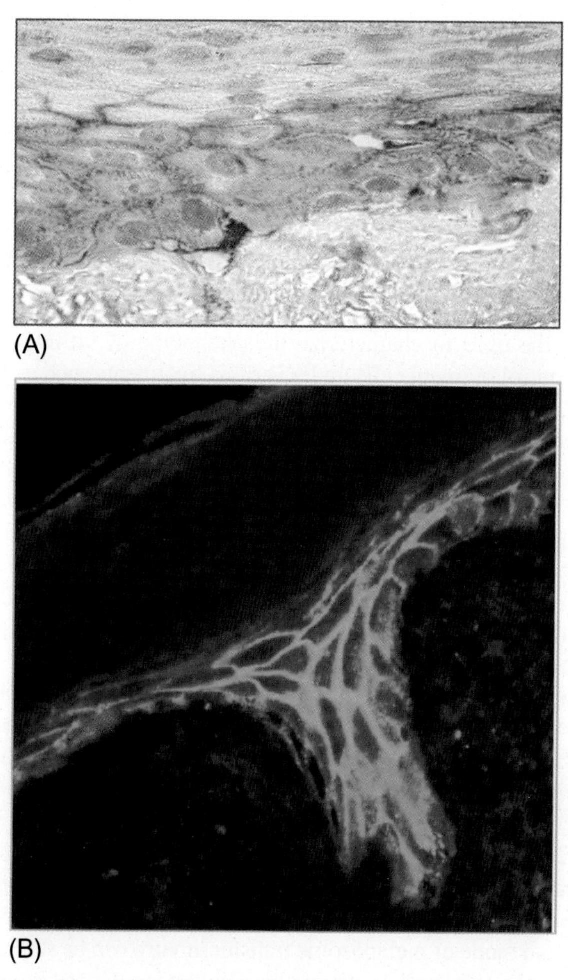

Fig. 5 Reconstructed pigmented epidermis using DDD: double staining melanA *(red)*-E-cadherin *(green)* on reconstructed epidermis using vitiligo melanocytes (A) or keratinocytes and melanocytes from vitiligo patient (B).

are standardized, and since the quality of cells can be managed before reconstruction, this model ensures production of reconstructs of similar architecture.

To date, no pigmented reconstructions on collagen gel without fibroblasts are available. Introduction of fibroblasts in DDD or in collagen (lattice) allows testing the influence of fibroblasts on pigmentation and differentiation. However, on DDD, the real number of colonizing fibroblasts is difficult to manage; thus, this model is only suitable for basic research. Fibroblasts have to be used carefully since passages [41] and age impact the secretion of soluble factors by fibroblasts, which may influence the reconstructions.

Therefore, having introduced the different techniques to allow isolation the different cell subsets involved in vitiligo pathogenesis and described methods for developing three-dimensional pigmented epidermis, these techniques can be used to better understand the functional interplay between each cell subsets involved in the pathogenesis of the disease.

6. Functional studies

6.1 Functional studies of vitiligo cells using monolayers: Melanocytes, keratinocytes, and fibroblasts

Cell cultures can be used to characterize the phenotype of vitiligo melanocytes as compared with control pigment cells under various treatments such as UV or pharmacological agents. The cells are usually seeded the day before to obtain 60%–70% confluency on the day of treatment. Various techniques can be used to observe melanocyte behavior. Direct observation by microscopy on fixed and specifically stained melanocytes (melan-A, DOPA, c-kit, S-100, and HMB-45) gives information on shape, dendricity, and pigmentation. DOPA staining is often used as alternative method to overcome the poor melanocyte number. Vitiligo melanocyte and keratinocyte cultures can be tested in vitro in order to determine their specific susceptibility to noxious stimuli or to physiological growth factors. UVB, cumene hydroperoxide, and *tert*-butyl-phenol are the most used stimuli [33,42–49].

Moreover, these cells can be also tested in vitro with pro- or antiinflammatory soluble factors. Recombinant cytokines identified as predominantly expressed in vitiligo, alone or in combination, could be added at various concentrations to stimulate vitiligo or normal melanocytes or keratinocytes [50]. Another way could be to test the culture supernatant of T cells isolated from vitiligo blood or skin following in vitro activation with anti-CD3/CD28.

Cell proliferation and mortality can be assessed through MTT test, manual cell count, annexin V/propidium staining, DNA ladder, or caspase profile [51]. In addition, the morphology and mode of melanosome transfer in vitro can be studied using atomic force microscopy, which allows to estimate and quantize the measure and the distribution of the dendrites including internal melanosomes distribution and arrangement [48,49,52]. Culture on 3T3 feeder layer induces the formation of colonies of keratinocytes that vary in size according to the state of cell proliferation, differentiation, or senescence. Lesional vitiligo keratinocytes are characterized by a lower proliferative potential, as indicated by a shorter in vitro life span. Moreover, the expression of p16, PCNA, p53, and p63 markers differs between lesional and nonlesional cells. Lesional keratinocytes show a lower level of the senescence marker p16 and a higher level of the melanocyte growth factor SCF [53].

Genetically modified cells in monolayer can be useful to verify if putative gene can reproduce some key features of diseased cells mostly adhesion and dendricity for melanocyte, but sometimes, in monolayer, cells do not express protein they express in tissue. For example, melanocytes in monolayer do not express E-cadherin, thus silencing of E-cadherin has no effect on their behavior in culture. Several genes are implicated in cell adhesion, but their silencing may have various effect; silencing of DDR1 (collagen IV receptor) has no effect on plastic adhesion, whereas silencing of CCN3 (melanocyte's homeostasis) is lethal for melanocytes [8].

6.2 Functional studies of vitiligo cells by next generation approach

Cultures can be also used to perform functional studies based on epigenome editing. This branch of biology refers to directed alteration of chromatin marks at specific genomic loci by EpiEffectors aiming to durable gene regulation with application in basic research and clinics. Targeted deposition or removal of chromatin modifications is a powerful approach for functional studies. It includes histone posttranslational modifications, DNA methylation, and hydroxymethylation, which in concert regulate the gene expression. The available profile of chromatin modifications yields thousands of global scale epigenomic maps. In preclinical studies, epigenome editing will give rise to a fine mapping of the regulatory mechanisms leading to health/disease process. It is an integrating part of interdisciplinary field of *synthetic biology*. It represents new and interesting avenues to treat metabolic diseases, possibly including vitiligo, characterized by aberrant signaling pathways through a causative therapy. However, some questions are still open: Which modifications have a causal role in governing processes such as transcription or alternative splicing? Which modifications lead to stable and heritable changes in chromatin status? Which chromatin behavior affects the maintenance of such modifications? Is the specificity per se able to provide clinical applications? [54]. Recently, single-cell epigenomic approach was developed by combining several different high-throughput techniques allowing studies on cytosine modification, protein-DNA interaction, chromatin structure, and 3-D organization [55]. Single-cell epigenomic studies might complement single-cell transcriptome and genome analysis in defining for example vitiligo melanocytes. Therefore, targeted therapeutics can be designed to better improve melanocytes regeneration.

6.3 Functional studies using reconstructed epidermis

Monolayer cultures and cocultures [56] are useful to study vitiligo melanocytes and keratinocytes direct (cell-cell contact) or indirect interaction (soluble factors/culture insert), but they do not reproduce the tridimensional interactions of cells of the EMU. Epidermal reconstructs, which reproduce the EMU and basal membrane attachment, are thus a handful "in vivo-like" model. Indeed, reconstructions of different levels of complexity can be prepared (reconstructs with keratinocytes alone, keratinocytes and melanocytes, keratinocytes, melanocytes, and fibroblasts), including chimeric reconstructs (normal vs pathological cells). According to the initial reflection, the analysis of the behavior of melanocytes and keratinocytes is done in a more physiological environment than that of monolayer cultures and cocultures and allows to test conveniently compounds that are suspected to be implicated in vitiligo etiology or susceptible to improve the attachment and survival of melanocytes upon the basal layer. An example of epidermal reconstructs tested with epinephrine, norepinephrine, dopamine, hydrogen

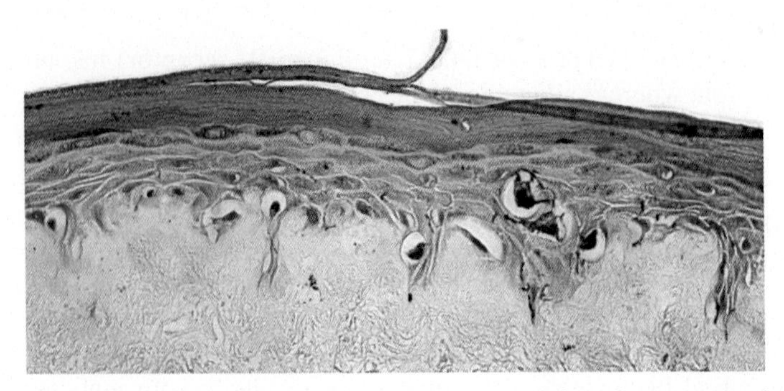

Fig. 6 Model of melanocytes migration: Fontana-Masson staining of reconstructed epidermis with melanocytes. Detached melanocyte *(white arrow)* after incubation with vitiligo serum during 17 h.

peroxide, or vitiligo sera is illustrated in Fig. 6. Our main results using this model [39] were the following: reconstructs made with melanocytes from nonlesional generalized vitiligo skin have a significantly reduced number of basal layer melanocytes, and the presence of vitiligo keratinocytes enhances this effect. Vitiligo sera may induce melanocyte detachment independently of the disease activity or extent. Hydrogen peroxide induces melanocyte detachment in reconstructs containing vitiligo melanocytes and normal keratinocytes, but not in normal controls. Epinephrine, but not norepinephrine, allows melanocyte detachment. Lastly, the effect of T cells or soluble factors produced by these cells in reconstructed pigmented epidermis would be important to decipher their precise role in the process leading to melanocyte loss in vitiligo.

Epidermal reconstructs are useful to address several research questions regarding the pathomechanism involved in melanocyte loss in vitiligo: Is the melanocyte primarily affected? Is the cellular environment important (keratinocytes and fibroblasts)? Are other (soluble) factors implicated in the development of vitiligo? However, this model has some limitation since we have not yet been able to introduce, for instance, LC or other immune cells to study their implication in vitiligo etiology; the study of topical molecules to improve vitiligo treatment is less easy than that of soluble factors and that long-term studies (>3 weeks) are not possible, since there is no renewal of the basal layer. Moreover, the analysis of intracellular signaling and membrane assessment per cell can be just on a single-cell-type culture, when cell sorting is not available.

Modelization of vitiligo by transgenesis has mostly to do with melanocyte. Since we supposed that genetic defect in vitiligo melanocyte does not induce per se detachment of melanocytes, experiments are designed in two steps. The first step consists to analyze the behavior of genetically modified melanocytes in reconstructed epidermis in the absence of stressor. In a second step, vitiligo known stressors are used to induce melanocyte detachment if the tested modification alone has not induced a mislocalization of melanocytes. Silencing of CCN3 in melanocyte induced a decrease in the number of melanocytes and the detachment of melanocyte according to the level of CCN3 inhibition [20]. On the other hand, hydrogen peroxide induces a detachment of E-cadherin silenced melanocyte as for vitiligo (Fig. 7) [21]. This model affords good

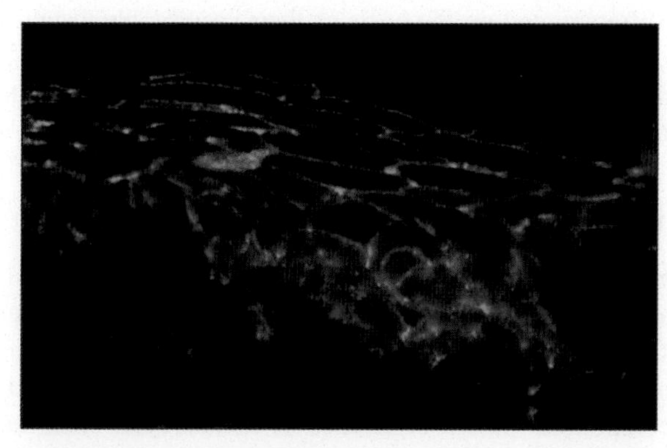

Fig. 7 Epidermal reconstruction using transgenic cells: double staining melanA (*red*)/E-cadherin *(green)* on reconstructed epidermis using normal keratinocytes and melanocytes inhibited for E-cadherin.

opportunities to study molecules on genetically altered cells; however, the effect of the molecules has to be fast enough to be seen in this model that can only be studied during 3 weeks.

7. Analytic techniques

7.1 Fluorescence-based assays

Fluorochrome-conjugated monoclonal or polyclonal antibodies are widely applied to visualize and quantify the expression of some surface or internal melanocyte markers [35,57–59]. The first data were derived from immunohistochemistry methods on frozen and paraffin-embedded sections, followed by fluorescence and confocal microscopy on slide-cultured cells. Phenotypic characterization of cultured vitiligo melanocytes with respect to control cells was carried out and revealed a reduced expression of c-Kit, endothelin 1 (ET-1) receptor, tyrosinase, and MITF in vitiligo melanocytes, with a pattern progressively varying from the edge of the white spot to the nonlesional area [58]. A flow cytometry approach permits both a qualitative and quantitative multiparametric and time-dependent analyses. Membrane and intracellular staining together with the analysis of the physical and biological parameters allows the structural and functional characterization of melanocytes and their sorting for further cultures. Beside the antibody-based approach, flow cytometry has been used to detect in vitiligo melanocytes levels of intracellular ROS production (DCFH-DA or dhRho123 staining), membrane lipo-peroxidation [60] (BODIPY581/591 staining), and content and transmembrane cardiolipin distribution (NAO fluorescence pattern) [42]. Several new approaches, based on "fluo world," are now available. Most of these innovative technologies have not yet been applied to the study of vitiligo, but they open promising perspectives. The laser scanning cytometer (LSC) permits slide-based cytometry (SBC) and the hyperchromatic approach. Its potential use in vitiligo study arises from its intrinsic features: nonconsumptive, iterative restaining, differential photobleaching (fluorochromes differentiated on the basis of their specific photostability), and photoactivation (for nanoparticles or photocaged dyes). A single cell can be reanalyzed, whereas the information gained per specimen is only limited by the number of available antibodies and sterical hindrance [61].

7.2 Proteomic

Mass spectrometry in conjunction with free-flow electrophoresis of sucrose density gradient has allowed the identification of early-stage melanosome proteins [62]. Two-dimensional differential image-gel electrophoresis (2D-DIGE) and liquid chromatography tandem mass spectrometry (LC-MS/MS) allow the analysis and identification of the protein components of the organelles of melanocytes with melanosomes at different maturation stages [63]. Even if proteomic assays have been so far carried out on murine healthy melanocytes, the scenario possibly designed by this approach may be crucial for the maturation process leading from nonpigmented melanocytes to pigmented melanocytes in vitiligo. Starting from the consideration of vitiligo as metabolic multifactorial disease, an analysis of the systemic proteomic profile will support its characterization. Recently, proteomic approach was applied to formalin-fixed paraffin-embedded tissue [64].

7.3 Metabolomic

Metabolomics is an emerging field of biological science and considers the high-throughput characterization of small compounds (<1500 Da) representing the final products of cellular metabolism that mirrors the chemical fingerprint of an organism at a precise point. It aims to identify and quantify the metabolites to correlate their changes with pathological conditions or with drug intake. Accordingly, metabolomics is one of the building blocks of biology systems. Metabolomic studies generate a complex set of data requiring specialized analyses based on cheminformatics, bioinformatics, and statistics expertise.

With regard to the specific vitiligo application, metabolomic research will provide a fingerprinting of the metabolic activities related to immune deregulation and degenerative process, possibly merging the vitiligo profile with those known and described in other diseases and health population. The separation techniques are based on gas chromatography, high-performance liquid chromatography, and capillary electrophoresis; the detection techniques usually are done by mass spectrometry or nuclear magnetic resonance [65].

7.4 Transcriptomic

Recently, oligonucleotide-based microarrays have been used to explore the pattern of gene expression of vitiligo melanocytes. Interestingly, the most upregulated genes are related to the network of endosome and lysosome organelles. The next step is to analyze the various clusters of the differentially expressed genes. This approach can represent the basis for further in-depth analyses to better clarify the complex vitiligo pathomechanisms [66].

The fields of genomics and transcriptomics have known a considerable progression with the next-generation sequencing (NGS) technology [67]. RNA-sequencing on samples or even on single cells represents an innovative approach for both mapping and quantifying a large number of transcribed RNA in a given sample. Such method is of interest to improve our understanding on vitiligo pathomechanisms through identification of genes important in depigmentation and melanocyte loss.

8. Concluding remarks

Cell culture models are useful to investigate the differences between normal and vitiligo cells and to test potential treatment options in a cell-focusing approach. New analytic techniques using a limited amount of biological material are very promising. Beyond the study of vitiligo pathomechanisms or therapies using unmodified patient's cells, in vitro approaches can be also adapted to generate melanocyte-specific silencing or overexpression of putative target genes or, a further step, to design animal models using melanocyte-specific expression of modified genes. The study of various issues pertaining to vitiligo, including the role and regulation of transcription factors, organelle genesis, intracellular transport, stem cell maintenance, and senescence, can be envisaged [68]. This step will allow the development of vitiligo models using normal cells bypassing the difficulty to culture vitiligo cells, which is currently the limiting factor for studying vitiligo in vitro. Combining these approaches with new generation tools (genomic and transcriptomic analyses and proteomic and/or metabolomic analyses) could be of major interest to precisely define primary defect of cells in the context of vitiligo pathogenesis. Moreover, the thorough characterization of the phenotype and function of immune cells involved in human vitiligo skin could emphasize the role of specific cell subsets or inflammatory soluble factors that could be added in these models to go more in depth in the interplay between epidermal (keratinocytes-melanocytes) or dermal cells (fibroblasts) with innate or adaptive immune cells.

Finally, one limit of our manual technique of reconstruction is that it does not ensure a homogenous deposition of melanocytes on dermis. 3-D printing could address this point in the future; yet, nobody has printed melanocytes so far, and nobody has been able to print a single cell, which represents a better way to study molecules in a similar environment. Indeed, such technology would allow on the same reconstruct printing of both unmodified and modified cells.

Conflict of interest

The authors have no conflict of interest to declare.

References

[1] Ezzedine K, Eleftheriadou V, Whitton M, van Geel N. Vitiligo. Lancet 2015; 386(9988):74–84.
[2] Taieb A, Picardo M. Clinical practice. Vitiligo. N Engl J Med 2009;360(2):160–9.

[3] Dell'anna ML, Picardo M. A review and a new hypothesis for non-immunological pathogenetic mechanisms in vitiligo. Pigment Cell Res 2006;19(5):406–11.
[4] Boniface K, Taieb A, Seneschal J. New insights into immune mechanisms of vitiligo. G Ital Dermatol Venereol 2016;151(1):44–54.
[5] Sandoval-Cruz M, Garcia-Carrasco M, Sanchez-Porras R, Mendoza-Pinto C, Jimenez-Hernandez M, Munguia-Realpozo P, et al. Immunopathogenesis of vitiligo. Autoimmun Rev 2011;10(12):762–5.
[6] Strassner JP, Harris JE. Understanding mechanisms of autoimmunity through translational research in vitiligo. Curr Opin Immunol 2016;43:81–8.
[7] Jin Y, Birlea SA, Fain PR, Gowan K, Riccardi SL, Holland PJ, et al. Variant of TYR and autoimmunity susceptibility loci in generalized vitiligo. N Engl J Med 2010;362(18):1686–97.
[8] Jin Y, Birlea SA, Fain PR, Mailloux CM, Riccardi SL, Gowan K, et al. Common variants in FOXP1 are associated with generalized vitiligo. Nat Genet 2010;42(7):576–8.
[9] Spritz RA. The genetics of vitiligo. J Invest Dermatol 2011;131(E1):E18–20.
[10] Jin Y, Birlea SA, Fain PR, Ferrara TM, Ben S, Riccardi SL, et al. Genome-wide association analyses identify 13 new susceptibility loci for generalized vitiligo. Nat Genet 2012;44(6):676–80.
[11] Jin Y, Andersen G, Yorgov D, Ferrara TM, Ben S, Brownson KM, et al. Genome-wide association studies of autoimmune vitiligo identify 23 new risk loci and highlight key pathways and regulatory variants. Nat Genet 2016;48(11):1418–24.
[12] Spritz RA, Andersen GH. Genetics of vitiligo. Dermatol Clin 2017;35(2):245–55.
[13] Badri AM, Todd PM, Garioch JJ, Gudgeon JE, Stewart DG, Goudie RB. An immunohistological study of cutaneous lymphocytes in vitiligo. J Pathol 1993;170(2):149–55.
[14] Le Poole IC, van den Wijngaard RM, Westerhof W, Das PK. Presence of T cells and macrophages in inflammatory vitiligo skin parallels melanocyte disappearance. Am J Pathol 1996;148(4):1219–28.
[15] Boniface K, Taieb A, Seneschal J. New insights into immune mechanisms of vitiligo. G Ital Dermatol Venereol 2016;151(1):44–54.
[16] Cichorek M, Wachulska M, Stasiewicz A, Tyminska A. Skin melanocytes: biology and development. Postepy Dermatol Alergol 2013;30(1):30–41.
[17] Medrano EE, Nordlund JJ. Successful culture of adult human melanocytes obtained from normal and vitiligo donors. J Invest Dermatol 1990;95(4):441–5.
[18] Dell'anna ML, Cario-Andre M, Bellei B, Taieb A, Picardo M. In vitro research on vitiligo: strategies, principles, methodological options and common pitfalls. Exp Dermatol 2012;21(7):490–6.
[19] Hunt G, Todd C, Cresswell JE, Thody AJ. Alpha-melanocyte stimulating hormone and its analogue Nle4DPhe7 alpha-MSH affect morphology, tyrosinase activity and melanogenesis in cultured human melanocytes. J Cell Sci 1994;107(Pt 1):205–11.
[20] Ricard AS, Pain C, Daubos A, Ezzedine K, Lamrissi-Garcia I, Bibeyran A, et al. Study of CCN3 (NOV) and DDR1 in normal melanocytes and vitiligo skin. Exp Dermatol 2012;21(6):411–6.
[21] Wagner RY, Luciani F, Cario-Andre M, Rubod A, Petit V, Benzekri L, et al. Altered E-cadherin levels and distribution in melanocytes precede clinical manifestations of vitiligo. J Invest Dermatol 2015;135(7):1810–9.
[22] Bertolotti A, Boniface K, Vergier B, Mossalayi D, Taieb A, Ezzedine K, et al. Type I interferon signature in the initiation of the immune response in vitiligo. Pigment Cell Melanoma Res 2014;27(3):398–407.
[23] Rashighi M, Agarwal P, Richmond JM, Harris TH, Dresser K, Su MW, et al. CXCL10 is critical for the progression and maintenance of depigmentation in a mouse model of vitiligo. Sci Transl Med 2014;6(223). 223ra23.

[24] Seneschal J, Clark RA, Gehad A, Baecher-Allan CM, Kupper TS. Human epidermal Langerhans cells maintain immune homeostasis in skin by activating skin resident regulatory T cells. Immunity 2012;36(5):873–84.
[25] Clark RA, Chong BF, Mirchandani N, Yamanaka K, Murphy GF, Dowgiert RK, et al. A novel method for the isolation of skin resident T cells from normal and diseased human skin. J Invest Dermatol 2006;126(5):1059–70.
[26] Caliari SR, Burdick JA. A practical guide to hydrogels for cell culture. Nat Methods 2016;13(5):405–14.
[27] Prunieras M, Regnier M, Schlotterer M. New procedure for culturing human epidermal cells on allogenic or xenogenic skin: preparation of recombined grafts. Ann Chir Plast 1979;24(4):357–62.
[28] Poumay Y, Coquette A. Modelling the human epidermis in vitro: tools for basic and applied research. Arch Dermatol Res 2007;298(8):361–9.
[29] Bell E, Sher S, Hull B, Merrill C, Rosen S, Chamson A, et al. The reconstitution of living skin. J Invest Dermatol 1983;81(1 Suppl):2s–10s.
[30] Black AF, Bouez C, Perrier E, Schlotmann K, Chapuis F, Damour O. Optimization and characterization of an engineered human skin equivalent. Tissue Eng 2005;11(5–6):723–33.
[31] Lee DY, Lee JH, Yang JM, Lee ES, Park KH, Mun GH. A new dermal equivalent: the use of dermal fibroblast culture alone without exogenous materials. J Dermatol Sci 2006;43(2):95–104.
[32] Rosdy M, Clauss LC. Terminal epidermal differentiation of human keratinocytes grown in chemically defined medium on inert filter substrates at the air-liquid interface. J Invest Dermatol 1990;95(4):409–14.
[33] Jimbow K, Chen H, Park JS, Thomas PD. Increased sensitivity of melanocytes to oxidative stress and abnormal expression of tyrosinase-related protein in vitiligo. Br J Dermatol 2001;144(1):55–65.
[34] Cario-Andre M, Bessou S, Gontier E, Maresca V, Picardo M, Taieb A. The reconstructed epidermis with melanocytes: a new tool to study pigmentation and photoprotection. Cell Mol Biol (Noisy-le-Grand) 1999;45(7):931–42.
[35] Regnier M, Staquet MJ, Schmitt D, Schmidt R. Integration of Langerhans cells into a pigmented reconstructed human epidermis. J Invest Dermatol 1997;109(4):510–2.
[36] van den Bogaard EH, Tjabringa GS, Joosten I, Vonk-Bergers M, van Rijssen E, Tijssen HJ, et al. Crosstalk between keratinocytes and T cells in a 3D microenvironment: a model to study inflammatory skin diseases. J Invest Dermatol 2014;134(3):719–27.
[37] Groeber F, Engelhardt L, Lange J, Kurdyn S, Schmid FF, Rucker C, et al. A first vascularized skin equivalent for as an alternative to animal experimentation. ALTEX 2016;33(4):415–22.
[38] Bessou S, Surleve-Bazeille JE, Sorbier E, Taieb A. Ex vivo reconstruction of the epidermis with melanocytes and the influence of UVB. Pigment Cell Res 1995;8(5):241–9.
[39] Cario-Andre M, Pain C, Gauthier Y, Taieb A. The melanocytorrhagic hypothesis of vitiligo tested on pigmented, stressed, reconstructed epidermis. Pigment Cell Res 2007;20(5):385–93.
[40] Salducci M, Andre N, Guere C, Martin M, Fitoussi R, Vie K, et al. Factors secreted by irradiated aged fibroblasts induce solar lentigo in pigmented reconstructed epidermis. Pigment Cell Melanoma Res 2014;27(3):502–4.
[41] Janson D, Saintigny G, Mahe C, El Ghalbzouri A. Papillary fibroblasts differentiate into reticular fibroblasts after prolonged in vitro culture. Exp Dermatol 2013;22(1):48–53.
[42] Dell'Anna ML, Ottaviani M, Albanesi V, Vidolin AP, Leone G, Ferraro C, et al. Membrane lipid alterations as a possible basis for melanocyte degeneration in vitiligo. J Invest Dermatol 2007;127(5):1226–33.

[43] Ivanova K, van den Wijngaard R, Gerzer R, Lamers WH, Das PK. Non-lesional vitiliginous melanocytes are not characterized by an increased proneness to nitric oxide-induced apoptosis. Exp Dermatol 2005;14(6):445–53.
[44] Kroll TM, Bommiasamy H, Boissy RE, Hernandez C, Nickoloff BJ, Mestril R, et al. 4-Tertiary butyl phenol exposure sensitizes human melanocytes to dendritic cell-mediated killing: relevance to vitiligo. J Invest Dermatol 2005;124(4):798–806.
[45] Lee AY, Kim NH, Choi WI, Youm YH. Less keratinocyte-derived factors related to more keratinocyte apoptosis in depigmented than normally pigmented suction-blistered epidermis may cause passive melanocyte death in vitiligo. J Invest Dermatol 2005;124(5):976–83.
[46] Maresca V, Roccella M, Roccella F, Camera E, Del Porto G, Passi S, et al. Increased sensitivity to peroxidative agents as a possible pathogenic factor of melanocyte damage in vitiligo. J Invest Dermatol 1997;109(3):310–3.
[47] Yang F, Sarangarajan R, Le Poole IC, Medrano EE, Boissy RE. The cytotoxicity and apoptosis induced by 4-tertiary butylphenol in human melanocytes are independent of tyrosinase activity. J Invest Dermatol 2000;114(1):157–64.
[48] Zhang RZ, Zhu WY, Xia MY, Feng Y. Morphology of cultured human epidermal melanocytes observed by atomic force microscopy. Pigment Cell Res 2004;17(1):62–5.
[49] Boissy RE, Liu YY, Medrano EE, Nordlund JJ. Structural aberration of the rough endoplasmic reticulum and melanosome compartmentalization in long-term cultures of melanocytes from vitiligo patients. J Invest Dermatol 1991;97(3):395–404.
[50] Wang S, Zhou M, Lin F, Liu D, Hong W, Lu L, et al. Interferon-gamma induces senescence in normal human melanocytes. PLoS One 2014;9(3):e93232.
[51] Yang L, Wei Y, Sun Y, Shi W, Yang J, Zhu L, et al. Interferon-gamma inhibits Melanogenesis and induces apoptosis in melanocytes: a pivotal role of CD8+ cytotoxic T lymphocytes in vitiligo. Acta Derm Venereol 2015;95(6):664–70.
[52] Bondanza S, Maurelli R, Paterna P, Migliore E, Giacomo FD, Primavera G, et al. Keratinocyte cultures from involved skin in vitiligo patients show an impaired in vitro behaviour. Pigment Cell Res 2007;20(4):288–300.
[53] van den Wijngaard RM, Aten J, Scheepmaker A, Le Poole IC, Tigges AJ, Westerhof W, et al. Expression and modulation of apoptosis regulatory molecules in human melanocytes: significance in vitiligo. Br J Dermatol 2000;143(3):573–81.
[54] Kungulovski G, Jeltsch A. Epigenome editing: state of the art, concepts, and perspectives. Trends Genet 2016;32(2):101–13.
[55] Clark SJ, Lee HJ, Smallwood SA, Kelsey G, Reik W. Single-cell epigenomics: powerful new methods for understanding gene regulation and cell identity. Genome Biol 2016;17:72.
[56] Eves PC, Beck AJ, Shard AG, Mac NS. A chemically defined surface for the co-culture of melanocytes and keratinocytes. Biomaterials 2005;26(34):7068–81.
[57] Graham A, Westerhof W, Thody AJ. The expression of alpha-MSH by melanocytes is reduced in vitiligo. Ann N Y Acad Sci 1999;885:470–3.
[58] Kitamura R, Tsukamoto K, Harada K, Shimizu A, Shimada S, Kobayashi T, et al. Mechanisms underlying the dysfunction of melanocytes in vitiligo epidermis: role of SCF/KIT protein interactions and the downstream effector, MITF-M. J Pathol 2004;202(4):463–75.
[59] Norris A, Todd C, Graham A, Quinn AG, Thody AJ. The expression of the c-kit receptor by epidermal melanocytes may be reduced in vitiligo. Br J Dermatol 1996;134(2):299–306.

[60] Dell'Anna ML, Urbanelli S, Mastrofrancesco A, Camera E, Iacovelli P, Leone G, et al. Alterations of mitochondria in peripheral blood mononuclear cells of vitiligo patients. Pigment Cell Res 2003;16(5):553–9.

[61] Tellez CS, Davis DW, Prieto VG, Gershenwald JE, Johnson MM, McCarty MF, et al. Quantitative analysis of melanocytic tissue array reveals inverse correlation between activator protein-2alpha and protease-activated receptor-1 expression during melanoma progression. J Invest Dermatol 2007;127(2):387–93.

[62] Ouvry-Patat SA, Torres MP, Quek HH, Gelfand CA, O'Mullan P, Nissum M, et al. Free-flow electrophoresis for top-down proteomics by Fourier transform ion cyclotron resonance mass spectrometry. Proteomics 2008;8(14):2798–808.

[63] Kawase A, Kushimoto T, Kawa Y, Ohsumi K, Nishikawa H, Kawakami T, et al. Proteomic analysis of immature murine melanocytes at different stages of maturation: a crucial role for calreticulin. J Dermatol Sci 2008;49(1):43–52.

[64] O'Rourke MB, Padula MP. Analysis of formalin-fixed, paraffin-embedded (FFPE) tissue via proteomic techniques and misconceptions of antigen retrieval. Biotechniques 2016;60(5):229–38.

[65] Cambiaghi A, Ferrario M, Masseroli M. Analysis of metabolomic data: tools, current strategies and future challenges for omics data integration. Brief Bioinform 2017;18(3):498–510.

[66] Stromberg S, Bjorklund MG, Asplund A, Rimini R, Lundeberg J, Nilsson P, et al. Transcriptional profiling of melanocytes from patients with vitiligo vulgaris. Pigment Cell Melanoma Res 2008;21(2):162–71.

[67] Wang Z, Gerstein M, Snyder M. RNA-Seq: a revolutionary tool for transcriptomics. Nat Rev Genet 2009;10(1):57–63.

[68] Goding CR. Melanocytes: the new black. Int J Biochem Cell Biol 2007;39(2):275–9.

Skin squamous cell carcinoma models: The role in combating the disease

7

Elizabeth Pavez Loriè, Hans-Jürgen Stark†, Manuel Berning†, Petra Boukamp*,†*
*IUF Leibniz Research Institute for Environmental Medicine, Düsseldorf, Germany,
†German Cancer Research Center, Heidelberg, Germany

Nonmelanoma skin cancer is the most prevalent and steadily increasing cancer worldwide. Cutaneous squamous cell carcinomas (cSCCs), the second most frequent tumor of that group, have gained strong metastatic potential and, in absolute numbers, are causing as many deaths as malignant melanoma. Immunosuppression of transplant recipients and novel cancer treatment modalities, such as the BRAF inhibitors, successfully applied for treating malignant melanoma, further increases the incidence of cSCC. Thus, understanding the cSCC's pheno- and genotype and mechanistic traits induced by the different compounds will be an absolute requirement to develop new treatment strategies. Therefore, the use of models that perfectly recapitulates this tumor type can thereby help to further unravel fundamental biological questions and to function as preclinical models to test, for example, novel treatment strategies.

1. Nonmelanoma skin cancers

Nonmelanoma skin cancer comprises several tumor types, with basal cell carcinoma (BCC) and cSCCs, the first and second most prevalent, respectively, representing 20% of all nonmelanoma skin cancers. While BCCs are locally destructive but rarely metastasize, cSCCs are more aggressive with a metastasis rate of up to 5%. As metastasis still lacks successful treatment, cSCCs contribute to the high cancer mortality rate of the present. It is estimated that one in every six North American will develop skin cancer during his/her lifetime, and the incidence of cSCC in Europe has increased up to 3%–8% each year since the 1960s (reviewed by Leiter et al. [1] and Perez et al. [2]).

The reason for this continuous increase is in part due to an altered attitude toward tanning, which has seen a rise in excessive sun exposure and use of tanning devices. Indoor tanning was first introduced in Europe in 1906, by the German company Heraeus, as sunlight therapy for patients suffering from bone diseases like rickets to help them develop more vitamin D. In the 1920s, being tanned became a trend, which lead to the development of sunbeds, a method by which to simulate sun by controlled UV exposure, by the 1970s. Ever since, the production of tanning beds is an evergrowing industry. However, due to the strong correlation with skin cancer, tanning

Skin Tissue Models. https://doi.org/10.1016/B978-0-12-810545-0.00007-3

salons and tanning bed manufacturers are now restricted by defined skin cancer prevention guidelines (IARC Working Group, 2007; WHO Artificial tanning sunbeds: risk and guidance, 2003). As skin cancer development typically requires 2–3 decades or more between the period of sun exposure and tumor onset, the consequence of frequent and/or regular use of sunbeds is already started to show and is likely to continue to increase the incidence of skin cancer in the future [3,4].

2. Risk factors for the development of cSCCs

2.1 Immunosuppresion and cSCC

While skin cancer is largely a disease of old age, due to the required accumulation of genetic events and a possible contribution of the age-dependent increase in impaired immunity (e.g., [5]), the immunosuppression required by transplant recipients has been suggested to influence UV sensitivity and strongly accelerates skin cancer development at younger age. This is particularly true for cSCC development. While immunocompetent patients show a general prevalence for BCCs with a ratio of 4:1 compared with cSCC, this ratio reverses in the immunosuppressed transplant recipients who bear a 100-fold risk for the development of cSCC (Scottish Intercollegiate Guidelines Network (SIGN). Edinburgh: SIGN, 2014). Further, these tumors that occur earlier in life are more aggressive, with a metastasis rate of up to 10%, and frequently occur as multiple cSCCs, also termed carcinomatous catastrophe. While this may support the notion that the immune system plays an important role in permitting cSCC development and/or progression, there is evidence that the risk conferred is dependent on the specific drug used and, thus, immune-independent factors may also play an important role [6]. Moreover, Abikhair and colleagues provided a link between the immunosuppressive agent cyclosporine A (CsA) and IL-22 in the tumors of the immunosuppressed cSCC patients. They found that "CsA drives T cell polarization toward IL-22-producing T22 cells, and CSA treatment increased IL-22 receptor in SCC cells" [7]. Since inhibition of IL-22 action reduced the tumor burden in their mouse model, they proposed anti-IL-22 treatment as a potential therapy to antagonize cSCC development in immunosuppressed transplant recipients. As the number of transplant recipients is steadily growing, there is a clear demand for better understanding and improve treatment modalities to combat this devastating type of skin cancer.

2.2 Smoking and cSCCs

While it was still a matter of debate whether smoking is a risk factor for developing cSCCs (see [8], and references therein), a recent study of as many as 43,794 skin cancer patients suggested that former smoking, irrespective of time and dose, did not influence the risk for skin cancer. However, for current smokers, the risk for developing a cSCC was significantly increased while that of developing a BCC was reduced [9].

2.3 Targeted therapy and cSCC

Upon targeted therapy, the skin is the tissue with the strongest adverse effects. As perfectly reviewed by Tang and Ratner [10], tyrosine kinase inhibitors frequently cause rash and perturb differentiation as particularly obvious for EGFRi (epidermal growth factor receptor inhibitor) and the tyrosine kinase inhibitor imatinib. For the multikinase inhibitor sorafenib, the adverse development of single or multiple keratoacanthomas (KAs) and cSCCs was reported. Also, the hedgehog pathway inhibitor vismodegib, applied for the treatment of BCCs, caused rapid appearance (within 4–12 weeks) of single or multiple cSCC at sites distant to the primary BCC. This is unexpected, since hedgehog/Gli1 signaling was long thought to be relevant for the regulation of hair follicles and its aberrant expression to be a characteristic for BCCs. Furthermore, it was recently shown that an aberrant hedgehog/Gli1 signal pathway facilitates proliferation, invasion, and migration of cutaneous SCC through regulating VEGF [11,12], thus further questioning the mechanism underlying the adverse event of cSCC development upon vismodegib treatment.

Also therapies with PD-1 or PDL-1 (programmed cell death-1 receptor and ligand) give rise to skin phenotypes. These immune-based checkpoint inhibitor therapies have demonstrated a distinct toxicity profile compared with chemotherapy, with the skin belonging to the most common target organ and erythema (rash) to the most common appearance of adverse events (reviewed by Maughan et al. [13]). It should be expected that when these drugs are more commonly used, the overall magnitude of adverse events would further increase.

The most severe example of skin-related adverse events relates to the BRAF inhibitors (vemurafenib and dabrafenib) used for the treatment of malignant melanoma. Vemurafenib treatment causes rapid development of KAs and cSCCs in up to 30% of the patients (for reference, see, e.g., [14]). Having identified adverse hyperactivation of the RAF-MEK-ERK pathway in the melanoma cells, BRAF inhibitors were combined with MEK inhibitors. Comparing monotherapy with combination therapy, Erfan et al. concluded that "combined treatment of MAPK inhibitors may prevent some of the skin toxicities, including the rapid development of cSCCs." Interestingly, the most frequent side effects, such as keratosis pilaris (KP)/hyperkeratinization of hair follicles; palmoplantar keratoderma (PPK)/hyperkeratosis, that is, perturbation of differentiation; and photosensitivity, were not prevented [15]. Thus, targeted anticancer therapies can cause cutaneous adverse events different from classical chemotherapeutic toxicities, which even include the rapid development of cSCCs.

3 What is known about the genetics of cSCCs

The etiology of cSCCs is multifactorial, and besides chemicals such as polycyclic aromatic hydrocarbons, UV radiation is believed to be the causal exogenous inductor. Accordingly, in 2009, the World Health Organization (WHO) accepted UV radiation into the group 1 of agents considered to be "carcinogenic in humans" [16]. In agreement with that, cSCCs predominantly develop at sun-exposed sites of the skin that also show other signs of sun damage. Interestingly, even for sun-exposed sites, the specific location of the primary tumor can be indicative of aggressiveness. It is suggested that cSCCs of the ear and lip are more prone to metastasize than cSCC from other sun-exposed sites [17]. In the guidelines of the National Comprehensive Cancer Network (NCCN), the list is extended to also other areas of the face including the eyelids, eyebrows, nose, and chin (NCCN, Guideline Version 1.2016, Squamous Cell Skin Cancer, 2016). In addition, the scalp was included as a high-risk site in the European guidelines [18]. Tumor diameter (>2 cm), tumor thickness, and anatomical depth have also been suggested as independent risk factor for tumor recurrence and metastasis; for example, invasion depths from 2 to 6 mm are correlated with an increase to 15% as summarized by Cheng et al. [19] and reviewed in a meta-analysis by Thompson and coworkers [20]. In addition, it has long been thought that histological grade and differentiation are of prognostic value, suggesting that poorly differentiated cSCCs have a higher potential for local recurrence and distant metastasis [20]. However, many cSCCs are moderately well differentiated, including the more aggressive cSCCs in the immunosuppressed transplant recipients [21], thus arguing against the requirement for the loss of differentiation in order to gain metastatic potential.

A number of studies have indicated that UV light initiates cSCC development by causing inactivating mutations in the p53 tumor suppressor gene. Accordingly, UV-indicative p53 mutations are still the most frequent recurrent aberrations found [22]. The loss of heterozygosity or loss-of-function mutations in CDKN2A are also reported as being frequent [23]. CDKN2A encodes two different, unrelated tumor suppressor proteins, p16INK4a and p14ARF. While p16INK4a contributes to the dysregulation of the Cdk4/6-cyclin-retinoblastoma (Rb) pathway and consequential loss of cell-cycle control, p14ARF mutations, although reported to be found in cSCC [24], are hardly followed up. This may be surprising as p14ARF is generally upregulated by oncogenic stimuli and serves as a critical regulator of p53 stability. The nucleolar localization of p14ARF may therefore be seen as a reserve to be able to cause rapid p53 induction or to coordinate effects on cell cycle, survival, and growth [25]. As CDKN2A mutations were already identified in actinic keratoses (AK), that is, precursor lesions of cSCC, it is suggested that they contribute to deregulating cell-cycle control but do not serve as tumorigenic drivers [26]. Similarly, NOTCH1, a gene involved in regulating differentiation, proliferation, and apoptosis, is one of the most frequently mutated genes in cSCCs [27]. Depending on the cellular context, NOTCH1 has been shown to act either as a tumor suppressor or as an oncogene [28].

Interestingly and similar to p53 mutations, NOTCH1 loss or downregulation is often found in normal-looking skin, thereby supporting a role for NOTCH1—as it was already done earlier for p53—in cancer initiation rather than tumorigenic progression [27,29]. Leigh and coworkers suggested NOTCH1 “as a gatekeeper in human cSCC” [27].

The role of oncogenic ras, on the other hand, is still controversially discussed (for review, see [30]). Even in the most recent sequence analysis, the rate of HRAS mutations was only 8%, a frequency similar to that listed in the COSMIC database (as discussed in [31]). However, the frequency of HRAS mutations may be enriched in cSCCs occurring in melanoma patients under BRAF-targeted therapy [32,33]. Under these specific conditions (vemurafenib-induced cellular and microenvironmental changes), the ras oncogene may act as a driver mutation for the tumorigenic conversion of human skin keratinocytes (Tham et al., in preparation).

To further increase the number of drug targets for the therapy of cSCC patients, a comprehensive genomic profiling was performed from 122 advanced-stage cSCCs [31]. This study further supported the strong mutational heterogeneity of cSCCs and, in addition, confirmed the presence of at least one clinically relevant (accessible for targeted therapy) genomic alteration in the majority of these cSCCs. Despite the genetic heterogeneity, also in this study, the most frequent and recurrent aberrations included p53, CDKN2A, and NOTCH1, thus confirming the previous findings with the cohort of cSCC from immunosuppressed patients [27] and making them the most prevalent aberrations for cSCCs.

Noteworthy and for a long time demonstrated for p53, a high mutation load was identified already for normal human skin. By ultradeep sequencing of 74 cancer genes in small biopsies of eyelid epidermis, Campbell and colleagues reported on a burden of somatic mutations as high as 2–6 mutations/megabase/cell, exhibiting characteristic signatures of UV exposure [34]. Positively selected "driver" mutations were found in 18%–32% of the normal skin cells, and the most frequently mutated genes were NOTCH1, 2, and 3; p53; and FGFR3. This again demonstrates that many of the cancer-related genes are already existent in their mutant form in normal human skin where they may still contribute to tissue homeostasis. Only in combination with a set of aberrant regulations these mutant genes get active and become able to confer tumorigenicity. The genetic heterogeneity of the cSCCs further suggests that this dysregulation does not necessarily occur via one "well-defined pathway" but that diverse regulatory combinations can lead to the same result of tumorigenic conversion, thus being consistent with the genetic complexity required for skin cancer development and progression.

Specifically addressing cSCC metastases, targeted sequencing of 504 cancer-associated genes demonstrated that the genomes of the lymph node metastases were as heterogeneous as the primary cSCCs, with large variations in the number of genomic alterations [11]. Consistent with the primary tumors, mutations in p53, CDKN2A, and NOTCH1/2 predominated, with UV-indicative C to T transitions being the dominant substitutions in the mutation spectra. In addition, they identified mutations in genes affecting RAS/RTK/PI3K, squamous differentiation, cell cycle, and chromatin remodeling pathways. Collectively, these findings indicate a high similarity between the aberration profile of primary tumor and metastasis and thereby underline the relevance of the current skin cancer cell models as excellent tools for unraveling cSCC-specific functional and responsive consequences.

Inconspicuous in most of the mutation studies are EGFR (epidermal growth factor receptor) and c-Myc, which appear to be altered more frequently by amplification or gene rearrangement. Reports of EGFR amplification in cSCC have ranged from as low as 1% (COSMIC database) up to 20% [11]. Nevertheless, advanced tumors often show upregulated EGFR expression without exhibiting EGFR or RAS mutations [35], and overexpression of EGFR was shown to be associated with poor outcome in cSCC [36]. The importance of EGFR overexpression was also demonstrated in a phase II trial where the EGFR antagonist gefitinib caused a complete response rate in 18% of the cSCC patients [37]. c-Myc aberrations were observed in 35% of AK lesions and 63% of the cSCCs analyzed, and its relevance for skin cancer was demonstrated by the significant association between copy number gains and protein expression ([38] and references therein). In addition, changes in the micro-RNA (miR) expression profile are depicted between the human healthy skin and cSCC [39]. The possible change in regulatory competence suggests that miRNAs may be promising targets as biomarkers for diagnosis and prediction of prognosis and/or targeted therapies (for review, see [40]). In this context, it was shown that miRNA-203 targets and regulates c-myc and functions as a tumor suppressor in cSCC [41].

As extensively reviewed by Rudnick et al. [42], so far, most patients are treated with surgical means. Additionally, retinoic acid or synthetic retinoid analogues in combination with interferon alpha and cisplatin are administered aiming for their differentiation, antiproliferative, and proapoptotic potential. Indeed, this combination has shown to produce objective response rated in locally advanced and metastatic cSCCs. In addition, an oral 5-FU off-label treatment for advanced cSCCs is used, and a prodrug of the inactive form of 5-FU is currently FAD approved. From the molecularly defined drugs, only the EGFR inhibitor has made it into clinical trials. Though not yet approved by the FDA for advanced cSCC failing standard treatment, they are aimed to stop proliferation and hinder EGF-dependent resistance to apoptosis. The inhibitory activity was proved by analyzing EGFR expression after a 14-day treatment, and for an observation time of 2 years, the disease-free survival rate was 60% and 73% of patients that remained recurrence-free [43]. Whether, for example, NOTCH1 and 2, as the most frequently mutated genes in cSCCs, will have the potential for targeted therapy remains to be seen.

4. Role of the microenvironment

It is, however, not only the cancer cells that drive tumor initiation and progression. Increasing evidence has pointed to the microenvironment, with its neighboring stromal extracellular matrix (ECM) and soluble growth factors as major drivers of tumorigenic behavior (for review, see [44]). Apart from the epidermal keratinocytes, UV radiation also causes genetic and epigenetic changes in the underlying dermal fibroblasts, thereby inducing changes that alter the interplay with the keratinocytes and allow premalignant lesions to become invasive or even metastatic cSCCs. Accordingly, alterations in the composition of basement membrane and dermal ECM are early events in cSCC progression. Proteases are required for matrix modulation and degradation with the consequence not only of altering tissue stiffness but also of releasing growth factors, cytokines, and chemokines that are preferentially stored in the ECM of the dermis, as perfectly reviewed by Nissinen et al. [45]. At present, the major focus remains on inflammatory cells as the main source for protease production. However, the dermal fibroblasts are also induced to modify their growth factor and protease profile and even to drastically modify their ECM profile in response to chronic UV exposure (Bauer et al., in preparation). We further showed that upon exogenous Wnt stimulation, it was not the keratinocytes but the stromal fibroblasts that became activated and expressed factors that, in a paracrine fashion, promoted keratinocyte growth and invasion [46].

Currently, the best documented example for a severe role of the microenvironment in cSCC development is provided by patients suffering from recessive dystrophic epidermolysis bullosa (RDEB), a hereditary skin disorder based on a mutation in the collagen VII (COL7A1) gene responsible for causing epidermal blistering and chronic wounds. These patients are afflicted with an increased risk for developing aggressive and rapidly metastasizing cSCCs. So far, it is still elusive whether RDEB cSCCs and sporadic UV-induced cSCCs are distinct tumor entities (for review, see [47]). However, as RDEB cSCCs preferentially arise in long-term skin wounds or cutaneous scars, it is suggested that the altered microenvironment may be of utmost importance and causally involved in the progression of these specific cSCCs. Bruckner-Tuderman and colleagues now suggest that the "inherent structural instability of the RDEB dermis, combined with repeated injury, increased the bioavailability of TGF-ß. TGF-ß, in turn, promotes ECM production, cross-linking, thickening of dermal fibrils, and tissue stiffening." This triggers myofibroblast activity in the modified dermis and consequently alters mechanosignaling in the tumor cells, thus supporting tumor cell invasion in this "preexisting injury-driven microenvironment" [48]. Together, these examples accentuate the increasing importance of the microenvironment in tumor progression of human cSCCs.

5. How to study skin cancer?

Taking these findings—the genotypic and phenotypic aspects of cSCC—into consideration, numerous approaches have been established to better understand their role in skin cancer development and progression and to allow for the establishment of models suitable to study these aspects and to evaluate for exogenous carcinogens and cancer treatment modalities.

5.1 Mouse models

For decades, the mouse has served as an important model to study skin carcinogenesis. Skin cancer is mostly induced by carcinogens—the two-stage skin carcinogenesis protocol—originating from the Berenblum/Mottram experiment [49]. In this model, skin cancer is based on the induction of ras mutations by the application of the carcinogen Dimethylbenzanthracene (DMBA) followed by long-term promotion with the tumor promoter 12-O-tetradecanoylphorbol-13-acetate (TPA). This skin cancer model is used to unravel the biological and genetic events and to determine factors that affect the process of skin carcinogenesis. Similarly, studies concerning the nature of the skin cancer-initiating cell, their malignant progression and metastasis, and the role of inflammation in this tumor type were successfully performed (for review, see [50]). As skin cancer risk is closely correlated with UV exposure, more recently, hairless mouse models have been utilized more frequently to induce skin carcinomas. Daily exposure to UVB radiation is predicted to provide a similar stimulus for tumor induction as UV does in human skin cancer (see, e.g., [51]).

Undoubtedly, mouse and human differ significantly, and this is particularly relevant for the anatomy, physiology, and regulation of the skin and skin cancer; as best expressed by Khavari, "the skin is illustrative of these differences" [52]. One of the most important differences is tissue architecture. The epithelium in not only fur-coated but also largely furless nude mice is characterized by an extensive and dense distribution of hair follicles, even when being defective as in the case of nude mice. They undergo regular synchronized cycles, that is, are frequently activated and thereby are likely to contribute to the epithelial pool that is the target of skin cancer initiation. In the human skin, hair follicles are sparse and replicate asynchronously and infrequently, thus making the interfollicular epidermis (IFE) the major target of skin cancer initiation. With the differences in architecture and regeneration time of the IFE, important differences in regulatory networks also exist between the mouse and human skin, including the fact that the identity of the human IFE stem cells is still elusive [52–54]. Similar differences account for the connective tissue, the dermis, of the mouse and human skin. As the interaction of the tumor cells with their environment (epidermal-dermal interaction) is increasingly appreciated [55], these profound cross species differences strongly demand for experimental models that also recapitulate the stromal component of the human skin more accurately.

Despite all differences/diversities, there is no question that genetic mouse models are invaluable and that, in its complexity, certain questions can yet still only be addressed in mouse models.

Accordingly, Kretzschmar and colleagues showed that the activation of β-catenin in different stem cells of the murine hair follicle (LGR5-, LRIG1-, and LGR6-positive epidermal stem cells) elicited different responses concerning the formation of ectopic hair follicles and therefrom derived tumors. They also demonstrated that the different stem cells caused distinct changes in the underlying dermis, concluding that compartmentalization of epidermal stem is responsible for tumor heterogeneity [56]. Similarly, important is the question concerning the cell of origin for BCCs and SCCs. Utilizing inducible Cre drivers to delete Ptch1, the gene known as being causal for BCC development, in different cell compartments of the mouse skin, Peterson et al. demonstrated that different stem cell populations from the hair follicle rapidly develop BCC-like tumors, while this was not the case with stem cells from the IFE ([57] and references therein). By determining the role of Wnt regulation in the different murine epidermal stem cell populations, Kretzschmar et al. showed "that different stem cells are associated with different tumor types and stromal responses" and postulated that "compartmentalization of epidermal stem cells underlies tumor heterogeneity" [56].

Xenograft mouse models are similarly indispensable. Immunodeficient mice can host human tumor cells, thus enabling the identification and functional characterization of genetic changes in the human skin cancer cells in a complex in vivo environment, thus allowing to dissect, for example, their role in angiogenic or inflammatory attraction. This said, we demonstrated that earlier unraveling overexpression of thrombospondin-1 acts as an important tumor suppressive mechanism for human skin cancer by halting tumor vascularization [58]. Furthermore, new humanized mouse models were established in which murine recipients are engrafted with human hematopoietic stem cells that develop into a functional human immune system. As these humanized mice can further be engrafted with other human tissues, including the skin and skin SCCs, they promise significant improvements in studying the role of immunity and cancer (for review, see [59]).

5.2 Human models

5.2.1 2D models; tumor-derived and genetically modified cell lines

The aim of understanding and functionally characterizing human skin cancer was first explored by establishing cell lines derived from skin cSCCs [60–64]. Later, in vitro-derived models, such as the HaCaT in vitro skin carcinogenesis model, were developed through selective genetic modification to recapitulate the multistep genetic development of skin cancer [65–70]. Of extreme value for skin cancer research is also the "MET model," comprising cell lines from the primary tumor, recurrence, and metastasis from the same patient [61,62]. These cells proved an excellent pendant to the frequently used HaCaT model as the MET cells are wild type for the tumor suppressor p53 and thus allow the dissection of p53-dependent versus p53-independent circuits in skin cancer. Of note, the origin of A431 cells, often referred to and commercially distributed as a human cutaneous epidermoid carcinoma cell line (Sigma Aldrich), is likely of vulvar origin (squamous cell carcinoma, NCIt: C4052 Cellosaurus) and thus may cause misleading interpretations when compared with and taken as a skin cancer cell.

5.2.2 3D models

Spheroids

While conventional 2D in vitro cultures have served as the basis for biological, biochemical, and molecular genetic discoveries and will remain an important tool, it is apparent that the cultivation of cells in monolayer cultures remains restrictive. Tissue organization not only requires a very specific overall regulatory network but also moves individual cells into different regulatory states. This similarly accounts for normal and cancerous tissues, creating great recognition for the need of 3D tissue-like models. The most frequent approach that particularly aims at recapitulating in vivo growth conditions of cancer is the spheroid models. As reviewed by Ishiguro et al. [71], spheroid models include tumor-derived organoids, tumor-derived spheroids, organotypic multicellular spheroids, and multicellular tumor spheroids—based on the source of the tumor cells, execution protocol, and timing. Hereby, the uniqueness of tumor-derived spheroids as perfect tools for the enrichment of cancer stem cells (CSCs) or cells with stem-cell-related characteristics is frequently emphasized. The hypothesis that CSCs are the basis for tumorigenicity and chemoresistance has quickened the use of tumor-derived spheroids for the research of many tumor entities, including cSCC [72], with the hope that they may prove instrumental for high-throughput screening approaches including the identification of regulatory cues that are drug targetable.

Organotypic cultures

However, spheroids do not fully recapitulate the three-dimensional (3D) organization of cells in a given tissue or tumor and, in particular, are missing the proper environment with the cells and ECM characteristic for the tissues and organs. Consequently, there is still a large gap between our detailed knowledge of subcellular processes and the understanding of mammalian biology at the tissue and organ level.

For the skin, the development of organotypic models followed the important work of Bell et al., which introduced the dermal collagen model, with incorporated fibroblasts as a dermal equivalent and the keratinocytes forming an epidermis on top (Bell et al. [73] and more extensively reviewed by Parenteau et al. [74]). Thereafter, many groups have been able to develop and use these models to, for example, study the formation of the basal membrane [75,76]. These models permitted the further understanding of the role of fibroblasts in epidermal regeneration and contributed substantially to the understanding of the development of the skin barrier [77–80]. To avoid some of the problems with collagen-based gels, like short life span and shrinkage, approaches to reinforce the models with scaffolds that provide more stability and alternate gel forms that offer a better environment for the fibroblasts to produce their own ECM have been explored [54,81–84]. The aim was to make the models more authentic and long-lived.

Organotypic skin cancer models

Human organotypic cancer models are usually constructed with the same basis as nonpathological models. Some will only focus on one tissue, for example, epidermis, and others will be more advanced and include different cell types, that is, are complemented with a stromal component. Thereby, the model's mesenchymal compartment either is based on an acellular gel made from animal or synthetic ECM proteins, the most common one being collagen I, or will include fibroblasts embedded in the gel. Alternatively, the mesenchymal cells make their own ECM, the so-called "cell-derived or fibroblast-derived" matrix, in order to study the function of tissue-tissue interaction in a more biologically accurate form [85-89]. One crucial aspect, particularly for cancer models, is to produce an oxygen gradient that resembles that seen in tumors, where the cycle of growth and vascularization is produced by the hypoxic center of the cancer mass. In scaffold-based systems, this can be achieved by increasing the density of collagen [90]. Here, the level of oxygen can easily be manipulated by changing the matrix density and/or cell amount. One could also reverse angiogenesis by means of perfusion by incorporating degradable phosphate-based glass fibers (PGFs) [91]. Another cancer model with a documented tumor hypoxia-like environment was made with a synthetic scaffold-based copolymers and oral squamous carcinoma cells [92].

Epithelia cancer models

Many organs in our body have similar tissue structures, one of them being epithelia, which means that the 3D in vitro model approaches that are used or developed for other epithelial tissues can be used or adapted to produce skin in vitro models and vice versa. The intestine is one of those organs. As for skin epithelia, the quest for finding an in vitro testing model and developing personalized cancer treatment approaches has given researchers a broad spectrum of in vitro techniques to choose from, from the simplest "organoids" to more sophisticated alternatives like "organ-on-a-chip" [93].

In epithelial models of colon cancer, the aim has been to target the polarization of the cells. This was approached by using a microporous filter membrane that allowed for an apical side, with a single polarized cell layer containing microvilli, and a basolateral side that either can be the source of nutrients and factors from media or, in the case of a 3D coculture, can "mimic" the living mucosa by including immune cells or myofibroblasts. By turning the polarized epithelial colon cells upside down and adding cells that are usually found in the loose connective tissue surrounding the epithelial tissue, researchers have improved the cell-cell interaction between different compartments, like immune cells, and have formed more in vivo-like models [94–97].

Another cancer type that has been modeled for quite some time is breast cancer. Starting in the 1970s, researchers have been investigating what we now take for granted in the field of in vitro modeling, like the use of different matrices as scaffolds and the effects of ECM components on the tumor or tumor cells [98,99]. Other important work showed the crucial role of stromal cells for cancer models [98,100–102], which applies for any model where the goal is to understand the tissue or organ-specific effects.

Skin cancer models

During the last decade, it has also become more common to utilize in vitro skin cancer models to answer questions regarding skin cancer biology, cancer progression, and treatment for cSCC. Following the footsteps of early cSCC in vitro models [60], other significant works were published at the beginning of this decade. In 2009, Commandeur and coworkers described and compared two existing skin-modeling techniques, an explant model and a skin equivalent model with SCC cell lines (SCC12B2 and SCC13) cocultured with a collagen scaffold populated with normal human fibroblasts [103]. Importantly, these models can be performed using defined skin model media containing L-serine and lipid supplements, among other factors [78]. This work clearly pointed out how the protein expression pattern of these two models differed and, even more so, how they diverged from what was seen in skin SCC biopsies. Furthermore, this study clearly demonstrated that even the explant tissue changed its characteristics from in vivo. Another aspect that needs to be taken into consideration for an explant model is the fact that several biopsies would be needed, which may be a not only ethical but also practical problem. In 2015, Danso et al. reported on the establishment of second- and third-generation explants, and indeed, their concept worked. However, the further away the explants were made, the more the new-generation models differed from the original [104].

When it comes to cell-line-based models, one argument is that cell lines have generally been artificially selected and represent a more homogenous view of SCC. Thus, the aspect of tumor heterogeneity is poorly recapitulated. Therefore, if the aim is to understand the mechanisms of SCC, it is of importance to include not only one cell line model. In the study mentioned above, two distinct and individual patterns could be observed, one invasive with SCC12B2 cells and the other leading to the detachment of the epidermis from the dermis with SCC13 cells [103].

A year later, Ridky and coworkers used the technique of deepidermized dermis (DED) from normal human skin. The key finding of this study was that the DED could maintain an intact basement membrane and being repopulated with foreskin fibroblasts produced a dermal equivalent, with a matrix similar to that seen during initiation of epithelial tumor cell invasion [105]. For this study, neoplastic tissues were generated from different kinds of normal epithelia, including the skin, by transforming the cells with the ras oncogene and cyclin-dependent kinase-4 (Cdk4). The neoplastic epithelial cells invaded into the stroma within 1 week. These models gave a rapid assessment tool of anticancer drugs and showed the effect that gene expression signature had on neoplastic invasion.

The importance of stromal cancer-associated fibroblasts (CAF) was addressed by Commandeur et al. [106]. This study showed that a collagen-based dermal compartment including CAFs indeed affected the way the epidermal cells (NHEK, SCC12B2, and SCC-13) behaved. By examining proliferation, differentiation, and invasion of the tumor cells and collagen fiber contraction in the stroma, they concluded that the CAFs in this collagen cancer model increased contractility of the dermal equivalent and, among others, contributed to a decreased proliferative activity of the epidermal cells, increased terminal differentiation, increased invasion of the SCC-12B2 cells, and increased dermal-epidermal detachment of the SCC-13 cancer cell line. Using the same model, the authors studied overexpression (addition of EGF) or inhibition (addition of tyrosine kinase inhibitor, erlotinib) of EGFR [107]. They concluded that erlotinib treatment produced a thinner epidermal compartment in both models and additionally showed epidermal scaling in one of the models, something that is a common side effect of this drug in vivo in the form of xerosis, giving insight in how these models can be used to evaluate the effects of drugs.

Another aspect of in vitro skin model application is to evaluate phototoxicity and sunscreen properties [108], thereby giving the industry a tool for evaluating new protective substances. A lot of work has been focused on the effects of UV radiation on normal human skin [109–114], which is a relevant point when trying to find what causes the damage that will persist and with time develop into cancer. UV-exposed models have also been useful to understand the role of key protein components of UV protection. Accordingly, a knockdown in vitro model of filaggrin showed that this protein is crucial for UVB protection [115].

Immunosuppression in transplant patients is strongly associated with an increase in the development of cSCC (for review, see Mittal et al. 2017) [116]. Thus, organotypic cultures have beneficially been used (besides hairless mice) to address the immediate role of different immunosuppressive drugs and additionally determined the role of acute UVB radiation on epidermal regeneration, proliferation, and apoptosis under "physiological" conditions [117]. Thereby, the authors could show that the immunosuppressive drug cyclosporin A inhibited apoptosis after UV and, on the contrary, rapamycin (m-TOR inhibitor) increased the apoptotic effect. Another key argument favoring work in organotypic models is that they give us an opportunity to manipulate and examine the role of different biological key players or discover new ones. By using this approach, it could, for instance, be shown that stromal Wnt activation caused the fibroblasts to express chemokines such as CCL2 and IL-8 and that these in turn act on the keratinocytes and stimulated proliferation of the HaCaT cells [46]. SCC models have also been used to understand a relatively unknown proteins' mode of action in the skin. A recent example is the protein flightless I (FLII), an actin remodeling protein. The authors showed its effect on cell adhesion, proliferation, and migration in the normal skin and its role in SCC progression by using the human MET-1 cell line cultured on a collagen: matrigel-based fibroblast-containing dermal equivalent [118].

All models that have been covered in this section have included a biological- or synthetic-based platform as the dermal compartment. However, these are limited in their sustainability due to either their composition or their short life span. Following the seminal work of Ahlfors and Billiar [119], many groups have used the approach of cell-derived matrix as the basis for a dermal equivalent [86–88,119–121]. Modifying the original protocol, we developed a fibroblast-derived matrix skin cancer model [85]. This model proved suitable to support normal epidermal regeneration for >20 weeks and, importantly, recapitulated invasion of tumor cells in its best sense. Starting from a superficial epithelium, the tumor cells rupture the BM, degrade the underlying collagen, and penetrate the dermal equivalent (Fig. 1). Moreover, each tumor cell line showed a unique and characteristic pattern of invasion, thus making this a relevant tool for studying early changes and mechanisms required for the initiation of tumor cell invasion and studying the role of tumor-stroma interaction in cancer progression. Due to its extended viability, this model also allows for chronic exposure of chemicals or UV radiation, thereby predestining it not only for experimental approaches but also for the use as preclinical models.

5.2.3 The future of skin cancer models

The progress in the field of 3D biology and bioengineering and the need to find human disease models in translational medicine to better understand a disease or to find new treatments have made the public and pharmacological and biomedical companies more open to the possibility to expand their use of 3D model. And although the existing models already have given us a lot of insights into the biology and chemistry behind tissues and disease, we still need to strive to make the models more biologically accurate and to include them in preclinical trials on a regular basis.

It should be the basis of any attempt to improve existing or design new models and modeling techniques to recapitulate the in vivo situation. One step toward this goal is clearly the complementation of the existing models with additional cell types. Important aspects to further develop skin cancer models are immune-related questions. The first approaches are promising. Based on collagen-glycosaminoglycan-chitosan dermal substrates, Dezutter-Dambuyant et al. were the first to present a living dermal equivalent containing fibroblasts, endothelial cells, and dendritic cells to study the differentiation of interstitial DC in a dermal equivalent [122]. The same group now presents an immunocompetent reconstructed skin model maintained in a microfluidic-based cell culture system [123]. Macrophages were incorporated into a collagen-based OTC model, which proved suitable to modulate macrophage activation toward a tumor-supporting phenotype [124]. Thus, incorporating the different kinds of immune cells into the different OTC models and determining their controlled distribution and participation in skin cancer progression will be the next challenge.

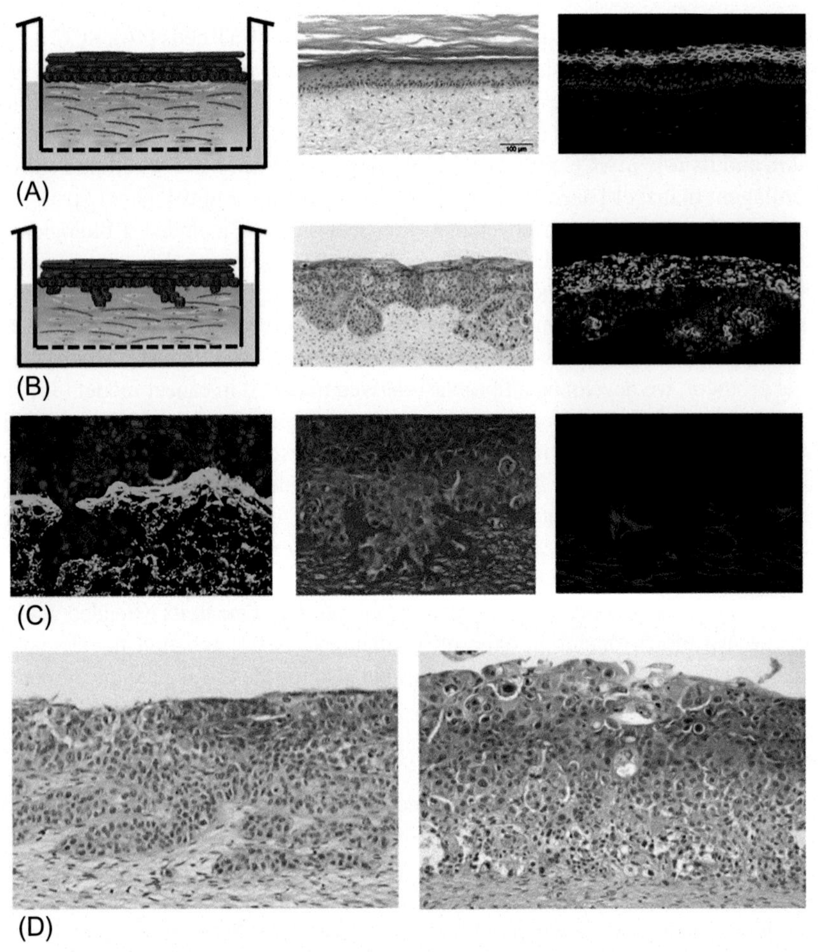

Fig. 1 Characterization of the 3D in vitro skin cancer model. *(A) 3D organotypic culture model*: schematic representation of the 3D organotypic culture (*left*); H&*E*-stained histological section of the skin equivalent (*middle*), immunofluorescence staining demonstrating the desmosomal network in the epidermis (staining for desmocollin); and the integrity of the basement membrane (BM) (staining collagen type IV) (*right*). *(B) 3D skin cancer invasion model*: schematic representation of the OTC (*left*); H&E-stained histological section representing invasively growing keratinocytes (*middle*), immunofluorescence staining demonstrating maintenance of the desmosomal network in the epidermis (staining for plakoglobin); and the in part disrupted BM (staining for collagen type IV) (*right*). *(C) Characterization of the invasion process of the skin cancer modes*: immunofluorescence staining demonstrates the invasion of the epithelial cells (stained for E-cadherin) into the dermal equivalent (stained for collagen type III) (left). Picrosirius red staining (middle) with its corresponding polarization (right) demonstrates collagen organization in the dermal equivalent, showing the loss of collagen fibers around and accumulation of collagen fibers between the invasive epithelial strands. *(D) Demonstration of tumor-cell-specific invasion.* H&E staining of histological sections from two different tumor cell lines, each shows a unique and highly reproducible invasion phenotype, supporting the relevance of this 3D organotypic model as an in vitro skin cancer model.

Vascularization is also an important trait that needs to be addressed in more depth. The first endothelialized skin model based on a chitosan cross-linked collagen-glycosaminoglycan scaffold has enabled the formation of capillary-like structures, which showed several morphological characteristics associated with the microvasculature in vivo [125]. Similarly, our approach of scaffold-based endothelialized OTCs allowed for the self-assembly and maturation of a microvasculature (Böhnke et al., in preparation). Based on a porcine vascular scaffold, a first bioreactor-based system has been described [126]. Besides these promising approaches, no model has yet proved suitable as a tumor model, and we are still far away from connecting them to a dynamic flow.

The first skin-on-a-chip model has been described [127], thereby paving the way to overcome the fluidity problems of the normal skin organotypic culture models. Furthermore, by improving the complexity, that is, by integrating this into the "human-on-a-chip" model, a new era of interactive studies will become possible. As this will clearly be extended to cancer cells, models are to be expected that, despite their great complexity, will feature reproducibility and the potential to learn about tissue interactive regulatory loops.

Presently, the most rapidly evolving approach to improve complexity, reproducibility, and in particular the quantity of specimen is the technique of 3D bioprinting. An automated cell printing system was used to bioprint a 3D coculture model of ovarian cancer cells and normal fibroblasts in Matrigel, thereby allowing for a highly controlled cell density and cell-cell distance [128]. First approaches for 3D bioprinted skin have also been described [129,130], though the quality of the epidermal tissue shown so far still leaves a lot to be desired. Furthermore, skin cancer models with their complex microenvironment (cells and ECM) have still to be shown. However, from the rapid advances in the field of bioprinting, we can expect that preclinical skin cancer models will be available soon (for review, see also [131]).

Independent on what the in vitro models will be used for, their future use and development lie in their flexibility, reliability, and reproducibility. Particularly in the field of personalized medicine, that is, dermatooncology, modeling techniques can play an important role in establishing patient-specific therapies also including clinical management methods and assessment, closing the gap between bench and clinic even more.

Finally, it should be mentioned that the question of biological authenticity in the models, be it tissue-specific, organ-specific, or organ-system-specific, will start to become more and more relevant due to the increasing understanding of how important the human body interplay between microenvironments, tissues, and organs is in health and disease.

References

[1] Leiter U, Eigentler T, Garbe C. Epidemiology of skin cancer. Adv Exp Med Biol 2014;810:120–40.

[2] Perez HC, Benavides X, Perez JS, Pabon MA, Tschen J, Maradei-Anaya SJ, et al. Basic aspects of the pathogenesis and prevention of non-melanoma skin cancer in solid organ transplant recipients: a review. Int J Dermatol 2017;56(4):370–8.

[3] Karagas MR, Stannard VA, Mott LA, Slattery MJ, Spencer SK, Weinstock MA. Use of tanning devices and risk of basal cell and squamous cell skin cancers. J Natl Cancer Inst 2002;94(3):224–6.
[4] Zuba EB, Francuzik W, Malicki P, Osmola-Mankowska A, Jenerowicz D. Knowledge about ultraviolet radiation hazards and tanning behavior of cosmetology and medical students. Acta Dermatovenerol Croat 2016;24(1):73–7.
[5] Grewe M. Chronological ageing and photoageing of dendritic cells. Clin Exp Dermatol 2001;26(7):608–12.
[6] Ingvar A, Smedby KE, Lindelof B, Fernberg P, Bellocco R, Tufveson G, et al. Immunosuppressive treatment after solid organ transplantation and risk of post-transplant cutaneous squamous cell carcinoma. Nephrol Dial Transplant 2010;25(8):2764–71.
[7] Abikhair M, Mitsui H, Yanofsky V, Roudiani N, Ovits C, Bryan T, et al. Cyclosporine a immunosuppression drives catastrophic squamous cell carcinoma through IL-22. JCI Insight 2016;1(8):e86434.
[8] McBride P, Olsen CM, Green AC. Tobacco smoking and cutaneous squamous cell carcinoma: a 16-year longitudinal population-based study. Cancer Epidemiol Biomark Prev 2011;20(8):1778–83.
[9] Dusingize JC, Olsen CM, Pandeya NP, Subramaniam P, Thompson BS, Neale RE, et al. Cigarette smoking and the risks of basal cell carcinoma and squamous cell carcinoma. J Investig Dermatol 2017;137(8):1700–8.
[10] Tang N, Ratner D. Managing cutaneous side effects from targeted molecular inhibitors for melanoma and nonmelanoma skin cancer. Dermatol Surg 2016;42(Suppl 1):S40–8.
[11] Li YY, Hanna GJ, Laga AC, Haddad RI, Lorch JH, Hammerman PS. Genomic analysis of metastatic cutaneous squamous cell carcinoma. Clin Cancer Res 2015;21(6):1447–56.
[12] Mohan SV, Chang J, Li S, Henry AS, Wood DJ, Chang AL. Increased risk of cutaneous squamous cell carcinoma after Vismodegib therapy for basal cell carcinoma. JAMA Dermatol 2016;152(5):527–32.
[13] Maughan BL, Bailey E, Gill DM, Agarwal N. Incidence of immune-related adverse events with program death receptor-1- and program death receptor-1 ligand-directed therapies in genitourinary cancers. Front Oncol 2017;7:56.
[14] Flaherty KT, Puzanov I, Kim KB, Ribas A, McArthur GA, Sosman JA, et al. Inhibition of mutated, activated BRAF in metastatic melanoma. N Engl J Med 2010;363(9):809–19.
[15] Erfan G, Puig S, Carrera C, Arance A, Gaba L, Victoria I, et al. Development of cutaneous toxicities during selective anti-BRAF therapies: preventive role of combination with MEK inhibitors. Acta Derm Venereol 2017;97(2):258–60.
[16] El Ghissassi F, Baan R, Straif K, Grosse Y, Secretan B, Bouvard V, et al. A review of human carcinogens—part D: radiation. Lancet Oncol 2009;10(8):751–2.
[17] Johnson TM, Rowe DE, Nelson BR, Swanson NA. Squamous cell carcinoma of the skin (excluding lip and oral mucosa). J Am Acad Dermatol 1992;26(3 Pt 2):467–84.
[18] Bonerandi JJ, Beauvillain C, Caquant L, Chassagne JF, Chaussade V, Clavere P, et al. Guidelines for the diagnosis and treatment of cutaneous squamous cell carcinoma and precursor lesions. J Eur Acad Dermatol Venereol 2011;25(Suppl 5):1–51.
[19] Cheng J, Yan S. Prognostic variables in high-risk cutaneous squamous cell carcinoma: a review. J Cutan Pathol 2016;43(11):994–1004.
[20] Thompson AK, Kelley BF, Prokop LJ, Murad MH, Baum CL. Risk factors for cutaneous squamous cell carcinoma recurrence, metastasis, and disease-specific death: a systematic review and meta-analysis. JAMA Dermatol 2016;152(4):419–28.

[21] Leufke C, Leykauf J, Krunic D, Jauch A, Holtgreve-Grez H, Bohm-Steuer B, et al. The telomere profile distinguishes two classes of genetically distinct cutaneous squamous cell carcinomas. Oncogene 2014;33(27):3506–18.
[22] Yilmaz AS, Ozer HG, Gillespie JL, Allain DC, Bernhardt MN, Furlan KC, et al. Differential mutation frequencies in metastatic cutaneous squamous cell carcinomas versus primary tumors. Cancer 2017;123(7):1184–93.
[23] Kusters-Vandevelde HV, Van Leeuwen A, Verdijk MA, de Koning MN, Quint WG, Melchers WJ, et al. CDKN2A but not TP53 mutations nor HPV presence predict poor outcome in metastatic squamous cell carcinoma of the skin. Int J Cancer 2010;126(9):2123–32.
[24] Blokx WA, Ruiter DJ, Verdijk MA, de Wilde PC, Willems RW, de Jong EM, et al. INK4-ARF and p53 mutations in metastatic cutaneous squamous cell carcinoma: case report and archival study on the use of Ink4a-ARF and p53 mutation analysis in identification of the corresponding primary tumor. Am J Surg Pathol 2005;29(1):125–30.
[25] Gallagher SJ, Kefford RF, Rizos H. The ARF tumour suppressor. Int J Biochem Cell Biol 2006;38(10):1637–41.
[26] Kanellou P, Zaravinos A, Zioga M, Stratigos A, Baritaki S, Soufla G, et al. Genomic instability, mutations and expression analysis of the tumour suppressor genes p14(ARF), p15(INK4b), p16(INK4a) and p53 in actinic keratosis. Cancer Lett 2008;264(1):145–61.
[27] South AP, Purdie KJ, Watt SA, Haldenby S, den Breems NY, Dimon M, et al. NOTCH1 mutations occur early during cutaneous squamous cell carcinogenesis. J Investig Dermatol 2014;134(10):2630–8.
[28] Wang NJ, Sanborn Z, Arnett KL, Bayston LJ, Liao W, Proby CM, et al. Loss-of-function mutations in Notch receptors in cutaneous and lung squamous cell carcinoma. Proc Natl Acad Sci U S A 2011;108(43):17761–6.
[29] Brash DE. Roles of the transcription factor p53 in keratinocyte carcinomas. Br J Dermatol 2006;154(Suppl 1):8–10.
[30] Boukamp P. Non-melanoma skin cancer: what drives tumor development and progression? Carcinogenesis 2005;26(10):1657–67.
[31] Al-Rohil RN, Tarasen AJ, Carlson JA, Wang K, Johnson A, Yelensky R, et al. Evaluation of 122 advanced-stage cutaneous squamous cell carcinomas by comprehensive genomic profiling opens the door for new routes to targeted therapies. Cancer 2016;122(2):249–57.
[32] Oberholzer PA, Kee D, Dziunycz P, Sucker A, Kamsukom N, Jones R, et al. RAS mutations are associated with the development of cutaneous squamous cell tumors in patients treated with RAF inhibitors. J Clin Oncol 2012;30(3):316–21.
[33] Su F, Viros A, Milagre C, Trunzer K, Bollag G, Spleiss O, et al. RAS mutations in cutaneous squamous-cell carcinomas in patients treated with BRAF inhibitors. N Engl J Med 2012;366(3):207–15.
[34] Martincorena I, Campbell PJ. Somatic mutation in cancer and normal cells. Science 2015;349(6255):1483–9.
[35] Mauerer A, Herschberger E, Dietmaier W, Landthaler M, Hafner C. Low incidence of EGFR and HRAS mutations in cutaneous squamous cell carcinomas of a German cohort. Exp Dermatol 2011;20(10):848–50.
[36] Canueto J, Cardenoso E, Garcia JL, Santos-Briz A, Castellanos-Martin A, Fernandez-Lopez E, et al. Epidermal growth factor receptor expression is associated with poor outcome in cutaneous squamous cell carcinoma. Br J Dermatol 2017;176(5):1279–87.
[37] Lewis CM, Glisson BS, Feng L, Wan F, Tang X, Wistuba II, et al. A phase II study of gefitinib for aggressive cutaneous squamous cell carcinoma of the head and neck. Clin Cancer Res 2012;18(5):1435–46.

[38] Toll A, Salgado R, Yebenes M, Martin-Ezquerra G, Gilaberte M, Baro T, et al. MYC gene numerical aberrations in actinic keratosis and cutaneous squamous cell carcinoma. Br J Dermatol 2009;161(5):1112–8.

[39] Bruegger C, Kempf W, Spoerri I, Arnold AW, Itin PH, Burger B. MicroRNA expression differs in cutaneous squamous cell carcinomas and healthy skin of immunocompetent individuals. Exp Dermatol 2013;22(6):426–8.

[40] Yu X, Li Z. The role of miRNAs in cutaneous squamous cell carcinoma. J Cell Mol Med 2016;20(1):3–9.

[41] Lohcharoenkal W, Harada M, Loven J, Meisgen F, Landen NX, Zhang L, et al. MicroRNA-203 inversely correlates with differentiation grade, targets c-MYC, and functions as a tumor suppressor in cSCC. J Investig Dermatol 2016;136(12):2485–94.

[42] Rudnick EW, Thareja S, Cherpelis B. Oral therapy for nonmelanoma skin cancer in patients with advanced disease and large tumor burden: a review of the literature with focus on a new generation of targeted therapies. Int J Dermatol 2016;55(3):249–58. quiz 256, 258.

[43] Heath CH, Deep NL, Nabell L, Carroll WR, Desmond R, Clemons L, et al. Phase 1 study of erlotinib plus radiation therapy in patients with advanced cutaneous squamous cell carcinoma. Int J Radiat Oncol Biol Phys 2013;85(5):1275–81.

[44] Bissell MJ, Hines WC. Why don't we get more cancer? A proposed role of the microenvironment in restraining cancer progression. Nat Med 2011;17(3):320–9.

[45] Nissinen L, Farshchian M, Riihila P, Kahari VM. New perspectives on role of tumor microenvironment in progression of cutaneous squamous cell carcinoma. Cell Tissue Res 2016;365(3):691–702.

[46] Sobel K, Tham M, Stark HJ, Stammer H, Pratzel-Wunder S, Bickenbach JR, et al. Wnt-3a-activated human fibroblasts promote human keratinocyte proliferation and matrix destruction. Int J Cancer 2015;136(12):2786–98.

[47] Kim M, Murrell DF. Update on the pathogenesis of squamous cell carcinoma development in recessive dystrophic epidermolysis bullosa. Eur J Dermatol 2015;25(Suppl 1):30–2.

[48] Mittapalli VR, Madl J, Loffek S, Kiritsi D, Kern JS, Romer W, et al. Injury-driven stiffening of the dermis expedites skin carcinoma progression. Cancer Res 2016;76(4):940–51.

[49] Schweizer J, Loehrke H, Goerttler K. Transmaternal modification of the Berenblum/Mottram experiment in mice. Bull Cancer 1978;65(3):265–70.

[50] Huang PY, Balmain A. Modeling cutaneous squamous carcinoma development in the mouse. Cold Spring Harb Perspect Med 2014;4(9):a013623.

[51] de Gruijl FR, Rebel H. Early events in UV carcinogenesis—DNA damage, target cells and mutant p53 foci. Photochem Photobiol 2008;84(2):382–7.

[52] Khavari PA. Modelling cancer in human skin tissue. Nat Rev Cancer 2006;6(4):270–80.

[53] Boehnke K, Falkowska-Hansen B, Stark HJ, Boukamp P. Stem cells of the human epidermis and their niche: composition and function in epidermal regeneration and carcinogenesis. Carcinogenesis 2012;33(7):1247–58.

[54] Muffler S, Stark HJ, Amoros M, Falkowska-Hansen B, Boehnke K, Buhring HJ, et al. A stable niche supports long-term maintenance of human epidermal stem cells in organotypic cultures. Stem Cells 2008;26(10):2506–15.

[55] Bhat R, Bissell MJ. Of plasticity and specificity: dialectics of the microenvironment and macroenvironment and the organ phenotype. Wiley Interdiscip Rev Dev Biol 2014;3(2):147–63.

[56] Kretzschmar K, Weber C, Driskell RR, Calonje E, Watt FM. Compartmentalized epidermal activation of beta-catenin differentially affects lineage reprogramming and underlies tumor heterogeneity. Cell Rep 2016;14(2):269–81.

[57] Peterson SC, Eberl M, Vagnozzi AN, Belkadi A, Veniaminova NA, Verhaegen ME, et al. Basal cell carcinoma preferentially arises from stem cells within hair follicle and mechanosensory niches. Cell Stem Cell 2015;16(4):400–12.
[58] Bleuel K, Popp S, Fusenig NE, Stanbridge EJ, Boukamp P. Tumor suppression in human skin carcinoma cells by chromosome 15 transfer or thrombospondin-1 overexpression through halted tumor vascularization. Proc Natl Acad Sci U S A 1999;96(5):2065–70.
[59] Walsh NC, Kenney LL, Jangalwe S, Aryee KE, Greiner DL, Brehm MA, et al. Humanized mouse models of clinical disease. Annu Rev Pathol 2017;12:187–215.
[60] Boukamp P, Tilgen W, Dzarlieva RT, Breitkreutz D, Haag D, Riehl RK, et al. Phenotypic and genotypic characteristics of a cell line from a squamous cell carcinoma of human skin. J Natl Cancer Inst 1982;68(3):415–27.
[61] Popp S, Waltering S, Holtgreve-Grez H, Jauch A, Proby C, Leigh IM, et al. Genetic characterization of a human skin carcinoma progression model: from primary tumor to metastasis. J Investig Dermatol 2000;115(6):1095–103.
[62] Proby CM, Purdie KJ, Sexton CJ, Purkis P, Navsaria HA, Stables JN, et al. Spontaneous keratinocyte cell lines representing early and advanced stages of malignant transformation of the epidermis. Exp Dermatol 2000;9(2):104–17.
[63] Rheinwald JG, Beckett MA. Tumorigenic keratinocyte lines requiring anchorage and fibroblast support cultured from human squamous cell carcinomas. Cancer Res 1981;41(5):1657–63.
[64] Tilgen W, Boukamp P, Breitkreutz D, Dzarlieva RT, Engstner M, Haag D, et al. Preservation of morphological, functional, and karyotypic traits during long-term culture and in vivo passage of two human skin squamous cell carcinomas. Cancer Res 1983;43(12 Pt 1):5995–6011.
[65] Boukamp P, Peter W, Pascheberg U, Altmeier S, Fasching C, Stanbridge EJ, et al. Step-wise progression in human skin carcinogenesis in vitro involves mutational inactivation of p53, rasH oncogene activation and additional chromosome loss. Oncogene 1995;11(5):961–9.
[66] Boukamp P, Petrussevska RT, Breitkreutz D, Hornung J, Markham A, Fusenig NE. Normal keratinization in a spontaneously immortalized aneuploid human keratinocyte cell line. J Cell Biol 1988;106(3):761–71.
[67] Boukamp P, Stanbridge EJ, Foo DY, Cerutti PA, Fusenig NE. c-Ha-ras oncogene expression in immortalized human keratinocytes (HaCaT) alters growth potential in vivo but lacks correlation with malignancy. Cancer Res 1990;50(9):2840–7.
[68] Buschke S, Stark HJ, Cerezo A, Pratzel-Wunder S, Boehnke K, Kollar J, et al. A decisive function of transforming growth factor-beta/Smad signaling in tissue morphogenesis and differentiation of human HaCaT keratinocytes. Mol Biol Cell 2011;22(6):782–94.
[69] Cerezo A, Kalthoff H, Schuermann M, Schafer B, Boukamp P. Dual regulation of telomerase activity through c-Myc-dependent inhibition and alternative splicing of hTERT. J Cell Sci 2002;115(Pt 6):1305–12.
[70] Lehman TA, Modali R, Boukamp P, Stanek J, Bennett WP, Welsh JA, et al. p53 mutations in human immortalized epithelial cell lines. Carcinogenesis 1993;14(5):833–9.
[71] Ishiguro T, Ohata H, Sato A, Yamawaki K, Enomoto T, Okamoto K. Tumor-derived spheroids: relevance to cancer stem cells and clinical applications. Cancer Sci 2017;108(3):283–9.
[72] Adhikary G, Grun D, Kerr C, Balasubramanian S, Rorke EA, Vemuri M, et al. Identification of a population of epidermal squamous cell carcinoma cells with enhanced potential for tumor formation. PLoS One 2013;8(12):e84324.

[73] Bell E, Ehrlich HP, Buttle DJ, Nakatsuji T. Living tissue formed in vitro and accepted as skin-equivalent tissue of full thickness. Science 1981;211(4486):1052–4.

[74] Parenteau NL, Nolte CM, Bilbo P, Rosenberg M, Wilkins LM, Johnson EW, et al. Epidermis generated in vitro: practical considerations and applications. J Cell Biochem 1991;45(3):245–51.

[75] Fleischmajer R, Utani A, MacDonald ED, Perlish JS, Pan TC, Chu ML, et al. Initiation of skin basement membrane formation at the epidermo-dermal interface involves assembly of laminins through binding to cell membrane receptors. J Cell Sci 1998;111(Pt 14):1929–40.

[76] Smola H, Stark HJ, Thiekotter G, Mirancea N, Krieg T, Fusenig NE. Dynamics of basement membrane formation by keratinocyte-fibroblast interactions in organotypic skin culture. Exp Cell Res 1998;239(2):399–410.

[77] Contard P, Bartel RL, Jacobs 2nd L, Perlish JS, MacDonald 2nd ED, Handler L, et al. Culturing keratinocytes and fibroblasts in a three-dimensional mesh results in epidermal differentiation and formation of a basal lamina-anchoring zone. J Investig Dermatol 1993;100(1):35–9.

[78] El Ghalbzouri A, Lamme E, Ponec M. Crucial role of fibroblasts in regulating epidermal morphogenesis. Cell Tissue Res 2002;310(2):189–99.

[79] El Ghalbzouri A, Ponec M. Diffusible factors released by fibroblasts support epidermal morphogenesis and deposition of basement membrane components. Wound Repair Regen 2004;12(3):359–67.

[80] Ponec M, Weerheim A, Kempenaar J, Mulder A, Gooris GS, Bouwstra J, et al. The formation of competent barrier lipids in reconstructed human epidermis requires the presence of vitamin C. J Investig Dermatol 1997;109(3):348–55.

[81] Boehnke K, Mirancea N, Pavesio A, Fusenig NE, Boukamp P, Stark HJ. Effects of fibroblasts and microenvironment on epidermal regeneration and tissue function in long-term skin equivalents. Eur J Cell Biol 2007;86(11–12):731–46.

[82] Cooper ML, Hansbrough JF. Use of a composite skin graft composed of cultured human keratinocytes and fibroblasts and a collagen-GAG matrix to cover full-thickness wounds on athymic mice. Surgery 1991;109(2):198–207.

[83] Murphy GF, Orgill DP, Yannas IV. Partial dermal regeneration is induced by biodegradable collagen-glycosaminoglycan grafts. Lab Investig 1990;62(3):305–13.

[84] Stark HJ, Willhauck MJ, Mirancea N, Boehnke K, Nord I, Breitkreutz D, et al. Authentic fibroblast matrix in dermal equivalents normalises epidermal histogenesis and dermoepidermal junction in organotypic co-culture. Eur J Cell Biol 2004;83(11–12):631–45.

[85] Berning M, Pratzel-Wunder S, Bickenbach JR, Boukamp P. Three-dimensional in vitro skin and skin cancer models based on human fibroblast-derived matrix. Tissue Eng Part C Methods 2015;21(9):958–70.

[86] El Ghalbzouri A, Commandeur S, Rietveld MH, Mulder AA, Willemze R. Replacement of animal-derived collagen matrix by human fibroblast-derived dermal matrix for human skin equivalent products. Biomaterials 2009;30(1):71–8.

[87] Ng YZ, Pourreyron C, Salas-Alanis JC, Dayal JH, Cepeda-Valdes R, Yan W, et al. Fibroblast-derived dermal matrix drives development of aggressive cutaneous squamous cell carcinoma in patients with recessive dystrophic epidermolysis bullosa. Cancer Res 2012;72(14):3522–34.

[88] Pouyani T, Ronfard V, Scott PG, Dodd CM, Ahmed A, Gallo RL, et al. De novo synthesis of human dermis in vitro in the absence of a three-dimensional scaffold. In Vitro Cell Dev Biol Anim 2009;45(8):430–41.

[89] Serebriiskii I, Castello-Cros R, Lamb A, Golemis EA, Cukierman E. Fibroblast-derived 3D matrix differentially regulates the growth and drug-responsiveness of human cancer cells. Matrix Biol 2008;27(6):573–85.

[90] Cheema U, Brown RA, Alp B, MacRobert AJ. Spatially defined oxygen gradients and vascular endothelial growth factor expression in an engineered 3D cell model. Cell Mol Life Sci 2008;65(1):177–86.

[91] Cheema U, Alekseeva T, Abou-Neel EA, Brown RA. Switching off angiogenic signalling: creating channelled constructs for adequate oxygen delivery in tissue engineered constructs. Eur Cell Mater 2010;20:274–80. [discussion 280-271].

[92] Fischbach C, Chen R, Matsumoto T, Schmelzle T, Brugge JS, Polverini PJ, et al. Engineering tumors with 3D scaffolds. Nat Methods 2007;4(10):855–60.

[93] Lancaster MA, Knoblich JA. Organogenesis in a dish: modeling development and disease using organoid technologies. Science 2014;345(6194):1247125.

[94] des Rieux A, Fievez V, Theate I, Mast J, Preat V, Schneider YJ. An improved in vitro model of human intestinal follicle-associated epithelium to study nanoparticle transport by M cells. Eur J Pharm Sci 2007;30(5):380–91.

[95] Fritsch C, Orian-Rousseaul V, Lefebvre O, Simon-Assmann P, Reimund JM, Duclos B, et al. Characterization of human intestinal stromal cell lines: response to cytokines and interactions with epithelial cells. Exp Cell Res 1999;248(2):391–406.

[96] Gullberg E, Leonard M, Karlsson J, Hopkins AM, Brayden D, Baird AW, et al. Expression of specific markers and particle transport in a new human intestinal M-cell model. Biochem Biophys Res Commun 2000;279(3):808–13.

[97] Kerneis S, Caliot E, Stubbe H, Bogdanova A, Kraehenbuhl J, Pringault E. Molecular studies of the intestinal mucosal barrier physiopathology using cocultures of epithelial and immune cells: a technical update. Microbes Infect 2000;2(9):1119–24.

[98] Wang CS, Goulet F, Auger F, Tremblay N, Germain L, Tetu B. Production of bioengineered cancer tissue constructs in vitro: epithelium-mesenchyme heterotypic interactions. In Vitro Cell Dev Biol Anim 2001;37(7):434–9.

[99] Weigelt B, Ghajar CM, Bissell MJ. The need for complex 3D culture models to unravel novel pathways and identify accurate biomarkers in breast cancer. Adv Drug Deliv Rev 2014;69–70:42–51.

[100] Bissell MJ, Radisky D. Putting tumours in context. Nat Rev Cancer 2001;1(1):46–54.

[101] Cunha GR, Hom YK. Role of mesenchymal-epithelial interactions in mammary gland development. J Mammary Gland Biol Neoplasia 1996;1(1):21–35.

[102] Holliday DL, Brouilette KT, Markert A, Gordon LA, Jones JL. Novel multicellular organotypic models of normal and malignant breast: tools for dissecting the role of the microenvironment in breast cancer progression. Breast Cancer Res 2009;11(1):R3.

[103] Commandeur S, de Gruijl FR, Willemze R, Tensen CP, El Ghalbzouri A. An in vitro three-dimensional model of primary human cutaneous squamous cell carcinoma. Exp Dermatol 2009;18(10):849–56.

[104] Danso MO, van Drongelen V, Mulder A, Gooris G, van Smeden J, El Ghalbzouri A, et al. Exploring the potentials of nurture: 2(nd) and 3(rd) generation explant human skin equivalents. J Dermatol Sci 2015;77(2):102–9.

[105] Ridky TW, Chow JM, Wong DJ, Khavari PA. Invasive three-dimensional organotypic neoplasia from multiple normal human epithelia. Nat Med 2010;16(12):1450–5.

[106] Commandeur S, Ho SH, de Gruijl FR, Willemze R, Tensen CP, El Ghalbzouri A. Functional characterization of cancer-associated fibroblasts of human cutaneous squamous cell carcinoma. Exp Dermatol 2011;20(9):737–42.

[107] Commandeur S, van Drongelen V, de Gruijl FR, El Ghalbzouri A. Epidermal growth factor receptor activation and inhibition in 3D in vitro models of normal skin and human cutaneous squamous cell carcinoma. Cancer Sci 2012;103(12):2120–6.
[108] Cohen C, Dossou KG, Rougier A, Roguet R. Episkin: an in vitro model for the evaluation of phototoxicity and sunscreen photoprotective properties. Toxicol in Vitro 1994;8(4):669–71.
[109] Bernerd F, Asselineau D. Successive alteration and recovery of epidermal differentiation and morphogenesis after specific UVB-damages in skin reconstructed in vitro. Dev Biol 1997;183(2):123–38.
[110] Bernerd F, Asselineau D. UVA exposure of human skin reconstructed in vitro induces apoptosis of dermal fibroblasts: subsequent connective tissue repair and implications in photoaging. Cell Death Differ 1998;5(9):792–802.
[111] Bernerd F, Asselineau D. An organotypic model of skin to study photodamage and photoprotection in vitro. J Am Acad Dermatol 2008;58(5 Suppl 2):S155–9.
[112] Marionnet C, Grether-Beck S, Seite S, Marini A, Jaenicke T, Lejeune F, et al. A broad-spectrum sunscreen prevents UVA radiation-induced gene expression in reconstructed skin in vitro and in human skin in vivo. Exp Dermatol 2011;20(6):477–82.
[113] Marionnet C, Pierrard C, Lejeune F, Sok J, Thomas M, Bernerd F. Different oxidative stress response in keratinocytes and fibroblasts of reconstructed skin exposed to non extreme daily-ultraviolet radiation. PLoS One 2010;5(8):e12059.
[114] Nelson D, Gay RJ. Effects of UV irradiation on a living skin equivalent. Photochem Photobiol 1993;57(5):830–7.
[115] Mildner M, Jin J, Eckhart L, Kezic S, Gruber F, Barresi C, et al. Knockdown of filaggrin impairs diffusion barrier function and increases UV sensitivity in a human skin model. J Investig Dermatol 2010;130(9):2286–94.
[116] Mittal A, Colegio OR. Skin cancers in organ transplant recipients. Am J Transplant 2017, https://doi.org/10.1111/ajt.14382.
[117] Voskamp P, Bodmann CA, Koehl GE, Rebel HG, Van Olderen MG, Gaumann A, et al. Dietary immunosuppressants do not enhance UV-induced skin carcinogenesis, and reveal discordance between p53-mutant early clones and carcinomas. Cancer Prev Res (Phila) 2013;6(2):129–38.
[118] Kopecki Z, Yang GN, Jackson JE, Melville EL, Calley MP, Murrell DF, et al. Cytoskeletal protein flightless I inhibits apoptosis, enhances tumor cell invasion and promotes cutaneous squamous cell carcinoma progression. Oncotarget 2015;6(34):36426–40.
[119] Ahlfors JE, Billiar KL. Biomechanical and biochemical characteristics of a human fibroblast-produced and remodeled matrix. Biomaterials 2007;28(13):2183–91.
[120] Chen CZ, Peng YX, Wang ZB, Fish PV, Kaar JL, Koepsel RR, et al. The scar-in-a-jar: studying potential antifibrotic compounds from the epigenetic to extracellular level in a single well. Br J Pharmacol 2009;158(5):1196–209.
[121] Michel M, L'Heureux N, Pouliot R, Xu W, Auger FA, Germain L. Characterization of a new tissue-engineered human skin equivalent with hair. In Vitro Cell Dev Biol Anim 1999;35(6):318–26.
[122] Dezutter-Dambuyant C, Black A, Bechetoille N, Bouez C, Marechal S, Auxenfans C, et al. Evolutive skin reconstructions: from the dermal collagen-glycosaminoglycan-chitosane substrate to an immunocompetent reconstructed skin. Biomed Mater Eng 2006;16(4 Suppl):S85–94.
[123] Dezutter-Dambuyant C, Durand I, Alberti L, Bendriss-Vermare N, Valladeau-Guilemond J, Duc A, et al. A novel regulation of PD-1 ligands on mesenchymal stromal cells through MMP-mediated proteolytic cleavage. Oncoimmunology 2016;5(3):e1091146.

[124] Linde N, Gutschalk CM, Hoffmann C, Yilmaz D, Mueller MM. Integrating macrophages into organotypic co-cultures: a 3D in vitro model to study tumor-associated macrophages. PLoS One 2012;7(7):e40058.

[125] Black AF, Hudon V, Damour O, Germain L, Auger FA. A novel approach for studying angiogenesis: a human skin equivalent with a capillary-like network. Cell Biol Toxicol 1999;15(2):81–90.

[126] Groeber F, Engelhardt L, Lange J, Kurdyn S, Schmid FF, Rucker C, et al. A first vascularized skin equivalent for as an alternative to animal experimentation. ALTEX 2016;33(4):415–22.

[127] Wufuer M, Lee G, Hur W, Jeon B, Kim BJ, Choi TH, et al. Skin-on-a-chip model simulating inflammation, edema and drug-based treatment. Sci Rep 2016;6:37471.

[128] Xu F, Celli J, Rizvi I, Moon S, Hasan T, Demirci U. A three-dimensional in vitro ovarian cancer coculture model using a high-throughput cell patterning platform. Biotechnol J 2011;6(2):204–12.

[129] Lee V, Singh G, Trasatti JP, Bjornsson C, Xu X, Tran TN, et al. Design and fabrication of human skin by three-dimensional bioprinting. Tissue Eng Part C Methods 2014;20(6):473–84.

[130] Pourchet LJ, Thepot A, Albouy M, Courtial EJ, Boher A, Blum LJ, et al. Human skin 3D bioprinting using scaffold-free approach. Adv Healthc Mater 2017;6(4):1–8.

[131] Albritton JL, Miller JS. 3D bioprinting: improving in vitro models of metastasis with heterogeneous tumor microenvironments. Dis Model Mech 2017;10(1):3–14.

Section C

Skin substitutes: Clinical demands and skin tissue equivalents

Strategies to promote the vascularization of skin substitutes after transplantation

Jennifer Bourland[*,†,‡], *Julie Fradette*[*,†,‡]

[*]Centre de recherche en organogénèse expérimentale de l'Université Laval/LOEX, Québec, QC, Canada, [†]Division of Regenerative Medicine, CHU de Québec – Université Laval Research Center, Québec, QC, Canada, [‡]Department of Surgery, Université Laval, Québec, QC, Canada

1 Introduction

Ensuring adequate vascularization remains one challenge of foremost importance for the engineering of thick tissues. Nutrient and oxygen diffusion ensured by the microvasculature remains a central issue for optimal tissue production in vitro and timely survival after in vivo transplantation. The importance of studying vascularization processes (angiogenesis, vasculogenesis, and intussusceptive angiogenesis) and designing appropriate vascularization strategies in the field of tissue engineering is reflected by the increased number of yearly publications on the topic (Fig. 1), which is also true for skin vascularization in particular, which represents up to 10% of the yearly publications in the last 5 years (Fig. 1).

Skin substitutes are among the first tissue-engineered constructs that have been grafted to patients and are still the focus of numerous clinical investigations. The well-defined hallmarks of healthy skin, including specific epithelial markers of keratinocyte differentiation, combined with the ease of visually monitoring grafted tissues, undoubtedly contribute to the intense scientific activity observed in the field of wound repair and skin replacement products. Indeed, novel technologies, cell-editing tools, and proof-of-concept studies are often investigated using the skin as a model system, including various vascularization strategies.

While not mutually exclusive, methods employed to promote vascularization can be subdivided into proangiogenic approaches and in vitro cell-based prevascularization techniques. Early achievements using the "angiogenic" approaches in the field of skin vascularization relied on gene transfer to stimulate the ingrowth of blood vessels into the grafted reconstructed skin, for example, by overexpressing VEGF [1–3], PDGF [4,5], or bFGF [6] using modified viral vectors (reviewed in [7,8]). Currently, the creation of bioinductive scaffolds is actively investigated, including the functionalization of materials by the incorporation of these proangiogenic growth factors (EGF, bFGF, and PDGF). Scaffold-based promotion of vascularization has been addressed in recent reviews describing how structural and physicochemical modifications and biological activation of dermal scaffolds can stimulate healing processes [9–11].

Skin Tissue Models. https://doi.org/10.1016/B978-0-12-810545-0.00008-5

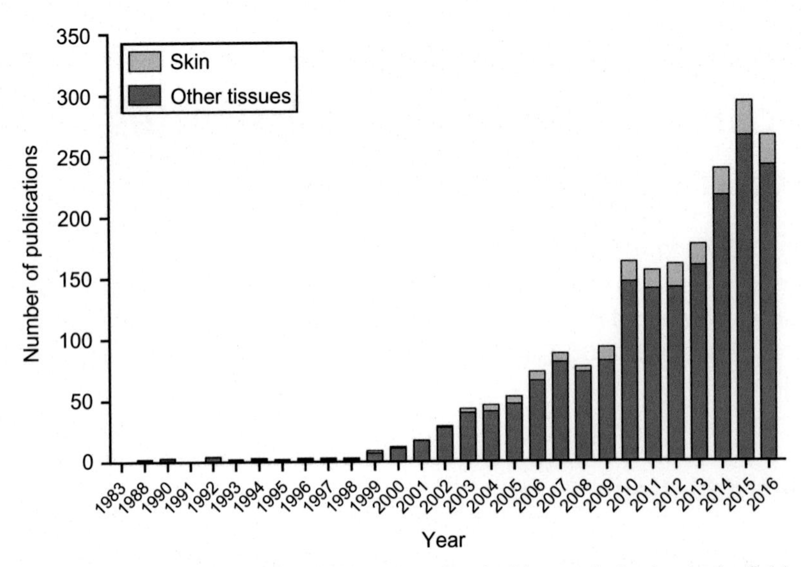

Fig. 1 Increase in the number of publications associated with vascularization in the field of tissue engineering. Consultation of the PubMed search engine with the keywords "tissue AND engineering AND vascularization" and "skin AND engineering AND vascularization." Performed December 12, 2016, 6:30pm.

Indeed, an increasing number of studies focus on the development of connective tissue constructs, devoid of keratinocytes, acting as fillers or bandages to stimulate resident skin cells at the wound periphery. A wide array of constructs with various specifications have been produced, and their impact on regenerative processes includes accelerated wound closure due to increased reepithelialization and/or contraction, with improved neovascularization being the most common functional outcome observed [12,13]. However, for more extensive and deeper tissue defects such as severe burn injuries, trauma, or tumor resection, stimulating peripheral keratinocyte migration, granulation tissue formation, and therapeutic angiogenesis is not sufficient. Rapid coverage with a substitute already featuring a dermis and an epidermis (referred here as reconstructed skin and bilamellar or bilayer skin) is required. Novel models incorporating an adipose compartment (hypodermis) to generate a trilayer skin substitute have also been developed. For all these types of constructs, a widely studied strategy to promote vascularization consists in incorporating vessel-forming cells that will reorganize to generate a preformed vascular network of capillary vessels in vitro. Upon grafting, inosculation between the host recipient bed and the substitute's network is therefore favored, resulting in timely blood perfusion [14–16].

This chapter will focus on recent cell-based vascularization strategies for the development of tissue-engineered constructs designed for skin replacement. Considering the diversity of the in vitro models available, we will procure an overview describing how the choice of scaffolding elements and the origin of the cell source (including progenitor cells and adult microvascular endothelial cells) can impact vascularized skin reconstruction. Particular emphasis will be given to studies that assessed functional parameters of the preendothelialized reconstructed tissues in vitro and their outcome after in vivo implantation. Lastly, recent investigations of lymphatic microvasculature, an important complement of blood microvasculature, will be briefly discussed.

2 Cell-based vascularization strategies

2.1 Promoting vascularization of dermal substitutes

The epidermis is an avascular compartment that is nourished from the underlying dermis. The active role of the dermis in supporting keratinocyte adhesion, migration, and differentiation is also essential during the reepithelialization phase and thus is key to wound healing. Therefore, many studies focused on dermal template vascularization as a first step toward bilayer skin reconstruction and assessed efficacy using various wound healing models. Most methods used to engineer prevascularized dermis have focused on synthetic or animal-derived matrices or the cell's ability to produce their own extracellular matrix. Some of these studies are listed in Table 1, which details the matrix and cell origin and the functional tests performed to assess these constructs (in vitro and in vivo).

Hydrogels based on collagen or fibrin materials are often used in combination with stromal cells and endothelial cells (EC) or endothelial progenitors to generate a vascularized dermis. For example, Kuo et al. extracted murine collagen to develop a hydrogel containing human blood-derived endothelial colony-forming cells (ECFC) or human bone-marrow-derived mesenchymal stem cells (BM-MSC), which can be injected into a full-thickness wound to form a dermal template [17]. A similar approach based on a hyaluronic acid-containing hydrogel and using human ECFC has shown that the human capillary network is remodeled but still present after 2 weeks of transplantation to nude mice. Murine smooth muscle cells wrapped around human vessels were identified in the transplant after hydrogel degradation during the inflammation period (day 3 after grafting) [18]. The vessels' functionality was assessed by UEA-1 lectin injections after grafting, and at days 5–7 post implantation, the human and murine vessels were successfully anastomosed with the host vasculature, with the presence of chimeric vessels. The connections and chimeric vessel number were reported to increase over time (at day 11) [18].

With a structure between a sponge and a typical hydrogel, gellan gum-hyaluronic acid hydrogels have also been used to produce dermal substitutes comprising human adipose-derived stromal/stem cells (ASC) and human adipose-derived endothelial cells [19]. Endothelial cells were characterized by the expression of surface markers such as CD31 and von Willebrand factor and their ability to uptake acetylated low-density lipoprotein (Dil-Ac-LDL) and to form tubes in Matrigel® [19]. The addition of endothelial cells was reported to improve tissue vascularization after grafting compared with the use of hydrogel with ASC only (quantified by CD31 staining).

Sponges are another type of biomaterials used for dermis reconstruction and usually feature more rigid scaffolding elements that are also conducive to the development of

Table 1 Examples of vascularized dermal substitutes

Scaffold/matrix	Matrix origin	Endothelial cells	Endothelial origin	Functionality assay	Animal model	References
Collagen gel	Murine	ECFC	h	–	Nude mice	[17]
HA hydrogel	Microbial fermentation	ECFC	h	UEA-1 lectin	Nude mice	[18]
Gellan gum-HA hydrogel	Microbial fermentation	AT-derived EC	h	Dil-AC-LDL uptake	Nude mice	[19]
Collagen-chitosan sponge		HMVEC	h	Pericyte recruitment	Nude mice	[20,22,43]
Integra®	Bovine and shark	AT-derived microvascular fragments	m	–	Nude mice	[24]
		Endothelial progenitors	r	–	Nude mice	[26]
Cell sheet nanofilms	Human, bovine, and porcine components	HUVEC	h	UEA-1 lectin	Nude mice, nude rats	[23]
Cell sheet/self-assembly	Human	HUVEC	h	CM-DiI staining	–	[30,35]
		HMVEC	h	CM-DiI staining	–	[30,35]
		HMVEC	h	Ang-2 secretion	–	[33]
		HBOEC	h	Ac-LDL uptake	–	[36]

h, human; m, murine; r, rat.

an in vitro preformed EC network [20,21]. In collagen-chitosan sponges, human microvascular endothelial cells (HMVEC) can form microvessels that have the capacity to recruit pericyte-like cells [22]. These α-smooth muscle actin-positive cells were identified around CD31-positive tubules when HMVEC were cocultured with human fibroblasts, but this was not observed with fibroblasts only. This illustrates the ability of in vitro preformed capillaries to be attractive to cell types involved in capillary stability such as pericytes [22,23].

The dermal substitute Integra® (Integra LifeSciences, Plainsboro, NJ), which is a Food and Drug Administration (FDA)-approved product, was used by different groups to study the effect of prevascularization [24,25]. It consists of a matrix made of cross-linked bovine collagen and shark glycosaminoglycan combined with a silicone layer mimicking a pseudoepidermis to protect the implant after grafting. This acellular off-the-shelf dermal substitute can be preseeded with endothelial cells (and stromal cells if desired) to facilitate its rapid integration and reduce the usual time required for vascularization (up to 3 weeks). For example, vascular resident endothelial progenitor cells extracted from rat hearts were differentiated toward endothelial cells and were then able to form tubes in a gel. These cells were then seeded in the Integra® matrix and grafted onto nude mice. When preseeded with EC, the implant showed a higher density of vessels 2 weeks after transplantation (analyzed by tissue transillumination and digital segmentation) [26]. Furthermore, the added endothelial progenitor cells displayed a proangiogenic secretory profile combined with chemoattractant secretion (such as MCP-1).

Using the same Integra® matrix, Frueh et al. studied the effect of prevascularization by incorporating adipose-tissue-derived microvascular fragments [24]. For this study, they used cells extracted from epididymal fat pads of GFP-expressing transgenic mice, and the implant was grafted on nude mice. This leads to the formation of a dense GFP^+ microvascular network in the tissue, primarily at its periphery. This network rapidly connected with the host vasculature and allowed quick perfusion (in 3–6 days), whereas nonprevascularized implants did not show perfusion after 14 days (only partial infiltration of the host vessels at the border periphery). During the first 2 weeks after transplantation, the preformed vessels showed signs of remodeling as assessed by microhemodynamic analyses [24].

Another approach to generate vascularized dermal substitutes is to create cell sheets, which slightly differs from methods based on cell seeding on or within a preformed scaffold. Different kinds of cell sheets can be produced, ranging from thin keratinocyte layers to cell sheets made of mesenchymal cells that are able to produce abundant extracellular matrix (ECM) (Fig. 2). Indeed, methods that reproduce with higher fidelity the native microenvironment are based on the ability of stromal cells to produce and assemble their own human extracellular matrix. Most of the recent developments we will describe relate to such cell sheets produced using either thermoresponsive culture substrates, the self-assembly approach (vitamin C supplementation) or a combination thereof. Cell sheets can be produced from a wide range of stromal cells using thermoresponsive plates [27]. These connective sheets can then be seeded with human umbilical vein endothelial cells (HUVEC) or HMVEC to coculture both cell types. This results in the formation of tube-like structures. For

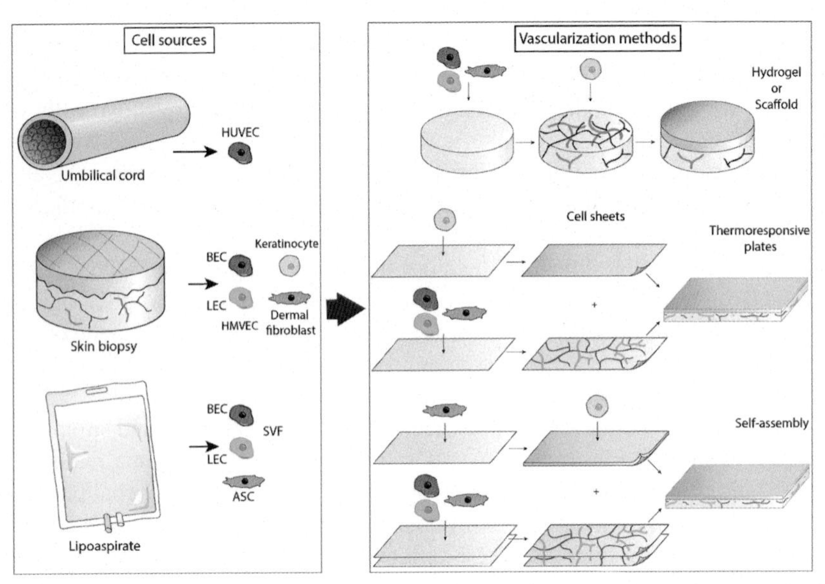

Fig. 2 Schematic summarizing common methods of tissue-engineered skin prevascularization using different cell sources. Endothelial cells can be extracted from different tissues to obtain either HUVEC, HMVEC, SVF, or progenitors (shown in *dark red*, BEC), as well as lymphatic endothelial cells (*green*, LEC). Stromal cells can also be isolated (skin fibroblasts or ASC, in *pink*). These primary cells can be incorporated into various scaffoldings or hydrogels or seeded on thermoresponsive plates (cell sheets) and/or for long-term culture in the presence of ascorbic acid (self-assembly approach) to produce prevascularized skin substitutes. Keratinocytes (in beige) are then isolated from skin biopsies and seeded at the surface of the dermal templates.

example, Sasagawa et al. showed that HUVEC could also form tube-like structures when cocultured with myoblasts [28]. They produced prevascularized tissues by using temperature-responsive culture dishes to form cell sheets. Myoblasts were cultured to confluence in a dish; the temperature was then brought to 20°C to detach the cell sheet. Cell sheets were manipulated using a hydrogel-coated plunger designed for this purpose. To add endothelial cells, a cell sheet was applied on a HUVEC monolayer, which was then detached using a lower temperature. Once transferred on the sheet, the HUVEC were sandwiched with another cell sheet [28]. This allows to produce thicker implants (about 100 μm thick) able to connect to the host vessels after transplantation, as observed with five-layer myoblasts and HUVEC sheets implanted in nude rats [28].

Another variation of a sheet method is the use of cell coating with fibronectin and gelatin to build up extracellular matrix nanofilms that can then be stacked [23]. Using this method, a dermal multilayered tissue (thickness of 30–50 μm but can go up to 100 μm with 20 layers of dermal fibroblasts) was reconstructed and prevascularized by seeding HUVEC on different layers [23].

Cell sheet formation based on ascorbic acid stimulation of ECM by the stromal cells is also an approach used to generate connective tissue constructs that can be further vascularized in vitro. Dermal extracellular matrix is composed of a complex combination of structural proteins and small molecules. This complexity is often not fully reproduced by hydrogels or sponges. At the LOEX center, the self-assembly approach of tissue engineering is the core technology for the reconstruction of multiple tissues. These types of cell sheets can be obtained when dermal fibroblasts (or other stromal cells) are cultured with ascorbic acid over many days, leading to tissues stable for many weeks in vitro (up to 3 months) [29,30]. Ascorbic acid is a cofactor of the prolyl 4-hydroxylase, an enzyme involved in the assembly of the triple helix of procollagen. This allows the formation of cell sheets enriched in ECM, on the surface of which endothelial cells are generally seeded, although endothelial cells can also be incorporated directly earlier in the process [31]. The cell and matrix sheets are then manipulated with forceps and stacked to produce a tissue of the desired thickness featuring human extracellular matrix favorable to endothelial cell reorganization into a capillary-like network. Using human cells, the extracellular matrix is therefore fully human and not altered by subsequent chemical or physical treatment, allowing to mimic more adequately the native microenvironment. With this method, HUVEC and HMVEC have been used to produce vascularized connective tissue substitutes comprising BM-MSC sheets [32], ASC sheets [33,34], and vascularized dermal substitutes using fibroblasts [30,35].

The self-assembly method can also be combined with endothelial progenitors such as human blood outgrowth endothelial cells (HBOEC) [36]. HBOEC can be extracted from peripheral blood and thus are easy to obtain. The expression of genes facilitating wound healing and implant vascularization was compared with those of HUVEC and early endothelial progenitor cells. HBOEC expressed higher levels of genes involved in the cross talk with cells from the wound bed such as resident endothelial cells, immune cells (e.g., through VEGF, MCP-1, and angiopoietin-2) and keratinocytes [36]. Like HUVEC, these cells could form tubes in vitro, and their functionality was assessed by Ac-LDL uptake and the binding of UEA-lectin. This reconstructed dermis featuring endothelial progenitor cells is another example of constructs that could be used in the future to treat wounds associated with impaired healing such as ulcers, using autologous cells displaying a therapeutic secretome.

2.2 Considerations for the choice of an endothelial cell source

The diversity of dermal substitutes shows that many different scaffolds and cells can be combined to generate a microvascularized tissue (Table 1), but the endothelial cell source is of uttermost importance for clinical translation. Existing methods investigating tissue prevascularization have used different types of endothelial cells: from HUVEC to HMVEC extracted from different tissues (adult skin and adipose tissue) or progenitors. Functional and transcriptomic studies have revealed differences between endothelial cells from different origins when maintained and assessed in culture. The first distinctions involve two subtypes: endothelial cells from large vessels (HUVEC, aortic endothelial cells, and vein endothelial cells) and microvascular endothelial cells. These cells can reorganize into vessels of various sizes when seeded into tissue constructs, but they also differ by their gene expression profile. Many variations between these cells are seen for genes involved in production and interaction with ECM. For example, HMVEC express genes linked to the production of basement membrane (laminin and collagen type IV), whereas macrovascular endothelial cells do not have this direct contact with ECM and basement membrane [37]. Microvascular endothelial cells are also characterized by genes associated with a phenotype more closely adapted to angiogenesis, with the expression of factors such as angiopoietin-2, LMO2, and transforming growth factor α. Even if this seems to correspond to their in vivo localization and function, it is particularly interesting for tissue engineering to consider these parameters, as they are maintained by cells in an in vitro context [37].

While more similar compared with macrovascular endothelial cells, differences were also observed between HMVEC originating from different tissues, reflecting the variations between vascular beds from different tissues. HMVEC have tissue-specific gene expression variations. For example, skin HMVEC were found to express bFGF and different genes involved in cholesterol biosynthesis. Molecules involved in immune trafficking and immune responses have also been reported to differ depending on the cell origin [37].

The choice of endothelial cells can be particularly important for clinical applications and needs to be further characterized. Endothelial cells are immunogenic and thus should be autologous. HMVEC are obtained in small quantities from skin extraction procedures and need to be expanded. HMVEC are obtained in small quantities from skin extractions and need to be expanded in vitro before tissue production.

This is why alternative sources such as adipose-tissue-derived cells and other progenitors are investigated [38–41]. A study recently compared the use of adipose-tissue-derived EC and dermal EC for their in vitro vascularization potency and found that even if some differences existed between these cells, they had a similar response and angiogenic potential. It confirmed that EC from an abundant and accessible source as subcutaneous adipose tissue can be an alternative to the expansion of dermal microvascular endothelial cells [38]. Further progress in these characterizations will likely improve autologous-based vascularization strategies.

2.3 Bilayer skin reconstruction models featuring vascular-like networks

To achieve optimal wound healing and skin regeneration, reestablishing the skin's essential roles is of the uttermost importance. Reepithelialization needs to occur early in the healing process to prevent infections and restore the barrier function. To ensure proper reepithelialization, different bilayer skin substitutes presenting an epidermis and a vascularized dermis are now being developed. Keratinocytes can be added to the different vascularized dermis described in Section 2.1 to produce bilamellar substitutes. Therefore, the main techniques used for skin reconstruction also include cell seeding in hydrogels, on decellularized matrix, or the use of the cell sheet/self-assembly methods (Fig. 2). Vascularized bilamellar skin reconstructed with hydrogels is often based on the use of animal-derived material such as bovine or rat collagen and combined with either allogenic or autologous cells [9]. Table 2 summarizes key studies including the ones that evaluated the reconstructed microvascularized skin in animal models.

Table 2 **Examples of vascularized skin substitutes**

Scaffold/matrix	Matrix origin	Endothelial cells	Endothelial origin	Animal model	References
Collagen-chitosan sponge	Bovine	HUVEC	h	–	[42]
Collagen-glycosaminoglycan		HMVEC	h	Nude mice	[44]
Collagen gel	Rat	HMVEC	h	Nude mice	[46]
Fibrin hydrogel	Bovine	AT-derived EC	m	Nude rat	[47]
Self-assembled generated ECM	Human	HMVEC	h	Nude mice	[30,48-50]
		HUVEC	h	Nude mice	[30,50]

h, human; m, murine.

Historically, the first vascularized bilamellar skin substitutes were produced using chitosan and bovine collagen biopolymers in which fibroblasts, HUVEC, and keratinocytes were cultured [42,43]. In these reconstructed tissues, HUVEC formed tubes with lumens featuring a basement membrane but only when grown in coculture with fibroblasts, thus demonstrating the critical role of stromal cells in skin vascularization. Another pioneering study developed a bilamellar skin substitute based on a collagen-glycosaminoglycan substrate [44]. They seeded fibroblasts and HMVEC in the substrate and then layered keratinocytes on the surface. In vitro, HMVEC seemed to affect keratinocyte stratification and alter their differentiation. HMVEC were also aggregated in the dermis near the dermoepidermal junction. No network developed or was observed in vitro compared with the previous work in collagen-chitosan sponges [45], but once grafted into athymic mice, the HMVEC in the substitute were reorganized to form vascular analogs and secreted collagen type IV, indicative of a basement membrane deposition, between 2 and 4 weeks after implantation [44].

More recently, rat collagen type I or bovine fibrin hydrogels were used to produce bilamellar skin featuring blood capillaries [46,47]. These skin substitutes are composed of exogenous matrix and human keratinocytes, fibroblasts, and HMVEC. This allows the production of vascularized skin substitutes in 3 weeks, and grafting on nu/nu rats showed improved wound healing and anastomosis with the host vasculature [46]. Of note, the extraction of sufficient quantities of HMVEC can represent a challenge, especially when autologous treatment of large skin surface areas is required. One alternative that has been proposed is to replace the HMVEC by white adipose-tissue-derived endothelial cells [47]. Using the same model based on bovine fibrin hydrogels, the authors reported that upon coculture in hydrogel with white adipose-tissue-derived stromal cells (ASC), the adipose-derived endothelial cells formed capillaries in the skin substitutes that were able to accelerate tissue perfusion after grafting (rat erythrocytes were observed in human capillaries at day 4 post implantation). A recent comparison of endothelial cells extracted from the dermis and adipose tissue also supports the use of adipose tissue endothelial cells for tissue engineering considering their availability and similarities with dermal HMVEC [38].

Similarly to dermal substitutes, vascularized skin can be produced by the self-assembly approach [30,48,49]. Using this method, the crucial role of cross talk between different cell types was shown by the effect of keratinocytes on HMVEC and HUVEC [30]. In this study, vascularized substitutes were produced using human dermal fibroblasts. The reconstructed tissues contained either HMVEC or HUVEC. Keratinocytes were seeded on top of the dermal compartment or not, and the formation of vascular network was assessed by time-lapse imaging using GFP-HUVEC or by subsequent immunostaining. Keratinocytes maintained a network of small diameter capillaries and were able to reduce the size of bigger capillaries seen in the stromal constructs devoid of keratinocytes. This effect was found to rely on a paracrine effect (in part VEGF-driven) and not dependent of direct contact between the keratinocytes and the endothelialized stroma. This model of self-assembled skin was later produced using GFP-transduced HMVEC and HUVEC to follow the substitutes after grafting in nude mice. HMVEC- and HUVEC-formed capillaries in the skin substitute in vitro, and those capillaries were remodeled within 4 days in mice [50]. At the same time, anastomosis between the human GFP+ endothelial cells and the murine capillary network showed the functionality of the human preformed capillaries.

2.4 Achieving the production of a trilayer skin substitute: Vascularization of reconstructed adipose tissue to recreate the hypodermis

While thinner skin substitutes can lead to effective wound coverage and graft take without the presence of a preformed capillary network [51–54], thicker skin substitutes will likely require more than proangiogenic cues from their cellular and scaffolding components in order to promote graft survival and integration without necrosis. Full-thickness skin constructs including the deepest layer of subcutaneous fat, also called hypodermis, represent emerging skin models. The precise contribution of the hypodermis to skin homeostasis is still unclear; however, adipose tissue is well vascularized, and both adipocytes and stromal vascular cells possess a secretome highly relevant for skin repair, including proproliferative, proangiogenic, and antiapoptotic actions [55–58]. Few studies investigated the direct role of adipocytes in wound healing [57,59], but Schmidt et al. reported an important functional role of adipocytes in recruiting fibroblasts to the wound bed during the proliferative phase of healing [60].

A limited number of studies addressed the technical engineering challenge of developing a human trilayer skin [61–64]. The first all-human trilayer skin was produced using the self-assembly approach of tissue engineering, resulting in a skin featuring a well-differentiated epidermis directly interacting with a stromal compartment made of either dermal fibroblasts or undifferentiated ASC, which itself was in contact with an hypodermal compartment [64,65]. This hypodermis was produced by differentiating ASC into adipocytes while stimulating matrix production and assembly, leading to the presence of rounded adipocytes embedded in a rich matrix made of human structural collagens and basement membrane components (laminin-1 and collagen type IV) present around the newly developed adipocytes [64,65]. This model features the advantage of being readily scalable for the production of large surface areas (at least 90 cm^2) that can be easily manipulated by the surgeons. Other models compared BM-MSC and ASC for the production of a plasma-based hydrogel hypodermis and reported that both cell types have the capacity to contribute to the epidermal differentiation program [61]. Silk scaffolds have also been used for the creation of the hypodermis based on in vitro differentiation of ASC within a reconstructed skin featuring a dermis made of fibroblasts and type I collagen [63]. Depending on the exogenous scaffolding components used, adipocyte-like cells can be difficult to identify on histological sections, but leptin and other secreted products usually support the presence of adipogenic cells in the reconstructed skin constructs. The most recent model used freshly isolated human adipocytes seeded into a rat collagen type I hydrogel to produce the hypodermis compartment of the trilayer skin constructs [62]. Different media formulations were tested to determine culture conditions optimal for these mixed cell types [62]. Indeed, for all models, the challenge of trilayer skin reconstruction is to find out culture conditions favorable to the differentiation/ maintenance of mature adipocytes since supplements present in either keratinocyte or endothelial cell media contain molecules known to influence the metabolic functions of adipocytes (adipogenesis and lipolysis) and able to trigger adipocyte dedifferentiation [33,66].

In the context of skin substitute production, the use of a trilayer skin could be beneficial for the treatment of deep lesions, especially at anatomical sites prone to scarring. The clinical application of such substitutes will likely depend on the prevascularization of both the dermal and hypodermal compartments to ensure efficient graft take and survival. To date, the production of a reconstructed skin featuring an epidermis supported by a collagen-gel-based dermis (prevascularized or not) under which a silk-based vascularized hypodermis was placed has been reported [63]. Investigating the fate of the network upon grafting will provide interesting insights into the potential benefits on skin quality of implanting such a construct in vivo.

Independently of skin reconstruction, numerous studies assessed the production of vascularized adipose tissue featuring mature adipocytes, using various reconstruction strategies [67]. Indeed, the characterization of the structural features and functionality of AT constructs before their incorporation into a skin construct comprising a dermis and an epidermis is an important step. Creating and maintaining a developed network of capillaries in vitro while differentiating cells into mature adipocytes require that special consideration is given to the culture protocol to determine the best sequence for cell seeding, induction of adipogenesis, and use of specialized culture media (such as EGM2 endothelial cell medium) to ensure an optimal composition supporting all cell types [33,66,68–70].

Table 3 In vitro-engineered adipose constructs featuring endothelial cells, natural scaffolding elements, and adipocytes differentiated from ASC[a]

Scaffold/matrix	Endothelial cells	EC assembly in vitro	Animal model	References
Silk fibroin	HUVEC	ns	–	[71,72]
Silk fibroin coated with VEGF and/or laminin	HMVEC	Yes (lumen)	–	[73]
ASC spheroids induced or not toward adipocytes	HUVEC	EC alignment	Nude mice	[78]
Fibrin gel or spheroids	HUVEC	ns	Nude mice/1 week	[79]
Collagen microparticles with or without fibrin and growth factors	HUVEC	ns	NOD SCID mice/12 days, 4 weeks, or 4 months	[75]
Collagen/alginate microspheres	HUVEC	Limited	–	[76]
Collagen/alginate microspheres coated with collagen fibrils	HUVEC	Limited	Injected in BALB/c-nu mice/4, 8, and 12 weeks	[77]
Self-assembly	HUVEC	Yes	–	[34]
	HMVEC	Yes	Nude mice [65]	[33]

[a] Since 2010; ns, not shown.

Among adipose tissue models described since 2010, constructs engineered from naturally-derived scaffolding elements and featuring adipocytes cocultured with endothelial cells are summarized in Table 3.

Self-assembled adipose sheets and thicker adipose tissues have been produced, and they efficiently supported EC reorganization into capillary-like networks into that highly natural microenvironment [33,34]. Our experience with in vivo implantation of such self-assembled adipose tissues [65] is that inosculation with vessels of the recipient site (muscular bed) is occurring as fast as reported for reconstructed dermis (within 3–4 days). Indeed, the preformed network in vitro within these adipose constructs (Fig. 3) rapidly connected with the murine vasculature, as supported by the presence of red blood cells within human CD31-expressing capillaries (Fig. 4).

Silk fibroin is a prominent material in this field of research [71,72], and coculture with endothelial cells has led to the formation of capillary-like networks in vitro [73,74]. Microtissues formed by collagen/alginate microspheres or microparticles [75–77] and from spheroids/fibrin [78,79] have also been investigated in combination with HUVEC, both in vitro and after implantation into nude mice (Table 3). Most studies demonstrated in vivo angiogenesis within the constructs.

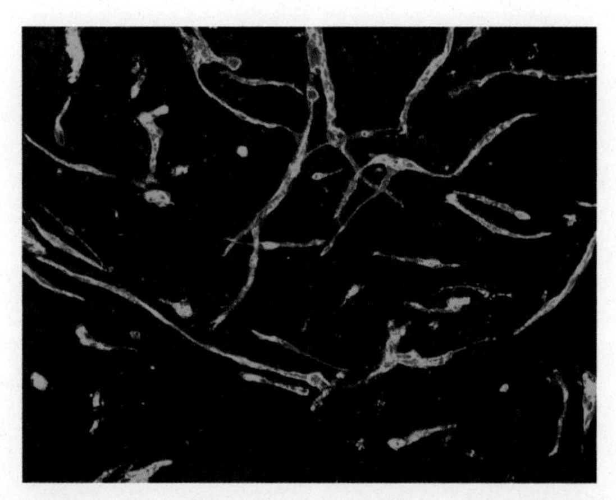

Fig. 3 Capillary network development in vitro. Confocal imaging reveals that HMVEC organize into CD-31$^+$ capillary-like structures *(green)* when seeded into a reconstructed adipose tissue produced from in vitro-differentiated ASC according to the self-assembly approach of tissue engineering [33]. HMVEC were cultured for a total of 13 days after seeding between two adipose sheets (adipogenically induced for 29 days) that were superposed for 6 days. Hoechst-labeled nuclei appear in *blue* (20×).

2.5 Considerations for vascularization assessment after grafting

Neovascularization is classically assessed following immunolabeling for blood vessel components on histological sections, a useful but time-consuming method that does not allow to easily distinguish nonperfused from functional vessels. Confocal and two-photon microscopy can reveal the architecture of the capillary networks, and a detailed characterization (e.g., network extent and branching) can be performed when used in conjunction with specialized software [80,81].

Systems suitable for the in vivo evaluation of angiogenesis are needed. Color-coded or crossover murine models have been developed to enable in situ imaging of anastomosis and to determine the origin of skin graft vasculature [82–84]. In preclinical models, these approaches in combination with dorsal skinfold chambers can establish the dynamics of the neovascularization process quite precisely [24,83,85]. The use of transillumination and computerized digital segmentation has also been reported as a simple, low-cost method allowing to visualize large tissue areas and to quantify the vasculature with better resolution than radioangiographic studies [86]. However, the need to excise the tissues to perform ex vivo analyses using transillumination methods is a limitation. Nonmicroscopic methods such as magnetic resonance imaging (MRI) with contrast agents [87,88], computerized tomography (CT), positron emission tomography (PET), and ultrasonography can determine tissue perfusion noninvasively, and new research designed at improving their limits in spatial resolution will likely result in more widespread applications in the clinical practice (reviewed in [81]). The development of a low-cost, accessible, reliable, and noninvasive procedure for the characterization and quantitative analysis of angiogenesis using methods directly applicable to human patients is still needed. The skin's easy access for such investigation will likely lead

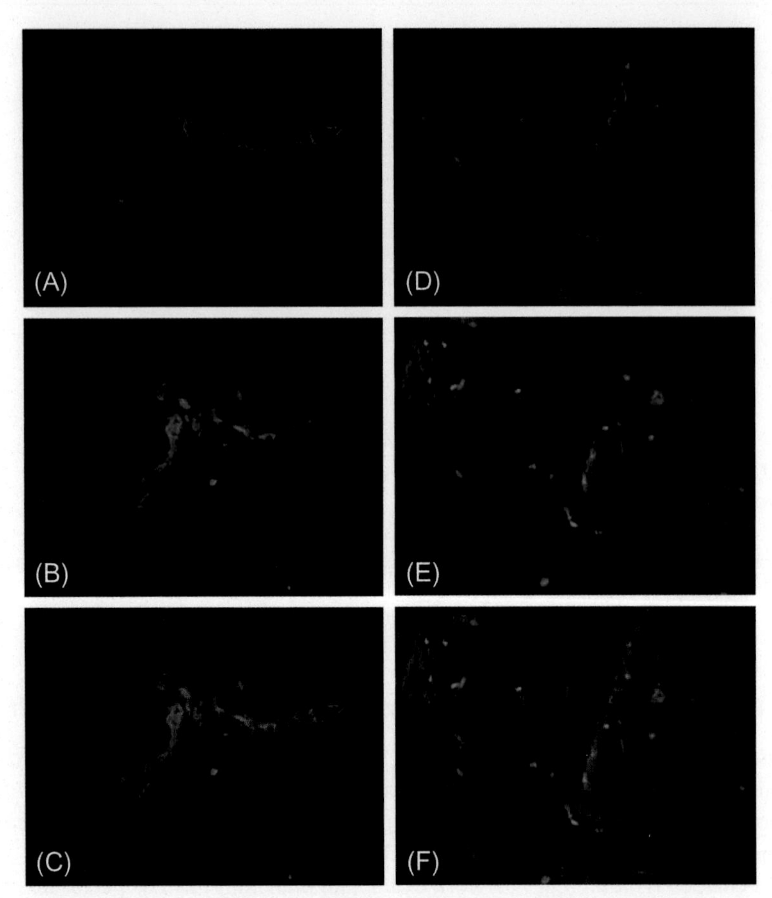

Fig. 4 Vascularization of adipose tissue constructs. Human capillaries within the microvascularized adipose tissues are stable and functional 14 days after grafting. Immunolabelings on cryosections reveal that human CD31-expressing capillary structures (A and D in *red*) are filled with murine red blood cells (B and E in *green*) indicating that functional inosculation occurred upon grafting. C and F are image superposition of A and D, with B and E images, respectively (40× magnification).

to new technologies and broader applicability of already existing methods such as laser Doppler flowmetry, transcutaneous oxygen tension measurements, and capillary microscopy. For example, capillary microscopy (also called videocapillaroscopy or intravital capillaroscopy) is a method typically used and validated to examine the morphological and functional changes in microcirculation in the skin affected by pathological conditions such as psoriasis and systemic sclerosis [89,90]. The noninvasive nature of the device for direct observation of the architecture of the local microcirculation is an advantage. This tool can determine capillary morphology including size, shape, and tortuosity in addition to capillary density; skin percentage occupied by capillaries (volume fraction); and changes in the velocity of the red blood cells within capillaries. This technique allows the assessment of the nutritive capillary bed, distinct from the thermoregulatory vascular bed, which contributes 85% of the total blood flow, usually evaluated using transcutaneous oxygen tension and Doppler laser. When capillaroscopy is performed in combination with the use of intravenous fluorescent dyes such as sodium fluorescein, the imaged capillaries allow the determination of perfusion indexes [90]. It would be interesting in the future to use similar approaches for the monitoring of prevascularized reconstructed skin after grafting to evaluate the evolution of neovascularization in comparison with normal skin. Recent state-of-the-art advancements achieved using multimodal imaging to evaluate blood flow will undoubtedly benefit the field of microcirculation and lead to a more widespread use of these modalities in the clinical realm (extensively reviewed in [91]).

2.6 Beyond blood vessel vasculature: Lymphatic-like networks revealed in vitro

As described in the previous sections, the need for nutrients and gas exchanges has driven extensive research on vascularization strategies with a focus on blood microvascularization. However, it is now recognized that lymphatic vascularization is also a key aspect to consider while performing tissue reconstruction, particularly for the clinical application of skin substitutes. The lymphatic capillaries are part of a unidirectional system and are characterized by blind ends and button-like junctions allowing fluid and cells to easily enter the capillaries [92].

The lymphatic vasculature allows to transport debris from the wound site and to drain interstitial fluid, but it also plays a major role in the immune response by allowing immune cells to access and leave an inflammation site. A quick drainage of the grafted site, for example, for burn or venous ulcer patients, is key to the successful integration of skin substitutes.

A small number of teams addressed the lack of lymphatic capillaries in skin substitutes generated by tissue engineering [41,46,47,50,93]. Lymphatic engineering mainly stems from the thorough characterization of human microvascular endothelial cells performed in recent years. Typical extraction methods for microvascular endothelial cells are based on cell extrusion and generate cell populations widely used to generate blood capillaries. These cells also include a subpopulation of lymphatic endothelial cells (LEC), which can form new dermal initial lymphatics [94–96]. LEC can be purified by flow cytometry or using immunomagnetic beads directed at podoplanin, lymphatic vessel endothelial hyaluronan receptor 1 (LYVE-1), and CD34 and then be incorporated into substitutes [94,97]. For example, purified human LEC seeded on top of fibroblast-derived cell sheets generated by the self-assembly approach, which were later stacked, produced a reconstructed dermis featuring a lymphatic capillary network [97]. This network displayed blind-ended lymphatic capillaries expressing prospero homeobox protein 1 (PROX1), LYVE-1, and podoplanin. Anchoring filaments were also present. Moreover, in vitro lymphangiogenesis could be modulated by VEGF-C and hepatocyte growth factor (HGF), which are able to induce lymphangiogenesis via VEGFR-3 and c-Met, respectively. In addition, this study demonstrated that when supplemented together, the impact of VEGF-C and HGF was more potent than for each molecule separately [97].

The first demonstration of the ability of lymphatic capillaries to be translated in vivo was reported in a bilayer skin substitute featuring blood and lymphatic capillaries produced in a rat collagen hydrogel, which was grafted to nu/nu rats [46]. In this study, it was shown that, as with blood capillaries, lymphatic vessels anastomosed with rat vessels by either direct or wrapping connection [46]. Direct connections happened at blunt ends, whereas wrapping connections formed hybrid vessels with rat and human cells. Rat capillaries were also often found in the proximity of human lymphatic vessels. The subsequent addition of melanocytes to the skin substitutes did not influence the formation and development of lymphatic vessels after grafting [98].

Skin equivalents comprising blood and lymphatic microvessels have also been generated by the use of fibronectin and gelatin cell coating with a mixture of HUVEC and human dermal lymphatic microvascular endothelial cells [93,99,100]. It will be interesting to determine the fate of the lymphatic microvessels in vivo, considering that similar dermal substitutes with blood vascularization only, generated by this method were grafted and have shown the persistence of the network in nude mice [23].

Similarly than for blood vascularization studies, other cell sources could provide an easier way to scale up autologous vascularized skin production by extracting more endothelial cells or progenitors than with classical skin biopsies. One alternative is the use of adipose-tissue-derived cells. The extraction can be done following a minimally invasive liposuction procedure. Human endothelial cells isolated from adipose tissue can form microvessels when seeded in a collagen type I hydrogel [41,47]. In addition, in order to assess the ability of microvascular fragments to produce lymphatic vascularization, murine GFP^+ cells were seeded in a cross-linked bovine collagen and shark glycosaminoglycan matrix (Integra®) and then applied on murine dorsal full-thickness skin defects [24]. After 14 days, all implants (with or without microvascular fragments) presented lymphatic vessels, but in the group enriched with fragments, the lymphatic network had a higher density, and more than 60% of the lymphatic vessels were GFP^+, indicating that they originated from the transplanted microvascular fragments. GFP^+ lymphatic vessels were also shown to invade the surrounding host tissue. A dense network of GFP^+ blood vessels was also obtained by preseeding of fragment-derived cells in Integra®, indicating the plasticity of these cells. This study demonstrated the implication of microvascular fragment-derived cells in a prevascularized implant for the development of new lymphatic vessels, expected to facilitate the graft vascularization and drainage.

Finally, in the future, skin tissue engineering could also benefit from the use of LEC derived from induced pluripotent stem cells. Lee et al. generated LEC from pluripotent stem cells and characterized them [101]. After different processes of differentiation, a mixed population of LEC and lymphatic progenitor cells was obtained. They expressed LYVE-1, VEGFR-3, and podoplanin. These cells are able to form lymphatic capillaries and can be produced on a large scale from the same population of fibroblasts that would be used to produce dermal substitutes. They also displayed a higher secretion of prolymphangiogenic and proangiogenic factors such as angiopoietins and VEGF-C compared with human normal LEC, making these cells a valuable tool to promote skin wound healing.

3. Conclusion

The field of skin tissue engineering has been progressing at great strides. The variety of reconstructed skin models being currently investigated in an experimental or clinical setting is very promising for the development of optimal skin replacement adapted to specific wound types (e.g., diabetic ulcers and severe burns). In addition to the advances mentioned for the reconstruction of a more complete reconstructed skin (the trilayer skin and lymphangiogenesis), robust but uncomplicated imaging platforms allowing vascularization assessment will be needed to further the field. Vascularization strategies will continue to evolve along with the more widespread use of new technologies such as bioprinting, microfluidics, nanotechnologies, and the production of microstructured templates [102–104]. These exciting technologies have the potential to generate perfusable microchannels or networks featuring predetermined organization and patterning resulting in improved spatial control of vessel distribution. We can speculate that thicker and customized skin substitutes could be designed in vitro in the future, with tissues optimized for distinct anatomical sites submitted to various mechanical loads, for example. In this regard, system-based approaches [105] and mathematical and in silico modelization [106,107] are also likely to gain more attention in the future in the design of new strategies. Finally, the emergence of cell-editing tools will likely result in a broader range of applications of tissue-engineering strategies, ranging from genetic skin diseases such as *epidermolysis bullosa*, to the manipulation of cutaneous and immune cells in order to generate a "universal" skin substitute. Altogether, the numerous applications of tissue-engineered skin substitutes for cosmetic and reconstructive surgery prompt continued efforts toward successful translation for broad clinical implementation.

Acknowledgments

The authors are grateful to all the members of the LOEX center and their collaborators that have contributed to the field of tissue vascularization over the years, as well as the funding agencies that have supported part of this research including the Canadian Institutes of Health Research (CIHR) grants # 84368 and #111233. We acknowledge the support of the Centre de recherche du CHU de Québec-Université Laval, of the Fonds de recherche du Québec-Santé (FRQS), and of the Quebec Network for cell and tissue therapies—ThéCell (a thematic network funded by the FRQS). The authors report that they have no conflict of interest.

Abbreviations

ASC adipose-derived stromal/stem cells
AT adipose tissue
BEC blood endothelial cells

BM-MSC bone marrow mesenchymal stem/stromal cells
EC endothelial cells
ECFC endothelial colony-forming cells
ECM extracellular matrix
GFP green fluorescent protein
HBOEC human blood outgrowth endothelial cells
HGF hepatocyte growth factor
HMVEC human microvascular endothelial cells
HUVEC human umbilical vein endothelial cells
LEC lymphatic endothelial cells
LYVE-1 lymphatic vessel endothelial hyaluronan receptor 1
MCP-1 monocyte chemoattractant protein 1
PROX1 prospero homeobox protein 1
SVF stromal vascular fraction
UEA-lectin *Ulex europaeus* agglutinin lectin
VEGF vascular endothelial growth factor
VEGFR vascular endothelial growth factor receptor

References

[1] Supp DM, Supp AP, Bell SM, Boyce ST. Enhanced vascularization of cultured skin substitutes genetically modified to overexpress vascular endothelial growth factor. J Invest Dermatol 2000;114(1):5–13.

[2] Supp DM, Boyce ST. Overexpression of vascular endothelial growth factor accelerates early vascularization and improves healing of genetically modified cultured skin substitutes. J Burn Care Rehabil 2002;23(1):10–20.

[3] Supp DM, Karpinski AC, Boyce ST. Vascular endothelial growth factor overexpression increases vascularization by murine but not human endothelial cells in cultured skin substitutes grafted to athymic mice. J Burn Care Rehabil 2004;25(4):337–45.

[4] Eming SA, Medalie DA, Tompkins RG, Yarmush ML, Morgan JR. Genetically modified human keratinocytes overexpressing PDGF-A enhance the performance of a composite skin graft. Hum Gene Ther 1998;9(4):529–39.

[5] Supp DM, Bell SM, Morgan JR, Boyce ST. Genetic modification of cultured skin substitutes by transduction of human keratinocytes and fibroblasts with platelet-derived growth factor-A. Wound Repair Regen 2000;8(1):26–35.

[6] Inoue S, Kijima H, Kidokoro M, Tanaka M, Suzuki Y, Motojuku M, et al. The effectiveness of basic fibroblast growth factor in fibrin-based cultured skin substitute in vivo. J Burn Care Res 2009;30(3):514–9.

[7] Andreadis ST. Gene-modified tissue-engineered skin: the next generation of skin substitutes. Adv Biochem Eng Biotechnol 2007;103:241–74.

[8] Jeschke MG, Shahrokhi S, Finnerty CC, Branski LK, Dibildox M, of Committee TA. Wound coverage technologies in burn care: established techniques. J Burn Care Res 2013;34(6):612–20.

[9] Frueh FS, Menger MD, Lindenblatt N, Giovanoli P, Laschke MW. Current and emerging vascularization strategies in skin tissue engineering. Crit Rev Biotechnol 2016;1–13.

[10] Novosel EC, Kleinhans C, Kluger PJ. Vascularization is the key challenge in tissue engineering. Adv Drug Deliv Rev 2011;63(4–5):300–11.
[11] Kim J, Mirando AC, Popel AS, Green JJ. Gene delivery nanoparticles to modulate angiogenesis. Adv Drug Deliv Rev 2016.
[12] Demidova-Rice TN, Durham JT, Herman IM. Wound healing angiogenesis: innovations and challenges in acute and chronic wound healing. Adv Wound Care *(New Rochelle)* 2012;1(1):17–22.
[13] Cerqueira MT, Pirraco RP, Marques AP. Stem cells in skin wound healing: are we there yet? Adv Wound Care (New Rochelle) 2016;5(4):164–75.
[14] Hendrickx B, Vranckx JJ, Luttun A. Cell-based vascularization strategies for skin tissue engineering. Tissue Eng Part B Rev 2011;17(1):13–24.
[15] Laschke MW, Vollmar B, Menger MD. Inosculation: connecting the life-sustaining pipelines. Tissue Eng Part B Rev 2009;15(4):455–65.
[16] Laschke MW, Menger MD. Prevascularization in tissue engineering: current concepts and future directions. Biotechnol Adv 2016;34(2):112–21.
[17] Kuo K-CC, Lin R-ZZ, Tien H-WW WP-YY, Li Y-CC, Melero-Martin JM, et al. Bioengineering vascularized tissue constructs using an injectable cell-laden enzymatically crosslinked collagen hydrogel derived from dermal extracellular matrix. Acta Biomater 2015;27:151–66.
[18] Hanjaya-Putra D, Shen Y-II, Wilson A, Fox-Talbot K, Khetan S, Burdick JA, et al. Integration and regression of implanted engineered human vascular networks during deep wound healing. Stem Cells Transl Med 2013;2(4):297–306.
[19] Cerqueira MT, da Silva LP, Santos TC, Pirraco RP, Correlo VM, Reis RL, et al. Gellan gum-hyaluronic acid spongy-like hydrogels and cells from adipose tissue synergize promoting neoskin vascularization. ACS Appl Mater Interfaces 2014;6(22):19668–79.
[20] Hudon V, Berthod F, Black AF, Damour O, Germain L, Auger FA. A tissue-engineered endothelialized dermis to study the modulation of angiogenic and angiostatic molecules on capillary-like tube formation in vitro. Br J Dermatol 2003;148(6):1094–104.
[21] Tremblay P-LL, Berthod F, Germain L, Auger FAA. In vitro evaluation of the angiostatic potential of drugs using an endothelialized tissue-engineered connective tissue. J Pharmacol Exp Ther 2005;315(2):510–6.
[22] Berthod F, Symes J, Tremblay N, Medin JA, Auger FAA. Spontaneous fibroblast-derived pericyte recruitment in a human tissue-engineered angiogenesis model in vitro. J Cell Physiol 2012;227(5):2130–7.
[23] Asano Y, Shimoda H, Okano D, Matsusaki M, Akashi M. Transplantation of three-dimensional artificial human vascular tissues fabricated using an extracellular matrix nanofilm-based cell-accumulation technique. J Tissue Eng Regen Med 2015;11(4):1303–7.
[24] Frueh FS, Später T, Lindenblatt N, Calcagni M, Giovanoli P, Scheuer C, et al. Adipose tissue-derived microvascular fragments improve vascularization, Lymphangiogenesis, and integration of dermal skin substitutes. J Invest Dermatol 2017;137(1):217–27.
[25] Zhang X, Yang J, Li Y, Liu S, Long K, Zhao Q, et al. Functional neovascularization in tissue engineering with porcine acellular dermal matrix and human umbilical vein endothelial cells. Tissue Eng Part C Methods 2011;17(4):423–33.
[26] Zhang Z, Ito WD, Hopfner U, Böhmert B, Kremer M, Reckhenrich AK, et al. The role of single cell derived vascular resident endothelial progenitor cells in the enhancement of vascularization in scaffold-based skin regeneration. Biomaterials 2011;32(17):4109–17.
[27] Yang J, Yamato M, Kohno C, Nishimoto A, Sekine H, Fukai F, et al. Cell sheet engineering: recreating tissues without biodegradable scaffolds. Biomaterials 2005;26(33):6415–22.

[28] Sasagawa T, Shimizu T, Sekiya S, Haraguchi Y, Yamato M, Sawa Y, et al. Design of prevascularized three-dimensional cell-dense tissues using a cell sheet stacking manipulation technology. Biomaterials 2010;31(7):1646–54.

[29] Aubin K, Safoine M, Proulx M, Audet-Casgrain M-AA, Côté J-FF, Têtu F-AA, et al. Characterization of in vitro engineered human adipose tissues: relevant adipokine secretion and impact of TNF-α. PLoS One 2015;10(9). e0137612.

[30] Rochon M-HH, Fradette J, Fortin V, Tomasetig F, Roberge CJ, Baker K, et al. Normal human epithelial cells regulate the size and morphology of tissue-engineered capillaries. Tissue Eng Part A 2010;16(5):1457–68.

[31] Chabaud S, Rousseau A, Marcoux T-LL, Bolduc S. Inexpensive production of near-native engineered stromas. J Tissue Eng Regen Med 2015;11(5):1377–89.

[32] Sorrell JM, Baber MA, Caplan AI. Influence of adult mesenchymal stem cells on in vitro vascular formation. Tissue Eng Part A 2009;15(7):1751–61.

[33] Aubin K, Vincent C, Proulx M, Mayrand D, Fradette J. Creating capillary networks within human engineered tissues: Impact of adipocytes and their secretory products. Acta Biomater 2015;11:333–45.

[34] Sorrell JM, Baber MA, Traktuev DO, March KL, Caplan AI. The creation of an in vitro adipose tissue that contains a vascular-adipocyte complex. Biomaterials 2011;32(36):9667–76.

[35] Sorrell JM, Baber MA, Caplan AI. A self-assembled fibroblast-endothelial cell co-culture system that supports in vitro vasculogenesis by both human umbilical vein endothelial cells and human dermal microvascular endothelial cells. Cells Tissues Organs 2007;186(3):157–68.

[36] Hendrickx B, Verdonck K, Van den Berge S, Dickens S, Eriksson E, Vranckx JJ, et al. Integration of blood outgrowth endothelial cells in dermal fibroblast sheets promotes full thickness wound healing. Stem Cells (Dayton, OH) 2010;28(7):1165–77.

[37] Chi JT, Chang HY, Haraldsen G, Jahnsen FL, Troyanskaya OG, Chang DS, et al. Endothelial cell diversity revealed by global expression profiling. Proc Natl Acad Sci U S A 2003;100(19):10623–8.

[38] Monsuur HN, Weijers EM, Niessen FB, Gefen A, Koolwijk P, Gibbs S, et al. Extensive characterization and comparison of endothelial cells derived from dermis and adipose tissue: potential use in tissue engineering. PLoS One 2016;11(11):e0167056.

[39] Haug V, Torio-Padron N, Stark GB, Finkenzeller G, Strassburg S. Comparison between endothelial progenitor cells and human umbilical vein endothelial cells on neovascularization in an adipogenesis mouse model. Microvasc Res 2015;97:159–66.

[40] Traktuev DO, Prater DN, Merfeld-Clauss S, Sanjeevaiah AR, Saadatzadeh MR, Murphy M, et al. Robust functional vascular network formation in vivo by cooperation of adipose progenitor and endothelial cells. Circ Res 2009;104(12):1410–20.

[41] Klar A, Güven S, Biedermann T, Luginbühl J, Böttcher-Haberzeth S, Meuli-Simmen C, et al. Tissue-engineered dermo-epidermal skin grafts prevascularized with adipose-derived cells. Biomaterials 2014;35(19):5065–78.

[42] Black AF, Berthod F, L'Heureux N, Germain L, Auger FA. In vitro reconstruction of a human capillary-like network in a tissue-engineered skin equivalent. FASEB J 1998;12(13):1331–40.

[43] Tremblay P-LL, Hudon V, Berthod F, Germain L, Auger FAA. Inosculation of tissue-engineered capillaries with the host's vasculature in a reconstructed skin transplanted on mice. Am J Transplant 2005;5(5):1002–10.

[44] Supp DM, Wilson-Landy K, Boyce ST. Human dermal microvascular endothelial cells form vascular analogs in cultured skin substitutes after grafting to athymic mice. FASEB J 2002;16(8):797–804.

[45] Black AF, Hudon V, Damour O, Germain L, Auger FA. A novel approach for studying angiogenesis: a human skin equivalent with a capillary-like network. Cell Biol Toxicol 1999;15(2):81–90.
[46] Marino D, Luginbühl J, Scola S, Meuli M, Reichmann E. Bioengineering dermo-epidermal skin grafts with blood and lymphatic capillaries. Science translational medicine 2014;6(221):221ra14.
[47] Klar AS, Güven S, Zimoch J, Zapiórkowska NA, Biedermann T, Böttcher-Haberzeth S, et al. Characterization of vasculogenic potential of human adipose-derived endothelial cells in a three-dimensional vascularized skin substitute. Pediatr Surg Int 2016;32(1):17–27.
[48] Liu Y, Luo H, Wang X, Takemura A, Fang YR, Jin Y, et al. In vitro construction of scaffold-free bilayered tissue-engineered skin containing capillary networks. Bio Med research international 2013;2013:561410.
[49] Cerqueira MT, Pirraco RP, Martins AR, Santos TC, Reis RL, Marques AP. Cell sheet technology-driven re-epithelialization and neovascularization of skin wounds. Acta Biomater 2014;10(7):3145–55.
[50] Gibot L, Galbraith T, Huot J, Auger FAA. A preexisting microvascular network benefits in vivo revascularization of a microvascularized tissue-engineered skin substitute. Tissue Eng Part A 2010;16(10):3199–206.
[51] Pouliot R, Larouche D, Auger FAA, Juhasz J, Xu W, Li H, et al. Reconstructed human skin produced in vitro and grafted on athymic mice. Transplantation 2002;73(11):1751–7.
[52] Beaudoin Cloutier C, Goyer B, Perron C, Guignard R, Larouche D, Moulin VJ, et al. In vivo evaluation and imaging of a bilayered self-assembled skin substitute using a decellularized dermal matrix grafted on mice. Tissue Eng Part A 2016;13:313–22.
[53] Bottcher-Haberzeth S, Biedermann T, Klar AS, Widmer DS, Neuhaus K, Schiestl C, et al. Characterization of pigmented dermo-epidermal skin substitutes in a long-term in vivo assay. Exp Dermatol 2015;24(1):16–21.
[54] Nyame TT, Chiang HA, Orgill DP. Clinical applications of skin substitutes. Surg Clin North Am 2014;94(4):839–50.
[55] Kapur SK, Katz AJ. Review of the adipose derived stem cell secretome. Biochimie 2013;95(12):2222–8.
[56] Bronckaers A, Hilkens P, Martens W, Gervois P, Ratajczak J, Struys T, et al. Mesenchymal stem/stromal cells as a pharmacological and therapeutic approach to accelerate angiogenesis. Pharmacol Ther 2014;143(2):181–96.
[57] Morissette Martin P, Maux A, Laterreur V, Mayrand D, L Gagné V, VJJ M, et al. Enhancing repair of full-thickness excisional wounds in a murine model: impact of tissue-engineered biological dressings featuring human differentiated adipocytes. Acta Biomater 2015;22:39–49.
[58] Dai M, Zhang Y, Yu M, Tian W. Therapeutic applications of conditioned medium from adipose tissue. Cell Prolif 2016;49(5):561–7.
[59] Chen YW, Scutaru TT, Ghetu N, Carasevici E, Lupascu CD, Ferariu D, et al. The effects of adipose-derived stem cell-differentiated adipocytes on skin burn wound healing in rats. J Burn Care Res 2017;38(1):1–10.
[60] Schmidt BA, Horsley V. Intradermal adipocytes mediate fibroblast recruitment during skin wound healing. Development 2013;140(7):1517–27.
[61] Monfort A, Soriano-Navarro M, Garcia-Verdugo JM, Izeta A. Production of human tissue-engineered skin trilayer on a plasma-based hypodermis. J Tissue Eng Regen Med 2013;7(6):479–90.
[62] Huber B, Link A, Linke K, Gehrke SA, Winnefeld M, Kluger PJ. Integration of mature adipocytes to build-up a functional three-layered full-skin equivalent. Tissue Eng Part C Methods 2016;22(8):756–64.

[63] Bellas E, Seiberg M, Garlick J, Kaplan DL. In vitro 3D full-thickness skin-equivalent tissue model using silk and collagen biomaterials. Macromol Biosci 2012;12(12):1627–36.

[64] Trottier V, Marceau-Fortier G, Germain L, Vincent C, Fradette J. IFATS collection: using human adipose-derived stem/stromal cells for the production of new skin substitutes. Stem Cells 2008;26(10):2713–23.

[65] Labbé B, Trottier V, Proulx M, Vincent C, Fradette J. Adipose stem cells, tissue engineering, and solid organ transplantation and regeneration. In: Illouz YG, Sterodimas A, editors. Adipose stem cells and regenerative medicine. Berlin Heidelberg: Springer-Verlag; 2011. p. 229–43.

[66] Huber B, Czaja AM, Kluger PJ. Influence of epidermal growth factor (EGF) and hydrocortisone on the co-culture of mature adipocytes and endothelial cells for vascularized adipose tissue engineering. Cell Biol Int 2016;40(5):569–78.

[67] Choi JH, Gimble JM, Lee K, Marra KG, Rubin JP, Yoo JJ, et al. Adipose tissue engineering for soft tissue regeneration. Tissue Eng Part B Rev 2010;16(4):413–26.

[68] Wittmann K, Dietl S, Ludwig N, Berberich O, Hoefner C, Storck K, et al. Engineering vascularized adipose tissue using the stromal-vascular fraction and fibrin hydrogels. Tissue Eng Part A 2015;21(7–8):1343–53.

[69] Volz AC, Huber B, Kluger PJ. Adipose-derived stem cell differentiation as a basic tool for vascularized adipose tissue engineering. Differentiation 2016;92(1–2):52–64.

[70] Huber B, Volz AC, Kluger PJ. Understanding the effects of mature adipocytes and endothelial cells on fatty acid metabolism and vascular tone in physiological fatty tissue for vascularized adipose tissue engineering. Cell Tissue Res 2015;362(2):269–79.

[71] Choi JH, Bellas E, Gimble JM, Vunjak-Novakovic G, Kaplan DL. Lipolytic function of adipocyte/endothelial cocultures. Tissue Eng Part A 2011;17(9–10):1437–44.

[72] Choi JH, Gimble JM, Vunjak-Novakovic G, Kaplan DL. Effects of hyperinsulinemia on lipolytic function of three-dimensional adipocyte/endothelial co-cultures. Tissue Eng Part C Methods 2010;16(5):1157–65.

[73] Bellas E, Marra KG, Kaplan DL. Sustainable three-dimensional tissue model of human adipose tissue. Tissue Eng Part C Methods 2013;19(10):745–54.

[74] Kang JH, Gimble JM, Kaplan DL. In vitro 3D model for human vascularized adipose tissue. Tissue Eng Part A 2009;15(8):2227–36.

[75] Frerich B, Winter K, Scheller K, Braumann UD. Comparison of different fabrication techniques for human adipose tissue engineering in severe combined immunodeficient mice. Artif Organs 2012;36(3):227–37.

[76] Yao R, Du Y, Zhang R, Lin F, Luan J. A biomimetic physiological model for human adipose tissue by adipocytes and endothelial cell cocultures with spatially controlled distribution. Biomed Mater 2013;8(4):045005.

[77] Yao R, Zhang R, Lin F, Luan J. Biomimetic injectable HUVEC-adipocytes/collagen/alginate microsphere co-cultures for adipose tissue engineering. Biotechnol Bioeng 2013;110(5):1430–43.

[78] Verseijden F, Posthumus-van Sluijs SJ, Farrell E, van Neck JW, Hovius SE, Hofer SO, et al. Prevascular structures promote vascularization in engineered human adipose tissue constructs upon implantation. Cell Transplant 2010;19(8):1007–20.

[79] Verseijden F, Posthumus-van Sluijs SJ, van Neck JW, Hofer SO, Hovius SE, Van Osch GJ. Comparing scaffold-free and fibrin-based adipose-derived stromal cell constructs for adipose tissue engineering: an in vitro and in vivo study. Cell Transplant 2012;21(10):2283–97.

[80] Mayrand D, Fradette J. High definition confocal imaging modalities for thick tissue-engineered substitutes. In: Gimble JM, Bunnell BA, editors. Methods in molecular biology. Springer; in press.
[81] Daly SM, Leahy MJ. 'Go with the flow': a review of methods and advancements in blood flow imaging. Journal of biophotonics 2013;6(3):217–55.
[82] Uehara F, Tome Y, Reynoso J, Mii S, Yano S, Miwa S, et al. Color-coded imaging of spontaneous vessel anastomosis in vivo. Anticancer Res 2013;33(8):3041–5.
[83] Calcagni M, Althaus MK, Knapik AD, Hegland N, Contaldo C, Giovanoli P, et al. In vivo visualization of the origination of skin graft vasculature in a wild-type/GFP crossover model. Microvasc Res 2011;82(3):237–45.
[84] Capla JM, Ceradini DJ, Tepper OM, Callaghan MJ, Bhatt KA, Galiano RD, et al. Skin graft vascularization involves precisely regulated regression and replacement of endothelial cells through both angiogenesis and vasculogenesis. Plast Reconstr Surg 2006;117(3):836–44.
[85] Lindenblatt N, Calcagni M, Contaldo C, Menger MD, Giovanoli P, Vollmar B. A new model for studying the revascularization of skin grafts in vivo: the role of angiogenesis. Plast Reconstr Surg 2008;122(6):1669–80.
[86] Egaña J, Condurache A, Lohmeyer J, Kremer M, Stöckelhuber BM, Lavandero S, et al. Ex vivo method to visualize and quantify vascular networks in native and tissue engineered skin. Langenbecks Arch Surg 2009;394(2):349–56.
[87] Proulx M, Aubin K, Lagueux J, Audet P, Auger M, Fortin M-AA, et al. Magnetic resonance imaging of human tissue-engineered adipose substitutes. Tissue Eng Part C Methods 2015;21(7):693–704.
[88] Sun W, Sun Y, Klar AS, Geutjes P, Reichmann E, Heerschap A, et al. Functional analysis of vascularized collagen/fibrin templates by MRI in vivo. Tissue Eng Part C Methods 2016;22(8):747–55.
[89] Gangemi EN, Carnino R, Stella M. Videocapillaroscopy in postburn scars: in vivo analysis of the microcirculation. Burns 2010;36(6):799–805.
[90] Hern S, Mortimer PS. Visualization of dermal blood vessels--capillaroscopy. Clin Exp Dermatol 1999;24(6):473–8.
[91] Lal C, Leahy MJ. An updated review of methods and advancements in microvascular blood flow imaging. Microcirculation 2016;23(5):345–63.
[92] Alitalo K, Tammela T, Petrova T. Lymphangiogenesis in development and human disease. Nature 2005;438(7070):946–53.
[93] Matsusaki M, Fujimoto K, Shirakata Y, Hirakawa S, Hashimoto K, Akashi M. Development of full-thickness human skin equivalents with blood and lymph-like capillary networks by cell coating technology. J Biomed Mater Res A 2015;103(10):3386–96.
[94] Podgrabinska S, Braun P, Velasco P, Kloos B, Pepper MS, Skobe M. Molecular characterization of lymphatic endothelial cells. Proc Natl Acad Sci U S A 2002;99(25):16069–74.
[95] Amatschek S, Kriehuber E, Bauer W, Reiniger B, Meraner P, Wolpl A, et al. Blood and lymphatic endothelial cell-specific differentiation programs are stringently controlled by the tissue environment. Blood 2007;109(11):4777–85.
[96] Kriehuber E, Breiteneder-Geleff S, Groeger M, Soleiman A, Schoppmann SF, Stingl G, et al. Isolation and characterization of dermal lymphatic and blood endothelial cells reveal stable and functionally specialized cell lineages. J Exp Med 2001;194(6):797–808.
[97] Gibot L, Galbraith T, Kloos B, Das S, Lacroix DA, Auger FAA, et al. Cell-based approach for 3D reconstruction of lymphatic capillaries in vitro reveals distinct functions of HGF and VEGF-C in lymphangiogenesis. Biomaterials 2016;78:129–39.

[98] Klar A, Böttcher-Haberzeth S, Biedermann T, Schiestl C, Reichmann E, Meuli M. Analysis of blood and lymph vascularization patterns in tissue-engineered human dermo-epidermal skin analogs of different pigmentation. Pediatr Surg Int 2014;30(2):223–31.

[99] Hikimoto D, Nishiguchi A, Matsusaki M, Akashi M. High-throughput blood- and lymph-capillaries with open-ended pores which allow the transport of drugs and cells. Adv Healthc Mater 2016;5(15):1969–78.

[100] Asano Y, Nishiguchi A, Matsusaki M, Okano D, Saito E, Akashi M, et al. Ultrastructure of blood and lymphatic vascular networks in three-dimensional cultured tissues fabricated by extracellular matrix nanofilm-based cell accumulation technique. Microscopy 2014;63(3):219–26.

[101] Lee S-JJ, Park C, Lee JY, Kim S, Kwon PJ, Kim W, et al. Generation of pure lymphatic endothelial cells from human pluripotent stem cells and their therapeutic effects on wound repair. Sci Rep 2015;5:11019.

[102] Michael S, Sorg H, Peck CT, Koch L, Deiwick A, Chichkov B, et al. Tissue engineered skin substitutes created by laser-assisted bioprinting form skin-like structures in the dorsal skin fold chamber in mice. PLoS One 2013;8(3):e57741.

[103] Kolesky DB, Homan KA, Skylar-Scott MA, Lewis JA. Three-dimensional bioprinting of thick vascularized tissues. Proc Natl Acad Sci U S A 2016;113(12):3179–84.

[104] Bae H, Puranik AS, Gauvin R, Edalat F, Carrillo-Conde B, Peppas NA, et al. Building vascular networks. Sci Transl Med 2012;4(160):160ps23.

[105] Buganza Tepole A, Kuhl E. Systems-based approaches toward wound healing. Pediatr Res 2013 Apr;73(4 Pt 2):553–63.

[106] Flegg JA, Menon SN, Maini PK, McElwain DL. On the mathematical modeling of wound healing angiogenesis in skin as a reaction-transport process. Front Physiol 2015;6:262.

[107] Perfahl H, Hughes BD, Alarcon T, Maini PK, Lloyd MC, Reuss M, et al. 3D hybrid modelling of vascular network formation. J Theor Biol 2016;414:254–68.

Wound healing and cutaneous scarring models of the human skin

Mohammed Ashrafi*, †, ‡, #, Adam Hague*, #, Mohamed Baguneid†, Teresa Alonso-Rasgado‡, Ardeshir Bayat*, †, ‡
*Plastic & Reconstructive Surgery Research, Centre for Dermatological Research, University of Manchester, Manchester, United Kingdom, †University Hospital of South Manchester NHS Foundation Trust, Manchester, United Kingdom, ‡Bioengineering Group, School of Materials, University of Manchester, Manchester, United Kingdom

1 Introduction

Wound healing is a multifactorial process, which commences after any form of cutaneous injury resulting in immediate vasoconstriction and coagulation [1]. The subsequent release of cytokines by keratinocytes and platelets leads to initial inflammation, with various immune cells, including neutrophils and monocytes, migrating into the wound bed [1]. Release of growth factors by macrophages stimulates angiogenesis, resident epithelial cells to proliferate, and fibroblast infiltration to occur, ultimately assisting with wound closure [2]. The final visual appearance of the wound-healing process is of upmost importance to both patients and physicians alike, with adverse scarring being a cause for concern. A major challenge currently facing researchers investigating this area is creating an ideal model to study this complex process, evaluate candidate therapies, and ultimately improve aesthetic and functional outcomes for patients with abnormal wounds and scars [3]. Many models currently exist, investigating all stages of wound healing including reepithelialization and scar remodeling [4].

Animal models have been widely used in attempts to replicate the human skin [5]. Pig skin is currently thought to be the most accurate animal model; however, no animal skin is an exact replica of the human skin, leading to significant differences in the wound-healing mechanisms between them [6,7]. Additionally, ethical issues surrounding their use also need to be taken into consideration [8].

Numerous human scar and wound-healing models have been developed, including in silico, in vitro, ex vivo, and in vivo. The aim of this review is to summarize and evaluate each of the above models including their advantages and limitations (Table 1).

2 In silico

The large variety of interactions that influence wound healing requires the need for computational models in order to accurately understand and investigate this complex process

#Authors contributed equally to the chapter and are joint first authors.

Skin Tissue Models. https://doi.org/10.1016/B978-0-12-810545-0.00009-7

Table 1 **Human wound models**

Model	Description/features	Application	Advantages	Disadvantages
In silico				
Differential equations Agent-based models	- The use of mathematical equations based on known cellular behaviors to simulate wound healing - A computational modeling technique that can simulate interactions between cells and their environment. Classically consist of three components: agent, region, and patch	- Investigation of all stages of wound healing, along with the impact that various cells, cytokines, and growth factors have on this process	- Can be used as an initial screening tool to help plan clinical trials/ research	- Can involve complex mathematical equations - The lack of standardized outcome measures - Remain experimental unless validated
In vitro				
Single cell Coculture Organ culture	- A defect simulating a wound is created in cells cultured in monolayer - Coculture of more than one cell type. The use of the Transwell system has classically been described for this purpose - Cells cultured on a three-dimensional (3-D) scaffold or grown in a 3-D format in order to create a skin "equivalent"	- Studying rates of cell migration and proliferation in a wound setting, for example, keratinocytes and fibroblasts - Investigation of specific cellular interactions and subsequent effects - Investigation of scar pathogenesis	- Single-cell models are relatively quick and inexpensive to conduct and allow the investigation of one specific cell type - Ability to include more than one cell type more accurately reflects the wound environment, particularly with organotypic models	- Models cannot accurately reflect cellular interactions in the wound environment, particularly single-cell and coculture models - Many models use serum, which is inflammatory when compared with plasma and interstitial fluid. Can therefore be difficult to translate findings into the in vivo setting - The lack of validated biomarkers - Dependence on excised tissue

ex vivo				
Scars Skin with the epidermis intact The skin with artificially created wounds	- The use of harvested cutaneous tissue - Common scar tissue used includes that from keloids and striae albae - Wounds created can be full thickness or partial thickness and include burns	- Investigation of scar pathogenesis and physiology of wound healing - Testing of potential therapeutic agents - Genotoxicity testing	- Provides a safe environment for the assessment of potential treatments - Creates an environment that more closely resembles normal skin	- Short investigative opportunity - Not subject to the same environmental factors that would be present in a patient - The lack of standardization in study design
In vivo				
Scars Wounds	- The use of live subjects. These can be either patients with wounds or scars already present or volunteers in which a wound is created	- Validation of wound and scar therapies	- Physiology of healing is identical to that in patients	- The lack of standardization in study design - Difficulties with recruitment

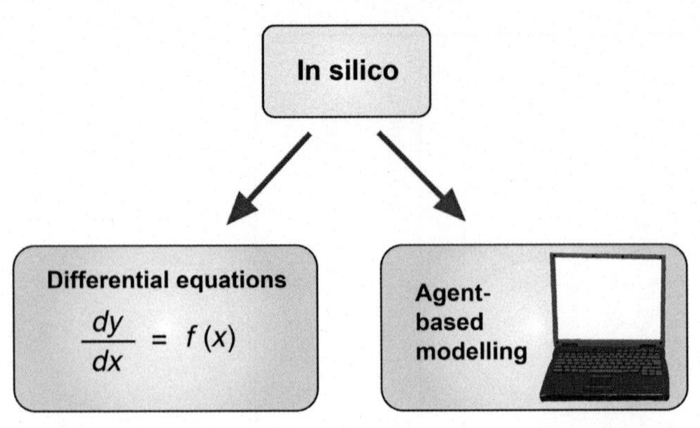

Fig. 1 Diagram illustrating examples of in silico human models of scar formation and wound healing.

(Fig. 1) [9]. Numerous work has been conducted into designing such models for various stages of the wound-healing processes [9]. These can then subsequently be validated experimentally and used to assist in various areas of tissue engineering research.

Differential equations have been widely used to investigate the differing phases of wound healing [9]. Models specifically looking at the inflammatory phase of healing have used ordinary differential equations to outline the effects of variability between patients in growth factors and structural proteins, along with macrophages and fibroblasts [10,11]. Equations have also been used to model wound closure [12,13]. Arciero et al. investigated the mechanisms that coordinate cell migration during wound healing by deriving a two-dimensional continuum model [12]. Evolution equations were then solved numerically using a level set method. Mathematical models were also utilized by Wearing et al. who looked into the effects of keratinocyte growth factor on reepithelialization [13]. The final stage of the wound-healing process, tissue remodeling, was investigated by McDougall et al. [14]. They applied mathematical models to demonstrate the influences that constituents of the extracellular matrix (ECM) have on the trajectories of fibroblasts and direction of collagen deposition.

Agent-based modeling is a computational modeling technique that can simulate complex behaviors and interactions between agents and their environment while taking into consideration the stochastic nature of biological processes [15]. Agent-based models (ABMs) typically consist of three components: agent, region, and patch [15]. The agent represents cells or molecules that move in the region, which consists of individual patches. The simulated behaviors of agents in their virtual environment are based on current knowledge regarding their known behavior [16]. They have several advantages including being easier to use for nonmathematicians while also simplifying the representation of tissue structure [16]. Adra et al. derived an ABM to assess the regulatory actions transforming growth factor-beta 1 (TGF-β1) has in tissue regeneration [17]. They subsequently validated the model by using it to create a virtual piece of epidermis and comparing keratinocyte behavior and the actions of TGF-β1 with the existing literature. Similarly, Li et al. used a three-dimensional (3-D) ABM of the epidermis to investigate epithelial renewal and basal keratinocyte regeneration [18].

ABMs have also been used to provide a framework for the creation of multiscale models for wound healing. Along with investigating the effects of interactions between keratinocytes and fibroblasts [19], such models have also been used to study the biological influences that regulate human keratinocyte organization [20]. Anderson et al. advanced this type of model to create a hybrid continuous-discrete model for deterministic systems [21], which has been utilized to study fluid flow through vascular networks [22,23].

3 In vitro

In vitro models have been widely used in the investigation of wound healing and scar pathogenesis [3]. They are generally inexpensive and require less ethical considerations when compared with their in vivo counterparts [24]. They vary in complexity ranging from single-cell and coculture (two different cell types) models to organotypic multicellular-layered constructs (Fig. 2) [24].

Single-cell models commonly involve culturing cells in monolayer directly on the surface of the culture dish itself or on substrates including collagen and fibrin [24]. A defect, simulating a wound, is then created in the cell monolayer through mechanical [25,26] or chemical methods [27]. The defect is then gradually repopulated by the surrounding cells [28]. Using microscopy, the rates at which this proliferation and migration occur can then be determined. Such models classically involve keratinocytes [29] or fibroblasts [28,30].

Although they are relatively quick and inexpensive to conduct, single-cell models do not accurately reflect the wound environment [3]. In order to overcome this, several other models have been created. Cocultures of fibroblasts and keratinocytes using Transwell systems (Corning Costar, MA) have been conducted, allowing the

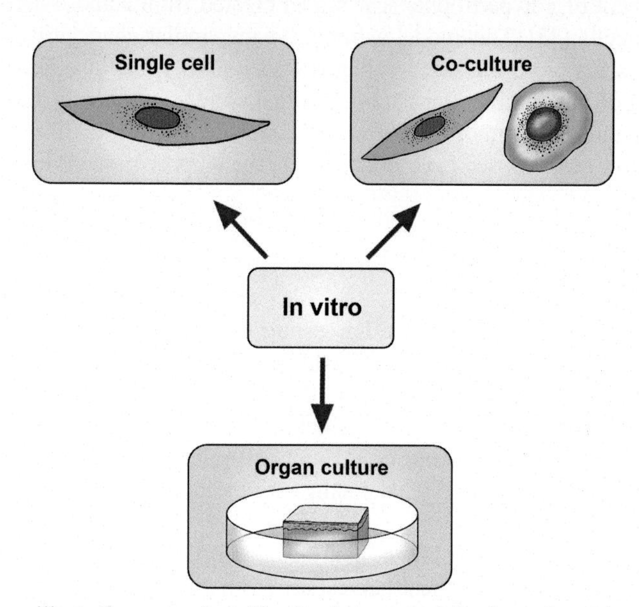

Fig. 2 Diagram illustrating examples of in vitro human models of scar formation and wound healing.

investigation of the interactions between these two cell types and the subsequent effects [31,32]. The Transwell assay is composed of two separate mediums divided by a porous membrane and can be used for various cell types including epithelial [33] and mesenchymal cells [34].

In vitro 3-D models have also been used in attempts to construct and mimic the ECM while also more accurately reflecting normal wound physiology [35]. It has already been shown that both collagen and collagenase production by fibroblasts varies when they are cultured in two and 3-D, presumably due to the action of integrins [36]. Nayak et al. utilized a construct of sericin to assess the proliferation and viability of cocultures of keratinocytes and fibroblasts, which were located on the upper and lower surfaces, respectively [37]. Similarly, collagen and fibrin have been used to create a 3-D environment, with both promoting fibroblasts to differentiate toward the scar phenotype when combined with a mechanical load [38]. Additionally, Karamichos et al. created a 3-D collagen matrix to study the effects of different culture conditions on matrix remodeling and fibroblast migration [39].

The knowledge that extensive interaction between keratinocytes and fibroblasts regulates the production of dermal ECM, led to the development and use of organotypic skin equivalents in scar research [40]. Bellemare et al. isolated keratinocytes from hypertrophic scars on a dermal matrix and created a fully differentiated epidermis [41]. This subsequently exhibited characteristics of adverse scarring and highlighted the contribution that keratinocytes have in hypertrophic scar formation. Likewise, Chiu et al. investigated the effects of photodynamic therapy on keloid scars through the use of an organotypic skin equivalent with keratinocytes and keloid fibroblasts [42]. The use of such models to investigate potential therapeutics is, however, limited due to the need for excised scar tissue along with a lack of validated biomarkers [3]. This led to the development of a hypertrophic scar model created from adipose-derived mesenchymal stem cells [43]. This model exhibited many similar characteristics to that of hypertrophic scars (e.g., increased collagen I secretion) while also allowing for the identification of measurable scar parameters that were subsequently validated using various treatments. The authors did, however, conclude that this model is only relevant for hypertrophic scars caused from injury where the adipose tissue is exposed.

4 Ex vivo

Ex vivo models initially involve the harvesting of cutaneous tissue from informed volunteers and with appropriate research ethical approvals in place. These models are critical in evaluating the pathophysiology of scars and wounds and their response to treatment, especially in diseases, such as keloid, where no animal models exist. The three main categories of cutaneous ex vivo models published are scars [44–47], skin with the epidermis intact [48–53], and skin with various artificially created wounds [48,52,54–58] (Fig. 3).

Scar models include keloid [44–47], striae alba [46], hypertrophic [46], and fine-line scars [46], with the site of scar harvest differing greatly between studies [46,47]. Biopsies of the scars used varied in shape and size; however, the majority were created

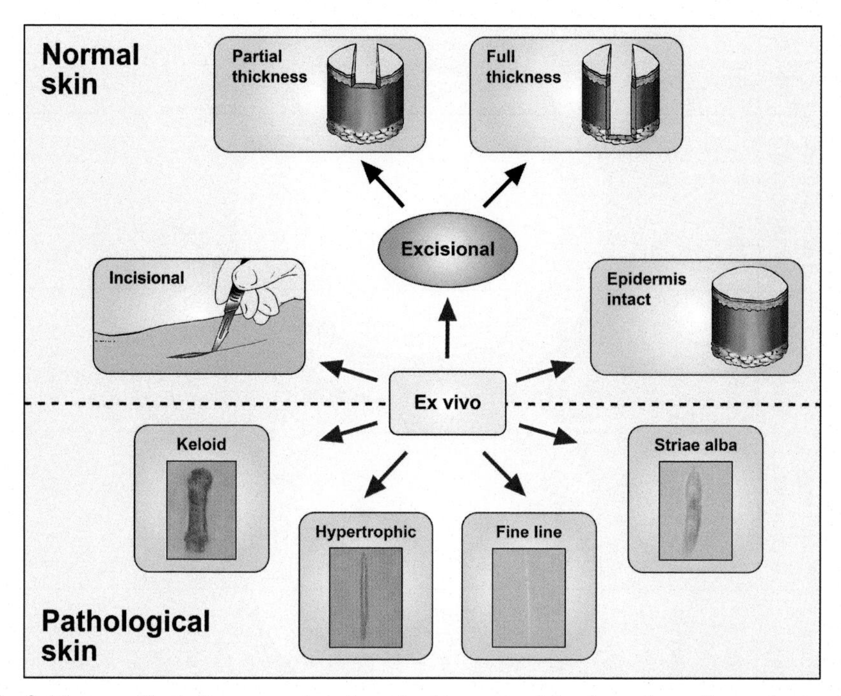

Fig. 3 Diagram illustrating examples of ex vivo human models of scar formation and wound healing.

using circular punch biopsies [44,46,47], although square-shaped explants have also been utilized [45]. The minimum diameter of biopsies used was 3 mm [44] with the largest reported at 6 mm [44,46]. Both Hanks buffer [44,45,47] and phosphate-buffered saline [46] have been used to wash the explants prior to embedding. Using multiple validation measures including histological assessment, cell proliferation and apoptotic assays, ELISA, and immunohistochemical and immunofluorescence staining, both Duong et al. [45] and Bagabir et al. [44] were able to preserve keloid ex vivo organ culture models for up to 6 weeks. Various media were used to support the explants (Table 2), with the use of supplemented William's E medium changed every 3 days, supporting superior proliferation with reduced cytotoxicity [44]. All explant models were embedded in collagen gel matrix with the epidermis either submerged in media [44,45] or air-exposed [44–47], with superior morphology being observed in the air-exposed explants [45]. Bagabir et al. [44] supplemented the media with dexamethasone, which is employed in the clinical management of keloid disease, and found shrinkage compared with untreated controls at 4 weeks, whereas at 3 weeks, no such macroscopic changes were apparent, highlighting the need for keloid models with longevity in order to accurately determine treatment output. The same model has also been used to evaluate photodynamic [46] and antifibrotic therapies [47].

Skin explant models with the epidermis intact are employed as a control to wounded explant models. They are typically harvested from the breast [49–52] and abdomen of patients undergoing elective surgery [51,53]. The size of the explants described

Table 2 Ex vivo models

	Model	Shape/size	Embedding techniques	Epidermis	Culture duration	Culture media	Validation techniques	Site	Therapeutics investigated	Disinfection
[44]	-Scar	-Circle -3–6 mm	-RTCG	-AE-S	-6 weeks	-DMEM -WEM -Changed every 3 days	-LDH -MTT -TUNEL -Ki67 -Col I–III ELISA -Histology -IHC-IF	-Sternum -Arm -Pubis -Scalp -Face -Ear -Abdomen	-DEX	-HB
[45]	-Scar	-Square -3–5 mm	-RTCG	-AE -S	-6 weeks	-DMEM -FBS -Changed every 24 -72 h	-IHC -TUNEL -ISH	-NS	-Nil	-HB
[46]	-Scar	-Circle -6 mm	-RTCG	-AE	-7 days	-WEM	-Histology -IHC -IF -qRT -PCR	-Abdomen -Arm -Breast -Thorax -Ear	-PDT	-PBS
[47]	-Scar	-Circle -4 mm	-RTCG	-AE	-4 weeks	-WEM -Changed every 3 days	-LDH -MTT -TUNEL -Ki67 -Col I–III ELISA -Histology -IHC -qRT -PCR	-Sternum -Arm -Pubis -Scalp -Face -Ear -Abdomen	-EGCG -PAI-1 -DEX	-HB

[48]	-Skin -Wound	-Circle -6 mm -Circle -6 mm -3 mm FTW	-RTCG -RTCG	-AE -AE	-3 weeks -3 weeks	-DMEM -Changed every 48 h	-LDH -Histology -IHC -VEGF ELISA	-NS	-IL 10	-HB -70% ethanol
[49]	-Skin	-Circle -3 mm	-TF	-AE	-24 h	-IMDM	-IHC	-Breast	-Cytokines -LPS	-NS
[50]	-Skin	-Square -1.5 cm^2	-TF	-AE	-15 days	-EpiLife with EDGS -Changed every 36 h	-Histology -TUNEL	-Breast	-Fixator pin -Biomaterial	-PBS
[51]	-Skin	-Circle -16 mm	-NI	-AE	-24 h	-DMEM -Changed every 24 h	-Comet assay -MTT	-Breast -Abdomen	-Toxins	-HB
[52]	-Skin -Wound	-Square -1 cm^2 -Square -1 cm^2 -Circle -4 mm FTW	-NS	-AE	-6 days	-KBM	-Planimetry -Histology	-Breast	-Toxins	-NS
[53]	-Skin	-NS	-CS	-AE	-14 days	-DMEM -Changed every 48 h	-Histology -TUNEL -Ki-67 -IF	-Abdomen	-Nil	-Betadine -70% ethanol
[54]	-Wound	-Circle -8 mm -4 mm FTW	-TF	-AE	-21 days	-DMEM -Changed NS	-Histology	-Abdomen	-Hyaluronan	-NS

(Continued)

Table 2 Continued

	Model	Shape/size	Embedding techniques	Epidermis	Culture duration	Culture media	Validation techniques	Site	Therapeutics investigated	Disinfection
[55]	-Wound	-Circle -6 mm -3 mm FTW -Square -Burn wound	-Nil	-S	-14 days	-DMEM -Changed every 48 h	-Histology -Cell culture	-Breast	-Nil	-NS
[56]	-Wound	-Circle -6 mm -3 mm PTW	-Culture disc	-AE	-7 days	-DMEM -Changed every 48 h	-IHC -ISH	-Trunk	-Keratinocytes	-NS
[57]	-Wound	-Circle -8 mm -4 mm FTW	-RTCG	-AE	-16 days	-DMEM -Changed every 48 h	-Histology -IHC -qRT-PCR -Western blotting -LDH -Flow cytometry	-Abdomen	-Electric stimulation	-PBS
[58]	-Wound	-Circle -8 mm -2 mm FTW	-TF	-S and AE	-8 days	-NS -Changed NS	-Histology -IHC	-Tissue-engineered human skin	-Dermal substitutes	-NS

FTW, full-thickness wound; *NS*, not specified; *PTW*, partial-thickness wound; *RTCG*, rat tail collagen gel; *TF*, Transwell filter; *NI*, Netwell inserts; *CS*, cell strainer; *AE*, air-exposed; *S*, submerged; *DMEM*, Dulbecco's modified eagle's medium; *WEM*, William's E medium; *FBS*, fetal bovine serum; *IMDM*, Iscove's modified Dulbecco's medium; *EDGS*, EpiLife defined growth supplement; *KBM*, keratinocyte basal medium; *LDH*, lactate dehydrogenase; *TUNEL*, terminal deoxynucleotidyl transferase dUTP nick end labeling; *Col*, collagen; *ELISA*, enzyme-linked immunosorbent assay; *IHC*, immunohistochemistry; *IF*, immunofluorescence; *ISH*, in situ hybridization; *qRT-PCR*, quantitative reverse transcription-polymerase chain reaction; *VEGF*, vascular endothelial growth factor; *DEX*, dexamethasone; *PDT*, photodynamic therapy; *EGCG*, (−)-epigallocatechin-3-gallate; *PAI-1*, plasminogen activator inhibitor-1; *IL*, interleukin; *LPS*, lipopolysaccharides; *HB*, Hanks buffer; *PBS*, phosphate-buffered saline.

varies in size and shape, ranging from 6 to 16 mm circular punches [48,49,51], uniform squares with a surface area of 1–1.5 cm^2 [52] or being irregularly shaped [50]. Numerous solutions have been used to disinfect the explants including 70% ethanol [48,53], betadine [53], Hanks buffer with [51] or without antibiotic supplementation [48], and phosphate-buffered saline supplemented with gentamicin and amphotericin B [50]. Explant viability ranged from 24 h [49] to 3 weeks [48] with the epidermis being air-exposed in all models [48–53]. Balaji et al. [48] embedded the skin explants in rat tail collagen 1 gel matrix cultured in supplemented serum-free Dulbecco's modified eagle's medium with the epidermal surface air-exposed. With the media changed every other day for 21 days, they showed detachment and decreasing thickness of the epidermis at day 14; however, they showed the dermis was similar to day 0 skin and lactate dehydrogenase (LDH) activity declined from day 1 to day 21. Companjen et al. used a different embedding technique with the skin explants inserted into Transwell filters, with the dermis suspended in culture medium and the epidermis also air-exposed [49]. These biopsies, taken from healthy volunteers and those with psoriatic plaque-type skin lesions, were incubated for only 24 h in standard conditions (5% carbon dioxide and 37°C) or special conditions, where explants were placed into Tedlar culture bags with an artificial atmosphere of 5% carbon dioxide and 95% oxygen at 32°C. They found morphological deterioration of psoriatic lesions in standard conditions that were abolished under special conditions. No such deterioration of normal skin explants was noted. They concluded the model allowed maintenance of the normal skin architecture without spontaneous induction of regenerative maturation markers, through measurement of cytokine release with ELISA and immunohistochemistry. However, they only cultured the model for a very short time period [49]. Xu et al. anchored a partial dermal thickness skin explant using sutures to a nylon mesh cell strainer and showed normal epidermal morphology, with keratinocyte proliferation up to 12 days and an intact basement membrane for up to 21 days. [53]. The epidermis was air-exposed, while the dermis was surrounded in media that was changed every other day. The skin explant models have been used to assess cytokine activity [48,49], glycosaminoglycans [50], and toxins [51,52].

Cutaneous explants with iatrogenic wounds were either harvested from the breast [52,55], the abdomen following abdominoplasty surgery [54,57], or the trunk [56]. All but one study used circular skin explants 6–8 mm in diameter [48,54–58]. Tomic-Canic et al. used 1 cm^2 explants with a 4 mm circular punch biopsy through the reticular dermis [52]. All other studies have also utilized circular wounds within the explants ranging in depth from 1 mm [55], the upper dermis [56], to full thickness [48,54,57,58]. Kratz et al. also created a deep dermal burn wound model by applying a heated brass string to the epidermal surface for 1 s at a temperature of 150°C [55]. Disinfection, embedding, length of culture, type of media, and validation methods varied between the studies (Table 2). Balaji et al. evaluated cytokine treatment on wound healing by embedding the wounded explants in rat tail collagen matrix, maintaining the model for up to 3 weeks with media changed every other day and monitoring viability using a LDH cytotoxicity assay and histological evaluation [48]. Kratz et al. found discrepancies in the wound-healing process between incisional and burn wounds, highlighting the vital role ex vivo models have in understanding the normal healing process of different types of wounds, and thus potentially allowing the investigation of therapeutics that are wound type-specific [55]. These cutaneous wounded explant models have also been used to assess toxins [52], glycosaminoglycans [54], and keratinocyte transplantation [56]. Dermal substitutes have also been assessed by inserting them into the full-thickness wounds created in the tissue explants [54,58]. Sebastian et al. used the donut-shaped full-thickness wounded explants to investigate the effect of electric stimulation on wound closure and found enhanced epidermal proliferation [57].

The use of these diverse scar and wound models described above in assessing cytokine activity, toxin effects, cell migration, and proliferation is extensively highlighted. These models have also allowed assessment of other features such as epidermal thickness [44], epithelialization [57], and the role of immune components in healing and response to treatment [5]. Other features of wound healing where these models could be used to investigate include the study of dermal thickness, angiogenesis, hair follicles, rete ridge formation, and ECM proteins.

5 In vivo

In vivo cutaneous scar and wound models are ideal for investigating the pathophysiology of healing in live subjects (Fig. 4). In vivo models are used to either assess normal scarring or wound-healing processes or used to investigate the effect of candidate therapeutics [59,60].

To assess the normal scarring process, the creation of a scar in healthy volunteers is a model that has been utilized. Dunkin et al. hypothesized that there is a critical depth of wounding beyond which a fibrous scar develops [61]. They confirmed this by creating an incisional wound of varying depth with deep dermal at one end and superficial dermal at the other in 113 healthy volunteers along the lateral aspect of the hip. Using digital images and high-frequency ultrasound, they found no detectable scar at the shallow end of the wound at the study end point of 9 months, confirming their hypothesis [61].

The impact of treatments on scars has been assessed in clinical studies. Ud-din et al. conducted a case series of 65 patients with keloid disease to assess steroid-sensitive responders using mostly subjective parameters [59]. Over a 3-month period, they found keloid scar steroid sensitivity was associated with anatomical location and frequency of injections. However, the lack of objective measures of response and short follow-up period are limitations to be considered. The same group conducted a similar case series on patients with keloid disease undergoing photodynamic therapy over a longer follow-up period of 9 months; however, on this occasion, they also included quantifiable non-invasive objective measures of response including spectrophotometric intracutaneous analysis and full-field laser perfusion imagery [60]. Human in vivo scars have also been used to assess less conventional treatments such as electric stimulation [62]. The application of electric stimulation on raised dermal scars found reduced pain and pruritus over a 2-month study period in 19 patients [62]. Although, objective measurements of scar melanin, collagen, hemoglobin, and blood flow were measured, the

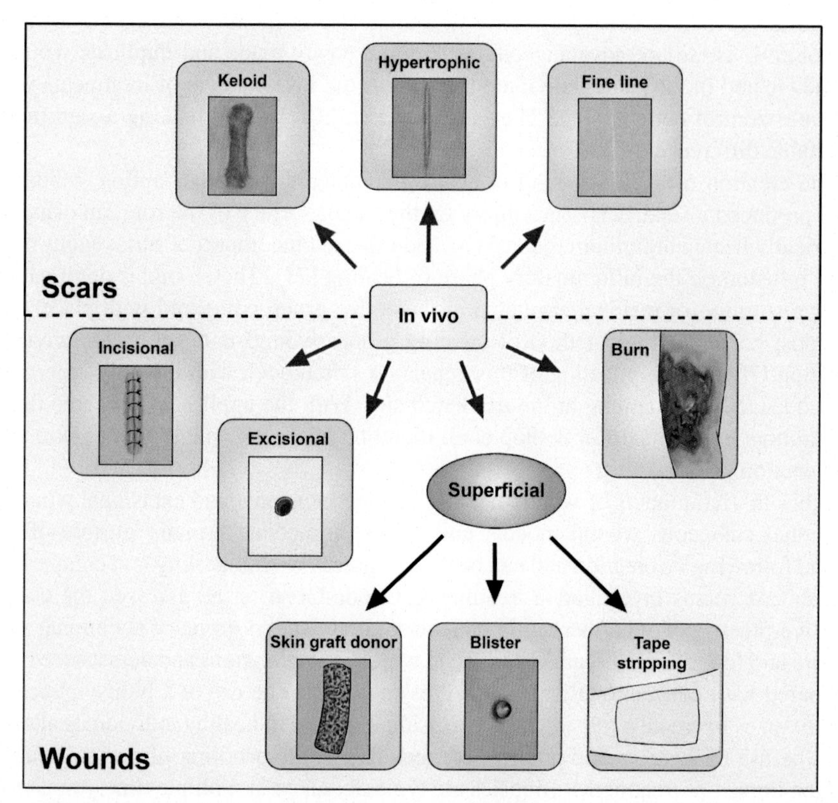

Fig. 4 Diagram illustrating examples of in vivo human models of scar formation and wound healing.

primary significant findings were subjective outcomes. As well as noninvasive objective measures, invasive biopsies of scars following treatment have also been done to assess morphological and immunohistochemical changes post treatment [63].

Various cutaneous wound models in patients and healthy volunteers have been investigated with regards to the healing process and the effect of various treatment modalities. These models include superficial [64–69], burn [70,71], incisional [72], and excisional wounds [73–77].

Superficial wounds include the blister model [64–66], tape-stripping method [67], and skin graft donor sites [68,69]. The blister model allows the creation of a standardized superficial wound that can be monitored for epidermal regeneration. This is created using various suction devices, and these models can be used to assess drug effects and drug absorption [65,66]. They are easily replicated; however, they are limited to only providing a useful tool for the investigation of superficial rather than deep dermal wounds. Less detrimental to the epidermis is the tape-stripping method. This involves the noninvasive application of adhesive tape to the skin with repeated stripping of the epidermis that leads to skin barrier loss [67]. Although offering the least side effects and proving useful in assessing treatment penetration, these advantages are limited to the stratum corneum. Skin graft donor sites are a further model of superficial wounding [68,69]. These are advantageous as they are easily made and duplicate wounds can be created in the same individual that allows the investigation of treatments with optimum control comparisons. They do, however, heal rapidly making a significant detectable difference difficult.

The creation of a burn wound in healthy volunteers is another option. Mattsson et al. produced a superficial burn injury on the flexor surface of the forearm using an electrically heated aluminum rod at 51°C and assessed the impact of intravenous lidocaine infusion on the inflammatory phase of healing [71]. They found a significantly faster restitution of residual erythema in the active group compared with placebo at 12h post burn. Another method of creating a burn wound is the use of ultraviolet radiation [70]. The application of this appears a safe model, with the only long-term consequence being tanning at the irradiated site. With the application of three times the minimal erythema dose, Bishop et al. found no blistering and no loss of skin barrier function [70].

Other in vivo models of wound healing include incisional and excisional wounds. Incisional cutaneous wound models involve the immediate primary closure of the wound following its creation and can be used to assess wound healing and comparison of standard versus investigative treatments. Conde-Green et al. assessed the use of negative-pressure wound therapy in patients who underwent primary abdominal wall closure and found a significant reduction in wound complications and dehiscence rates compared with conventional dry gauze dressings [72]. The use of a biopsy punch of various sizes to create a full-thickness excisional wound in healthy individuals allows, with the use of invasive and noninvasive techniques, the profiling of normal healing and the impact of treatments. Biopsies are typically taken at multiple time points usually from the inner aspect of the upper arms under local anesthetic [73,77]. This makes the process convenient for the subjects and the location being relatively well hidden. Ud-din et al. used objective noninvasive devices such as spectrophotometric intracutaneous analysis, full-field laser perfusion, and 3-D imaging to ascertain melanin, hemoglobin, collagen, blood flow, and wound size up to 14days post wounding [74]. They found corroboration of these findings with immunohistochemical techniques, thus supporting the use of these noninvasive devices in wound theranostics. The excisional punch biopsy model has also been used to assess treatments such as skin substitutes [75,76] and also less common treatments such as electric stimulation [73,77]. Greaves et al. assessed the use of skin substitutes including Decellularised Dermis (NHSBT, Watford, the United Kingdom), Integra Matrix Wound Dressing (Integra LifeSciences Inc., Plainsboro, NJ), and autografting of the skin and found enhanced angiogenesis [75] with acute wounds treated with Decellularised Dermis and reduced dermal fibrosis compared with control wounds [76]. Ud-din et al. identified increased blood flow in this excisional wound model following electric stimulation of the wound using noninvasive objective devices [77] and later confirmed this using immunohistochemistry and western blotting [73]. They also found increased angiogenic marker response in electrically stimulated wounds [73].

In vivo models using animals have also been widely described in the investigation of wound healing. Perhaps the most economical of these is the murine model.

Numerous studies have used mice, with wounds being created on the head, ear, tail, and back [78,79]. Mice also have the additional benefit of being amenable to genetic manipulation [80]. These genetically modified mice allow for specific genes and their function to be investigated during the wound-healing process [80]. Despite their benefits, wound healing in mice primarily occurs through contraction and is therefore significantly different from that in humans [79,80]. Attempts have been made to try and overcome this limitation, however, through the use of back splints to limit the degree of wound contraction [81]. The animal model that currently offers the most accurate replication of human wound healing is that of the pig [82]. They have a thick dermis, which is structurally similar to that in humans [83], and have sparse hair and skin that is adherent to underlying structures, which is also observed in humans [84]. The porcine model has been used to study various wound-healing pathologies including chronic ischemic wounds [84] and burns [85]. They have also been instrumental in the development of epidermal and dermal skin substitutes during preclinical studies [86,87]. However, despite their benefits, there are several limitations to their use. Despite being similar to the human skin, there are significant structural differences observed in porcine skin, such as a less vascular dermis [88]. Pigs are also more expensive and difficult to manage when compared with their murine counterparts. Additionally, as with all animal models, there are numerous ethical considerations that must also be taken into account before their use.

6 Conclusion

Human cutaneous scar and wound models are vital in helping researchers better understand the physiological and pathological processes of normal and abnormal healing, and in certain scenarios can be considered to be superior to animal models in the absence of an exact replica. In silico, in vitro, ex vivo, and in vivo models also provide relevant and realistic models to assess the therapeutic effects of experimental treatments aimed at improving the wound-healing processes.

The use of in silico models are currently limited by the absence of standardization and the lack of validation and therefore remain in the experimental phase [89]. The need for transferability of methods across different laboratories and inter- and intralaboratory reproducibility are crucial before such models are deemed reliable [8]. In vitro models are beneficial as they are relatively less expensive and their complexity is broad ranging from single-cell models to organ cultures. However, translation of findings into an in vivo model is challenging given the multiple confounding factors encountered in clinical situations. To some extent, ex vivo human models aim to solve these difficulties in clinical translatability. They provide an excellent means of assessing the complex wound-healing processes and providing a platform to assess the impact of treatments in a safe environment. However, certain environmental factors patients with wounds would be subjected to are not reproducible, and these models only provide a short investigative opportunity into a process that continues for several months to years [90]. In vivo models of scarring and wound healing are ideal for investigative purposes as they provide the closest resemblance, if not the same, to patients. However, limitations include difficulties in recruitment and current studies lacking consistency in study design, assessment techniques, and follow-up [82]. Tissue-engineered models are emerging to closely resemble the in vivo environment through regulation of environmental factors that would help resolve limitations with current wound-healing models [91,92].

Human cutaneous scar and wound models are critical in improving our understanding of the pathophysiology of cutaneous healing and scarring. They are additionally vital to aiding with the safe assessment of current and emerging wound therapeutics, with the ultimate aim of creating a scarring-free environment leading to optimal skin regeneration.

Abbreviations

ECM extracellular matrix
ABM agent-based model
TGF-β1 transforming growth factor-beta 1
LDH lactate dehydrogenase

References

[1] Broughton G, Janis JE, Attinger CE. The basic science of wound healing. Plast Reconstr Surg 2006;117(7 Suppl):12S–34S.
[2] Barrientos S, Stojadinovic O, Golinko MS, Brem H, Tomic-Canic M. Growth factors and cytokines in wound healing. Wound Repair Regen 2008;16(5):585–601.
[3] van den Broek LJ, Limandjaja GC, Niessen FB, Gibbs S. Human hypertrophic and keloid scar models: principles, limitations and future challenges from a tissue engineering perspective. Exp Dermatol 2014;23(6):382–6.
[4] Greenhalgh DG. Models of wound healing. J Burn Care Rehabil 2005;26(4):293–305.
[5] Sidgwick GP, McGeorge D, Bayat A. Functional testing of topical skin formulations using an optimised ex vivo skin organ culture model. Arch Dermatol Res 2016;308(5):297–308.
[6] Harunari N, Zhu KQ, Armendariz RT, Deubner H, Muangman P, Carrougher GJ, et al. Histology of the thick scar on the female, red duroc pig: final similarities to human hypertrophic scar. Burns 2006;32(6):669–77.
[7] Simon GA, Maibach HI. The pig as an experimental animal model of percutaneous permeation in man: qualitative and quantitative observations—an overview. Skin Pharmacol Appl Skin Physiol 2000;13(5):229–34.
[8] Pfuhler S, Fautz R, Ouedraogo G, Latil A, Kenny J, Moore C, et al. The cosmetics Europe strategy for animal-free genotoxicity testing: project status up-date. Toxicol In Vitro 2014;28(1):18–23.
[9] Ziraldo C, Mi Q, An G, Vodovotz Y. Computational modeling of inflammation and wound healing. Adv Wound Care (New Rochelle) 2013;2(9):527–37.
[10] Waugh HV, Sherratt JA. Modeling the effects of treating diabetic wounds with engineered skin substitutes. Wound Repair Regen 2007;15(4):556–65.
[11] Waugh HV, Sherratt JA. Macrophage dynamics in diabetic wound dealing. Bull Math Biol 2006;68(1):197–207.
[12] Arciero JC, Mi Q, Branca MF, Hackam DJ, Swigon D. Continuum model of collective cell migration in wound healing and colony expansion. Biophys J 2011;100(3):535–43.

[13] Wearing HJ, Sherratt JA. Keratinocyte growth factor signalling: a mathematical model of dermal-epidermal interaction in epidermal wound healing. Math Biosci 2000;165(1):41–62.
[14] McDougall S, Dallon J, Sherratt J, Maini P. Fibroblast migration and collagen deposition during dermal wound healing: mathematical modelling and clinical implications. Philos Trans A Math Phys Eng Sci 1843;2006(364):1385–405.
[15] Mi Q, Rivière B, Clermont G, Steed DL, Vodovotz Y. Agent-based model of inflammation and wound healing: insights into diabetic foot ulcer pathology and the role of transforming growth factor-beta1. Wound Repair Regen 2007;15(5):671–82.
[16] An G, Mi Q, Dutta-Moscato J, Vodovotz Y. Agent-based models in translational systems biology. Wiley Interdiscip Rev Syst Biol Med 2009;1(2):159–71.
[17] Adra S, Sun T, MacNeil S, Holcombe M, Smallwood R. Development of a three dimensional multiscale computational model of the human epidermis. PLoS One 2010;5(1):e8511.
[18] Li X, Upadhyay AK, Bullock AJ, Dicolandrea T, Xu J, Binder RL, et al. Skin stem cell hypotheses and long term clone survival—explored using agent-based modelling. Sci Rep 2013;3:1904.
[19] Sun T, McMinn P, Holcombe M, Smallwood R, MacNeil S. Agent based modelling helps in understanding the rules by which fibroblasts support keratinocyte colony formation. PLoS One 2008;3(5):e2129.
[20] Sun T, McMinn P, Coakley S, Holcombe M, Smallwood R, Macneil S. An integrated systems biology approach to understanding the rules of keratinocyte colony formation. J R Soc Interface 2007;4(17):1077–92.
[21] Anderson AR, Chaplain MA. Continuous and discrete mathematical models of tumor-induced angiogenesis. Bull Math Biol 1998;60(5):857–99.
[22] McDougall SR, Anderson AR, Chaplain MA, Sherratt JA. Mathematical modelling of flow through vascular networks: implications for tumour-induced angiogenesis and chemotherapy strategies. Bull Math Biol 2002;64(4):673–702.
[23] Chaplain MA, McDougall SR, Anderson AR. Mathematical modeling of tumor-induced angiogenesis. Annu Rev Biomed Eng 2006;8:233–57.
[24] Gottrup F, Agren MS, Karlsmark T. Models for use in wound healing research: a survey focusing on in vitro and in vivo adult soft tissue. Wound Repair Regen 2000;8(2):83–96.
[25] Cha D, O'Brien P, O'Toole EA, Woodley DT, Hudson LG. Enhanced modulation of keratinocyte motility by transforming growth factor-alpha (TGF-alpha) relative to epidermal growth factor (EGF). J Invest Dermatol 1996;106(4):590–7.
[26] Bürk RR. A factor from a transformed cell line that affects cell migration. Proc Natl Acad Sci U S A 1973;70(2):369–72.
[27] Buisson AC, Zahm JM, Polette M, Pierrot D, Bellon G, Puchelle E, et al. Gelatinase B is involved in the in vitro wound repair of human respiratory epithelium. J Cell Physiol 1996;166(2):413–26.
[28] Calderon M, Lawrence WT, Banes AJ. Increased proliferation in keloid fibroblasts wounded in vitro. J Surg Res 1996;61(2):343–7.
[29] Henemyre-Harris CL, Adkins AL, Chuang AH, Graham JS. Addition of epidermal growth factor improves the rate of sulfur mustard wound healing in an in vitro model. Eplasty 2008;8:e16.
[30] Kim WS, Lee JS, Bae GY, Kim JJ, Chin YW, Bahk YY, et al. Extract of Aneilema keisak inhibits transforming growth factor-β-dependent signalling by inducing Smad2 downregulation in keloid fibroblasts. Exp Dermatol 2013;22(1):69–71.
[31] Butler PD, Ly DP, Longaker MT, Yang GP. Use of organotypic coculture to study keloid biology. Am J Surg 2008;195(2):144–8.

[32] Phan TT, Lim IJ, Aalami O, Lorget F, Khoo A, Tan EK, et al. Smad3 signalling plays an important role in keloid pathogenesis via epithelial-mesenchymal interactions. J Pathol 2005;207(2):232–42.

[33] Chen WL, Kuo KT, Chou TY, Chen CL, Wang CH, Wei YH, et al. The role of cytochrome c oxidase subunit Va in non-small cell lung carcinoma cells: association with migration, invasion and prediction of distant metastasis. BMC Cancer 2012;12:273.

[34] Harisi R, Kenessey I, Olah JN, Timar F, Babo I, Pogany G, et al. Differential inhibition of single and cluster type tumor cell migration. Anticancer Res 2009;29(8):2981–5.

[35] Xie Y, Rizzi SC, Dawson R, Lynam E, Richards S, Leavesley DI, et al. Development of a three-dimensional human skin equivalent wound model for investigating novel wound healing therapies. Tissue Eng Part C Methods 2010;16(5):1111–23.

[36] Eckes B, Krieg T, Nusgens BV, Lapière CM. In vitro reconstituted skin as a tool for biology, pharmacology and therapy: a review. Wound Repair Regen 1995;3(3):248–57.

[37] Nayak S, Dey S, Kundu SC. Skin equivalent tissue-engineered construct: co-cultured fibroblasts/keratinocytes on 3D matrices of sericin hope cocoons. PLoS One 2013;8(9):e74779.

[38] Derderian CA, Bastidas N, Lerman OZ, Bhatt KA, Lin SE, Voss J, et al. Mechanical strain alters gene expression in an in vitro model of hypertrophic scarring. Ann Plast Surg 2005;55(1):69–75 [discussion].

[39] Karamichos D, Lakshman N, Petroll WM. An experimental model for assessing fibroblast migration in 3-D collagen matrices. Cell Motil Cytoskeleton 2009;66(1):1–9.

[40] Ghaffari A, Kilani RT, Ghahary A. Keratinocyte-conditioned media regulate collagen expression in dermal fibroblasts. J Invest Dermatol 2009;129(2):340–7.

[41] Bellemare J, Roberge CJ, Bergeron D, Lopez-Vallé CA, Roy M, Moulin VJ. Epidermis promotes dermal fibrosis: role in the pathogenesis of hypertrophic scars. J Pathol 2005;206(1):1–8.

[42] Chiu LL, Sun CH, Yeh AT, Torkian B, Karamzadeh A, Tromberg B, et al. Photodynamic therapy on keloid fibroblasts in tissue-engineered keratinocyte-fibroblast co-culture. Lasers Surg Med 2005;37(3):231–44.

[43] van den Broek LJ, Niessen FB, Scheper RJ, Gibbs S. Development, validation and testing of a human tissue engineered hypertrophic scar model. ALTEX 2012;29(4):389–402.

[44] Bagabir R, Syed F, Paus R, Bayat A. Long-term organ culture of keloid disease tissue. Exp Dermatol 2012;21(5):376–81.

[45] Duong HS, Zhang Q, Kobi A, Le A, Messadi DV. Assessment of morphological and immunohistological alterations in long-term keloid skin explants. Cells Tissues Organs 2005;181(2):89–102.

[46] Mendoza-Garcia J, Sebastian A, Alonso-Rasgado T, Bayat A. Ex vivo evaluation of the effect of photodynamic therapy on skin scars and striae distensae. Photodermatol Photoimmunol Photomed 2015;31(5):239–51.

[47] Syed F, Bagabir RA, Paus R, Bayat A. Ex vivo evaluation of antifibrotic compounds in skin scarring: EGCG and silencing of PAI-1 independently inhibit growth and induce keloid shrinkage. Lab Invest 2013;93(8):946–60.

[48] Balaji S, Moles CM, Bhattacharya SS, LeSaint M, Dhamija Y, Le LD, et al. Comparison of interleukin 10 homologs on dermal wound healing using a novel human skin ex vivo organ culture model. J Surg Res 2014;190(1):358–66.

[49] Companjen AR, van der Wel LI, Wei L, Laman JD, Prens EP. A modified ex vivo skin organ culture system for functional studies. Arch Dermatol Res 2001;293(4):184–90.

[50] Peramo A, Marcelo CL, Goldstein SA, Martin DC. Improved preservation of the tissue surrounding percutaneous devices by hyaluronic acid and dermatan sulfate in a human skin explant model. Ann Biomed Eng 2010;38(3):1098–110.

[51] Reus AA, Usta M, Krul CA. The use of ex vivo human skin tissue for genotoxicity testing. Toxicol Appl Pharmacol 2012;261(2):154–63.

[52] Tomic-Canic M, Mamber SW, Stojadinovic O, Lee B, Radoja N, McMichael J. Streptolysin O enhances keratinocyte migration and proliferation and promotes skin organ culture wound healing in vitro. Wound Repair Regen 2007;15(1):71–9.

[53] Xu W, Jong Hong S, Jia S, Zhao Y, Galiano RD, Mustoe TA. Application of a partial-thickness human ex vivo skin culture model in cutaneous wound healing study. Lab Invest 2012;92(4):584–99.

[54] Hodgkinson T, Bayat A. In vitro and ex vivo analysis of hyaluronan supplementation of Integra dermal template on human dermal fibroblasts and keratinocytes. J Appl Biomater Funct Mater 2016;14(1):e9–18.

[55] Kratz G. Modeling of wound healing processes in human skin using tissue culture. Microsc Res Tech 1998;42(5):345–50.

[56] Moll I, Houdek P, Schmidt H, Moll R. Characterization of epidermal wound healing in a human skin organ culture model: acceleration by transplanted keratinocytes. J Invest Dermatol 1998;111(2):251–8.

[57] Sebastian A, Iqbal SA, Colthurst J, Volk SW, Bayat A. Electrical stimulation enhances epidermal proliferation in human cutaneous wounds by modulating p53-SIVA1 interaction. J Invest Dermatol 2015;135(4):1166–74.

[58] van Kilsdonk JW, van den Bogaard EH, Jansen PA, Bos C, Bergers M, Schalkwijk J. An in vitro wound healing model for evaluation of dermal substitutes. Wound Repair Regen 2013;21(6):890–6.

[59] Ud-Din S, Bowring A, Derbyshire B, Morris J, Bayat A. Identification of steroid sensitive responders versus non-responders in the treatment of keloid disease. Arch Dermatol Res 2013;305(5):423–32.

[60] Ud-Din S, Thomas G, Morris J, Bayat A. Photodynamic therapy: an innovative approach to the treatment of keloid disease evaluated using subjective and objective non-invasive tools. Arch Dermatol Res 2013;305(3):205–14.

[61] Dunkin CS, Pleat JM, Gillespie PH, Tyler MP, Roberts AH, McGrouther DA. Scarring occurs at a critical depth of skin injury: precise measurement in a graduated dermal scratch in human volunteers. Plast Reconstr Surg 2007;119(6):1722–32 [discussion 33-4].

[62] Ud-Din S, Giddings Dip P, Colthurst J, Whiteside S, Morris J, Bayat A. Significant reduction of symptoms of scarring with electrical stimulation: evaluated with subjective and objective assessment tools in a prospective noncontrolled case series. Wounds 2013;25(8):212–24.

[63] Ud-Din S, McAnelly SL, Bowring A, Whiteside S, Morris J, Chaudhry I, et al. A double-blind controlled clinical trial assessing the effect of topical gels on striae distensae (stretch marks): a non-invasive imaging, morphological and immunohistochemical study. Arch Dermatol Res 2013;305(7):603–17.

[64] Kiistala U. Dermal-epidermal separation. II. External factors in suction blister formation with special reference to the effect of temperature. Ann Clin Res 1972;4(4):236–46.

[65] Lévy JJ, von Rosen J, Gassmüller J, Kleine Kuhlmann R, Lange L. Validation of an in vivo wound healing model for the quantification of pharmacological effects on epidermal regeneration. Dermatology 1995;190(2):136–41.

[66] Svedman P, Lundin S, Höglund P, Hammarlund C, Malmros C, Pantzar N. Passive drug diffusion via standardized skin mini-erosion; methodological aspects and clinical findings with new device. Pharm Res 1996;13(9):1354–9.

[67] Jeong SH, Kim JH, Yi SM, Lee JP, Sohn KH, Park KL, et al. Assessment of penetration of quantum dots through in vitro and in vivo human skin using the human skin equivalent model and the tape stripping method. Biochem Biophys Res Commun 2010;394(3):612–5.

[68] Brown GL, Nanney LB, Griffen J, Cramer AB, Yancey JM, Curtsinger LJ, et al. Enhancement of wound healing by topical treatment with epidermal growth factor. N Engl J Med 1989;321(2):76–9.
[69] Greenhalgh DG, Rieman M. Effects of basic fibroblast growth factor on the healing of partial-thickness donor sites. A prospective, randomized, double-blind trial. Wound Repair Regen 1994;2(2):113–21.
[70] Bishop T, Ballard A, Holmes H, Young AR, McMahon SB. Ultraviolet-B induced inflammation of human skin: characterisation and comparison with traditional models of hyperalgesia. Eur J Pain 2009;13(5):524–32.
[71] Mattsson U, Cassuto J, Tarnow P, Jönsson A, Jontell M. Intravenous lidocaine infusion in the treatment of experimental human skin burns—digital colour image analysis of erythema development. Burns 2000;26(8):710–5.
[72] Condé-Green A, Chung TL, Holton LH, Hui-Chou HG, Zhu Y, Wang H, et al. Incisional negative-pressure wound therapy versus conventional dressings following abdominal wall reconstruction: a comparative study. Ann Plast Surg 2013;71(4):394–7.
[73] Ud-Din S, Sebastian A, Giddings P, Colthurst J, Whiteside S, Morris J, et al. Angiogenesis is induced and wound size is reduced by electrical stimulation in an acute wound healing model in human skin. PLoS One 2015;10(4):e0124502.
[74] Ud-Din S, Greaves NS, Sebastian A, Baguneid M, Bayat A. Noninvasive device readouts validated by immunohistochemical analysis enable objective quantitative assessment of acute wound healing in human skin. Wound Repair Regen 2015;23(6):901–14.
[75] Greaves NS, Lqbal SA, Morris J, Benatar B, Alonso-Rasgado T, Baguneid M, et al. Acute cutaneous wounds treated with human decellularised dermis show enhanced angiogenesis during healing. PLoS One 2015;10(1):e0113209.
[76] Greaves NS, Iqbal SA, Hodgkinson T, Morris J, Benatar B, Alonso-Rasgado T, et al. Skin substitute-assisted repair shows reduced dermal fibrosis in acute human wounds validated simultaneously by histology and optical coherence tomography. Wound Repair Regen 2015;23(4):483–94.
[77] Ud-Din S, Perry D, Giddings P, Colthurst J, Zaman K, Cotton S, et al. Electrical stimulation increases blood flow and haemoglobin levels in acute cutaneous wounds without affecting wound closure time: evidenced by non-invasive assessment of temporal biopsy wounds in human volunteers. Exp Dermatol 2012;21(10):758–64.
[78] Reid RR, Said HK, Mogford JE, Mustoe TA. The future of wound healing: pursuing surgical models in transgenic and knockout mice. J Am Coll Surg 2004;199(4):578–85.
[79] Falanga V, Schrayer D, Cha J, Butmarc J, Carson P, Roberts AB, et al. Full-thickness wounding of the mouse tail as a model for delayed wound healing: accelerated wound closure in Smad3 knock-out mice. Wound Repair Regen 2004;12(3):320–6.
[80] Grose R, Werner S. Wound-healing studies in transgenic and knockout mice. Mol Biotechnol 2004;28(2):147–66.
[81] Dunn L, Prosser HC, Tan JT, Vanags LZ, Ng MK, Bursill CA. Murine model of wound healing. J Vis Exp 2013;75:e50265.
[82] Ud-Din S, Bayat A. Non-animal models of wound healing in cutaneous repair: in silico, in vitro, ex vivo and in vivo models of wounds and scars in human skin. Wound Rep Reg 2017;25:164–76. https://doi.org/10.1111/wrr.12513.
[83] Lindblad WJ. Considerations for selecting the correct animal model for dermal wound-healing studies. J Biomater Sci Polym Ed 2008;19(8):1087–96.
[84] Roy S, Biswas S, Khanna S, Gordillo G, Bergdall V, Green J, et al. Characterization of a preclinical model of chronic ischemic wound. Physiol Genomics 2009;37(3):211–24.

[85] Singer AJ, McClain SA, Taira BR, Romanov A, Rooney J, Zimmerman T. Validation of a porcine comb burn model. Am J Emerg Med 2009;27(3):285–8.
[86] Navarro FA, Stoner ML, Park CS, Huertas JC, Lee HB, Wood FM, et al. Sprayed keratinocyte suspensions accelerate epidermal coverage in a porcine microwound model. J Burn Care Rehabil 2000;21(6):513–8.
[87] Elgharably H, Roy S, Khanna S, Abas M, Dasghatak P, Das A, et al. A modified collagen gel enhances healing outcome in a preclinical swine model of excisional wounds. Wound Repair Regen 2013;21(3):473–81.
[88] Montagna W, Yun JS. The skin of the domestic pig. J Invest Dermatol 1964;42:11–21.
[89] Vermolen FJ, Javierre E. Computer simulations from a finite-element model for wound contraction and closure. J Tissue Viability 2010;19(2):43–53.
[90] Hewitt NJ, Edwards RJ, Fritsche E, Goebel C, Aeby P, Scheel J, et al. Use of human in vitro skin models for accurate and ethical risk assessment: metabolic considerations. Toxicol Sci 2013;133(2):209–17.
[91] Ataç B, Wagner I, Horland R, Lauster R, Marx U, Tonevitsky AG, et al. Skin and hair on-a-chip: in vitro skin models versus ex vivo tissue maintenance with dynamic perfusion. Lab Chip 2013;13(18):3555–61.
[92] Wagner I, Materne EM, Brincker S, Süssbier U, Frädrich C, Busek M, et al. A dynamic multi-organ-chip for long-term cultivation and substance testing proven by 3D human liver and skin tissue co-culture. Lab Chip 2013;13(18):3538–47.

Preclinical models for wound-healing studies

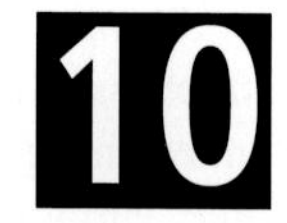

Irena Pastar, Liang Liang, Andrew P. Sawaya, Tongyu Cao Wikramanayake, George D. Glinos, Stefan Drakulich, Vivien Chen, Olivera Stojadinovic, Stephen C. Davis, Marjana Tomic-Canic
Department of Dermatology and Cutaneous Surgery, University of Miami Miller School of Medicine, Miami, FL,United States

1. Introduction

Wound healing is an evolutionarily conserved process aimed to efficiently restore cutaneous barrier upon injury. It consists of the sequential yet overlapping phases of hemostasis, inflammation, proliferation, and tissue remodeling [1]. Previous basic and preclinical research utilizing in vitro and in vivo wound models has greatly advanced understanding of distinct molecular and cellular processes involved in wound healing. However, attempts to translate these findings into the clinical setting had only limited success. Here, we outline the most commonly utilized in vitro, ex vivo, and in vivo wound-healing models, highlighting their applications in both mechanistic and preclinical studies.

Preclinical research remains the gold standard to evaluate safety, toxicology, and efficacy in a dose-response fashion. In addition, it provides great advantage to study specific mechanism(s) that contribute to the healing process. In vitro wound-healing assays such as wound scratch or transwell migration assays are useful when investigating the effect of a novel therapeutic agent on particular cell types or a specific factor that could be important during tissue repair [2–4]. However, two-dimensional in vitro assays are unable to completely reproduce biological conditions such as immune response, cell-cell, and cell-matrix interaction and pathology of wound-healing disorders. Therefore, organotypic cultures that recreate dermal layer and the stratified epidermis have been developed and successfully utilized to evaluate keratinocyte-fibroblast cross talk during wound healing and repose to topical treatments in three dimensions [5,6]. Similar human ex vivo wound model has been utilized successfully to study mechanism of wound epithelialization under a wide variety of experimental conditions [7]. In vivo animal wound models are the next step in the preclinical testing. Both small and large animal models are widely used in the wound-healing field as they illuminate the mechanisms of cutaneous wound repair under systemic conditions [8–11]. Importantly, in vivo wound models mimicking complex conditions such as diabetes, infection, or ischemia have been established and extensively utilized [2,12–18]. In vitro, ex vivo, and in vivo models have greatly advanced our understanding of the wound-healing process and molecular mechanisms; however, they often fail to completely reproduce human condition, particularly the complexity of nonhealing chronic wounds.

Skin Tissue Models. https://doi.org/10.1016/B978-0-12-810545-0.00010-3

Chronic wounds are the result of a dysregulated tissue repair process to restore a structurally and functionally intact skin barrier. They are frequently associated with underlying conditions that include vascular disease, diabetes, and aging [1]. This multiple, variable systemic contributing factors along with local tissue impairments make it challenging to model and replicate. Yet, estimates are that up to 2% of the world population will develop chronic wound, with the United States alone accounting for 6.5 million patients [1,19]. Treatment of chronic wounds has become a tremendous economic burden with costs estimated up to $25 billion a year in the United States alone [19]. Despite the high mortality rate and severe morbidities, mechanisms that result in the development of chronic wounds remain poorly understood, which, together with limitations in experimental models of these diseases, resulted in fewer effective therapies [1]. In addition, current treatment options are often not fully efficacious in restoring healing. The complexity of heterogeneous patient population with multifactorial etiology and lack of animal models that recapitulate the human condition hindered the research on chronic wounds and the development of new therapies [1]. Although experimental chronic wound models equivalent to human in laboratory animals do not exist, some animals develop nonhealing or slow healing wounds [20] with a low incidence.

Here, we describe the most current in vitro and in vivo models for wound-healing studies referring to their advantages and disadvantages. Recent advances in "microengineered" models of the functional human skin, known as organs-on-chip [21], that could provide the basis for preclinical assays with greater predictive power are also discussed. Although well-controlled, randomized clinical trials are the final testimonial of product efficacy, in vitro and in vivo models remain essential for the evaluation of wound-healing therapeutics.

2. In vitro wound healing models

In vitro models are important tools for preclinical wound-healing research. They not only aid in understanding the mechanisms of cellular response to wound-initiated migration and cell-cell interactions but also provide a tool to evaluate the effects of novel therapeutic agents at the cellular level in a controlled environment. These easily implemented, affordable assays can be employed to assess migratory capacity of primary cells (keratinocytes and fibroblasts) grown from patients' healthy or diseased skin specimens or from genetically manipulated animals. In the light of high-throughput and high-content imaging systems, these models provide an effective platform for drug discovery and screening.

2.1 In vitro two-dimensional wound assays

2.1.1 In vitro wound scratch assay

The scratch assay is typically utilized to quantify cellular migration on two-dimensional (2-D) surfaces over time upon different treatments. It is one of the most commonly used in vitro wound-healing assays, allowing determination of the optimal dose of the agents being tested. Cells are grown to confluency in a monolayer, and a scratch is made with a pipette tip to create an incision-like gap. The "wounded" area is photographed immediately after wounding and at defined time points thereafter, and cell migration is quantified and expressed as average percentage of closure of the scratch area [2,3]. Alternatively, laser ablation or hydrogel-coated dishes can be used in order to create a well-defined gap for the cells to move across [22,23]. For longer-term wound-healing assays (>24 h), mitosis inhibitors such as mitomycin C treatment are required to eliminate the contribution of cell proliferation to gap closure, allowing the assay to specifically assess migration of the given cell population. Scratch assay is compatible with time-lapse microscopy including live cell imaging, imaging specific patterns of cell migration behavior, and moreover a single-cell movement [24]. Live cell imaging also allows mechanistic studies during cell migration including those focusing on intracellular signaling events, such as visualization of fluorescently tagged proteins and their subcellular localization, or fluorescent resonance energy transfer (FRET) for protein-protein interactions [3]. Multiple studies have utilized wound scratch assays to evaluate the effects of major wound-healing growth factors, chemokines, and cytokines on cell migration [25–27]. Major promotility factors within human serum have been identified to stimulate keratinocyte migration in the following order of potency: TGFα > insulin > EGF > HB-EGF > IGF-1 > FGF > IL-8 > HGF > IL-1 > KGF > TGFβ [4]. These factors have been utilized as positive controls in in vitro wound assays [27,28].

2.1.2 Transwell migration assay

Another commonly used in vitro wound-healing assay utilized for cell migration assessment is the transwell assay, also known as the Boyden chamber assay. This method is ideal for quantitative analysis of single-cell migratory capacity toward certain chemoattractants. It consists of two medium-containing chambers separated by a porous membrane that allows cells to migrate through. By adjusting the cell numbers and concentrations of chemoattractant in the upper and lower chambers, it allows studies focusing on wound healing, relevant chemotaxis, and chemokinesis [29]. Generally, cells are seeded in the upper compartment and cultured to allow migration. After nonmigratory cells are removed by a cotton swab from the upper side of the membrane, the membrane is stained for cells that have migrated through the pores to the bottom of the membrane. The number of migratory cells is quantified per imaging field [30]. Utilizing transwell migration assay, Kroeze and colleagues demonstrated that chemotaxis of human primary keratinocytes is attracted by an array of cytokines including CCL14, CCL22, CCL27, CXCL1, and CXCL10, whereas keratinocyte chemokinesis is stimulated by CXCL8 [31].

Although highly relevant for the initial preclinical screening, the physiological relevance of in vitro migration assays is relatively low and cannot fully simulate the complexity of the cutaneous wound-healing process (Table 1). Due to the limitations of 2-D monolayer cell culture, these assays lack the cross talk between different cell types involved in the wound-healing process, such as keratinocyte-fibroblast interactions and

Table 1 Summary of the frequently used in vitro and ex vivo models for wound-healing studies

In vitro and ex vivo wound-healing models			
Model	**Advantages**	**Disadvantage**	**Reference**
In vitro—2-D cellular assays			
Wound scratch assay	- Scratching monolayer cell culture mimics "wounding" in vitro - Low-cost, high-throughput, easily implemented assays to measure cell migration	- The lack of cross talk with other cell types - The lack of ECM, low physiological relevance	[2-4,22-28]
Boyden chamber assay	- Measures cell motility such as chemotaxis and chemokinesis	- Assesses single-cell motility	[29-31]
In vitro—3-D skin equivalent			
Human skin organotypic wound model	- Standardized functional skin equivalent - Assesses reepithelialization and interaction between keratinocytes and fibroblasts; may include lymph and blood vessels - Ability to mimic chronic wound phenotype with primary or genetically modified cells	- Simplified ECM - The lack of a functional dermal-epidermal junction	[5,32-40]
Organ-on-a-chip skin model	- Microfluidic chip system provides perfusion to skin culture - Compatible with ex vivo skin model	- In development	[21,45]
Ex vivo wound models			
Human acute wound ex vivo model	- Ex vivo skin culture obtained from reduction surgery that maintains healing capabilities in vitro - Human skin full-thickness epidermis and dermis; Langerhans cells, pigment cells and nerve endings	- Absence of immune cells and blood supply - Variation in wound depth can affect healing outcome	[7,28,46, 48-54,56,57]
Murine skin explant model	- Assess keratinocytes migration from the full-thickness murine skin	- The lack of immune cells and blood supply	[59]

Table 1 Continued

In vitro and ex vivo wound-healing models			
Model	**Advantages**	**Disadvantage**	**Reference**
Human pressure ulcer ex vivo model	- Human skin full-thickness epidermis and dermis; Langerhans cells, pigment cells, and nerve endings - Utilizes mechanical explant test system to create chronic wounds	- Variation in skin thickness may affect outcomes - The absence of immune cells	[58,60-62]
Porcine skin explant wound infection model	- Full-thickness porcine skin that maintains healing capability in vitro - Allows for evaluation of a variety of antimicrobial dressings and topical treatments	- The lack of immune response and blood supply - Does not account for effects of host microbiome	[63-66]

interactions of resident cells with infiltrating immune cells, neurons, and other cell types. Therefore, it is often necessary to validate the results of in vitro assays with other preclinical tests that allow such interactions.

2.2 *In vitro three-dimensional wound models*

2.2.1 *Organotypic human wound model*

Organotypic cultures are in vitro systems that recreate rudimentary skin layers and the stratified epidermis and can assess keratinocyte response to topical treatments in three dimensions (3-D) before moving to in vivo studies. Similar to ex vivo wound models, this model can evaluate the effects of a wide variety of molecules and agents on keratinocyte migration and keratinocyte-fibroblast interaction during the process of wound healing. The organotypic culture model consists of human keratinocytes—HaCaT (immortalized cell line) being commonly used—grown on a matrix of type I collagen with fibroblasts [5,32] or devitalized human dermis [33]. Wounding is performed on this cellular bilayer using a biopsy punch, and after removal of the punched tissue, the remaining tissue is placed on top of a second premade type I collagen matrix and allowed to reepithelialize (Fig. 1) [6]. Other wounding methods include CO_2 laser-induced burns [34] and needle punctures [35]. Upon wounding, topical treatments may be applied. Reepithelialization can be measured by histomorphometric analysis or keratin immunostaining. Organotypic cultures use primary human cells and provide a standardized platform to study the pathogenesis of wound healing and its related disorders and development of more efficacious therapeutics.

This model has several advantages compared with traditional primary cell culture. First, this biomimetic model allows for the study of cell behavior in a 3-D environment

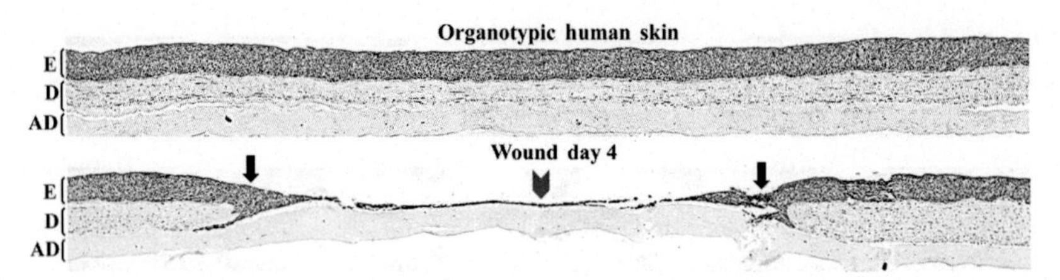

Fig. 1 Organotypic human wound model. H&E-stained section of the organotypic human model generated from HaCaT keratinocyte cell line and human fibroblasts grown on a matrix of type I collagen. Keratinocytes migrate over the matrix to fully reepithelialize wounding 4 days upon wounding. Black arrows indicate the edges of the initial wound area, and red arrow indicates wound closure. E, epidermis; D, dermis; AD, acellular dermis.

as opposed to the 2-D monolayer culture. The cells in organotypic cultures can maintain complex behaviors such as migration and even hair follicle formation that requires the interactions between different cell types in 3-D [5]. Of particular interest in the study of wound healing is that organotypic cell cultures have a basal keratinocyte proliferation rate similar to that found in vivo, a well-differentiated epithelium, and complete reepithelialization following wounding [6,36]. Second, this model allows for tight control of experimental conditions. Because well-defined lineages of cells are used, the keratinocytes in this model are without any interpatient or locational variability. Third, organotypic culture allows for genetic manipulations that are not possible or are unethical in vivo [5]. More importantly, 3-D organotypic model allows the incorporation of primary cells that are isolated from patient biopsy to study and compare a variety of disease-like scenarios [37]. Furthermore, keratinocytes derived from induced pluripotent stem cell (iPSC) isolated from both the normal and patient skin have been successfully incorporated into 3-D cultures and were fully functional [38]. Recently, full-thickness human skin constructs equipped with functional lymph and blood capillaries have been developed and tested on animal models demonstrating their potential to improve the survival of engrafted tissue [39].

Despite these advantages, the organotypic culture system has its own limitations, including a simplified anatomy and cell types compared with that seen in vivo. Therefore, it cannot replicate some complex cellular interactions that occur during the wound-healing process such as inflammation and angiogenesis [5]. Additionally, the lack of a functional dermal-epidermal junction leads to incomplete differentiation of cultured cells and affects their stability and the nature of their interactions [40]. Furthermore, organotypic cell behavior can be affected dramatically by the culture conditions, and thus proper media choice is paramount [5].

2.2.2 3D organotypic cultures mimicking wound healing disorders

Currently, there are chronic wound and hypertrophic scar organotypic models established using diseased and/or healthy human keratinocytes and fibroblasts. Three-dimensional organotypic cultures assembled with diabetic foot ulcer (DFU)-derived fibroblasts and healthy keratinocytes maintained pathological phenotype of DFUs, such as increased proliferation of basal keratinocytes, impaired reepithelialization, decreased extracellular matrix (ECM) production, and reduced neoangiogenesis [41].

The most advanced organotypic fibrosis model is the hypertrophic scar formation model with a reconstructed epidermis on a dermal matrix populated with adipose-derived mesenchymal stem cells (ASC). This organotypic model with ASC-populated matrices represents hypertrophic scar both macroscopically and microscopically and has been utilized for therapeutic evaluations. This organotypic model with ASC-populated matrices can reproduce a number of epidermal and dermal scar characteristics, mimics hypertrophic scar both macroscopically and microscopically, and has been validated with five known therapeutics [42]. For 3-D keloid models, either combinations of keloid fibroblasts with normal skin-derived keratinocytes [43] or a fully differentiated epidermis constructed from keratinocytes isolated from hypertrophic scars on a healthy fibroblast-populated dermal matrix was able to exhibit characteristics of an adverse scar (e.g., dermal thickness, epidermal thickness, and collagen I) suggesting the pathological changes in both cell types during hypertrophic scar formation [44]. In all, organotypic culture is a valuable model in the field of cutaneous wound healing because of its potential preclinical applications and assessment of novel therapeutics.

2.2.3 Organ-on-a-chip model

Organ-on-a-chip technologies that can mimic multilevel organ functions are being developed to screen drugs and treatments in a way that circumvents the need for live animals [45]. These biomimetic, microfluidic devices are generally small in size and are being developed with increased complexity. Organ-on-a-chip applications for skin-centered models have been developed and focus on accommodating a bioengineered skin equivalent that is continuously perfused through a microvascular channel. Atac et al. have shown that human skin equivalents, human ex vivo skin, and hair follicle cultures all had increased longevity in the organ-on-chip, as compared with static 3-D cultures [21]. Additionally, tissue disintegration was decreased in the chip cultures [21]. While there are several barriers to overcome before a full multilevel, organ-on-a-chip skin equivalent can be fully utilized for wound-healing studies, the power of this technology has already been made evident. What remains is to create and uphold the native skin's three-layered architecture, with sensory organs, appendages including the hair follicles and sebaceous and sweat glands, and a full vascular network, all on a single organ-on-a-chip. Once created, this can then be integrated onto multiorgan chips, mimicking a full organism's native environment, thus opening doors to more powerful treatment screening methods for wound-healing disorders.

3. Ex vivo wound models

3.1 Human ex vivo wound model

Ex vivo wound models are time- and cost-efficient to determine the effectiveness of topical or injected treatments before using in vivo models [7]. This model system can be used to assess a wide variety of molecules and agents on epithelialization. The skin

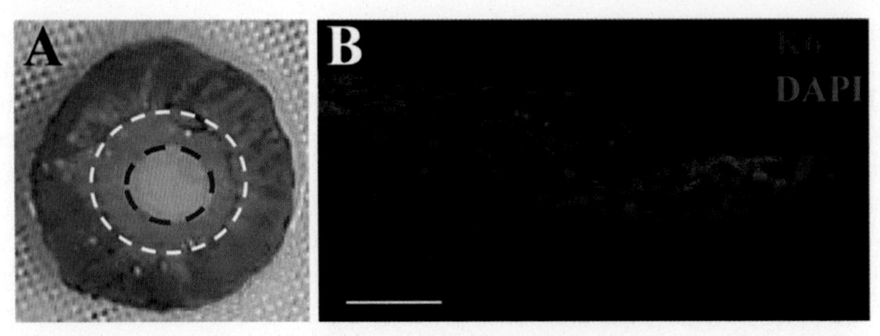

Fig. 2 Human ex vivo wound-healing model. (A) Gross photo of human ex vivo wound. White dashed line indicates wound edge; red dashed line indicates visual signs of epithelialization 4 days after wounding. (B) Keratin 6-stained ex vivo wound edge. Red color depicts keratin 6, and blue color depicts nuclei. Scale bar = 50 μm.

utilized in these models is typically acquired from reduction surgeries. The skin is cleaned from underlying fat to allow for adequate adherence of the dermis to the plate, and an excisional wound is created with a 3–4 mm punch. A larger, full-thickness 8–10 mm biopsy punch around the wound is then used to separate the wound from the rest of the skin (Fig. 2). The donutlike wounds are placed at air-liquid interface with media [7]. Treatments can be applied directly to the wound and the media to mimic either topical or "systemic" effect. Reepithelialization can be measured and quantified using histomorphometric analysis and keratin immunostaining [2,7,28,46,47].

Advantages of this model are numerous (Table 1). Foremost, it allows for assessment of reepithelialization in the human skin with uniform wound sizes under standardized conditions. A good amount of the skin can usually be obtained from reduction surgeries, allowing for multiple ex vivo wounds to be generated from a single patient source, allowing experimental replicates, and eliminating variability due to skin differences between individuals [7]. Furthermore, basement membrane integrity is preserved in the ex vivo culture, while it is not in the in vitro models such as organotypic culture [48]. However, whereas some resident inflammatory cells are retained, a major limitation of the ex vivo model is the lack of blood supply, thus limiting studies involving immune cell infiltrates.

To this date, the human ex vivo skin wound model has been successfully used to test the effects of growth factors like epidermal growth factor (EGF), microRNAs, and various pharmacological agents including glucocorticoids and beta-adrenergic antagonists [28,47,49-53]. A great amount of information can be obtained using this model including changes in gene expression pattern, protein levels, signaling events, and tissue morphology in order to comprehensively evaluate novel therapeutic agents. Importantly, the human ex vivo wound model has been shown to have comparable gene expression patterns to in vivo models, thus supporting its implementation in the functional assessment of human epidermal healing [54,55]. In all, this model has been implemented successfully in the literature to assess epithelialization under a wide variety of experimental conditions [52,56-58], and therefore, it is an extremely useful tool for the exploration of wound-healing epithelialization of the human skin and evaluation of novel treatments.

3.2 Murine skin explant model

In addition to human ex vivo wound model, murine skin explant culture can be used to evaluate keratinocyte migration. When fascia is removed from mouse skin samples, the skin is able to adhere to tissue culture plates while submerged in culture media. Keratinocyte migration out of skin explants can be observed over the course of up to 7 days [59]. This model represents a useful tool to study collective cell migration in the transgenic or knockout mouse skin or in response to pharmacological treatments to determine the effects of the overexpressed or knocked out gene or the pharmacological treatments on keratinocyte migration [59].

3.3 Pressure ulcer ex vivo model

In addition to studying acute wound healing, the human ex vivo wound model can be applied to study chronic wounds such as pressure ulcers [58]. Stojadinovic et al. demonstrated inflammasome induction in the ex vivo skin that was loaded with static stress at different time points using a well-described mechanical explant test system [58,60,61]. These specimens were analyzed for protein and RNA expression levels and morphological changes in response to mechanical load. A similar model has been developed in a porcine model. Yeung et al. applied pressure to porcine ex vivo wounds to assess the effectiveness of novel pressure-reducing devices [62]. These results demonstrate the flexibility of the ex vivo wound model to mimic both acute and chronic wound condition and its durability under various environmental and experimental influences. In summary, ex vivo models are useful to explore molecular mechanisms of epithelialization, as well as evaluation of novel pharmacological treatments and devices.

In vitro cell culture, ex vivo skin explants, skin-on-a-chip, and reconstituted organotypic cultures have their respective advantages in wound-healing studies. Single-cell type cultures, grown in vitro, allow individual repair processes—such as proliferation or cell migration—to be investigated at the lowest costs. Three-dimensional "skin equivalent" organotypic cultures generally use keratinocytes grown on a feeder layer of fibroblasts within a collagen matrix, thus recreating rudimentary skin layers and stratified epidermis. When wounded, this model allows studies pertaining to interactions between keratinocytes and fibroblasts and matrix, even though important wound-healing processes such as the inflammatory response and angiogenesis are missing. Human skin explants cultured ex vivo, on the other hand, contain all the local cell types including some resident inflammatory cells, thus mimicking the in vivo situation more closely. Taken together, all the in vitro and ex vivo models described represent powerful tools for wound-healing study, allowing us to gain insights in molecular events pertaining to this sophisticated process while also serving as valuable tools for preclinical testing.

3.4 Porcine ex vivo wound infection model

Porcine skin explants have been recently used as a model for wound infection and to evaluate novel antimicrobial therapies. Porcine skin was sterilized with chlorine gas in order to remove commensal bacteria while preserving histological properties of epidermis and dermis [63,64]. Partial-thickness wounds can be generated with a dermatome, which are then inoculated with the bacteria of interest [63-65]. Importantly, this ex vivo model simulated growth of *Pseudomonas aeruginosa* and *Staphylococcus aureus* biofilms on wounds, mimicking chronic wound environment [63-65]. After allowing time for a mature biofilm to form, the wound may be treated with a variety of dressings and other topical antimicrobial compounds to evaluate their efficacy [63-66]. For example, this model has been used to evaluate iodine-containing dressings, silver-containing dressings, surfactant-based dressings, and honey ointments, among many others [64,65]. The similarity of cutaneous physiology of the porcine skin as compared with the human skin is an advantage of this model. Additionally, this tissue can be easily obtained from meat processing facilities. Limitations of this model include a limited host immune response as would be seen in vivo and an absent commensal microbiome due to sterilization.

4. In vivo wound healing models

In vivo wound models help illuminate the mechanisms and pathophysiology of cutaneous wound repair under systemic conditions that allow paracrine and autocrine interactions between local cells and recruitment of immune cells through circulation. To this end, various in vivo wound models have been developed, and the animals most commonly used are mice, rats, rabbits, and swine (Table 2). Despite their vastly varied appearances, the mammalian skins share many common features in both anatomy and physiology. These models have been successfully used to delineate the basic mechanisms of wound healing and provide preclinical testing of novel interventions and optimizing treatment regimens.

All animal procedures need to be approved by institutional animal care and use committees prior to work and must be performed under aseptic conditions in anesthetized animals. Hair is often removed from the areas to be wounded by clipping or depilatory chemicals.

4.1 Rodent wound healing models

Mice and rats are the most frequently used animal models for wound-healing studies for several reasons. They are relatively inexpensive to acquire and maintain, develop rapidly, are of small body size that translates to small amounts of reagents needed, and have accelerated modes of healing (so that experiments usually last for days instead of weeks). The many inbred strains and large litters offer genetic homogeneity and allow for quantification and reproducibility, and the growing number of natural mutants and genetically modified strains (transgenic or various knockouts/knockins) facilitates studies of particular molecular pathways. The broad use of mice and rats in other biological and biomedical studies means that reagents (e.g., antibodies, ELISA, shRNA, and microRNAs) for a wide array of analyses are readily available. Additionally, mice and rats are easily manipulated to mimic specific wound conditions such as diabetes, infection, malnutrition, immunodeficiency, local tissue hypoxia, and ischemia [12].

Table 2 Summary of the frequently used in vivo models for wound-healing studies

In vivo wound-healing models			
Model	**Advantages**	**Disadvantages**	**References**
Rodent wound-healing models			
Murine incisional wound model	- Allows studying the interactions and influences of different cell types - Elucidates biomechanics of wound healing and scarring - Use of wound splinting can minimize contraction - Uncomplicated standardization and multiple replicates, repeated sampling as well	- Not optimal for assessment of epithelialization - Differences in skin anatomy and physiology in rodents versus humans - Influence of the hair follicle cycling on epithelialization	[8,67-69, 71-76]
Murine excisional wound model	- Allows studying the interactions and influences of different cell types on contraction - Can be used to evaluate treatment effects	- Not optimal for assessment of epithelialization	[76,78]
Murine full-thickness splinting model	- Can attenuate the effect of contraction, thus modeling human epithelialization process more closely - More reproducible than nonsplinted wounds	- Variation in skin anatomy and physiology in rodents versus humans	[11,84-87]
Parabiosis model	- Effective for study of circulation and infiltration of cells and factors in the context of wound healing - Demonstrates involvement of factors and various cell types in models of delayed wounding	- Not optimal for assessment of epithelialization	[10,88]
Dead space wound model	- Connective tissue formation isolated from other events of wound healing - Interstitial fluid accumulation provides valuable information about the wound environment	- Not useful for assessment of epithelialization - Variation in skin anatomy and physiology in rodents versus humans	[89-91]
Dorsal skinfold chamber model	- Enables vasculature microscopic imaging in vivo - In combination with intravital fluorescence microscopy enables pathophysiological imaging in real time	- Variation in skin anatomy and physiology in rodents versus humans	[10,92]

Continued

Table 2 Continued

In vivo wound-healing models			References
Model	**Advantages**	**Disadvantages**	
Tape stripping	- Partial removal of epidermis - Several levels of barrier disruption can be induced	- Variation in depth of barrier disruption	[94]
Diabetic wound-healing models			
Diabetes type 2 murine wound-healing model	- Not significantly influenced by other diabetes-related parameters - Permits application of bacterial biofilm to mimic chronic wound phenotype	- Variation in skin anatomy and physiology in rodents versus humans	[13-15,95,96]
Streptozotocin-induced diabetic wound model	- Permits study of wound healing in the context of type 1 diabetes	- Nonspecific toxicity in other organs; no significant delay in wound closure	[95,97-101]
Porcine model of wound healing			
Porcine partial/full-thickness wound model	- Greatest structural, biomechanical, immune, and most importantly wound-healing process similarity to humans - Multiple replicates within single animal, repeated sampling, and multiple concomitant assessment treatments	- Sweat gland locations dissimilar to humans - The lack of transgenic animals and reagents for analyses - High maintenance costs	[9,110]
Rabbit ear wound model			
Rabbit ear wound model	- Decreased blood flow - Absent wound contraction - Multiple replicates within single animal	- Dermis firmly attached to underlying cartilage - Avascular wound base	[117,118]
Ischemic wound model			
Ischemic wound models	- Accurate and reproducible model to depict characteristics of ischemic tissue with age-dependent response	- Labor-intensive and costly	[18,125]
Denervation wound model			
Denervation wound models	- Can be used to test potential therapeutics to restore sensation and improve wound healing	- The total lack of sensation does not fully mimic neuropathy in chronic wounds	[1,102,126]

Although contraction plays a major role in murine wound healing as compared with human skin wound-healing process, many molecular pathways involved in wound healing are well conserved between mice, rats, and humans.

The rat model is used less frequently than the mouse model largely due to the smaller number of genetically manipulated (transgenic and knockouts/knockins) rat strains, their larger size, and higher costs to maintain. However, the rat model also provides a variety of strains with different characteristics, and their larger size may allow comparison of different treatments in the same animal.

Murine hair follicles undergo cycles of growth and regression, and all the follicles on the dorsal skin are synchronized in their postnatal morphogenesis and the first post-natal hair cycle (catagen/regression phase, telogen/resting phase, and anagen/growth phase) [67]. A wounded skin shows accelerated healing when the follicles are in the growth phase (anagen) because follicular keratinocytes migrate to the wound site to contribute to healing [68]. Therefore, the age of mice and rats has to be carefully planned in wound-healing experiments, and the hair growth is closely monitored. To minimize the contribution to wound healing by keratinocyte migration from the hair follicles to the wound sites, wounds are ideally induced during telogen. For C57BL/6, the most commonly used mouse strain in biomedical research, all the dorsal skin hair follicles are in telogen 7 weeks after birth. This is evidenced by the homogeneous pink/gray skin color. Because follicular melanogenesis is strictly coupled with hair follicle cycling [69], anagen is associated with characteristic increase in skin pigmentation in C57BL/6 mice, with the skin turning dark gray at anagen entry, then completely black at the peak of anagen, and then turning gray at the follicles entry catagen. Additionally, synchronized anagen entry in mice results in profound increase in the thickness of almost all skin compartments [67], making anagen skin black and thick and "rubbery" compared with pink/gray and thin skin of telogen. Besides naturally occurring synchronized hair cycling in early postnatal life, synchronized anagen and subsequent spontaneous catagen/telogen can also be induced by depilation [70].

Samples collected at different time points and at the end of the wound-healing studies can be analyzed for the rate of wound closure by histology, breaking load and tensile strength, biochemistry (e.g., collagen deposition), and gene expression. Noninvasive methods such as bioluminescence in transgenic mice and laser Doppler imaging (to monitor blood flow) make it possible to monitor wound-healing progression in the same animals over time [71]. By comparing wound healing in experimental animals with that in control animals, one can determine the role of the mutated gene product, physiological condition, or therapeutic agent administered during acute wound healing. Here, we discuss incisional wound, full-thickness excisional wound including splinting, dead space wound, and tape-stripping murine models.

4.1.1 Incisional murine wounds

Cutting the skin with a sharp blade results in rapid disruption of tissue integrity with minimal collateral damage [61,62]. A fibrin clot quickly forms to bridge the injury margins. The size of the gape depends on the species, the size of the incision, the amount of subcutaneous fat, and the tensional forces on the wound site determined by the location and orientation of the incision. The loose skin in mice and rats can slide and retract over the subcutaneous fascia and produce a relatively large gape initially. In "primary intention closure," closure of incisional wounds by suture, staple, clip, or bandage reduces the tissue gap to allow rapid and efficient bridging of the wounded edges by granulation tissue and new epithelium and provides a useful model for biomechanical studies of healing [63,72-74]. In "secondary intention closure," an incisional wound is intentionally left open, and a much more extensive fibrin clot in the wound space is eventually replaced by ample granulation tissue and the surface gap closed by epithelialization, as well as contraction [8,75]. This model is more useful for the study of scarring over long periods of time and is not suitable for the study of epithelialization [76].

4.1.2 Full-thickness excisional murine wounds

In mice and rats, excisional wounds are created by removal of the full-thickness dorsal skin using 5–8 mm biopsy punch (wound area of 20–50 mm^2), and the entire wound is left open (Fig. 3). Alternatively, full-thickness excisional wounds can be generated on the mouse tail resulting in delayed healing [77]. Excisional wound model is used for the study of rate of contraction and epithelialization and factors that influence these processes. Agents that may accelerate or delay healing (including infectious agents) [78] may be applied topically to the wound bed before a scab is formed. When excisional wounds are covered with occlusive dressings, the exudate (wound fluid) is retained and can be used for assessing various soluble factors in the wound environment, such as cytokines, growth factors, nutrients, proteinases, and tissue degradation products. Both contraction and epithelialization contribute to wound closure in this model. Contrary to intuition, the size of the scab does not reflect the rate of epithelialization, which is measured reliably only by histology. The excision sites including

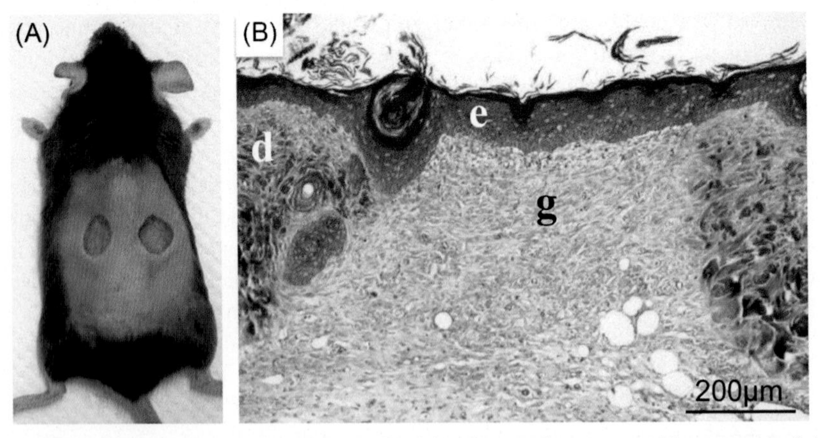

Fig. 3 Mouse excisional wound model. (A) Excisional wounds are created by removal of the full-thickness skin using biopsy punch, and the entire wound is left open. (B) Representative trichrome-Masson stained section of murine full-thickness wound at day 15 post wounding. Morphological features are indicated as E, epidermis; D, dermis; and G, granulation tissue.

wound edge are harvested at different time points for the analysis of histology (e.g., granulation tissue, epithelialization, contraction, and leukocyte infiltration), biochemistry (e.g., exudate analysis), and gene expression (e.g., RNA, microRNA, or protein). This model is the most widely used in murine wound studies and has led to the identification of important signaling pathways in wound healing, such as Wnt/β-catenin, TGF-β, Notch, hedgehog, and BMPs [79-83].

It has been further shown that variability is substantially lower in histological measurements of incisional and excisional wounds than planimetry [76]. Additionally, wound planimetry is less variable in excisional wounds than incisional wounds (secondary intention closure), but correlation between planimetric and histological measurements is greater in incisional wounds than excisional wounds [76].

4.1.3 Full-thickness splinting murine wounds

In addition to epithelialization, contraction makes a significant contribution to full-thickness wound closure in the murine models. The murine skin has an extensive layer of striated muscle just beneath the subcutaneous fat, called the panniculus carnosus, which is largely absent in humans. Upon wounding, the panniculus carnosus allows the skin to move independently from the deeper muscles and is largely responsible for the rapid skin contraction. To reduce the influence of contraction so that murine wound models can more closely mimic the human processes of epithelialization and new tissue formation, a silicon splint can be applied to the full-thickness wound [11,84-86]. Wound healing is then dependent on epithelialization, cell proliferation, and angiogenesis, which closely mirror the biological processes of human wound healing. Furthermore, splinted wounds were found to be more reproducible than nonsplinted ones [87].

4.1.4 Parabiosis model

The parabiosis model has been an effective model used to study circulation and infiltration of cells and factors in the context of wound healing. The model consists of two mice, a donor mouse (e.g., a wild-type mouse or a mouse carrying a genetic marker such as EGFP) and a recipient mouse (e.g., a mouse carrying a genetic mutation such as Lepr- or a wild-type mouse), surgically conjoined at their flank skin in which cross circulation is established between the joined mice several days post surgery through vascular anastomoses [10]. A wound can then be created on the recipient mouse, permitting the study of circulating factors and systemically derived cells from the donor mouse important for wound healing [10]. The parabiosis model has been utilized to highlight the importance of systemic factors and cell types in delayed wounding in diabetic mice and inflammatory cells and mesenchymal derived cells that function in skin homeostasis and wound repair [10,88].

4.1.5 Dead space wound model

In this model, connective tissue formation is isolated from other events of wound healing, such as contraction and epithelialization. This is achieved by inserting porous implants into the subcutaneous space, into which plasma infuses. This infusion leads to the development of a fibrin clot and subsequent formation of granulation tissue. The implants are made of relatively inert and nonbiodegradable materials to minimize inflammation. The implants allow interstitial fluid to accumulate, which can be extracted for analysis of metabolites, cytokines, growth factors, and nonadherent cell populations. Agents of interest can be orally administered or injected into the implants (such as growth factors, chemoattractants, and neutralizing antibodies) [89,90]. Because the implants have defined dimensions, this model is ideal for biochemical analysis once the granulation tissue is removed. The most common analyses include the hydroxyproline assay to estimate collagen content and hexosamine assay to estimate the stabilization of collagen and hence the strength of the repaired tissue [91].

4.1.6 Dorsal skin fold chamber model

The dorsal skinfold chamber utilizes an implanted chamber over exposed skin that allows for the microscopic imaging through an observation window embedded in the chamber [10]. It is a well-established model that permits the in vivo study of vasculature physiology in the context of wound healing under normal and pathophysiological states [10]. In addition, the use of intravital fluorescence microscopy allows for the capture of high-resolution images in real time of cutaneous physiology that includes vessel growth, microvascular perfusion, vascular permeability, and cell interactions [92]. Combined with intravital fluorescence microscopy is then performed over a period of time lasting up to 3 weeks allowing for the analyses of the early inflammatory response and angiogenic host response in real time after wounding.

4.1.7 Tape stripping wound model

The tape-stripping wound model involves the partial removal of the epidermis with adhesive tape. Depending on the adhesive used and the number of times it is applied, one can disrupt the epidermal barrier by removing the stratum corneum and stratum granulosum. Although the basement membrane is left intact and there is no blood loss, the epidermal permeability barrier is disrupted and leads to increased transepidermal water loss and activates epidermal repair [93,94]. This model is the least invasive type of all murine wound models discussed here and is useful to study genes and pathways involved in epidermal barrier differentiation and function.

4.2 Diabetic wound healing models

4.2.1 Diabetes type 2 murine wound healing model

Since it is expected that the incidence of chronic wounds in human population will increase due to a rise in diabetes prevalence [1], significant research effort has been employed to develop diabetic models of wound healing. Diabetic db/db mice are the most commonly used model for type 2 diabetes and wound-healing studies. A deletion mutation renders the leptin receptor (LepR, db/db) in the hypothalamus defective, causing the mouse to lose control of satiation, overeat, and develop hyperphagia, obesity, hyperlipidemia, hyperinsulinemia, and insulin resistance, thus closely resembling type 2 diabetes mellitus in humans [13,95]. Diabetic db/db mice exhibit significantly impaired wound healing, for both excisional and incisional wounds [13-15]. Furthermore, synergism of advanced age and db/db genotype can further delay wound healing [96]. One benefit of using the db/db mouse model is that the wound-healing process is not significantly influenced by other diabetes-related parameters. Variability in the rate of wound closure among individual animals does not correlate with the severity of obesity, hyperglycemia, hyperinsulinemia, or insulin resistance, helping to rule out possible confounding variables in the study [14]. A variation of the db/db model involves application of bacterial biofilm, such as *P. aeruginosa*, on excisional wounds. Introduction of the bacterial biofilm with the goal to mimic chronic wound phenotype results in extensive inflammatory cell infiltration, tissue necrosis, epidermal hyperplasia adjacent to the challenged wounds, and a severe delay in wound healing [13].

4.2.2 Streptozotocin-induced diabetic rodent wound healing models

Streptozotocin-induced diabetic murine models develop type 1 diabetes, due to the cytotoxic glucose analogue streptozotocin (STZ) that is toxic to pancreatic β-cells and causes insulin deficiency. STZ methylates DNA, causing DNA fragmentation and killing pancreatic β-cells [95,97]. In addition to the pancreas, STZ may also have toxic effects on other organs. To reduce nonspecific toxicity, diabetes is often induced through multiple injections of low doses of STZ that cause repetitive low-grade β-cell damage. The Animal Models of Diabetic Complications Consortium protocol recommends daily intraperitoneal injections of 40–50 mg/kg of STZ for 5 consecutive days [95]. STZ-induced diabetic mice may exhibit significant weight loss, due to insulin deficiency and severe hyperglycemia, but do not develop hypertension. Excisional or incisional wounds can be created in STZ-induced diabetic mice as described above. However, there is no significant delay in wound closure for STZ-treated mice as there is in db/db mice, which also demonstrate decreased granulation tissue formation, decreased wound bed vascularity, and diminished proliferation during the wound-healing process [98,99]. Therefore, compared with db/db mice, STZ-induced diabetic mouse model may not be the most appropriate model to study impaired wound healing. Other diabetic animal models include STZ-induced diabetic rats, in which STZ is injected intraperitoneally for a few consecutive days (2–5 days) at 45–70 mg/kg per injection [100]. A variation of the STZ-induced diabetic rat model is by adding an 8-week high-fat diet prior to the multiple STZ injections. The addition of a high-fat diet establishes an ischemic diabetic rat model. Ischemia—restricted blood supply to the body's organs—is one of the main characteristics of chronic wounds. Excisional wounds in this modified model resemble the interrupted healing process including delayed wound closure, increased inflammatory response, and excessive and prolonged production of reactive oxygen species (ROS) and matrix metalloproteinases (MMPs) [101]. However, in spite significant efforts to modify db/db or STZ wound models to more closely resemble chronic wound phenotype, wounds generated in these animals eventually heal resembling all phases of acute healing process.

In summary, murine models are the most frequently used animal models for wound-healing studies. The ease to acquire and maintain, relatively low costs, genetic homogeneity reproducibility, and the many manipulations already developed have facilitated the investigation of in vivo wound-healing processes that mimic a wide array of physiological/pathological conditions. In addition, the many mutant strains (both spontaneous and induced) enable the dissection of specific signaling pathways. However, there are significant anatomical and physiological differences between the mouse, rat, and human skin. Mice and rats have loose skin with much higher hair follicle density compared with the human skin, and the mouse skin has very thin epidermis. Caution must be taken when one tries to extrapolate results obtained in these models to clinical applications [2].

Besides those discussed above, skin grafting and human skin xenograft in mice have also made it possible to study the wound-healing processes under special circumstances (e.g., early lethality of mutant mouse strains and patient skin biopsies). In vivo wound models have also been developed that include partial-thickness wounds, blisters, burn and caustic agents (such as alkalies), and infections. However, the murine skin, especially the mouse skin, has very thin epidermis and abundant hair follicles, making it technically difficult to induce such wounds with consistency and reproducibility. The porcine model is more suitable for some of these applications.

4.3 Porcine models of wound healing

The porcine wound-healing model was developed by Mertz and Eaglstein in 1978 [9] and used since then to obtain a better understanding of wound-healing processes including scar formation, regeneration, and wound infection. Among all the in vivo wound-healing models commonly used, the porcine skin is structurally most similar to the human skin, including parameters such as epidermal thickness and dermal-epidermal thickness ratios and similar patterns of hair follicles and blood vessels [9,102]. The histological location of epidermal keratins 10 and 16 and dermal collagen IV, vimentin, and fibronectin is also similar to humans [103]. Biochemically, the porcine skin contains dermal collagen and a dermal elastic content more similar to the human skin than other commonly used mammals and has similar biomechanical properties [104]. Moreover, the resident immune cells in the porcine skin, including dendritic cells, are similar to those in the human skin [105]. In addition to these significant similarities between the porcine and human skin, there are also important differences. For example, the porcine skin has only apocrine sweat glands, while the human skin has eccrine sweat glands distributed over the entire body that also serve as essential sources of keratinocytes during epithelialization [106]. Importantly, pig wounds, similarly to human, heal primarily by epithelialization, whereas murine, rabbit, and canine wounds heal primarily by contraction, making the porcine model the most appropriate for certain aspects of wound-healing studies (Fig. 4) [9,107]. Sullivan and colleagues compared studies of wound therapy assessment in different animal models and found that porcine wound models were 78% concordant with human studies [9]. This far exceeded other animal wound models, which were only 53% concordant with human studies and more similar to in vitro studies [9]. Yorkshire specific pathogen-free (SPF)

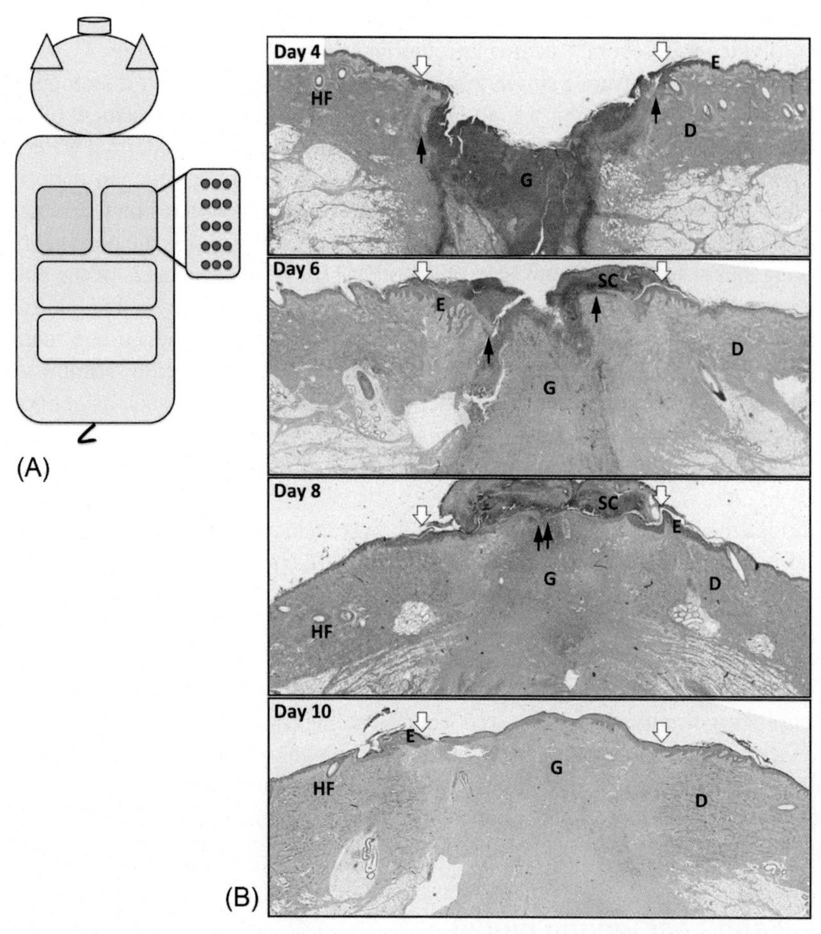

Fig. 4 Porcine full-thickness wound healing. (A) Schematic illustration showing the layout of experimental wounds in the pig model. Each quadrant can include multiple wounds and represents independent treatment conditions (color coded). Representative H&E-stained transverse sections of porcine full-thickness wounds at 4, 6, 8, and 10 days post wounding. White arrows indicate the edges of the initial wound area. Progression of reepithelialization is indicated by black arrows. By day 4, granulation tissue is formed in the wound bed. Meanwhile, keratinocytes migrate across granulation tissue to close the open wound. Reepithelialization is complete by day 8, and scab is cleared by day 10. Morphological features are indicated as E, epidermis; D, dermis; G, granulation tissue; SC, scab; and HF, hair follicle.

crossbred pigs are usually chosen for wound-healing studies due to their analogous repair mechanisms to humans as elaborated, and other strains such as the red Duroc pigs are used as models of hypertrophic scarring [108,109]. The anatomical location of wounding is important when performing scar studies since the wounds near tension areas scar differently.

The ample skin area of pigs allows for multiple wounds in the same animal that can be divided into several treatment groups and assessed over time (Fig. 4). Multiple biopsies can be taken from a single wound to assess wound closure by histology or for RNA/protein and microbiology analyses, allowing simultaneous assessment of multiple cellular processes including epithelialization, erythema, transepidermal water loss, scar formation, inflammatory cell infiltrate, bacterial infection, collagen deposition, and stem cell recruitment. The ability to compare various treatments on the same animal and to perform assessments during the course of healing makes the porcine model advantageous over rodent wound-healing models [107]. Disadvantages of the porcine model are their high costs for purchase and maintenance, their large size that makes them difficult to manage, and that in pigs it is challenging to induce and maintain diabetes [110] or to mimic a nonhealing wound model. In addition, administering anesthesia to a pig requires a skilled veterinarian. Some reagents, such as antibodies, are less readily available than for other animal species. Lastly, transgenic pigs are more difficult to produce than other animals used in wound-healing studies such as mice and rats.

Various models to study wound healing in pigs include incisional, excisional (deep partial or full thickness), wound infection, second-degree burn, keloid, laser injury, UV radiation injury, and skin hypoxia [111-116]. These models have been published and validated in studies testing different formulations including a variety of commonly used wound care products.

Overall, pigs provide a versatile vertebrate model that can reproduce the numerous etiologies of normal and impaired wound-healing process. Porcine wound models have increased our understanding of normal and pathological wound-healing process and have provided ample preclinical data, justifying continued development. Taken together, wound-healing studies conducted on large animals are extremely valuable and allow for initial preclinical evaluation for testing potential treatment modalities.

4.4 Rabbit ear wound model

The rabbit ear model has been extensively used for studying wound healing in the context of ischemia and infection. Since ischemia and infection play a significant role in the development of chronic wounds, the rabbit ear model has contributed to a better understanding of how these processes are deregulated in chronic wounds. The ischemic wound is created by suturing off the arterial blood supply to the ear, and then, the wound is created using a punch biopsy, creating a full-thickness wound down to the cartilage of the ear [117,118]. A benefit of this model has been its use to test effects of bacteria on wound healing and potential therapeutic treatments. It has been shown using the rabbit ear model that wounds infected with *P. aeruginosa* resulted in impairment of healing through biofilm [16,119]. In addition, it was found that topical application of growth factors that include platelet-derived growth factor B (PDGF-B), keratinocyte growth factor 2 (KGF-2), and vascular endothelial growth factor (VEGF) resulted in improved healing, demonstrating that this model can be used for effective testing of potential therapeutic treatments that could be translated to the clinic [17,120,121]. Advantages of this model include that rabbits are relatively inexpensive to maintain, their wounds heal primarily by epithelialization due to the underlying cartilage, and multiple replicates within a single animal can be made. As such, the rabbit ear model fails to fully recapitulate the human setting due to the dermis being firmly attached to the underlying cartilage and an avascular wound base—features that are not present in human wounds.

4.5 Ischemic wound models

Ischemic models of wound healing have been developed to study the effects of inadequate blood supply to tissue, which is characteristic of chronic nonhealing wounds. Several animal models have been developed to study ischemia with varying degrees of efficiency. The primary porcine model for ischemia is by means of a full-thickness bipedicle flap, whereby the dermis is prevented from readhering to the subcutaneous tissue, thus preventing revascularization [18]. The impaired reepithelialization, delayed recruitment of the adaptive immune response, and diminished organization and abundance of endothelial cells, characteristic of ischemic tissue, are produced in this model with accuracy and reliability [18,122]. A rabbit ear ischemic wound model has also been developed in which the main arteries to the rabbit ear are ligated, leaving the ear devoid of blood flow [123,124]. Wound-healing times are lengthened in the ischemic ear and show other signs characteristic of ischemia [118]. Another primary model, the murine limb-ligation model, was developed to mimic ischemia that occurs in the elderly population who often develop ischemia in their extremities. The murine limb-ligation model produces severe, acute ischemia [125]. Ischemia occurred to a similar extent in older mice as compared with younger ones, but the older mice experienced substantially impaired arteriogenesis and functional recovery upon wounding [125]. While other models to study ischemia do exist, these three primary models can depict with relatively significant accuracy the different aspects of chronic wound ischemia.

4.6 Denervation wound models

Neuropathy is a characteristic typical of many chronic wounds that can result from diabetes or sustained pressure load [1]. As a result, patients lose the protective sensation of pain and become unaware of the development of a wound. Denervation models are used to study wound healing in the context of neuropathy. These models can be created in vivo using mice and porcine models by surgically exposing the T9 to L1 vertebrae and transecting the nerve endings [102,126]. The animals are then allowed to recover, and wounds can then be created by excision. The response to sensation is determined by stimulating with a needle at the wound site to verify successful nerve resection. An advantage of this model is that they resemble DFU and pressure ulcer neuropathy and can be used to test potential therapeutics to improve sensation and wound healing [102,126].

4.7 In vivo models of wound infection

Recent advances in high-throughput sequencing technologies have facilitated studies of the complex microbial communities residing on the human skin [127]. These novel methods have characterized the diversity of cutaneous microbiome at specific body locations, including animal models and more importantly human chronic wounds [127-130]. Microbiomes of chronic wounds play significant roles in impaired wound healing, even in the presence or absence of clinical signs of infection [129,131]. The fact that chronic wound bioburden exists predominantly in the form of a biofilm [132] has imposed development of polymicrobial biofilm infection models. Biofilms facilitate genetic exchange of virulence factors, whereby bacterial resistant genes can be transferred to susceptible bacteria [133,134], while bacteria within biofilms are more resistant to antimicrobial agents than those in the free-floating planktonic state [135], thus more closely mimicking chronic wound environment. In one of the first polymicrobial wound infection models, the authors used four aerobic and anaerobic bacterial species commonly detected in chronic wounds (*S. aureus*, *P. aeruginosa*, *Enterococcus faecalis*, and *Finegoldia magna*) to develop biofilms *in vitro* and then transferred them into full-thickness wounds in mice [136]. Although not statistically significant at all time points, polymicrobial biofilms in this infected wound model had a clear trend of delayed wound closure, indicating that multibacterial species biofilms promote a more pathogenic wound phenotype. Remarkably, the polymicrobial biofilms were more resistant to antimicrobial agents (bleach and gentamicin) than monospecies biofilms, suggesting that synergistic interactions between bacterial species impart protection from therapeutic agents [136]. Another valuable in vivo biofilm model in New Zealand rabbit ear was developed to assess wound chronicity. The monospecies wound biofilms of *Klebsiella pneumoniae*, *S. aureus*, and *P. aeruginosa* in rabbit ear wounds lead to impaired wound closure compared with uninfected wounds [137]. *S. aureus* and *P. aeruginosa* were also used to form polymicrobial biofilms in the rabbit model, which showed enhanced delay in wound closure compared with their monospecies counterparts, confirming a synergistic interaction between both bacterial species. Lastly, pig models for studies focusing on wound biofilms provide major advantages [115,138] such as their similarity to the human skin and its large surface area that allows the creation of multiple wounds that can be assessed for both microbial load and host response simultaneously (Fig. 4A). Partial-thickness or full-thickness infection models in pigs have been extensively used to understand host-pathogen interaction in a polymicrobial wound environment. Importantly, porcine infected wound models have been extensively utilized for preclinical testing of topical antimicrobials [114,139-141]. Further development of polymicrobial wound models in conjunction with characterization of the chronic wound microbiome will aid the development of more specific and efficient biofilm-targeted therapies for chronic wounds.

5. Conclusion

Preclinical testing using in vitro and in vivo wound-healing models remains indispensable for addressing safety, toxicology, and efficacy in a dose-response fashion in development new therapies. However, rising incidence of nonhealing wounds and limitations of currently available therapeutics underscore the need for improved effectiveness of preclinical predictions of human treatment responses. Previous progress was hampered by the lack of understanding regarding the limitations of current preclinical wound models, along with gaps in understanding of the complex molecular and cellular network of processes that guide wound healing and their dysregulation in patients. Recent data suggest that studies focusing on mechanism of healing impairment in patients can provide the rationale and hypothesis driving future preclinical and basic research [142]. In addition, advances of iPSC technologies that provide opportunity to generate multiple cell types and tissues from patients' somatic cells coupled with progress in tissue engineering that allow building complex 3-D tissue structures that closely resembles human skin may prove advantageous in future preclinical testing. The most recent advances in cell biology and microfabrication have already enabled the development of microengineered models of the functional human skin—known as organ-on-chips. Future developments in microengineered wound-on-chip models incorporating primary cells from patients and adding multiple organs-on-chip may provide the basis for preclinical models that more accurately may represent human condition with greater predictive power.

Abbreviations

FDA Food and Drug Administration
PDGF platelet-derived growth factor
FRET fluorescent resonance energy transfer
TGFα transforming growth factor α
EGF epidermal growth factor
HB-EGF heparin binding-epidermal growth factor
IGF-1 insulin-like growth factor 1
FGF fibroblast growth factor
IL-8 interleukin-8
HGF hepatocyte growth factor
IL-1 interleukin-1
KGF keratinocyte growth factor
TGFβ transforming growth factor β
CCL14 C–C motif chemokine ligand 14
CCL22 C–C motif chemokine ligand 22
CCL27 C–C motif chemokine ligand 27
CXCL1 C–X–C motif chemokine ligand 1
CXCL10 C–X–C motif chemokine ligand 10
CXCL8 C–X–C motif chemokine ligand 8
ECM extracellular matrix
iPSC induced pluripotent stem cell
BMPs bone morphogenetic protein
STZ streptozotocin
ROS reactive oxygen species
MMPs matrix metalloproteinases
SPF specific pathogen-free
VEGF vascular endothelial growth factor

References

[1] Eming SA, Martin P, Tomic-Canic M. Wound repair and regeneration: mechanisms, signaling, and translation. Sci Transl Med 2014;6(265):265sr6.
[2] Pastar I, Stojadinovic O, Yin NC, Ramirez H, Nusbaum AG, Sawaya A, et al. Epithelialization in wound healing: a comprehensive review. Adv Wound Care (New Rochelle) 2014;3(7):445–64.
[3] Liang CC, Park AY, Guan JL. In vitro scratch assay: a convenient and inexpensive method for analysis of cell migration in vitro. Nat Protoc 2007;2(2):329–33.
[4] Li Y, Fan J, Chen M, Li W, Woodley DT. Transforming growth factor-alpha: a major human serum factor that promotes human keratinocyte migration. J Invest Dermatol 2006;126(9):2096–105.
[5] Oh JW, Hsi TC, Guerrero-Juarez CF, Ramos R, Plikus MV. Organotypic skin culture. J Invest Dermatol 2013;133(11):e14.
[6] Garlick JA, Taichman LB. Fate of human keratinocytes during reepithelialization in an organotypic culture model. Lab Invest 1994;70(6):916–24.
[7] Stojadinovic O, Tomic-Canic M. Human ex vivo wound healing model. Methods Mol Biol 2013;1037:255–64.
[8] Ashcroft GS, Horan MA, Ferguson MW. The effects of ageing on wound healing: immunolocalisation of growth factors and their receptors in a murine incisional model. J Anat 1997;190(Pt 3):351–65.
[9] Sullivan TP, Eaglstein WH, Davis SC, Mertz P. The pig as a model for human wound healing. Wound Repair Regen 2001;9(2):66–76.
[10] Wong VW, Sorkin M, Glotzbach JP, Longaker MT, Gurtner GC. Surgical approaches to create murine models of human wound healing. J Biomed Biotechnol 2011;2011:969618.
[11] Dunn L, Prosser HC, Tan JT, Vanags LZ, Ng MK, Bursill CA. Murine model of wound healing. J Vis Exp 2013;75:e50265.
[12] Davidson JM. Animal models for wound repair. Arch Dermatol Res 1998;290(Suppl):S1–11.
[13] Zhao G, Hochwalt PC, Usui ML, Underwood RA, Singh PK, James GA, et al. Delayed wound healing in diabetic (db/db) mice with *Pseudomonas aeruginosa* biofilm challenge: a model for the study of chronic wounds. Wound Repair Regen 2010;18(5):467–77.
[14] Trousdale RK, Jacobs S, Simhaee DA, Wu JK, Lustbader JW. Wound closure and metabolic parameter variability in a db/db mouse model for diabetic ulcers. J Surg Res 2009;151(1):100–7.
[15] Bitto A, Altavilla D, Pizzino G, Irrera N, Pallio G, Colonna MR, et al. Inhibition of inflammasome activation improves the impaired pattern of healing in genetically diabetic mice. Br J Pharmacol 2014;171(9):2300–7.
[16] Seth AK, Geringer MR, Gurjala AN, Hong SJ, Galiano RD, Leung KP, et al. Treatment of *Pseudomonas aeruginosa* biofilm-infected wounds with clinical wound care strategies: a quantitative study using an in vivo rabbit ear model. Plast Reconstr Surg 2012;129(2):262e–74e.
[17] Liechty KW, Crombleholme TM, Quinn TM, Cass DL, Flake AW, Adzick NS. Elevated platelet-derived growth factor-B in congenital cystic adenomatoid malformations requiring fetal resection. J Pediatr Surg 1999;34(5):805–9 [discussion 9-10].
[18] Roy S, Biswas S, Khanna S, Gordillo G, Bergdall V, Green J, et al. Characterization of a preclinical model of chronic ischemic wound. Physiol Genomics 2009;37(3):211–24.
[19] Sen CK, Gordillo GM, Roy S, Kirsner R, Lambert L, Hunt TK, et al. Human skin wounds: a major and snowballing threat to public health and the economy. Wound Repair Regen 2009;17(6):763–71.

[20] Hakkinen L, Koivisto L, Gardner H, Saarialho-Kere U, Carroll JM, Lakso M, et al. Increased expression of beta 6-integrin in skin leads to spontaneous development of chronic wounds. Am J Pathol 2004;164(1):229–42.
[21] Atac B, Wagner I, Horland R, Lauster R, Marx U, Tonevitsky AG, et al. Skin and hair on-a-chip: in vitro skin models versus ex vivo tissue maintenance with dynamic perfusion. Lab Chip 2013;13(18):3555–61.
[22] Tamada M, Perez TD, Nelson WJ, Sheetz MP. Two distinct modes of myosin assembly and dynamics during epithelial wound closure. J Cell Biol 2007;176(1):27–33.
[23] Fu X, Xu M, Liu J, Qi Y, Li S, Wang H. Regulation of migratory activity of human keratinocytes by topography of multiscale collagen-containing nanofibrous matrices. Biomaterials 2014;35(5):1496–506.
[24] Timm DM, Chen J, Sing D, Gage JA, Haisler WL, Neeley SK, et al. A high-throughput three-dimensional cell migration assay for toxicity screening with mobile device-based macroscopic image analysis. Sci Rep 2013;3:3000.
[25] Barrientos S, Stojadinovic O, Golinko MS, Brem H, Tomic-Canic M. Growth factors and cytokines in wound healing. Wound Repair Regen 2008;16(5):585–601.
[26] Barrientos S, Brem H, Stojadinovic O, Tomic-Canic M. Clinical application of growth factors and cytokines in wound healing. Wound Repair Regen 2014;22(5):569–78.
[27] Liang L, Stone RC, Stojadinovic O, Ramirez H, Pastar I, Maione AG, et al. Integrative analysis of mi RNA and mRNA paired expression profiling of primary fibroblast derived from diabetic foot ulcers reveals multiple impaired cellular functions. Wound Repair Regen 2016;24(6):943–53.
[28] Jozic I, Vukelic S, Stojadinovic O, Liang L, Ramirez HA, Pastar I, et al. Stress signals, mediated by membranous glucocorticoid receptor, activate PLC/PKC/GSK-3beta/beta-catenin pathway to inhibit wound closure. J Invest Dermatol 2016;137(5):1144–54.
[29] Boyden S. The chemotactic effect of mixtures of antibody and antigen on polymorphonuclear leucocytes. J Exp Med 1962;115:453–66.
[30] Marshall J. Transwell((R)) invasion assays. Methods Mol Biol 2011;769:97–110.
[31] Kroeze KL, Boink MA, Sampat-Sardjoepersad SC, Waaijman T, Scheper RJ, Gibbs S. Autocrine regulation of re-epithelialization after wounding by chemokine receptors CCR1, CCR10, CXCR1, CXCR2, and CXCR3. J Invest Dermatol 2012;132(1):216–25.
[32] Stark HJ, Szabowski A, Fusenig NE, Maas-Szabowski N. Organotypic cocultures as skin equivalents: a complex and sophisticated in vitro system. Biol Proced Online 2004;6:55–60.
[33] Li J, Sen GL. Generation of genetically modified organotypic skin cultures using devitalized human dermis. J Vis Exp 2015;106:e53280.
[34] Collawn SS, Banerjee NS, de la Torre J, Vasconez L, Chow LT. Adipose-derived stromal cells accelerate wound healing in an organotypic raft culture model. Ann Plast Surg 2012;68(5):501–4.
[35] Monslow J, Sato N, Mack JA, Maytin EV. Wounding-induced synthesis of hyaluronic acid in organotypic epidermal cultures requires the release of heparin-binding egf and activation of the EGFR. J Invest Dermatol 2009;129(8):2046–58.
[36] Garlick JA, Taichman LB. Effect of TGF-beta 1 on re-epithelialization of human keratinocytes in vitro: an organotypic model. J Invest Dermatol 1994;103(4):554–9.
[37] Maione AG, Brudno Y, Stojadinovic O, Park LK, Smith A, Tellechea A, et al. Three-dimensional human tissue models that incorporate diabetic foot ulcer-derived fibroblasts mimic in vivo features of chronic wounds. Tissue Eng Part C Methods 2015;21(5):499–508.

[38] Itoh M, Kiuru M, Cairo MS, Christiano AM. Generation of keratinocytes from normal and recessive dystrophic epidermolysis bullosa-induced pluripotent stem cells. Proc Natl Acad Sci U S A 2011;108(21):8797–802.
[39] Marino D, Luginbuhl J, Scola S, Meuli M, Reichmann E. Bioengineering dermo-epidermal skin grafts with blood and lymphatic capillaries. Sci Transl Med 2014;6(221):221ra14.
[40] Benny P, Badowski C, Lane EB, Raghunath M. Making more matrix: enhancing the deposition of dermal-epidermal junction components in vitro and accelerating organotypic skin culture development, using macromolecular crowding. Tissue Eng Part A 2015;21(1–2):183–92.
[41] Maione AG, Smith A, Kashpur O, Yanez V, Knight E, Mooney DJ, et al. Altered ECM deposition by diabetic foot ulcer-derived fibroblasts implicates fibronectin in chronic wound repair. Wound Repair Regen 2016;24(4):630–43.
[42] van den Broek LJ, Niessen FB, Scheper RJ, Gibbs S. Development, validation and testing of a human tissue engineered hypertrophic scar model. ALTEX 2012;29(4):389–402.
[43] Butler PD, Ly DP, Longaker MT, Yang GP. Use of organotypic coculture to study keloid biology. Am J Surg 2008;195(2):144–8.
[44] Bellemare J, Roberge CJ, Bergeron D, Lopez-Valle CA, Roy M, Moulin VJ. Epidermis promotes dermal fibrosis: role in the pathogenesis of hypertrophic scars. J Pathol 2005;206(1):1–8.
[45] Chan CY, Huang PH, Guo F, Ding X, Kapur V, Mai JD, et al. Accelerating drug discovery via organs-on-chips. Lab Chip 2013;13(24):4697–710.
[46] Pastar I, Stojadinovic O, Sawaya AP, Stone RC, Lindley LE, Ojeh N, et al. Skin metabolite, farnesyl pyrophosphate, regulates epidermal response to inflammation, oxidative stress, and migration. J Cell Physiol 2016;231(11):2452–63.
[47] Pastar I, Khan AA, Stojadinovic O, Lebrun EA, Medina MC, Brem H, et al. Induction of specific micro RNAs inhibits cutaneous wound healing. J Biol Chem 2012;287(35):29324–35.
[48] Xu W, Jong Hong S, Jia S, Zhao Y, Galiano RD, Mustoe TA. Application of a partial-thickness human ex vivo skin culture model in cutaneous wound healing study. Lab Invest 2012;92(4):584–99.
[49] Lee B, Vouthounis C, Stojadinovic O, Brem H, Im M, Tomic-Canic M. From an enhanceosome to a repressosome: molecular antagonism between glucocorticoids and EGF leads to inhibition of wound healing. J Mol Biol 2005;345(5):1083–97.
[50] Stojadinovic O, Lee B, Vouthounis C, Vukelic S, Pastar I, Blumenberg M, et al. Novel genomic effects of glucocorticoids in epidermal keratinocytes: Inhibition of apoptosis, interferon-gamma pathway, and wound healing along with promotion of terminal differentiation. J Biol Chem 2007;282(6):4021–34.
[51] Stojadinovic O, Brem H, Vouthounis C, Lee B, Fallon J, Stallcup M, et al. Molecular pathogenesis of chronic wounds: the role of beta-catenin and c-myc in the inhibition of epithelialization and wound healing. Am J Pathol 2005;167(1):59–69.
[52] Sivamani RK, Pullar CE, Manabat-Hidalgo CG, Rocke DM, Carlsen RC, Greenhalgh DG, et al. Stress-mediated increases in systemic and local epinephrine impair skin wound healing: potential new indication for beta blockers. PLoS Med 2009;6(1):e12.
[53] Tomic-Canic M, Mamber SW, Stojadinovic O, Lee B, Radoja N, McMichael J. Streptolysin O enhances keratinocyte migration and proliferation and promotes skin organ culture wound healing in vitro. Wound Repair Regen 2007;15(1):71–9.
[54] Roupe KM, Nybo M, Sjobring U, Alberius P, Schmidtchen A, Sorensen OE. Injury is a major inducer of epidermal innate immune responses during wound healing. J Invest Dermatol 2010;130(4):1167–77.

[55] Andrew Chan KL, Zhang G, Tomic-Canic M, Stojadinovic O, Lee B, Flach CR, et al. A coordinated approach to cutaneous wound healing: vibrational microscopy and molecular biology. J Cell Mol Med 2008;12(5B):2145–54.

[56] Vukelic S, Stojadinovic O, Pastar I, Rabach M, Krzyzanowska A, Lebrun E, et al. Cortisol synthesis in epidermis is induced by IL-1 and tissue injury. J Biol Chem 2011;286(12):10265–75.

[57] Vukelic S, Stojadinovic O, Pastar I, Vouthounis C, Krzyzanowska A, Das S, et al. Farnesyl pyrophosphate inhibits epithelialization and wound healing through the glucocorticoid receptor. J Biol Chem 2010;285(3):1980–8.

[58] Stojadinovic O, Minkiewicz J, Sawaya A, Bourne JW, Torzilli P, de Rivero Vaccari JP, et al. Deep tissue injury in development of pressure ulcers: a decrease of inflammasome activation and changes in human skin morphology in response to aging and mechanical load. PLoS One 2013;8(8):e69223.

[59] Harsha A, Stojadinovic O, Brem H, Sehara-Fujisawa A, Wewer U, Loomis CA, et al. ADAM12: a potential target for the treatment of chronic wounds. J Mol Med (Berl) 2008;86(8):961–9.

[60] Torzilli PA, Bhargava M, Chen CT. Mechanical loading of articular cartilage reduces IL-1-induced enzyme expression. Cartilage 2011;2(4):364–73.

[61] Torzilli PA, Grigiene R, Huang C, Friedman SM, Doty SB, Boskey AL, et al. Characterization of cartilage metabolic response to static and dynamic stress using a mechanical explant test system. J Biomech 1997;30(1):1–9.

[62] Yeung CC, Holmes DF, Thomason HA, Stephenson C, Derby B, Hardman MJ. An ex vivo porcine skin model to evaluate pressure-reducing devices of different mechanical properties used for pressure ulcer prevention. Wound Repair Regen 2016;24(6):1089–96.

[63] Yang Q, Phillips PL, Sampson EM, Progulske-Fox A, Jin S, Antonelli P, et al. Development of a novel ex vivo porcine skin explant model for the assessment of mature bacterial biofilms. Wound Repair Regen 2013;21(5):704–14.

[64] Yang Q, Larose C, Della Porta AC, Schultz GS, Gibson DJ. A surfactant-based wound dressing can reduce bacterial biofilms in a porcine skin explant model. Int Wound J 2017;14(2):408–13.

[65] Phillips PL, Yang Q, Davis S, Sampson EM, Azeke JI, Hamad A, et al. Antimicrobial dressing efficacy against mature *Pseudomonas aeruginosa* biofilm on porcine skin explants. Int Wound J 2015;12(4):469–83.

[66] Phillips PL, Yang Q, Schultz GS. The effect of negative pressure wound therapy with periodic instillation using antimicrobial solutions on *Pseudomonas aeruginosa* biofilm on porcine skin explants. Int Wound J 2013;10(Suppl 1):48–55.

[67] Muller-Rover S, Handjiski B, van der Veen C, Eichmuller S, Foitzik K, McKay IA, et al. A comprehensive guide for the accurate classification of murine hair follicles in distinct hair cycle stages. J Invest Dermatol 2001;117(1):3–15.

[68] Ansell DM, Kloepper JE, Thomason HA, Paus R, Hardman MJ. Exploring the "hair growth-wound healing connection": anagen phase promotes wound re-epithelialization. J Invest Dermatol 2011;131(2):518–28.

[69] Slominski A, Paus R. Melanogenesis is coupled to murine anagen: toward new concepts for the role of melanocytes and the regulation of melanogenesis in hair growth. J Invest Dermatol 1993;101(1 Suppl):90S–7S.

[70] Paus R, Handjiski B, Czarnetzki BM, Eichmuller S. A murine model for inducing and manipulating hair follicle regression (catagen): effects of dexamethasone and cyclosporin a. J Invest Dermatol 1994;103(2):143–7.

[71] Paul DW, Ghassemi P, Ramella-Roman JC, Prindeze NJ, Moffatt LT, Alkhalil A, et al. Noninvasive imaging technologies for cutaneous wound assessment: a review. Wound Repair Regen 2015;23(2):149–62.
[72] Rayner TE, Cowin AJ, Robertson JG, Cooter RD, Harries RC, Regester GO, et al. Mitogenic whey extract stimulates wound repair activity in vitro and promotes healing of rat incisional wounds. Am J Physiol Regul Integr Comp Physiol 2000;278(6):R1651–60.
[73] Muller-Decker K, Hirschner W, Marks F, Furstenberger G. The effects of cyclooxygenase isozyme inhibition on incisional wound healing in mouse skin. J Invest Dermatol 2002;119(5):1189–95.
[74] Chen L, Nagaraja S, Zhou J, Zhao Y, Fine D, Mitrophanov AY, et al. Wound healing in mac-1 deficient mice. Wound Repair Regen 2017;.
[75] Routley CE, Ashcroft GS. Effect of estrogen and progesterone on macrophage activation during wound healing. Wound Repair Regen 2009;17(1):42–50.
[76] Ansell DM, Campbell L, Thomason HA, Brass A, Hardman MJ. A statistical analysis of murine incisional and excisional acute wound models. Wound Repair Regen 2014;22(2):281–7.
[77] Falanga V, Schrayer D, Cha J, Butmarc J, Carson P, Roberts AB, et al. Full-thickness wounding of the mouse tail as a model for delayed wound healing: accelerated wound closure in Smad 3 knock-out mice. Wound Repair Regen 2004;12(3):320–6.
[78] Mihu MR, Roman-Sosa J, Varshney AK, Eugenin EA, Shah BP, Ham Lee H, et al. Methamphetamine alters the antimicrobial efficacy of phagocytic cells during methicillin-resistant *Staphylococcus aureus* skin infection. MBio 2015;6(6):15e01622.
[79] Shi Y, Shu B, Yang R, Xu Y, Xing B, Liu J, et al. Wnt and notch signaling pathway involved in wound healing by targeting c-Myc and Hes 1 separately. Stem Cell Res Ther 2015;6:120.
[80] Han G, Li F, Ten Dijke P, Wang XJ. Temporal smad 7 transgene induction in mouse epidermis accelerates skin wound healing. Am J Pathol 2011;179(4):1768–79.
[81] Yang RH, Qi SH, Shu B, Ruan SB, Lin ZP, Lin Y, et al. Epidermal stem cells (ESCs) accelerate diabetic wound healing via the Notch signalling pathway. Biosci Rep 2016;36(4).
[82] Suh HN, Han HJ. Sonic hedgehog increases the skin wound-healing ability of mouse embryonic stem cells through the micro RNA 200 family. Br J Pharmacol 2015;172(3):815–28.
[83] Lewis CJ, Mardaryev AN, Poterlowicz K, Sharova TY, Aziz A, Sharpe DT, et al. Bone morphogenetic protein signaling suppresses wound-induced skin repair by inhibiting keratinocyte proliferation and migration. J Invest Dermatol 2014;134(3):827–37.
[84] Chen J, Chen Y, Chen Y, Yang Z, You B, Ruan YC, et al. Epidermal CFTR suppresses MAPK/NF-kappa B to promote cutaneous wound healing. Cell Physiol Biochem 2016;39(6):2262–74.
[85] Cho H, Balaji S, Hone NL, Moles CM, Sheikh AQ, Crombleholme TM, et al. Diabetic wound healing in a MMP9−/− mouse model. Wound Repair Regen 2016;24(5):829–40.
[86] Fujiwara T, Duscher D, Rustad KC, Kosaraju R, Rodrigues M, Whittam AJ, et al. Extracellular superoxide dismutase deficiency impairs wound healing in advanced age by reducing neovascularization and fibroblast function. Exp Dermatol 2016;25(3):206–11.
[87] Park SA, Covert J, Teixeira L, Motta MJ, DeRemer SL, Abbott NL, et al. Importance of defining experimental conditions in a mouse excisional wound model. Wound Repair Regen 2015;23(2):251–61.
[88] Song G, Nguyen DT, Pietramaggiori G, Scherer S, Chen B, Zhan Q, et al. Use of the parabiotic model in studies of cutaneous wound healing to define the participation of circulating cells. Wound Repair Regen 2010;18(4):426–32.

[89] Mekonnen A, Sidamo T, Asres K, Engidawork E. In vivo wound healing activity and phytochemical screening of the crude extract and various fractions of Kalanchoe petitiana A. Rich (Crassulaceae) leaves in mice. J Ethnopharmacol 2013;145(2):638–46.
[90] Mukherjee H, Ojha D, Bharitkar YP, Ghosh S, Mondal S, Kaity S, et al. Evaluation of the wound healing activity of Shorea robusta, an Indian ethnomedicine, and its isolated constituent (s) in topical formulation. J Ethnopharmacol 2013;149(1):335–43.
[91] Oryan A, Mohammadalipour A, Moshiri A, Tabandeh MR. Avocado/soybean unsaponifiables: a novel regulator of cutaneous wound healing, modelling and remodelling. Int Wound J 2015;12(6):674–85.
[92] Laschke MW, Menger MD. The dorsal skinfold chamber: a versatile tool for preclinical research in tissue engineering and regenerative medicine. Eur Cell Mater 2016;32:202–15.
[93] Wojcik SM, Bundman DS, Roop DR. Delayed wound healing in keratin 6a knockout mice. Mol Cell Biol 2000;20(14):5248–55.
[94] McCormack RM, de Armas LR, Shiratsuchi M, Fiorentino DG, Olsson ML, Lichtenheld MG, et al. Perforin-2 is essential for intracellular defense of parenchymal cells and phagocytes against pathogenic bacteria. Elife 2015;4.
[95] Kitada M, Ogura Y, Koya D. Rodent models of diabetic nephropathy: their utility and limitations. Int J Nephrol Renovasc Dis 2016;9:279–90.
[96] Brem H, Tomic-Canic M, Entero H, Hanflik AM, Wang VM, Fallon JT, et al. The synergism of age and db/db genotype impairs wound healing. Exp Gerontol 2007;42(6):523–31.
[97] Al-Awar A, Kupai K, Veszelka M, Szucs G, Attieh Z, Murlasits Z, et al. Experimental diabetes mellitus in different animal models. J Diabetes Res 2016;2016:9051426.
[98] Desmet CM, Lafosse A, Veriter S, Porporato PE, Sonveaux P, Dufrane D, et al. Application of electron paramagnetic resonance (EPR) oximetry to monitor oxygen in wounds in diabetic models. PLoS One 2015;10(12):e0144914.
[99] Michaels J, Churgin SS, Blechman KM, Greives MR, Aarabi S, Galiano RD, et al. Db/db mice exhibit severe wound-healing impairments compared with other murine diabetic strains in a silicone-splinted excisional wound model. Wound Repair Regen 2007;15(5):665–70.
[100] Muhammad AA, Arulselvan P, Cheah PS, Abas F, Fakurazi S. Evaluation of wound healing properties of bioactive aqueous fraction from Moringa oleifera Lam on experimentally induced diabetic animal model. Drug Des Devel Ther 2016;10:1715–30.
[101] Yang P, Pei Q, Yu T, Chang Q, Wang D, Gao M, et al. Compromised wound healing in ischemic type 2 diabetic rats. PLoS One 2016;11(3):e0152068.
[102] Seaton M, Hocking A, Gibran NS. Porcine models of cutaneous wound healing. ILAR J 2015;56(1):127–38.
[103] Wollina U, Berger U, Mahrle G. Immunohistochemistry of porcine skin. Acta Histochem 1991;90(1):87–91.
[104] Heinrich W, Lange PM, Stirtz T, Iancu C, Heidemann E. Isolation and characterization of the large cyanogen bromide peptides from the alpha 1- and alpha 2-chains of pig skin collagen. FEBS Lett 1971;16(1):63–7.
[105] Summerfield A, Meurens F, Ricklin ME. The immunology of the porcine skin and its value as a model for human skin. Mol Immunol 2015;66(1):14–21.
[106] Rittie L, Sachs DL, Orringer JS, Voorhees JJ, Fisher GJ. Eccrine sweat glands are major contributors to reepithelialization of human wounds. Am J Pathol 2013;182(1):163–71.
[107] Gordillo GM, Bernatchez SF, Diegelmann R, Di Pietro LA, Eriksson E, Hinz B, et al. Preclinical models of wound healing: Is man the model? Proceedings of the wound healing society symposium. Adv Wound Care (New Rochelle) 2013;2(1):1–4.

[108] Sood RF, Muffley LA, Seaton ME, Ga M, Sirimahachaiyakul P, Hocking AM, et al. Dermal fibroblasts from the red Duroc pig have an inherently fibrogenic phenotype: an in vitro model of fibroproliferative scarring. Plast Reconstr Surg 2015;136(5):990–1000.
[109] Tejiram S, Zhang J, Travis TE, Carney BC, Alkhalil A, Moffatt LT, et al. Compression therapy affects collagen type balance in hypertrophic scar. J Surg Res 2016;201(2):299–305.
[110] Singer AJ, Taira BR, McClain SA, Rooney J, Steinhauff N, Zimmerman T, et al. Healing of mid-dermal burns in a diabetic porcine model. J Burn Care Res 2009;30(5):880–6.
[111] Davis SC, Mertz PM, Bilevich ED, Cazzaniga AL, Eaglstein WH. Early debridement of second-degree burn wounds enhances the rate of epithelization—an animal model to evaluate burn wound therapies. J Burn Care Rehabil 1996;17(6 Pt 1):558–61.
[112] Davis SC, Badiavas E, Rendon-Pellerano MI, Pardo RJ. Histological comparison of postoperative wound care regimens for laser resurfacing in a porcine model. Dermatol Surg 1999;25(5):387–91 [discussion 92-3].
[113] Davis SC, Eaglstein WH, Cazzaniga AL, Mertz PM. An octyl-2-cyanoacrylate formulation speeds healing of partial-thickness wounds. Dermatol Surg 2001;27(9):783–8.
[114] Davis SC, Cazzaniga AL, Ricotti C, Zalesky P, Hsu LC, Creech J, et al. Topical oxygen emulsion: a novel wound therapy. Arch Dermatol 2007;143(10):1252–6.
[115] Pastar I, Nusbaum AG, Gil J, Patel SB, Chen J, Valdes J, et al. Interactions of methicillin resistant *Staphylococcus aureus* USA300 and *Pseudomonas aeruginosa* in polymicrobial wound infection. PLoS One 2013;8(2):e56846.
[116] Kaiser MR, Davis SC, Mertz BA. Effect of ultraviolet radiation-induced inflammation on epidermal wound healing. Wound Repair Regen 1995;3(3):311–5.
[117] Nunan R, Harding KG, Martin P. Clinical challenges of chronic wounds: searching for an optimal animal model to recapitulate their complexity. Dis Model Mech 2014;7(11):1205–13.
[118] Chien S. Ischemic rabbit ear model created by minimally invasive surgery. Wound Repair Regen 2007;15(6):928–35.
[119] Karna SL, D'Arpa P, Chen T, Qian LW, Fourcaudot AB, Yamane K, et al. RNA-Seq Transcriptomic responses of full-thickness dermal excision wounds to *Pseudomonas aeruginosa* acute and biofilm infection. PLoS One 2016;11(10):e0165312.
[120] Xia YP, Zhao Y, Marcus J, Jimenez PA, Ruben SM, Moore PA, et al. Effects of keratinocyte growth factor-2 (KGF-2) on wound healing in an ischaemia-impaired rabbit ear model and on scar formation. J Pathol 1999;188(4):431–8.
[121] Corral CJ, Siddiqui A, Wu L, Farrell CL, Lyons D, Mustoe TA. Vascular endothelial growth factor is more important than basic fibroblastic growth factor during ischemic wound healing. Arch Surg 1999;134(2):200–5.
[122] Kerrigan CL, Zelt RG, Thomson JG, Diano E. The pig as an experimental animal in plastic surgery research for the study of skin flaps, myocutaneous flaps and fasciocutaneous flaps. Lab Anim Sci 1986;36(4):408–12.
[123] Steinberg JP, Hong SJ, Geringer MR, Galiano RD, Mustoe TA. Equivalent effects of topically-delivered adipose-derived stem cells and dermal fibroblasts in the ischemic rabbit ear model for chronic wounds. Aesthet Surg J 2012;32(4):504–19.
[124] Chien S, Wilhelmi BJ. A simplified technique for producing an ischemic wound model. J Vis Exp 2012;63:e3341.
[125] Westvik TS, Fitzgerald TN, Muto A, Maloney SP, Pimiento JM, Fancher TT, et al. Limb ischemia after iliac ligation in aged mice stimulates angiogenesis without arteriogenesis. J Vasc Surg 2009;49(2):464–73.
[126] Shu B, Xie JL, Xu YB, Lai W, Huang Y, Mao RX, et al. Effects of skin-derived precursors on wound healing of denervated skin in a nude mouse model. Int J Clin Exp Pathol 2015;8(3):2660–9.

[127] Group NHW, Peterson J, Garges S, Giovanni M, McInnes P, Wang L, et al. The NIH human microbiome project. Genome Res 2009;19(12):2317–23.

[128] McIntyre MK, Peacock TJ, Akers KS, Burmeister DM. Initial characterization of the pig skin bacteriome and its effect on in vitro models of wound healing. PLoS One 2016;11(11):e0166176.

[129] Loesche M, Gardner SE, Kalan L, Horwinski J, Zheng Q, Hodkinson BP, et al. Temporal stability in chronic wound microbiota is associated with poor healing. J Invest Dermatol 2017;137(1):237–44.

[130] Grice EA, Snitkin ES, Yockey LJ, Bermudez DM, Program NCS, Liechty KW, et al. Longitudinal shift in diabetic wound microbiota correlates with prolonged skin defense response. Proc Natl Acad Sci U S A 2010;107(33):14799–804.

[131] Gardner SE, Hillis SL, Heilmann K, Segre JA, Grice EA. The neuropathic diabetic foot ulcer microbiome is associated with clinical factors. Diabetes 2013;62(3):923–30.

[132] James GA, Swogger E, Wolcott R, Pulcini E, Secor P, Sestrich J, et al. Biofilms in chronic wounds. Wound Repair Regen 2008;16(1):37–44.

[133] Molin S, Tolker-Nielsen T. Gene transfer occurs with enhanced efficiency in biofilms and induces enhanced stabilisation of the biofilm structure. Curr Opin Biotechnol 2003;14(3):255–61.

[134] Lewis K. Riddle of biofilm resistance. Antimicrob Agents Chemother 2001;45(4):999–1007.

[135] Evans RC, Holmes CJ. Effect of vancomycin hydrochloride on *Staphylococcus epidermidis* biofilm associated with silicone elastomer. Antimicrob Agents Chemother 1987;31(6):889–94.

[136] Dalton T, Dowd SE, Wolcott RD, Sun Y, Watters C, Griswold JA, et al. An in vivo polymicrobial biofilm wound infection model to study interspecies interactions. PLoS One 2011;6(11):e27317.

[137] Seth AK, Geringer MR, Galiano RD, Leung KP, Mustoe TA, Hong SJ. Quantitative comparison and analysis of species-specific wound biofilm virulence using an in vivo, rabbit-ear model. J Am Coll Surg 2012;215(3):388–99.

[138] Davis SC, Ricotti C, Cazzaniga A, Welsh E, Eaglstein WH, Mertz PM. Microscopic and physiologic evidence for biofilm-associated wound colonization in vivo. Wound Repair Regen 2008;16(1):23–9.

[139] Bionda N, Fleeman RM, de la Fuente-Nunez C, Rodriguez MC, Reffuveille F, Shaw LN, et al. Identification of novel cyclic lipopeptides from a positional scanning combinatorial library with enhanced antibacterial and antibiofilm activities. Eur J Med Chem 2016;108:354–63.

[140] Martineau L, Davis SC, Peng HT, Hung A. Controlling methicillin resistant Staphyloccocus aureus and *Pseudomonas aeruginosa* wound infections with a novel biomaterial. J Invest Surg 2007;20(4):217–27.

[141] Mertz PM, Oliveira-Gandia MF, Davis SC. The evaluation of a cadexomer iodine wound dressing on methicillin resistant *Staphylococcus aureus* (MRSA) in acute wounds. Dermatol Surg 1999;25(2):89–93.

[142] Stone RC, Stojadinovic O, Rosa AM, Ramirez HA, Badiavas E, Blumenberg M, et al. A bioengineered living cell construct activates an acute wound healing response in venous leg ulcers. Sci Transl Med 2017;9(371).

The importance of targeting inflammation in skin regeneration 11

Megan Schrementi, Lin Chen, Luisa A. DiPietro
Center for Wound Healing and Tissue Regeneration, College of Dentistry, University of Illinois at Chicago, Chicago, IL, United States

1. Introduction

Wound healing is a multistep process that often concludes with scar formation. Healing occurs through the four overlapping phases of hemostasis, inflammation, proliferation, and tissue remodeling. These phases are interactive and orchestrated primarily through the release of soluble mediators that stimulate cells at the wound edges to migrate, proliferate, and synthesize the components necessary to restore the protective function of the skin. While all stages of wound healing are vital for a desired outcome of adequate wound closure, inflammation is particularly critical to skin repair and can regulate scarring outcomes. In addition to reducing microbial content in the wound, immune cells are now known to influence and direct many of the proliferative events that are needed for wound closure and tissue remodeling [1-3]. Thus, a regulated and appropriate inflammatory response supports proper repair [4,5]. Together, the accumulated evidence suggests that any attempts to regenerate skin must consider the complexity and role of the immune system in skin repair. This chapter focuses on the role of inflammation in wound healing and the potential to target the inflammatory response to improve healing outcomes and skin regeneration.

2. Stages of wound healing

As mentioned above, effective skin wound healing generally proceeds through the four intersecting phases of hemostasis, inflammation, proliferation, and remodeling (or resolution) (Fig. 1). Hemostasis and inflammation are the first two stages of the wound-healing process, and they occur in quick and overlapping succession. Tissue injury damages capillaries, and hemostasis is necessary to prevent excessive hemorrhage. The control of blood loss begins immediately after injury when platelets aggregate and adhere to the exposed extracellular matrix. The clotting cascade, which involves the successive conversion of clotting enzymes into their active forms, is also initiated, ultimately resulting in a fibrin clot. As platelets are activated, they degranulate, releasing growth factors and chemotactic factors [6,7]. Resident mast cells also degranulate quickly in response to trauma, releasing an additional set of inflammatory mediators [8]. Dermal keratinocytes at the site of injury further promote the inflammatory response, responding quickly to tissue damage by producing pro-inflammatory

Skin Tissue Models. https://doi.org/10.1016/B978-0-12-810545-0.00011-5

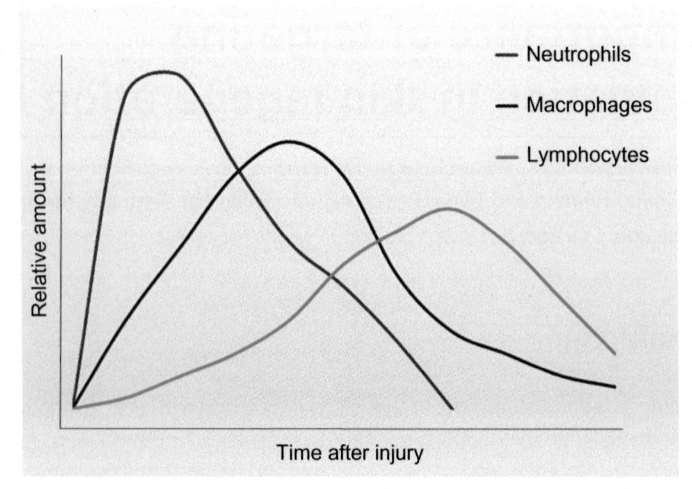

Fig. 1 Pattern of infiltration of immune cells in healing wounds. In response to wounding, inflammatory cells infiltrate the wound bed to control infection and stimulate downstream events vital to healing. The relative density of neutrophils, macrophages, and lymphocytes is depicted over time. Neutrophils infiltrate first, followed by macrophages and lymphocytes. Proper repair involves the coordinated response and interplay of these cells.

cytokines [9]. Overall, the release of inflammatory mediators and chemoattractants by multiple cell types results in the vascular changes that are typical of an acute inflammatory response along with recruitment of leukocytes from the circulation. Leukocyte infiltration, particularly neutrophils, is important to effective wound decontamination. As discussed below, wound leukocytes can also provide signals that initiate the downstream proliferative events of wound healing and influence scar formation.

As microbial content is controlled and inflammation subsides, leukocytes slowly exit the wound or undergo apoptosis [5]. The resolution of inflammation overlaps with the beginning of the proliferative phase of repair. Growth factors released from leukocytes, keratinocytes, fibroblasts, and other cell types stimulate growth and movement of the resident cells to repair the wound [5,10,11]. One to 2 days post injury, keratinocytes at the wound margins begin to migrate and proliferate. Once a single layer of cells covers the wound, keratinocytes stop migrating and begin to anchor to the basement membrane, forming a newly stratified epidermis. Blood vessel formation (angiogenesis) also occurs, supplying nutrients and oxygen [12]. Fibroblasts within the dermis produce collagen and repair the extracellular matrix [13,14]. Myofibroblasts, specialized contractile fibroblasts, initiate contraction of the wound, pulling the margins closer together, while collagen is synthesized [13]. During the final remodeling phase of repair, the assembled collagen undergoes degradation and resynthesis into more organized fibers. Collagen also undergoes cross-linking to stabilize the fibrils and add strength to the wound [10]. This constant cycle of collagen remodeling can last up to 20 years after wounding, although the skin will never achieve a completely normal structure. Because dermal repair in adults is always imperfect, the final strength of the remodeled wound is generally only 70%–80% of the original tissue [15].

3. Inflammation and skin regeneration

The prominent inflammatory response that occurs in wounds has demonstrable downstream effects on tissue repair, and well-regulated inflammation has been suggested to be critical to optimal skin healing [1,5]. Almost immediately following injury, immune cells initiate an inflammatory response at the site of the wound [16]. While this response is essential to the prevention of infection, studies in multiple models suggest that inflammation also supports skin regeneration and more perfect repair outcomes [5,17,18,116]. A balanced level of inflammation is critical to healing, as disequilibrium of the response can lead to untoward outcomes. Of relevance to tissue regeneration, aberrancies of wound inflammation have been linked to both impaired healing, such as chronic skin ulcers, and to excessive scar formation and skin fibrosis.

Evidence that inflammation can influence scar formation is abundant. This evidence derives both from the study of scar-forming wounds and from studies of nonscarring wounds. In particular, inappropriate inflammation has been linked to the formation of hypertrophic scars, scars that are raised and characterized by excessive formation of disorganized collagen. Keloids are another type of scar that has been suggested to have immunologic pathogenesis [5,18]. Keloids are large skin scars that grow beyond the bounds of the original injury site and contain disorganized collagen in whorl-like architecture [19]. Once excised, keloids frequently recur, creating a difficult clinical problem. Individual susceptibility to keloid formation appears to have a genetic component, although the specific genetics that underlies this pathology remains poorly understood [20].

Insight into the role of inflammation in wound healing and skin regeneration also comes from wounds that heal better than normal and without a scar. Early gestation fetal wounds (within the first and second trimester) heal rapidly in the absence of an inflammatory response and without a scar. By the third trimester, wounding invokes an increased inflammatory response, and fetal wounds at this stage are marked by inflammation and scar formation. Studies of privileged healing inside and outside of the uterus suggests that healing properties of the fetus are intrinsic and independent on the environment, thus providing additional support for the concept that inflammatory reduction results in less scarring. When inflammatory cells are introduced in the fetal wound, the result is delayed healing and scar formation [21]. Another tissue that exhibits a privileged healing phenotype is oral mucosa. Like fetal wounds, oral wounds heal rapidly and have minimal scar formation. Studies in at least three different models of oral mucosal healing demonstrate that this special healing pattern is associated with reduced inflammation [8]. Taken together, studies from the rapidly healing, scar-free fetal and oral mucosal tissues give a clear indication that reduced inflammation leads to faster, reduced scar healing. Such studies might be interpreted to suggest that adult skin repair might be improved by eliminating the inflammatory response. However, most studies of adult skin do suggest that a certain level of inflammation (in particular wound macrophages) is required for adequate wound repair [22].

Below, we review the immune cells, mediators, and receptors that are prominent in wounds, with an emphasis on what is known about their role in scar formation and in supporting skin regeneration (summarized in Table 1).

Table 1 **Effects of immune cells on healing outcomes**

Cell type	Beneficial	Detrimental
Neutrophils	Phagocytosis of microbes	Release of harmful enzymes and oxygen radicals, damaging normal tissue
Mast cells	Stimulation of angiogenesis and proliferation of keratinocytes/fibroblasts	Promote scar formation
Macrophages		
M1		Enhanced inflammation—impaired healing
M2	Stimulation of angiogenesis and collagen synthesis	
T cells		
CD4	Controversial	
CD8		Controversial
DETC	Stimulation of reepithelialization Stimulation of macrophage migration	
NKT	Controversial	Controversial
Treg	Downregulation of inflammation. Stimulation of reepithelialization	

Functions of immune cells involved in wound healing. A summary of the role of each immune cell involved in the wound-healing response is shown. The benefits of inflammatory cell infiltration such as controlling infection and stimulating growth and other downstream factors necessary for healing are further discussed in the text. Excessive numbers of inflammatory cells can lead to damage and detrimental effects on healing including chronic inflammation and increased scarring.

3.1 Mast cells

Mast cells are widely distributed in the body and are prominent near surfaces exposed to the environment, including the skin. Mast cells are often the first immune cell to respond to trauma and initiate an immune response [23]. These cells respond to a variety of stimuli: physical trauma, chemical exposure, allergens, and microbes. All of these challenges activate mast cells to degranulate and quickly release preformed biological mediators. Often thought of as primarily part of a hypersensitivity reaction, mast cells unmistakably influence the tissue repair process [8,23]. Factors released from mast cells can influence the activation, migration, and proliferation of endothelial cells, keratinocytes, and fibroblasts within skin [24,25]. Mast cells also secrete proangiogenic factors that can promote blood vessel formation. Evidence for an essential role of mast cells in normal wound closure is mixed, with some studies suggesting that wounds heal normally without mast cells and others demonstrating a slight deficit in healing in mast cell-deficient animals [8]. In contrast, though, excessive levels of mast cells have linked to the development of hypertrophic scars and keloids [8]. Experimental studies in both pigs and mice suggest that pharmacological inhibition of mast cell activation yields normal wound closure with reduced scar formation [26,27]. The finding that mast cell content is also reduced in the privileged healing of the oral mucosa further supports the association of mast cells with scar formation (Schrementi and DiPietro, unpublished data). Finally, increased mast cell content and maturity have been linked to the switch from nonscarring to scarring phenotype in fetal wounds [8]. Together, the accumulated data suggest that mast cell activity can spur a fibrotic response in skin wounds.

Many mechanisms by which mast cells might influence scar formation have been suggested. Firstly, activated mast cells release known profibrotic mediators such as transforming growth factor-β (TGF-β) and platelet-derived growth factor (PDGF) [28]. Other factors from mast cells that can influence fibroblast function and might promote scar formation include histamine, tryptase, and chymase. Direct mast cell to fibroblast contact via gap junctions has also been shown to stimulate fibroblast proliferation and myofibroblast differentiation [29]. As multiple experimental studies demonstrate a role for mast cells in scar formation, mast cell function seems likely to be an important variable that requires consideration for any skin regeneration strategies.

3.2 Neutrophils

Due to their abundance in circulation and the ability to mobilize quickly, neutrophils are the first infiltrating cell to arrive at the injury site from circulation, and they continue to increase in number for several days after injury. Neutrophils respond to products from platelet degranulation and clot formation, as well as bacterial components [10]. Because of their longevity at the wound site, neutrophils have several functions in early and late phases of inflammation [30]. During the early response, the main role of the neutrophil is to kill microbes and prevent infection. They are able to bind to microorganisms, ingest them through phagocytosis, and use enzymes and oxygen radicals to destroy them [31]. The activity of many enzymes and chemicals used by neutrophils to control microbes is not pathogen-specific and can cause bystander tissue damage. During phagocytosis, leakage of toxic lysosomal contents into the surrounding normal tissue can occur. Moreover, the release of oxygen free radicals and proteolytic enzymes can damage nearby normal tissue [32]. Notably, proteases released by neutrophils can be problematic for healing wounds when neutrophil counts are high. The additional tissue destruction then causes persistent inflammation and more tissue damage, which prevents the wound from progressing through the proper stages of repair. Thus, the consequences of damage to the surrounding tissue can represent a negative outcome of their antimicrobial function.

Research into the exact role of the neutrophil in wound repair is a continuing field. Some studies demonstrate that neutrophils are important for efficient wound repair, especially when the host is under microbial pressure. Clinical observations support this idea, as patients with neutrophil impairments such as defects in trafficking or granzyme function are more likely to develop a wound infection, often leading to a slow healing or chronic wound [33]. In normal patients, an idiopathic persistent neutrophil response increases the overall inflammatory response and can lead to poor wound healing, fibrosis, and excess scarring [34]. High levels of neutrophil proteases, such as serine proteases and matrix metalloproteinases (MMPs), have also been reported in chronic, nonhealing wounds [35,36]. However, in animal models, administration of neutrophil antisera does not alter wound repair as long as the wound is kept sterile [37]. Additionally, in a diabetic strain of mice noted for delayed repair, neutropenia accelerated wound closure by 50% [30]. Finally, in wounds that heal faster with less scarring, such as fetal wounds, neutrophils are physiologically different compared with adult neutrophils and produce fewer cytokines, thus lowering the inflammatory response [38]. Oral mucosal wounds, which also exhibit rapid scarless healing, exhibit a diminished neutrophil infiltration after wounding that leads to a diminished cytokine response [18]. Taken together, these studies are consistent with the notion that too much inflammation negatively affects healing. While a number of studies have demonstrated an important role for neutrophils in promoting wound healing and contributing to the resolution of inflammation after injury, a complete picture of the role of the neutrophil in the healing process has yet to be elucidated.

3.3 Macrophages

Macrophage content in skin wounds peaks several days after injury and is composed of both resident and newly arriving cells. Following wounding, monocytes extravasate into the wound bed from the blood stream and mature into macrophages, greatly increasing the number of macrophages in the tissue. In the wound, macrophages phagocytose microorganisms and apoptotic cells while attaching themselves to extracellular matrix within the tissue. The phagocytosis of apoptotic cells is a key signal that is believed to alter the macrophage phenotype from primarily pro-inflammatory to primarily reparative, a phenotype that supports tissue regrowth [12,39]. A continuum of macrophage phenotypes exists in wounds, and among these, the mature reparative phenotype has received a great deal of experimental attention [40]. Reparative macrophages, sometimes called M2, produce growth factors that stimulate angiogenesis and fibrogenesis, and in adult wounds, this phenotype is essential for proper repair [34,41]. More specifically, recent studies using genetically modified mice that are specifically void of macrophages or mice that are depleted of macrophages show that the loss of macrophages yields a significant delay in wound healing [22,42,43].

Because macrophages can stimulate fibroblast proliferation and collagen synthesis, pathological overactivity of macrophages might be expected to cause a fibrotic scar response. In support of this idea, macrophages do secrete many profibrotic factors including TGF-β and connective tissue growth factor (CTGF) [44]. Recent studies also strongly suggest that the overexpression of stromal-derived factor 1 (SDF1/CXCR4), a factor that recruits and stimulates macrophages, might contribute to hypertrophic scar formation, and the use of CXCR4 antagonist has been shown to inhibit scar formation in a mouse model of hypertrophic scars [45]. Similarly, the macrophage chemoattractant monocyte chemoattractant protein-1 (MCP-1/CCL2) has been linked to skin fibrosis [46]. This idea that macrophage dysfunction might underlie scar formation in skin is complemented by many studies in other tissues such as liver, lung, and kidney that suggest that macrophages can stimulate fibrosis [47]. Additional direct evidence for a role of macrophages in scar formation in skin wounds is needed, leaving the door open for future investigations.

3.4 T cells

T cells are an important part of innate and adaptive immune system. T-cell subsets including CD4+, CD8+, NKT, γδT cells (dendritic epidermal T cells, DETC), and regulatory T cells (Tregs, CD4+CD25+Fox-p3+) have been reported to play significant roles in wound healing. Following migration of macrophages and neutrophils into the wound, T cells including CD4+ helper cells and CD8+ cytotoxic T cells infiltrate the wound at around day 7 in an acute mouse wound model [48]. CD4+ cells from day 7 wounds express immunoregulatory or inflammatory cytokines including IFN-γ, TGF-β1, IL-10, and IL-17, while CD8+ cells express IFN-γ and TGF-β1 [48]. The functions of CD4+ and CD8+ T cells in skin wound healing and tissue regeneration, however, are not well understood, as there are several conflicting reports. The absence of CD4+ or CD8+ T cells has been reported to have no effect on wound closure [49]. However, wound breaking strength and collagen synthesis were shown to be significantly increased in the absence of CD8+ cells in one study [49] but not in another [48]. In contrast, CD4+ cell deficiency has been consistently shown to have no effect on collagen content and wound breaking strength [48,49].

An additional subset of T cells, natural killer T (NKT) cells, infiltrate into early skin wounds at a time similar to neutrophils. NKT cells seem to inhibit skin repair, as NKT cell-deficient mice exhibit accelerated wound closure, an augmented production of neutrophil and macrophage chemokines (CXCL-1, CXCL-2, CCL-3, and CCL-4) and TGF-β, and increased collagen deposition [50,51]. Some questions about NKT cells and skin repair remain, though, as another study, using a different NKT cell-deficient mouse strain demonstrated different results including decreased TGF-β production, myofibroblast differentiation, collagen deposition, and wound breaking strength [52].

DETC, which are resident cells, are very early and rapid responders to tissue damage. After T-cell receptor (TCR) stimulation, DETC can be activated by damaged keratinocytes to produce to critical growth factors responsible for keratinocyte proliferation and migration [53-55]. DETC-deficient mice display a significantly impaired skin wound closure [53]. Additionally, DETC−/− mice have delayed macrophage infiltration into burn wounds suggesting that DETC affect macrophage migration [56]. Therefore, DETC not only play a role in reepithelization but also have an impact on the inflammatory response during the healing process.

Tregs have an indispensable role in maintaining immune homeostasis by generally inhibiting the over reactive immune response. Activated Tregs are known to accumulate in early skin wounds. They impede IFN-γ production and pro-inflammatory macrophage infiltration. The depletion of Tregs results in delayed wound reepithelialization [57].

While multiple studies demonstrate a role for many types of T cells in skin wound repair, the role of T cells in scar formation has received less attention. In hypertrophic scar tissue of burn patients, there are increased numbers of TGF-β-producing CD4+ cells. Fibroblasts treated with conditioned medium of burn patient CD4+ cells demonstrate an increased rate of proliferation and collagen synthesis, suggesting CD4+ cells have an indirect immunoregulatory role in hypertrophic scar or fibrosis by interfering with the functions of fibroblasts [58]. Another T-cell type that seems likely to influence scar formation is the regulatory T cell. The immunoregulatory roles of Tregs and the connection between inflammation and scar formation suggests that Tregs may influence scar formation, a function that needs to be explored in future studies. More investigations are also needed to understand whether other T cell types that are known to modulate healing, including DETC and NKT cells, influence scar formation and skin regeneration capacity.

3.5 Keratinocytes

Although they are not myeloid-derived, skin keratinocytes are an important part of the immune system [59]. The skin epithelium provides a protective barrier and as such is an important part of innate immunity. Keratinocytes also produce numerous antimicrobial molecules, further contributing to a strong defensive blockade. Following skin injury, though, keratinocytes take on immune roles well beyond barrier function. Keratinocytes express Toll-like receptors (TLRs) and so are able to respond to a variety of external stimuli. TLRs, which are described further below, allow keratinocytes to respond to endogenous molecular patterns that are associated with tissue damage. In response to tissue damage, keratinocytes produce numerous chemokines and cytokines that assist in the recruitment of inflammatory cells to the wound [60]. Keratinocytes can also produce growth factors that stimulate tissue repair, such as TGF-β and vascular endothelial growth factor (VEGF) [61]. Some mediators produced by activated keratinocytes, such as IL-1, influence healing outcomes via cross talk with dermal fibroblasts [62].

4. Inflammatory mediators and receptors

A very large number of inflammatory mediators and receptors have been implicated as critical to wound healing. Below, we review some of the factors that have been shown to be most important to scar formation in wounds and thus important to the modulation of skin regeneration.

4.1 IL-1

IL-1 is produced by numerous cells including neutrophils, macrophages, keratinocytes, and fibroblasts. In response to injury, preformed IL-1 is quickly released and activated and in turn upregulates adhesion molecules in capillaries, facilitating macrophage and neutrophil infiltration. IL-1 also induces proliferation of keratinocytes and endothelial cells [11,63]. Thus, IL-1 plays a pivotal role in promoting host defense at the wound site by amplifying the immune response and can also directly affect tissue repair [64]. The activity of IL-1 in wounds is modulated by the presence of a receptor antagonist (IL-1ra), which inhibits IL-1 cellular signaling, reducing its proinflammatory effects [65].

An overabundance of IL-1 has been suggested to impair healing, as IL-1 levels are high in chronic wounds [66]. IL-1 receptor antagonist-deficient mice exhibit delayed wound healing, accompanied by an altered cytokine secretion profile and collagen deposition [67]. Interestingly, though, IL-1 production by epithelia has been linked to scar formation in skin wounds. In the skin, the epidermis produces IL-1, and the administration of an IL-1 receptor antagonist has been shown to block scar formation in a rabbit model of hypertrophic scars [62]. Overall, then, a balanced production of IL-1 seems essential to appropriate repair and skin regeneration.

4.2 IL-6

In the skin, IL-6 not only is mainly produced by keratinocytes but also can be secreted by neutrophils, fibroblasts, and macrophages [68,69]. IL-6 is a pro-inflammatory cytokine that has also been shown to cause keratinocyte proliferation [70]. IL-6-deficient mice display a significant postponement in healing including delayed reepithelialization and angiogenesis [71]. These deficiencies can be rescued via reconstitution with recombinant IL-6. IL-6 production has also been associated with scar formation. In particular, the overproduction IL-6 by macrophages and fibroblasts is associated with keloid scar formation [72,73]. Inhibition of both IL-6 and IL-6r by antibody neutralization reduces collagen and fibronectin gene expression, indicating that IL-6 can regulate collagen and ECM synthesis [74]. These studies indicate that like IL-1, a careful balance of IL-6 is necessary for proper wound repair.

4.3 Transforming growth factor-β

TGF-β has a pivotal role in regulating the immune response to wounding [75,76]. There are three isoforms of TGF-β in mammals: TGF-β1, TGF-β2, and TGF-β3. These isoforms can influence cell proliferation, differentiation, migration, and survival within the wound bed. Although the three isoforms appear to have similar action through cellular signal transduction, they are differently regulated and expressed during wound healing [77]. Immediately after injury, platelets release TGF-β1 that is chemotactic for inflammatory cells and mitogenic for collagen-producing fibroblasts. During healing, TGF-β1 continues to be produced and causes an increase in matrix protein synthesis while decreasing matrix protein degradation, thereby serving as an important regulator of scar formation [78]. TGF-β1 has also been implicated in the development of hypertrophic scarring and keloid formation [79]. TGF-β1 is produced at higher levels within keloid scars. The addition of TGF-β1 neutralizing antibodies to adult wounds has been shown to reduce scarring. Conversely, the addition of TGF-β1 to nonscarring fetal wounds results in scar formation [78,80]. TGF-β2, mainly secreted by wound fibroblasts, is also involved in the stages of wound healing, mainly in the recruitment of inflammatory cells and fibroblasts to the wound area. Application of TGF-β2 to a wound increases angiogenesis and the rate of reepithelialization [81]. Although it has a similar structure, the role of TGF-β3 in healing is very different from the other isoforms. TGF-β3, which is synthesized by keratinocytes and fibroblasts, has potential antiscarring effects [82]. In mice, TGF-β3 is present at high levels in scarless fetal wounds while at lower levels in adult wounds. TGF-β3 knockout mice exhibit scarring of normally scarless fetal wounds [83]. In the adult wound, the application of exogenous TGF-β3 reduces scarring [78,84]. The ratio of TGF-β3/TGF-β1 has been suggested to dictate healing outcomes. In a mouse model of dermal healing, higher ratios lead to scar reduction, and lower ratios lead to scar formation [79]. The scarless healing of the oral mucosa also is marked by a higher level of TGF-β3 expression compared with TGF-β1 [85]. The utility of altering the levels of and proportions of TGF-β1 and TGF-β3 to reduce scars has been examined in human studies. In human studies, Ferguson et al. reported the results of three double-blind, randomized controlled trials that evaluated the effects of adding active, human TGF-β3 to human wounds. The results of phase I and II trials showed that the application of TGF-β3 improved the appearance of the scars as compared with controls. However, in phase III studies, the application of TGF-β3 failed to significantly alter scarring [86].

4.4 Tumor necrosis factor-α

TNF-α is another early mediator of the inflammatory response that plays an important role in healing. TNF-α amplifies inflammation through cellular signaling pathways, promotes leukocyte infiltration, and contributes to microbial defense [87]. The effect of TNF-α on wound healing seems to be dependent on several factors and is dose-dependent. Small doses of TNF-α aid in protection from host infection, stimulate inflammation, and cause macrophages to release growth factors that promote healing [11,88]. However, at higher levels, TNF-α has been shown to suppress the synthesis of extracellular matrix, vital to proper wound repair, and to cause prolonged inflammation and slower healing [11]. Consistent with this observation, high levels of TNF-α are found in chronic wounds [89]. In human patients, the drug infliximab (a monoclonal antibody) has been used to inhibit TNF-α in chronic wounds. In a clinical study on a group of nine patients total, application of this inhibitor resulted in complete (100%) resolution in five patients and 75% resolution in the remaining four patients [90].

4.5 Secretory leukocyte protease inhibitor

SLPI is an inhibitor of serine proteases and is expressed near the end of the inflammatory phase of wound healing. SLPI-deficient mice show impaired cutaneous (healing?) marked by constitutive production of inflammatory cytokines and abundant leukocyte activation. Consistent with the known effects of excess inflammation, these mice exhibit delayed wound closure and increased scar formation [91]. Thus, SLPI has been suggested as a potential candidate for enhancing wound healing and reducing scar formation.

4.6 IL-10

The cytokine IL-10 is produced by a variety of immune cells such as T cells, monocytes, and macrophages [92]. IL-10 can also be produced by skin cells such as keratinocytes and fibroblasts [93,94]. IL-10 downregulates the expression of collagen I, collagen III, and α-SMA and upregulates the expression of MMP1 and MMP8 in TGF-β-treated dermal fibroblasts [95]. Significant information suggests that IL-10 production influences scar formation in skin wounds. IL-10 is abundantly produced in nonscarring fetal wounds, and fetal wounds produced in IL-10 knockout mice exhibit adultlike scar formation [96]. Studies in adult mice show that the treatment of skin wounds with IL-10 leads to improved collagen deposition along with decreased pro-inflammatory mediators such as IL-6, MCP-1/CCL2, MIP-2, and MIP-1α /CCL3 [94,97]. In another study, adult murine skin wounds injected with IL-10 show smaller scars, reduced deposition of collagens including collagen I and collagen III, and decreased α-SMA-positive fibroblasts [95]. These studies suggest that IL-10 not only downregulates the inflammatory response, creating an environment favorable for tissue regeneration, but also has a direct effect on collagen synthesis and degradation. A role for IL-10 in improving healing outcomes by reducing scar formation has been shown in sheep models as well [98]. Observations in human normal skin wounds further support the idea that IL-10 can reduce scar formation. In humans, midgestational fetal skin secretes high levels of IL-10 in keratinocytes and fibroblasts, while IL-10 is absent in neonatal skin [94]. Human hypertrophic scar formation has been shown to coincide with a prolonged decrease in the expression of IL-10 [99]. Finally, a clinical trial has now shown that treatment of incisional skin wounds with low concentrations of IL-10 yields a significantly improved macroscopic scar appearance [100]. Together, these studies strongly suggest that IL-10 is a critical factor that negatively regulates inflammation and scar formation during wound healing. Therefore, IL-10 has potential as a therapeutic agent to prevent or reduce scar formation and to promote skin regeneration.

5. Toll-like receptors

Toll-like receptors are a group of pattern recognizing molecules that are present on the surface or endosomal vesicles of many cells including leukocytes and skin-resident cells such as keratinocytes. TLRs recognize small segments of pathogens known as pathogen-associated molecular patterns (PAMPs). There are 13 known TLRs at this time. The cell surface receptors (TLR1, TLR2, TLR4, and TLR6) recognize bacterial and fungal components. Intracellular TLRs such as TLR3, TLR7, TLR8, and TLR9 recognize internal PAMPs such as microbial DNAs or RNAs [101]. Although TLRs do not normally recognize most host molecules, they can interact with ligands produced by injured or damaged cells; these ligands are termed damage-associated molecular patterns (DAMPs). Thus, TLRs present on the surface of sentinel skin cells provide a recognition system for tissue damage and possible infection [101,102]. In the skin, TLR2 and TLR4 are the most abundant and are found on the surface of keratinocytes, fibroblasts, dendritic cells, mast cells, and macrophages [103]. When a TLR binds with a cognate PAMP or DAMP, a conformational change occurs that results in a complex signaling cascade, leading to induction of important mediators such as transcription factors, cytokines, chemokines, and growth factors [101-103]. Until recently, little was known about the function of TLRs in skin wound healing.

Recent studies in mouse wound models have shown that TLR production is modulated during healing and the expression of TLR1–9 and 13 mRNAs are all greatly increased in early skin wounds [104]. The functional significance of the TLR expression in wounds has been examined for a few specific TLRs. TLR3 activation probably occurs through binding of RNAs released from necrotic or damaged cells during injury. TLR3 production is significantly elevated in mouse skin wounds, and TLR3 colocalizes with keratinocytes, fibroblasts, neutrophils, macrophages, and endothelial cells [105,106]. TLR3 knockout mice exhibit delayed skin wound closure, reduced angiogenesis, and impaired recruitment inflammatory cells [105]. Treatment of murine wounds with the TLR3 agonist poly I/C significantly promotes healing [106]. Experiments in genetically deficient mice have also demonstrated a role for TLR4 and TLR9 in wound healing that is similar to that described for TLR3 [107,108]. In the context of poorly healing diabetic wounds, however, TLRs seem to play a different role. More specifically, activation of TLRs, especially TLR2 and TLR4, seems to contribute to the prolonged inflammation and healing impairment seen in diabetic skin wounds [109,110]. The role of TLRs in wound healing is a new but promising field of study, and most of the published studies focus on the role of TLRs in inflammatory regulation in acute wound healing. However, a recent study discovered that fibroblasts isolated from human hypertrophic scar express significantly higher levels of TLR1, TLR2, TLR3, TLR5, TLR6, TLR7, TLR8, TLR9, and TLR signaling pathway adapter molecule, MyD88. These cells also exhibited increased expression of inflammatory mediators including IL-6, IL-8/CXCL8, MCP-1/CCL-2, and PGE-2 when compared with fibroblasts from normal skin [111]. The authors speculate that increased expression of the pro-inflammatory mediators recruits leukocytes, which produce TGF-β, stimulating fibroblasts to produce ECM. Therefore, TLRs especially TLR4 activation may be partially responsible for excessive ECM production during HTS formation. Further exploration is needed to find the direct evidence of the functions and involvement of TLR signaling in tissue regeneration, especially collagen synthesis and scar formation. Taken together, the accumulated studies suggest that TLRs may play important roles in the regulation of inflammatory response and ECM production by binding to PAMPS and/or DAMPS during the wound-healing process.

6. Summary and conclusions

As described above, many studies now implicate immune cells and mediators as causative agents in enhancing scar formation and preventing complete skin regeneration (Fig. 2). These studies suggest that the modulation of inflammation might be a promising therapeutic to encourage skin regeneration and minimize scar formation following injury. At the current time, several common treatments that are used to reduce scar formation are in fact known to cause a reduction in wound inflammation. These include the injection of steroids to prevent or treat hypertrophic scars and the use of moist or occlusive dressings. In particular, the creation of a moist environment mimics the creation of a mature epidermis, promotes reepithelialization, and reduces scar formation as compared with a dry wound environment [112]. In animal models too, the use of

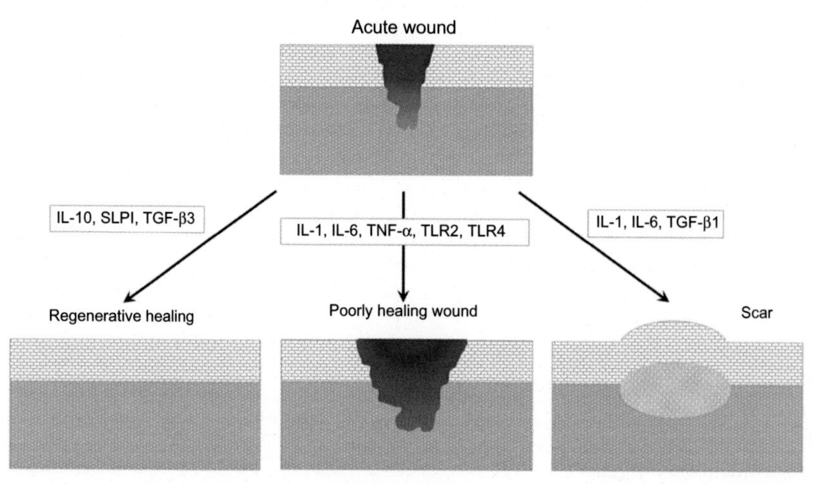

Fig. 2 Dominant effects of immune mediators and receptors on healing outcomes. Mediators and receptors with well-described influence on healing outcomes are shown. The function of specific mediators in wounds is often dependent upon the level of production. IL-10 not only downregulates the inflammatory response but also has a direct effect on collagen synthesis and degradation, with an overall effect of supporting regenerative repair. SLPI inhibits inflammation, thus supporting improved repair. High levels of TGF-β3, such as found in fetal wounds, foster scarless healing. In contrast, overexpression of IL-1, IL-6, and TNF-α or overstimulation of TLR2 and TLR4 has been shown to result in excessive inflammation, thus leading to poorly healing or chronic wounds. Overexpression of IL-1, IL-6, and TGF-β1 is associated with abnormal scar formation such as hypertrophic scars and keloids.

global anti-inflammatory treatments, such as celecoxib, has been shown to decrease inflammation and reduce scar formation [113].

Despite the considerable amount of information now available on how specific immune mediators might influence scarring and skin regeneration, the therapeutic development and use of targeted immunomodulators at sites of injury has been quite limited. There are several likely reasons for this situation. First, the appropriate therapeutic immunomodulation of skin regeneration will most certainly be challenging, as any dampening of the immune response must consider the likelihood of increased wound infection. Second, although several specific approaches (such as alterations of TGF-β1/3 ratios or introduction of exogenous IL-10) have shown great success in animal models, results in human subjects have often been less compelling. These caveats to development suggest that novel approaches will be needed to predict which types of immunomodulation will be most successful in the human skin. One possible approach might be the use of computational modeling of the inflammatory response in skin wounds [114,115]. The ability to model outcomes might permit a more rational selection of the key mediators that can be safely targeted to improve skin regeneration. Such an approach will require the development of increasingly sophisticated and complex models to allow successful and accurate clinical predictions.

Abbreviations

CTGF connective tissue growth factor
DAMP damage-associated molecular pattern
DETC dendritic epidermal T cell
ECM extracellular matrix
IL interleukin
MCP macrophage chemoattractant protein
PAMP pathogen-associated molecular pattern
PDGF platelet-derived growth factor
NKT natural killer T cell
SLPI secretory leukocyte protease inhibitor
TGF transforming growth factor
TLR Toll-like receptor
TNF tumor necrosis factor
Treg regulatory T cell

References

[1] Leibovich SJ, Ross R. The role of the macrophage in wound repair. A study with hydrocortisone and antimacrophage serum. Am J Pathol 1975; 78(1):71–100. Pub Med PMID: 1109560. Pubmed Central PMCID: 1915032.
[2] DiPietro LA, Burdick M, Low QE, Kunkel SL, Strieter RM. MIP-1alpha as a critical macrophage chemoattractant in murine wound repair. J Clin Invest 1998;101(8):1693–8.
[3] Hubner G, Brauchle M, Smola H, Madlener M, Fassler R, Werner S. Differential regulation of pro-inflammatory cytokines during wound healing in normal and glucocorticoid-treated mice. Cytokine 1996; 8(7):548–56. Pub Med PMID: 8891436.
[4] Szpaderska AM, Egozi EI, Gamelli RL, DiPietro LA. The effect of thrombocytopenia on dermal wound healing. J Invest Dermatol 2003; 120(6):1130–7. Pub Med PMID: 12787144.
[5] Martin P. Wound healing—aiming for perfect skin regeneration. Science 1997; 276 (5309):75–81. Pub Med PMID: 9082989.
[6] Ross R, Bowen-Pope DF, Raines EW. Platelet-derived growth factor and its role in health and disease. Philos Trans R Soc Lond Ser B Biol Sci. 1990; 327 (1239):155–69. Pub Med PMID: 1969656.
[7] Weksler BB. Platelets and the inflammatory response. Clin Lab Med 1983; 3(4):667–76. Pub Med PMID: 6360503.
[8] Wulff BC, Parent AE, Meleski MA, DiPietro LA, Schrementi ME, Wilgus TA. Mast cells contribute to scar formation during fetal wound healing. J Invest Dermatol 2012; 132(2):458–65. Pub Med PMID: 21993557. Pubmed Central PMCID: 3258379.
[9] Kupper TS, Fuhlbrigge RC. Immune surveillance in the skin: mechanisms and clinical consequences. Nat Rev Immunol 2004; 4(3):211–22. Pub Med PMID: 15039758.
[10] Singer AJ, Clark RA. Cutaneous wound healing. N Engl J Med 1999; 341 (10):738–46. Pub Med PMID: 10471461.
[11] Barrientos S, Stojadinovic O, Golinko MS, Brem H, Tomic-Canic M. Growth factors and cytokines in wound healing. Wound Repair Regener Off Publ Wound Heal Soc Eur Tissue Repair Soc. 2008 Sep-Oct; 16(5):585–601. Pub Med PMID: 19128254.

[12] Sciubba JJ, Waterhouse JP, Meyer J. A fine structural comparison of the healing of incisional wounds of mucosa and skin. J Oral Pathol 1978; 7(4):214–27. Pub Med PMID: 99502.

[13] Herzhoff K, Sollberg S, Huerkamp C, Krieg T, Eckes B. Fibroblast expression of collagen integrin receptors alpha 1beta1 and alpha 2beta1 is not changed in systemic scleroderma. Br J Dermatol 1999; 141(2):218–23. Pub Med PMID: 10468791.

[14] Woodley DT, Stanley JR, Reese MJ, O'Keefe EJ. Human dermal fibroblasts synthesize laminin. J Invest Dermatol 1988; 90(5):679–83. Pub Med PMID: 3283250.

[15] Doillon CJ, Dunn MG, Bender E, Silver FH. Collagen fiber formation in repair tissue: development of strength and toughness. Coll Relat Res 1985; 5(6):481–92. Pub Med PMID: 3833451.

[16] Doillon CJ, Dunn MG, Berg RA, Silver FH. Collagen deposition during wound repair. Scan Electron Microsc. 1985 (Pt 2):897–903. Pub Med PMID: 4048854.

[17] Wong JW, Gallant-Behm C, Wiebe C, Mak K, Hart DA, Larjava H, et al. Wound healing in oral mucosa results in reduced scar formation as compared with skin: evidence from the red Duroc pig model and humans. Wound Repair Regener Off Publ Wound Heal Soc Eur Tissue Repair Soc. 2009; 17(5):717–29. Pub Med PMID: 19769724.

[18] Szpaderska AM, Zuckerman JD, DiPietro LA. Differential injury responses in oral mucosal and cutaneous wounds. J Dent Res 2003; 82(8):621–6. Pub Med PMID: 12885847.

[19] Seifert O, Mrowietz U. Keloid scarring: bench and bedside. Arch Dermatol Res 2009; 301(4):259–72. Pub Med PMID: 19360429.

[20] Shih B, Bayat A. Genetics of keloid scarring. Arch Dermatol Res 2010; 302(5):319–39. Pub Med PMID: 20130896.

[21] Beanes SR, Hu FY, Soo C, Dang CM, Urata M, Ting K, et al. Confocal microscopic analysis of scarless repair in the fetal rat: defining the transition. Plast Reconstr Surg 2002; 109(1):160–70. Pub Med PMID: 11786808.

[22] Mizrahi M, Ben-Chetrit E. Relapsing macrophage activating syndrome in a 15-year-old girl with Still's disease: a case report. J Med Case Rep 2009; 3:138 .Pub Med PMID: 20062775. Pubmed Central PMCID: 2803809.

[23] Artuc M, Hermes B, Steckelings UM, Grutzkau A, Henz BM. Mast cells and their mediators in cutaneous wound healing—active participants or innocent bystanders? Exp Dermatol 1999;8(1):1–16.

[24] Maurer M, Theoharides T, Granstein RD, Bischoff SC, Bienenstock J, Henz B, et al. What is the physiological function of mast cells? Exp Dermatol 2003; 12(6):886–910. Pub Med PMID: 14719507.

[25] Persinger MA. Degranulation of brain mast cells in young albino rats. Behav Neural Biol 1983; 39(2):299–306. Pub Med PMID: 6200100.

[26] Chen L, Schrementi ME, Ranzer MJ, Wilgus TA, DiPietro LA. Blockade of mast cell activation reduces cutaneous scar formation. PLoS One 2014; 9(1):e85226. Pub Med PMID: 24465509. Pubmed Central PMCID: PMC3898956 [Epub 2014/01/28.eng].

[27] Gallant-Behm CL, Hildebrand KA, Hart DA. The mast cell stabilizer ketotifen prevents development of excessive skin wound contraction and fibrosis in red Duroc pigs. Wound Repair Regener Off Publ Wound Heal Soc Eur Tissue Repair Soc. 2008; 16(2):226–33. Pub Med PMID: 18318808.

[28] Gruber BL. Mast cells in the pathogenesis of fibrosis. Curr Rheumatol Rep 2003; 5(2):147–53. Pub Med PMID: 12628046.

[29] Moyer KE, Saggers GC, Ehrlich HP. Mast cells promote fibroblast populated collagen lattice contraction through gap junction intercellular communication. Wound Repair Regener Off Publ Wound Heal Soc Eur Tissue Repair Soc. 2004; 12(3):269–75. Pub Med PMID: 15225205.

[30] Velnar T, Bailey T, Smrkolj V. The wound healing process: an overview of the cellular and molecular mechanisms. J Int Med Res 2009; 37(5):1528–42. Pub Med PMID: 19930861.
[31] Broughton G, 2nd, Janis JE, Attinger CE. Wound healing: an overview. Plast Reconstr Surg. 2006; 117 (7 Suppl):1e-S-32e-S. Pub Med PMID: 16801750.
[32] Broughton G, 2nd, Janis JE, Attinger CE. The basic science of wound healing. Plast Reconstr Surg 2006; 117 (7 Suppl):12S–34S. Pub Med PMID: 16799372.
[33] Dinauer MC, Lekstrom-Himes JA, Dale DC. Inherited neutrophil disorders: molecular basis and new therapies. Hematol Am Soc Hematol Edu Prog. 2000:303–18. Pub Med PMID: 11701548.
[34] Satish L, Kathju S. Cellular and molecular characteristics of scarless versus fibrotic wound healing. Dermatol Res Pract 2010; 2010:790234. Pub Med PMID: 21253544. Pubmed Central PMCID: 3021858.
[35] Bullen EC, Longaker MT, Updike DL, Benton R, Ladin D, Hou Z, et al. Tissue inhibitor of metalloproteinases-1 is decreased and activated gelatinases are increased in chronic wounds. J Invest Dermatol 1995; 104(2):236–40. Pub Med PMID: 7829879.
[36] Danielsen PL, Holst AV, Maltesen HR, Bassi MR, Holst PJ, Heinemeier KM, et al. Matrix metalloproteinase-8 overexpression prevents proper tissue repair. Surgery 2011; 150(5):897–906. Pub Med PMID: 21875735.
[37] Dovi JV, Szpaderska AM, DiPietro LA. Neutrophil function in the healing wound: adding insult to injury? Thromb Haemost 2004; 92(2):275–80. Pub Med PMID: 15269822.
[38] Olutoye OO, Yager DR, Cohen IK, Diegelmann RF. Lower cytokine release by fetal porcine platelets: a possible explanation for reduced inflammation after fetal wounding. J Pediatr Surg 1996; 31(1):91–5. Pub Med PMID: 8632294.
[39] Savill J. Apoptosis in resolution of inflammation. J Leukoc Biol 1997; 61(4):375–80. Pub Med PMID: 9103222.
[40] Brancato SK, Albina JE. Wound macrophages as key regulators of repair: origin, phenotype, and function. Am J Pathol 2011; 178(1):19–25. Pub Med PMID: 21224038. Pubmed Central PMCID: 3069845.
[41] Novak ML, Koh TJ.Macrophage phenotypes during tissue repair J Leukoc Biol 2013; 93 (6):875–81. Pub Med PMID: 23505314. Pubmed Central PMCID: 3656331.
[42] Goren I, Allmann N, Yogev N, Schurmann C, Linke A, Holdener M, et al. A transgenic mouse model of inducible macrophage depletion: effects of diphtheria toxin-driven lysozyme M-specific cell lineage ablation on wound inflammatory, angiogenic, and contractive processes. Am J Pathol 2009; 175(1):132–47. Pub Med PMID: 19528348. Pubmed Central PMCID: 2708801.
[43] Lucas T, Waisman A, Ranjan R, Roes J, Krieg T, Muller W, et al. Differential roles of macrophages in diverse phases of skin repair. J Immunol 2010; 184(7):3964–77. Pub Med PMID: 20176743.
[44] Koh TJ, DiPietro LA. Inflammation and wound healing: the role of the macrophage. Expert Rev Mol Med. 2011 Jul 11; 13. Pub Med PMID: WOS:000292612800001. [English].
[45] Ding J, Ma Z, Liu H, Kwan P, Iwashina T, Shankowsky HA, et al. The therapeutic potential of a C-X-C chemokine receptor type 4 (CXCR-4) antagonist on hypertrophic scarring in vivo. Wound Repair Regener Off Publ Wound Heal Soc Eur Tissue Repair Soc. 2014; 22(5):622–30. Pub Med PMID: 25139227.
[46] Ferreira AM, Takagawa S, Fresco R, Zhu X, Varga J, DiPietro LA. Diminished induction of skin fibrosis in mice with MCP-1 deficiency. J Invest Dermatol 2006; 126(8):1900–8. Pub Med PMID: 16691201.

[47] Lech M, Anders HJ. Macrophages and fibrosis: how resident and infiltrating mononuclear phagocytes orchestrate all phases of tissue injury and repair. Biochim Biophys Acta 2013; 1832(7):989–97. Pub Med PMID: 23246690.

[48] Chen L, Mehta ND, Zhao Y, DiPietro LA. Absence of CD4 or CD8 lymphocytes changes infiltration of inflammatory cells and profiles of cytokine expression in skin wounds, but does not impair healing. Exp Dermatol 2014; 23(3):189–94. Pub Med PMID: 24521099. Pubmed Central PMCID: PMC3989939 [Epub 2014/02/14.eng].

[49] Barbul A, Breslin RJ, Woodyard JP, Wasserkrug HL, Efron G. The effect of in vivo T helper and T suppressor lymphocyte depletion on wound healing. Ann Surg 1989; 209(4):479–83. Pub Med PMID: 2522759. Pubmed Central PMCID: 1493975.

[50] Schneider DF, Palmer JL, Tulley JM, Speicher JT, Kovacs EJ, Gamelli RL, et al. A novel role for NKT cells in cutaneous wound repair. J Surg Res 2011; 168(2):325–33 e1. Pub Med PMID: 20089261. Pubmed Central PMCID: 3324973.

[51] Schneider DF, Palmer JL, Tulley JM, Kovacs EJ, Gamelli RL, Faunce DE. Prevention of NKT cell activation accelerates cutaneous wound closure and alters local inflammatory signals. J Surg Res 2011; 171(1):361–73. Pub Med PMID: 21067780. Pubmed Central PMCID: 3324976.

[52] Tanno H, Kawakami K, Ritsu M, Kanno E, Suzuki A, Kamimatsuno R, et al. Contribution of invariant natural killer T cells to skin wound healing. Am J Pathol 2015; 185 (12):3248–57. Pub Med PMID: 26468976.

[53] Jameson J, Ugarte K, Chen N, Yachi P, Fuchs E, Boismenu R, et al. A role for skin gammadelta T cells in wound repair. Science 2002; 296 (5568):747–9. Pub Med PMID: 11976459.

[54] Jameson JM, Cauvi G, Witherden DA, Havran WL. A keratinocyte-responsive gamma delta TCR is necessary for dendritic epidermal T cell activation by damaged keratinocytes and maintenance in the epidermis. J Immunol 2004; 172(6):3573–9. Pub Med PMID: 15004158.

[55] Toulon A, Breton L, Taylor KR, Tenenhaus M, Bhavsar D, Lanigan C, et al. A role for human skin-resident T cells in wound healing. J Exp Med 2009; 206(4):743–50. Pub Med PMID: 19307328. Pubmed Central PMCID: 2715110.

[56] Rani M, Zhang Q, Schwacha MG. Gamma delta T cells regulate wound myeloid cell activity after burn. Shock 2014; 42(2):133–41. Pub Med PMID: 24667623. Pubmed Central PMCID: 4101017.

[57] Nosbaum A, Prevel N, Truong HA, Mehta P, Ettinger M, Scharschmidt TC, et al. Cutting edge: regulatory t cells facilitate cutaneous wound healing. J Immunol 2016; 196(5):2010–4. Pub Med PMID: WOS:000372336300005 [English].

[58] Wang J, Jiao H, Stewart TL, Shankowsky HA, Scott PG, Tredget EE. Improvement in postburn hypertrophic scar after treatment with IFN-alpha 2b is associated with decreased fibrocytes. J Interferon Cytokine Res Off J Int Soc Interferon Cytokine Res. 2007; 27 (11):921–30. Pub Med PMID: 18052725.

[59] Strbo N, Yin N, Stojadinovic O. Innate and adaptive immune responses in wound epithelialization. Adv Wound Care 2014; 3(7):492–501. Pub Med PMID: 25032069. Pubmed Central PMCID: PMC4086194 [Epub 2014/07/18.eng].

[60] Seeger MA, Paller AS. The roles of growth factors in keratinocyte migration. Adv Wound Care 2015; 4(4):213–24. Pub Med PMID: 25945284. Pubmed Central PMCID: PMC4397993 [Epub 2015/05/07.eng].

[61] Mirza RE, Koh TJ. Contributions of cell subsets to cytokine production during normal and impaired wound healing. Cytokine 2015; 71(2):409–12. Pub Med PMID: 25281359. Pubmed Central PMCID: PMC4297569 [Epub 2014/10/05.eng].

[62] Gallant-Behm CL, Du P, Lin SM, Marucha PT, DiPietro LA, Mustoe TA. Epithelial regulation of mesenchymal tissue behavior. J Invest Dermatol 2011; 131(4):892–9. Pub Med PMID: 21228814. Pubmed Central PMCID: 3137131.

[63] Kupper TS. Interleukin 1 and other human keratinocyte cytokines: molecular and functional characterization. Adv Dermatol 1988; 3:293–307. Pub Med PMID: 3152825.

[64] Angele MK, Knoferl MW, Ayala A, Albina JE, Cioffi WG, Bland KI, et al. Trauma-hemorrhage delays wound healing potentially by increasing pro-inflammatory cytokines at the wound site. Surgery 1999; 126(2):279–85. Pub Med PMID: 10455895.

[65] Arend WP, Malyak M, Guthridge CJ, Gabay C. Interleukin-1 receptor antagonist: role in biology. Annu Rev Immunol 1998; 16:27–55. Pub Med PMID: 9597123.

[66] Trengove NJ, Bielefeldt-Ohmann H, Stacey MC. Mitogenic activity and cytokine levels in non-healing and healing chronic leg ulcers. Wound Repair Regener Off Publ Wound Heal Soc Eur Tissue Repair Soc. 2000; 8(1):13–25. Pub Med PMID: 10760211.

[67] Ishida Y, Kondo T, Kimura A, Matsushima K, Mukaida N. Absence of IL-1 receptor antagonist impaired wound healing along with aberrant NF-kappa B activation and a reciprocal suppression of TGF-beta signal pathway. J Immunol 2006; 176(9):5598–606. Pub Med PMID: 16622029.

[68] Paquet P, Pierard GE. Interleukin-6 and the skin. Int Arch Allergy Immunol 1996; 109(4):308–17. Pub Med PMID: 8634514.

[69] Sugawara T, Gallucci RM, Simeonova PP, Luster MI. Regulation and role of interleukin 6 in wounded human epithelial keratinocytes. Cytokine 2001; 15(6):328–36. Pub Med PMID: 11594800.

[70] Gallucci RM, Sloan DK, Heck JM, Murray AR, O'Dell SJ. Interleukin 6 indirectly induces keratinocyte migration. J Invest Dermatol 2004; 122(3):764–72. Pub Med PMID: 15086564.

[71] Zhang QZ, Nguyen AL, Yu WH, Le AD. Human oral mucosa and gingiva: a unique reservoir for mesenchymal stem cells. J Dent Res 2012; 91 (11):1011–8. Pub Med PMID: 22988012. Pubmed Central PMCID: 3490281.

[72] Xue H, McCauley RL, Zhang W. Elevated interleukin-6 expression in keloid fibroblasts. J Surg Res 2000; 89(1):74–7. Pub Med PMID: 10720455.

[73] Tosa M, Ghazizadeh M, Shimizu H, Hirai T, Hyakusoku H, Kawanami O. Global gene expression analysis of keloid fibroblasts in response to electron beam irradiation reveals the involvement of interleukin-6 pathway. J Invest Dermatol 2005; 124(4):704–13. Pub Med PMID: 15816827.

[74] Ghazizadeh M, Tosa M, Shimizu H, Hyakusoku H, Kawanami O. Functional implications of the IL-6 signaling pathway in keloid pathogenesis. J Invest Dermatol 2007; 127(1):98–105. Pub Med PMID: 17024100.

[75] Wahl SM, Hunt DA, Wakefield LM, McCartney-Francis N, Wahl LM, Roberts AB, et al. Transforming growth factor type beta induces monocyte chemotaxis and growth factor production. Proc Natl Acad Sci U S A 1987; 84 (16):5788–92. Pub Med PMID: 2886992. Pubmed Central PMCID: 298948.

[76] Letterio JJ, Roberts AB. Regulation of immune responses by TGF-beta. Annu Rev Immunol 1998; 16:137–61. Pub Med PMID: 9597127.

[77] Frank S, Madlener M, Werner S. Transforming growth factors beta 1, beta 2, and beta 3 and their receptors are differentially regulated during normal and impaired wound healing. J Biol Chem 1996; 271 (17):10188–93. Pub Med PMID: 8626581.

[78] Shah M, Foreman DM, Ferguson MW. Neutralisation of TGF-beta 1 and TGF-beta 2 or exogenous addition of TGF-beta 3 to cutaneous rat wounds reduces scarring. J Cell Sci 1995; 108 (Pt 3):985–1002. Pub Med PMID: 7542672.

[79] Russell SB, Trupin KM, Rodriguez-Eaton S, Russell JD, Trupin JS. Reduced growth-factor requirement of keloid-derived fibroblasts may account for tumor growth. Proc Natl Acad Sci U S A 1988; 85(2):587–91. Pub Med PMID: 3422443. Pubmed Central PMCID: 279596.

[80] Lanning DA, Nwomeh BC, Montante SJ, Yager DR, Diegelmann RF, Haynes JH. TGF-beta 1 alters the healing of cutaneous fetal excisional wounds. J Pediatr Surg 1999; 34(5):695–700. Pub Med PMID: 10359166.

[81] Cox DA, Kunz S, Cerletti N, McMaster GK, Burk RR. Wound healing in aged animals--effects of locally applied transforming growth factor beta 2 in different model systems. EXS 1992; 61:287–95. Pub Med PMID: 1377545.

[82] O'Kane S, Ferguson MW. Transforming growth factor beta s and wound healing. Int J Biochem Cell Biol 1997; 29(1):63–78. Pub Med PMID: 9076942.

[83] Occleston NL, Laverty HG, O'Kane S, Ferguson MW. Prevention and reduction of scarring in the skin by Transforming Growth Factor beta 3 (TGFbeta3): from laboratory discovery to clinical pharmaceutical. J Biomater Sci Polym Ed 2008; 19(8):1047–63. Pub Med PMID: 18644230.

[84] Proetzel G, Pawlowski SA, Wiles MV, Yin M, Boivin GP, Howles PN, et al. Transforming growth factor-beta 3 is required for secondary palate fusion. Nat Genet 1995; 11(4):409–14. Pub Med PMID: 7493021. Pubmed Central PMCID: 3855390.

[85] Schrementi ME, Ferreira AM, Zender C, DiPietro LA. Site-specific production of TGF-beta in oral mucosal and cutaneous wounds. Wound Repair Regener Off Publ Wound Heal Soc Eur Tissue Repair Soc. 2008; 16(1):80–6. Pub Med PMID: 18086295.

[86] Ferguson MW, Duncan J, Bond J, Bush J, Durani P, So K, et al. Prophylactic administration of avotermin for improvement of skin scarring: three double-blind, placebo-controlled, phase I/II studies. Lancet 2009; 373 (9671):1264–74. Pub Med PMID: 19362676.

[87] Kohno T, Brewer MT, Baker SL, Schwartz PE, King MW, Hale KK, et al. A second tumor necrosis factor receptor gene product can shed a naturally occurring tumor necrosis factor inhibitor. Proc Natl Acad Sci U S A 1990; 87 (21):8331–5. Pub Med PMID: 2172983. Pubmed Central PMCID: 54949.

[88] Brauchle M, Angermeyer K, Hubner G, Werner S. Large induction of keratinocyte growth factor expression by serum growth factors and pro-inflammatory cytokines in cultured fibroblasts. Oncogene 1994; 9(11):3199–204. Pub Med PMID: 7936642.

[89] Wallace HJ, Stacey MC. Levels of tumor necrosis factor-alpha (TNF-alpha) and soluble TNF receptors in chronic venous leg ulcers--correlations to healing status. J Invest Dermatol 1998; 110(3):292–6. Pub Med PMID: 9506452.

[90] Streit M, Beleznay Z, Braathen LR. Topical application of the tumour necrosis factor-alpha antibody infliximab improves healing of chronic wounds. Int Wound J 2006; 3(3):171–9. Pub Med PMID: 16984574.

[91] Ashcroft GS, Lei K, Jin W, Longenecker G, Kulkarni AB, Greenwell-Wild T, et al. Secretory leukocyte protease inhibitor mediates non-redundant functions necessary for normal wound healing. Nat Med 2000; 6(10):1147–53. Pub Med PMID: 11017147.

[92] Ouyang W, Rutz S, Crellin NK, Valdez PA, Hymowitz SG. Regulation and functions of the IL-10 family of cytokines in inflammation and disease. Annu Rev Immunol 2011; 29:71–109. Pub Med PMID: 21166540.

[93] Grone A. Keratinocytes and cytokines. Vet Immunol Immunopathol. 2002; 88 (1–2):1–12. Pub Med PMID: 12088639.

[94] Gordon A, Kozin ED, Keswani SG, Vaikunth SS, Katz AB, Zoltick PW, et al. Permissive environment in postnatal wounds induced by adenoviral-mediated overexpression of the anti-inflammatory cytokine interleukin-10 prevents scar formation. Wound Repair Regener Off Publ Wound Heal Soc Eur Tissue Repair Soc. 2008; 16(1):70–9. Pub Med PMID: 18086289.

[95] Shi JH, Guan H, Shi S, Cai WX, Bai XZ, Hu XL, et al. Protection against TGF-beta 1-induced fibrosis effects of IL-10 on dermal fibroblasts and its potential therapeutics for the reduction of skin scarring. Arch Dermatol Res 2013; 305(4):341–52. Pub Med PMID: 23321694.

[96] Liechty KW, Kim HB, Adzick NS, Crombleholme TM. Fetal wound repair results in scar formation in interleukin-10-deficient mice in a syngeneic murine model of scarless fetal wound repair. J Pediatr Surg 2000;35(6):866–72 [discussion 72-3].

[97] Peranteau WH, Zhang L, Muvarak N, Badillo AT, Radu A, Zoltick PW, et al. IL-10 overexpression decreases inflammatory mediators and promotes regenerative healing in an adult model of scar formation. J Invest Dermatol 2008; 128(7):1852–60. Pub Med PMID: 18200061.

[98] Morris MW, Jr., Allukian M, 3rd, Herdrich BJ, Caskey RC, Zgheib C, Xu J, et al. Modulation of the inflammatory response by increasing fetal wound size or interleukin-10 overexpression determines wound phenotype and scar formation. Wound Repair Regener Off Publ Wound Heal Soc Eur Tissue Repair Soc. 2014; 22(3):406–14. Pub Med PMID24844340.

[99] van den Broek LJ, van der Veer WM, de Jong EH, Gibbs S, Niessen FB. Suppressed inflammatory gene expression during human hypertrophic scar compared to normotrophic scar formation. Exp Dermatol 2015; 24(8):623–9. Pub Med PMID: 25939875.

[100] Kieran I, Knock A, Bush J, So K, Metcalfe A, Hobson R, et al. Interleukin-10 reduces scar formation in both animal and human cutaneous wounds: results of two preclinical and phase II randomized control studies. Wound Repair Regener Off Publ Wound Heal Soc Eur Tissue Repair Soc. 2013; 21(3):428–36. Pub Med PMID: 23627460.

[101] O'Neill LA, Golenbock D, Bowie AG. The history of Toll-like receptors—redefining innate immunity. Nat Rev Immunol 2013; 13(6):453–60. Pub Med PMID: 23681101 [Epub 2013/05/18.eng].

[102] Kondo T, Kawai T, Akira S. Dissecting negative regulation of Toll-like receptor signaling. Trends Immunol 2012; 33(9):449–58. Pub Med PMID: 22721918 [Epub 2012/06/23.eng].

[103] Miller LS, Modlin RL. Toll-like receptors in the skin. Semin Immunopathol 2007; 29(1):15–26. Pub Med PMID: 17621951.

[104] Chen L, Arbieva ZH, Guo S, Marucha PT, Mustoe TA, DiPietro LA. Positional differences in the wound transcriptome of skin and oral mucosa. BMC Genomics 2010; 11:471. Pub Med PMID: 20704739.

[105] Lin Q, Fang D, Fang J, Ren X, Yang X, Wen F, et al. Impaired wound healing with defective expression of chemokines and recruitment of myeloid cells in TLR3-deficient mice. J Immunol (Baltimore, MD: 1950) 2011; 186(6):3710–7. Pub Med PMID: 21317384 [Epub 2011/02/15.eng].

[106] Lin Q, Wang L, Lin Y, Liu X, Ren X, Wen S, et al. Toll-like receptor 3 ligand polyinosinic: polycytidylic acid promotes wound healing in human and murine skin. J Invest Dermatol 2012; 132(8):2085–92. Pub Med PMID: 22572822 [Epub 2012/05/11.eng].

[107] Chen L, Guo S, Ranzer MJ, DiPietro LA. Toll-like receptor 4 has an essential role in early skin wound healing. J Invest Dermatol 2013; 133(1):258–67. Pub Med PMID: 22951730. Pubmed Central PMCID: PMC3519973 [Epub 2012/09/07.eng].

[108] Sato T, Yamamoto M, Shimosato T, Klinman DM. Accelerated wound healing mediated by activation of Toll-like receptor 9. Wound Repair Regener Off Publ Wound Heal Soc Eur Tissue Repair Soc. 2010; 18(6):586–93. Pub Med PMID: 20946144. Pubmed Central PMCID: PMC3010290 [Epub 2010/10/16.eng].

[109] Dasu MR, Thangappan RK, Bourgette A, DiPietro LA, Isseroff R, Jialal I. TLR2 expression and signaling-dependent inflammation impair wound healing in diabetic mice. Lab Investig J Tech Methods Pathol. 2010; 90 (11):1628–36. Pub Med PMID: 20733560 [Epub 2010/08/25.eng].

[110] Dasu MR, Jialal I. Amelioration in wound healing in diabetic Toll-like receptor-4 knockout mice. J Diabetes Complicat 2013; 27(5):417–21. Pub Med PMID: 23773694. Pubmed Central PMCID: PMC3770740 [Epub 2013/06/19.eng].

[111] Wang J, Hori K, Ding J, Huang Y, Kwan P, Ladak A, et al. Toll-like receptors expressed by dermal fibroblasts contribute to hypertrophic scarring. J Cell Physiol 2011; 226(5):1265–73. Pub Med PMID: 20945369 [Epub 2010/10/15.eng].

[112] Junker JPE, Caterson EJ, Eriksson E. The Microenvironment of Wound Healing. J Craniofac Surg 2013; 24(1):12–6. Pub Med PMID: WOS:000314853300047 [English].

[113] Wilgus TA, Vodovotz Y, Vittadini E, Clubbs EA, Oberyszyn TM. Reduction of scar formation in full-thickness wounds with topical celecoxib treatment. Wound Repair Regener Off Publ Wound Heal Soc Eur Tissue Repair Soc. 2003; 11(1):25–34. Pub Med PMID: 12581424.

[114] Nagaraja S, Wallqvist A, Reifman J, Mitrophanov AY. Computational approach to characterize causative factors and molecular indicators of chronic wound inflammation. J Immunol 2014; 192(4):1824–34. Pub Med PMID: WOS:000331267200050 [English].

[115] Nagaraja S, Chen L, Zhou J, Zhao Y, Fine D, DiPietro A, et al. Mechanistic drivers and molecular indicators of delayed inflammation resolution in traumatic wounds. Wound Repair Regen. 2016; 24(2):A19-A. Pub Med PMID: WOS:000375157700081 [English].

[116] Ashcroft GS, Yang X, Glick AB, et al. Mice lacking Smad 3 show accelerated wound healing and an impaired local inflammatory response. Nat Cell Biol 1999;1(5):260–6. https://doi.org/10.1038/12971.

Section D

In Vitro Models as study platforms of skin Biology

In vitro models to study hair follicle generation

12

Ana Korosec, Beate M. Lichtenberger
Skin & Endothelium Research Division, Department of Dermatology, Medical University of Vienna, Vienna, Austria

1. Introduction

In its simplest form, the mammalian skin comprises two layers, the ectoderm-derived epidermis and a layer of mesenchyme, the dermis. A closer look however reveals a striking complexity of the skin including an intricate network of vasculature and innervation and elaborate structures arising from the epidermis during embryonic development such as hair follicles (HFs) and sebaceous and sweat glands [1]. The HF is a complex miniorgan that is unique in its remarkable ability to undergo cycles of regeneration for a lifetime [2,3]. Despite this high regenerative capacity, under normal circumstances, HFs develop during embryogenesis but never in adult tissues [4], leaving patients suffering from alopecia or severe injuries and burns in despair.

HFs are composed of cells originating from three distinct stem cell pools: epithelial and mesenchymal stem cells and neural crest cells giving rise to melanocytes [5,6]. HF development and hair growth involves not only reciprocal interactions between epidermal stem cells and specialized niche cells in the dermis [7], which makes HFs an exceptional model system to study stem cell (SC) self-renewal and their function in tissue homeostasis [8,9] and regeneration [10,11], but also the interaction of SCs with their niche [12–14]. Apart from SC biology, HF research involves various fields including immunology [15,16], chronobiology [17], metabolism [18], neurobiology [19], pigmentation research [20], and endocrinology [21]. Furthermore, as the HF cycles between a state of relative quiescence (telogen) and a state of rapid growth (anagen) [1], the HF and its entire microenvironment undergo transformations. In telogen, the HF has its minimal length, which elongates and generates a pigmented hair shaft (HS) in anagen. This proliferative phase is followed by a short period of apoptosis-driven organ involution (catagen) leading back to telogen. Not only the morphology of the HF and its associated sebaceous glands changes but also its immune system, the expression and secretory profile, and the surrounding dermis including the adipose tissue, innervation, and perifollicular vasculature [3,22–25], making the HF an ideal model for systems biology analyses.

With the development of a method to isolate and culture epidermal cells in the presence of fibroblasts or fibroblast-derived growth factors by Rheinwald and Green in the 1970s [26–29] and to grow hair ex vivo by Philpott and colleagues in 1990 [30–35], skin and HF research has thrived enormously. Detailed protocols for the establishment of in vitro skin equivalents (SEs) have enabled an expanding network of scientists to

Skin Tissue Models. https://doi.org/10.1016/B978-0-12-810545-0.00012-7

construct artificial skin in their laboratories. However, these bioengineered SEs still lack the complexity of mammalian skin including HFs, vasculature, or hypodermis. Recently, different approaches of introducing additional skin-derived cell types into classical SEs harboring epidermal cells and fibroblasts only have been developed, in order to resemble the complexity of the human skin more closely.

Based on the current knowledge on epidermal and dermal cells and their interactions during HF morphogenesis and hair cycling in adult tissue, many researchers have attempted to generate mature HFs with different experimental strategies. These mostly involve embryonic or neonatal rodent cells. Besides, patent applications aiming at de novo HF generation both by companies and academic researchers have been increasing lately [36], but until recently, all approaches to generate human HFs in vitro have failed.

In this chapter, we will discuss how HFs develop during embryogenesis and why new follicles do not grow in adult tissue under normal circumstances. We will describe the approaches developed in the last quarter of a century to study HFs in vitro from monolayer cell culture systems to ex vivo human HF organ cultures and three-dimensional artificial skin models and exemplify recent innovative attempts to generate HFs in the dish.

2. Hair follicle morphogenesis

HF organogenesis occurs early during embryonic development with the appearance of dermal cells toward the end of the first trimester [3,37]. In mice, HFs develop in waves from embryonic day 12.5 to 18.5 and result in a number of different hair types: primary or tylotrich (guard) HFs, characterized by huge hair bulbs, long straight hair, and two sebaceous glands; secondary or nontylotrich (awl, auchene, and zigzag) HFs with thinner and shorter HSs and one sebaceous gland; and vibrissae HFs with specialized sensory functions [37]. HF morphogenesis depends on reciprocal signaling between epidermal stem cells and specialized cells of the underlying mesenchyme, which aggregate forming the dermal condensate (DC) and will later become the dermal papilla (DP, Fig. 1) of the HF [7]. Dermal signals induce a focal thickening in the basal layer of the epidermis (hair germ or hair placode) and stimulate epidermal stem cells to grow downward and invaginate into the dermis, thereby engulfing the cells of the DC, which leads to the formation of the DP (Figs. 1 and 2A) and further drives HF formation [37]. As the follicle grows downward, it is encapsuled by the highly proliferative, transit-amplifying matrix cells at the leading edge. The formation of the HF lumen is mediated by early outward movement of keratin 79 (K79) epidermal cells from within the cores of developing hair buds and into the epidermis [38]. The inner layers of the HF differentiate into concentric cylinders and generate the inner root sheath (IRS) and the central HS, while the outer root sheath (ORS) is continuous with the basal layer of the interfollicular epidermis (IFE). The IRS cells migrate outward from the base of the hair buds. Subsequently, $K79^{+}$ cells are lost from the epidermis, thereby leaving behind a gap above the future site of the hair canal [38]. Once the HF reaches the bottom of the dermis, the HF becomes fully mature. However, matrix cells continue dividing, and their successors terminally differentiate to form the growing hair that exits the skin surface. In the mouse back skin, HF morphogenesis is completed between postnatal days 6 and 8 [37,39,40]. The ability

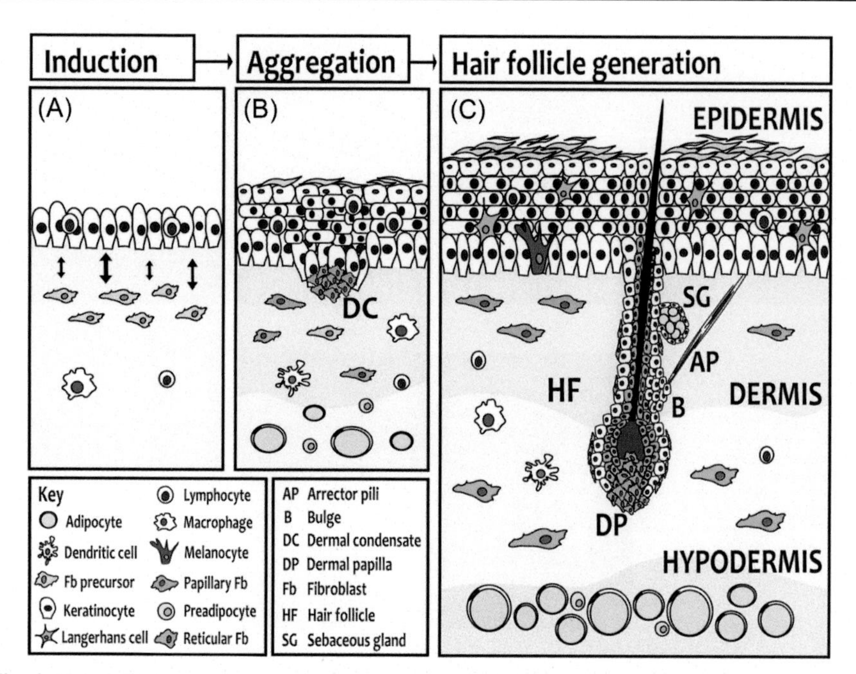

Fig. 1 Hair follicle morphogenesis depends on epithelial-mesenchymal interactions. (A) Reciprocal signaling between cells of the developing epidermis and the underlying mesenchyme induces hair follicle (HF) formation during embryogenesis. (B) Mesenchymal cells subsequently form a dermal condensate (DC), which will later become the dermal papilla (DP) of the hair follicle. (C) The dermal papilla instructs the surrounding epithelial cells that ensheathe the DC during the initial down growth, to proliferate and differentiate into complex HF structures and produce the hair shaft, and controls hair follicle cycling.

of the HF to regenerate is due to the presence of a pool of stem cells located in the bulge region of the HF, which forms a special niche that is critical for controlling SC self-renewal and thus the process of regeneration [41]. Bulge stem cell identity specification occurs early during HF morphogenesis in the embryo [41].

3. Hair follicle cycle

Matrix cells are transit-amplifying cells that undergo only a limited number of divisions before they terminally differentiate. Once the supply of matrix cells declines, HFs enter the so-called hair growth cycle, which starts with a degenerative phase called catagen. In murine skin, catagen occurs between postnatal days 14 and 18 in waves starting from the head caudally toward the tail and from the spinal cord down to the sides of the mouse [42].

During catagen, apoptosis of epithelial cells in the bulb and ORS leads to a complete regression of the lower "cycling" portion of the HF. This translocates the DP upward to rest below the permanent, noncycling part of the follicle [42,43] (Fig. 2B).

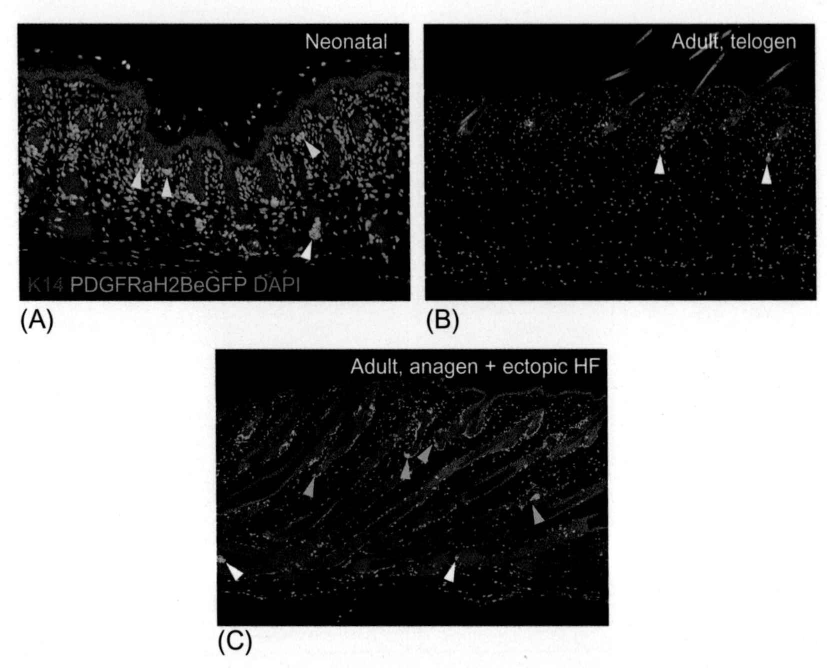

Fig. 2 Hair follicle neogenesis does not occur in adult skin. (A) Murine neonatal skin contains follicles in different stages of morphogenesis including hair germ stages *(yellow arrow heads)*, hair pegs, and hair follicles with enveloped DPs *(white arrow head)*. (B) During catagen, epithelial cells at the base of fully developed HFs undergo apoptosis and regress, while the DP *(white arrow heads)* moves upward until it comes to rest beneath the stem cells of the hair follicle bulge. (C) De novo HFs never develop in the adult skin unless Wnt/β-catenin signaling is activated in epidermal stem cells, which drives existing follicles into anagen and induces the formation of ectopic follicles (EF) in the interfollicular epidermis and from existing hair follicles. Like anagen HFs *(white arrow heads)*, these EFs have associated DPs *(orange arrow heads)*.

Upon degeneration, HFs enter a quiescent, resting stage called telogen, which lasts only 2 days in the first and more than 2 weeks in the second hair cycle in mice. Once quiescent epidermal SCs located in the lowest permanent portion of the HF, the bulge region, are activated, their progeny become transit-amplifying cells and divide rapidly to regenerate the lower portion of the HF in a phase called anagen, which resembles HF morphogenesis [4,42]. While in the mouse, anagen lasts for 2 weeks, human HFs may remain in anagen up to 6 years [44].

While transit-amplifying cells ultimately commit to terminal differentiation and produce the HS, the early progeny of stem cells on the way to becoming transit-amplifying cells maintains their slow-cycling properties and stemness. These cells home back to the bulge SC niche when hair growth terminates and serve as the primary SCs in the next hair growth cycle [12]. In contrast, initial bulge SCs reside as a resource to replenish the tissue upon injury [12]. HF and melanocyte SCs are miscellaneous in the bulge and hair germ. During hair cycle initiation, melanocyte SCs divide to produce proliferative and committed melanocyte progenitors. These melanocytes reside within the HF matrix of anagen follicles and generate melanin pigment, which is transferred to differentiating hair cells, and degenerate during catagen [1]. SCs of the sebaceous gland lineage terminally differentiate into sebocytes filled with lipids, which lubricate the skin's surface once they degenerate to release their contents [45].

In order to maintain skin homeostasis and to control HF growth and regression, proliferation and differentiation of epithelial cells must be tightly regulated and coordinated. Several signaling pathways have been implicated in HF development and cycling, including the PDGF, Wnt, FGF, Notch, Bmp, Shh, and TGFβ pathways, with essential instructive signals coming from fibroblasts forming the DP [7,37]. HF growth can be affected not only by deregulated paracrine and endocrine factors but also by the surrounding dermal niche, thereby leading to hair loss.

4. Dermal papilla cells instruct hair follicle morphogenesis and control hair cycling

Classical tissue recombination and transplantation experiments have revealed a pivotal role for dermal fibroblasts in regulating various aspects of HF biology, such as HF formation and regenerative hair cycling. While the contribution of epidermal cells to HF morphogenesis has been extensively investigated, our knowledge about the development of the DP and its functions in HF morphogenesis, growth, and HF cycling is limited. Novel DP-specific genetic tools now permit studying epidermal-dermal interactions and the signaling pathways involved during development and adulthood [46–49].

Several signaling pathways have been implicated in mesenchyme instruction during embryogenesis and subsequent DP formation as well as DP-derived instructive signaling that is essential to activate epithelial progenitors and initiate HF regeneration and growth [7,50]. While interdependent interactions between Wnt/β-catenin and ectodysplasin signaling pathways have been shown to determine DC formation before epithelial placodes appear [51], a recent study suggests that the commitment to the DP lineage likely occurs once the embryonic mesenchyme is exposed to Fgf20 derived from the epithelial placode, and cells of the early DC may already have committed to the DP fate [52]. Once committed, these dermal cells undergo lineage maturation and progressively acquire terminally differentiated characteristics. While paracrine PDGF signaling stimulates dermal mesenchyme proliferation essential for the assembly of (normal and ectopic) DPs [53,54], recent studies have shown that mesenchymal PDGF signaling is dispensable for HF induction and formation [48]. In contrast, epidermis-derived sonic hedgehog (Shh) is a crucial hair-inductive signal controlling the expression of DC- and DP-specific genes and thus DP maturation and maintenance of its function, thereby driving HF morphogenesis [40,55,56], whereas dermal deletion of the hedgehog effector smoothened inhibits follicle formation [57].

DP cells are specialized mesenchymal cells derived from the papillary dermal fibroblast lineage [58], which differ from interfollicular dermal fibroblasts by unique properties such as the formation of aggregates in culture and the ability to induce HFs [58,59]. As mentioned previously, HF morphogenesis depends on reciprocal cross talk between epidermal cells and fibroblasts and occurs during late embryonic and early postnatal development but never in adult tissues [7], and thus, in adult skin, wound healing results in a lack of HFs at the wound site in contrast to scar-free healing in fetal or neonatal tissue [60]. Only in rabbits and mice wounding-induced HF neogenesis has been described in adult tissues [61–63]. This is either due to a decline in papillary fibroblasts in the adult skin or because fibroblasts change their gene signature and lose their hair-inductive properties with age [57,64]. However, activation of Wnt/β-catenin signaling in the epidermis drives stem cells of the IFE and sebaceous glands into the HF lineage and reprograms adult dermal fibroblasts via paracrine signals to a neonatal state, thereby inducing formation of de novo follicles with associated DPs in the adult skin [53,57,65] (Fig. 2C). Interestingly, the newly induced follicles are not guard hairs but zigzag follicles corresponding to the third wave of HF development, likely because the dermal fibroblasts are reprogrammed to a neonatal but not embryonic state. These findings suggest that remodeling dermal signaling in adult fibroblasts may restore their hair-inductive capabilities. Indeed, reestablishing a neonatal gene signature in adult fibroblasts and the subsequent increase in the number of papillary fibroblasts facilitate HF neogenesis during skin regeneration upon injury of the adult skin [57,58]. Furthermore, β-catenin ablation in dermal fibroblasts promoted HF regeneration in neonatal and adult wounds, while dermal β-catenin stabilization inhibited HF neogenesis in neonatal wounds [64], suggesting that modulating dermal Wnt signaling may support in vitro HF regeneration as well.

In contrast to other fibroblasts of the papillary lineage, DP cells retain their capacity to induce HF neogenesis. This was shown in elegant studies where dissected rat vibrissae DPs were implanted into afollicular skin or beneath amputated vibrissae follicles, thereby inducing de novo HF formation in an ectopic site or hair fiber formation, respectively [66–69]. Interestingly, if the DP is removed surgically, lower follicle DS cells pitch in for the DP cells and restore functional hair growth [70–72], and the germinative epithelium transforms DS into DP cells, as shown in an in vitro assay [73]. Similarly, dissected human DPs induced HF formation when implanted into transected athymic mouse vibrissae follicles [74] and human microdissected DS cells from the base of male scalp skin follicles resulted in new HFs when grafted into small wounds of an immunologically incompatible female recipient forearm [75]. This indicates that HF-derived dermal cells have hair-inductive abilities and that DS and DP cells belong to a restricted group of immune-privileged tissues that can be transplanted to foreign sites without being rejected [76].

Recent studies have demonstrated that cells of the DP rarely divide [47] and that HF dermal stem cells repopulate the DP and DS and modulate HF type [77]. Furthermore, while it was shown that DP cell number specifies hair size, morphology, and HF cycling [78], an increase in DP cell numbers subsequent to activation of Wnt/β-catenin signaling did not result in hair type switching [47,79,80], despite the fact that only few DP genes seem to define HF type [81,82].

In conclusion, DP cells are specialized cells that are essential to induce HFs in vivo, and maintaining their gene signature and thus their hair-inductive properties is key to generating de novo follicles in vitro. Importantly, DP cells may have a much broader therapeutic potential than inducing new HFs for regenerative medicine. As DP cells can be reprogrammed into induced pluripotent (iPS) SCs more readily than many other cell types [83,84] and the noninvasive isolation method of HF plucking represents a huge advantage, DP cells may be used as a source to study or correct various diseases. Besides, recent studies have revealed that scalp HFs from patients suffering from androgenic alopecia have normal bulge stem cell populations [85], but their DPs display altered expression patterns compared with healthy donor HF as recently shown in immortalized DP cell lines derived from balding and nonbalding scalp [86], suggesting that modulating gene expression in diseased DPs might serve as a therapeutic intervention to restore hair growth.

5. Experimental models for hair follicle research

Current models to study HF growth range from normal and genetically modified mice, which are reviewed elsewhere [87,88] or are published on the Sanger Mouse Resources Portal (http://www.sanger.ac.uk/mouseportal) [89], in vivo reconstitution and human skin transplantation assays in immunocompromised mice, to monolayer cultures of specific cell types isolated from human HFs, hair follicle organ cultures (HFOCs), and three-dimensional coculture systems. Each of these model systems has advantages to address specific HF functions and may provide novel therapies for HF disorders. Although studies in animals have provided important insights in HF development and cycling and the signaling cascades involved, in vitro studies with human HFs are essential, not only because of recent legal restrictions for the use of animal models for the cosmetic industry. Human HF growth differs from that of rodents and other mammals as (i) it does not occur in synchronized waves, (ii) anagen of human scalp HF may last up to 6 years [44], (iii) human hair has a reduced function in thermoregulation [90], but most importantly (iv) due to the more dramatic effect of androgens in humans, which stimulate HF growth in some body parts but inhibit hair growth on the scalp and other body sites [91].

6. Human follicle organ cultures

An important milestone in the history of hair research was achieved by Philpott and colleagues, who developed ex vivo human HFOCs under serum-free conditions in 1990, which not only maintained HFs alive but also allowed HF elongation at similar rates to that seen in vivo [30]. As reviewed in great detail by Langan et al. [92], a series of ex vivo studies of intact human HFs grown in submerged culture systems or at the air-liquid interface have led to major advances in hair research. Although hair cycling in vivo cannot be recapitulated entirely in HFOCs, this simple, easily accessible, and translationally relevant model system is very versatile and allows a wide range of assays to address the impact of physical or chemical cues on HF elongation, proliferation, apoptosis, cytotoxicity, pigmentation, and regenerative HF cycling (e.g., HF staging). Novel methods allow gene knockdown by siRNA technology or in situ fluorescence labeling of specific HF SC populations in HFOCs, thereby permitting the study of new aspects and biological functions with different scientific approaches.

7. Monolayer cell culture systems of distinct hair follicle cell types

The establishment of a method to isolate and culture epidermal cells in the 1970s [26–29] was key to studying the effect of various growth factors and chemicals on epidermal cell proliferation, apoptosis and expression, and secretory profiles in two-dimensional (2-D) culture systems. Meanwhile, various companies sell defined, serum-free media for maintaining epidermal cells in undifferentiated and proliferative states and for inducing terminal differentiation. The identification of markers for specific HF epidermal (stem) cell populations especially in the mouse allowed the isolation and functional analysis of specific subsets of HF epidermal cells in vitro [93,94].

The most commonly used technique to isolate DP (and also adjacent DS) cells is surgical microdissection of human HFs or rodent vibrissae follicles or dispase and subsequent collagenase treatment of the lower portion of anagen follicles [59,67]. Larger numbers of DP cells can be isolated from the mouse skin by fluorescence-activated cell sorting (FACS) of dissociated dermal tissue [58] with DP-specific markers such as prominin-1 (CD133) [47]. However, as discussed below, DP cells are difficult to propagate and analyze in 2-D cultures as they almost immediately lose their innate properties in the dish [95]. Other HF-derived cells investigated in monolayer cultures or cocultures with epidermal and dermal cells include melanocytes [96] and sebocytes [97].

Although both epidermal and DP cells change their properties in vitro, important information about their interactions can be gained when coculturing these two cell types in vitro or by adding conditioned medium from either cell type to the other [98]. For example, epidermal stem cells cocultured with DP cells display increased proliferative and migratory potential [99] and induced hair-specific differentiation in transwell coculture systems [100] or by using DP cell-conditioned medium [101].

8. Bioengineered skin equivalents

Since Rheinwald and Green managed to isolate and grow skin epithelial cells in the presence of fibroblasts or fibroblast-derived growth factors [26–29], the development of three-dimensional (3-D) bioengineered SEs has thrived enormously. These days, human SEs are used not only for research and drug screening like EpiDerm (MatTek) or EpiSkin (L'Oreal) but also in regenerative medicine (e.g., the FDA approved bilayer skin substitutes Integra and Apligraf) to provide an artificial skin barrier and epithelial covering for large or chronic wounds to promote the healing process [102–104]. Such SEs created from autologous epidermal cells in vitro have been successfully used to treat chronic wounds [105]. Scientists have been searching for new scaffolds especially replacing animal-derived dermal matrix [106] or alternative tissue and cell sources as well as specific subsets of HF epidermal and dermal cells [107–111] in an effort to improve human skin substitutes and their production, especially for the generation of autologous skin grafts. For example, Higgins et al. reported increased support of SEs if human interfollicular fibroblasts are replaced by human HF dermal

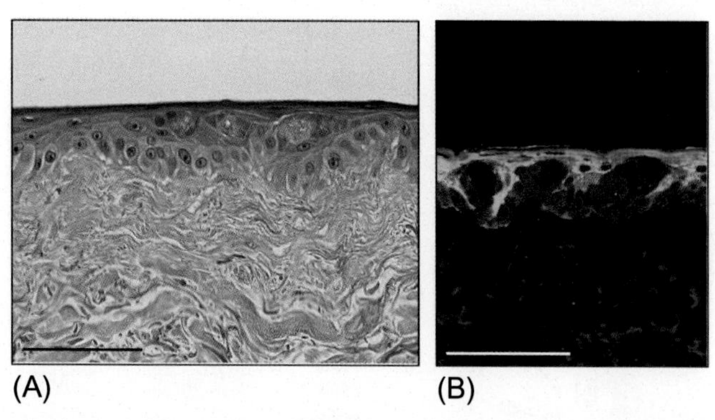

Fig. 3 Classical skin equivalents lack hair follicles. (A) Hematoxylin and eosin staining of a human skin equivalent (SE) section. Classical SEs lack hair follicles and other complex structures found in vivo. (B) Immunofluorescent staining of SE section with antibodies targeting keratin 14 (basal epidermal cells, *red*) and involucrin (differentiation marker, *green*) (and DAPI counterstaining to visualize nuclei, *blue*) reveals stratification of the artificial epidermis. Epidermal cells were seeded onto devitalized human dermis substrates submerged in a medium containing a feeder layer of irradiated fibroblasts, which provide essential growth factors.

cells (DP and DS) [112]. However, bioengineered human SEs available at present are simplistic and devoid of HFs, blood vessels, and nerves, which together make the skin a highly complex tissue (Fig. 3).

Recently, different innovative approaches of introducing additional skin-derived cell types into classical SEs harboring epidermal cells and fibroblasts only have been developed, in order to resemble the complexity of the human skin more closely (Fig. 4). These include addition of endothelial cells and adipose-derived mesenchymal SCs as well as bioengineered vascular networks and adipose tissue [113–119], which would support epidermal cell growth and differentiation and improve the survival of grafted tissue. Naturally, incorporating HFs in such bioengineered skin substitutes is a major goal, both because of the appearance of the transplanted artificial skin and because several studies indicate that HFs promote wound healing: wounds heal faster if HFs of rodent skin are in an active anagen growth phase rather than in the resting telogen, and transplantation of terminal anagen HFs into chronic leg ulcers promoted wound healing [120,121].

9. Hair follicle generation in in vivo reconstitution and SE transplantation assays

In vivo hair reconstitution assays were developed to study the early events of HF development. In these assays, a mixture of individually prepared epidermal cells or specific HFSC populations and mesenchymal cells from rodent embryonic or neonatal skin is grafted onto dorsal skin of immunocompromised recipient mice within a silicon chamber or by intradermal injection [122–125], which subsequently regenerate intact

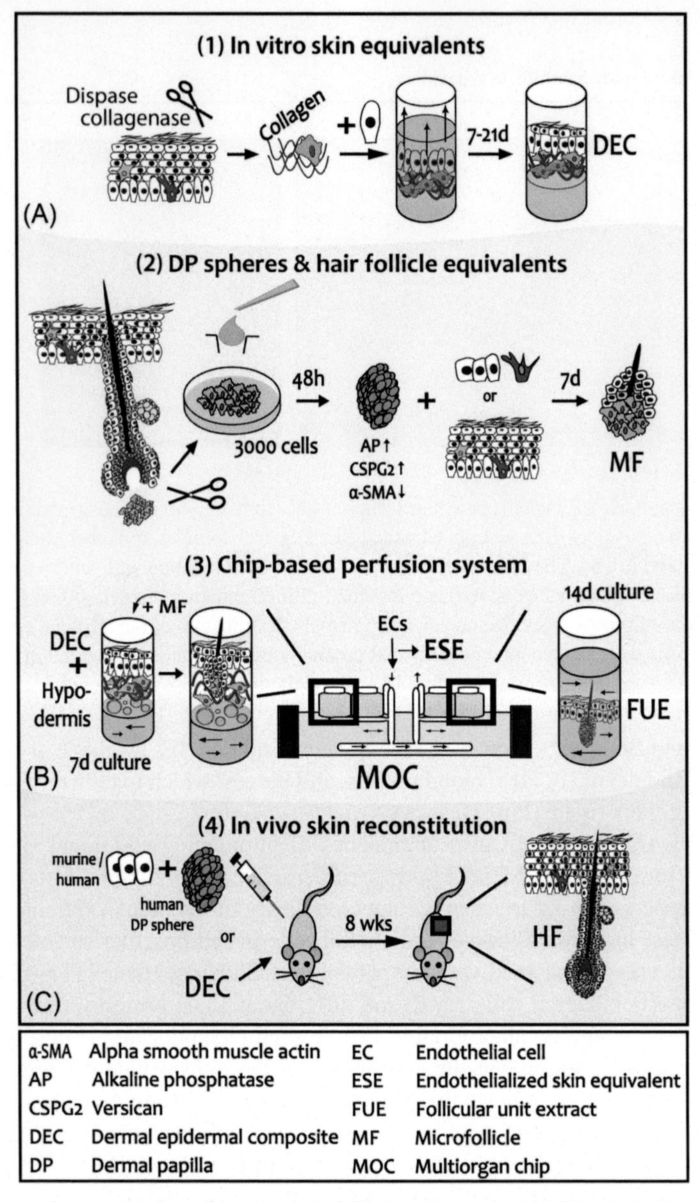

Fig. 4 Innovative approaches of in vitro hair follicle neogenesis. (A) Classical skin equivalents comprise fibroblasts in a collagen matrix and are overlayed with epidermal cells that stratify once they are raised to the air-liquid phase. (A) Higgins et al. (2013) managed to maintain the hair-inductive properties of DP cells by culturing human DP cells in hanging drop culture systems. The size of the DP-like spheres is similar to DPs in situ. (C) Recently, a group of scientists generated microfollicles from human cultured DP and epidermal cells and melanocytes entirely in vitro and introduced them into skin equivalents grown under dynamic perfusion on multiorgan chips (MOCs). To further improve the complexity of these SEs, subcutaneous tissue or other cell types could be included. These next-generation skin equivalents comprising in vitro-generated microfollicles can be used to study HF biology and for drug testing. (D) Hair follicles can be generated in vivo in skin reconstitution assays by injecting cultured human DP spheres in combination with murine or human epidermal cells into grafting chambers on top of immunocompromised mice or by transplanting in vitro-generated dermal-epidermal composites (DECs) onto the back skin of nude mice.

skin epithelium comprising normal-appearing HFs (Fig. 4). This ability of DP cells to induce de novo HF development, when combined with epithelial cells, has long fascinated HF researchers. Furthermore, generating new HFs from patient-derived DPs and competent epithelial cells has long been considered a viable therapeutic intervention for restoring hair growth. However, while murine DP cells often display impressive inductive capabilities in reconstitution assays [126], human DP cells independent of the patient age are less prone to induce HF neogenesis. Unlike cultured rat DPs, cultured human DPs do not form papilla-like aggregates within the dermis, highlighting species-specific differences [67]. While reconstitution assays with homo-specific human epidermal and dermal cells had not been successful, in 2007, Ehama et al. reported the generation of "chimeric" HF-like structures consisting of human epithelial and murine DP cells in a chamber assay [127]. However, these HF-like structures had no bulge region, nor all epithelial layers found in intact HFs, and the DP-like tissue failed to express DP markers. Human epidermal ORS and DPs grown in 3-D collagen matrices have been shown to form tubular structures in vitro [128] and organize themselves into epidermoid cyst-like spheroids [128,129], but do not result in intact HFs. Sriwiriyanont et al. reported HF neogenesis in SEs harboring mouse DPs and human epidermal cells grafted onto nude mice [130]. Others managed to induce HF neogenesis by combining human scalp DP cells and mouse epidermal cells in flap grafts [131] or by injecting human DP cells grown as spheroids, mixed with mouse keratinocytes in "patch" reconstitution assays [132].

Finally, in 2013, Higgins et al. managed to grow human hair in vivo in a hybrid reconstitution/recombination assay by placing DP-like aggregates formed of human scalp DP cells grown as spheroids in hanging drop culture systems between separated epidermis and dermis, which was recombined and transplanted onto immunocompromised mice [95]. In another study, skin substitutes were generated by seeding human neonatal foreskin keratinocytes onto a dermal substrate composed of collagen gel and human scalp DP cells cultured at the air-liquid interface [133]. Two months post transplantation onto immunocompromised mice, HFs of human origin with hair bulbs, DS, concentric layers of IRS and ORS, HR, and sebaceous glands were detected. Leiros et al. developed SEs from cultured human scalp HF epidermal SCs and DP cells seeded onto a devitalized dermal matrix, which were then transplanted onto nude mice. Although no fully mature HF was observed, HF-like structures with concentric layers of human origin were detected [134], indicating that cultured epidermal and dermal cells isolated from the adult human scalp are capable of producing new HFs in vivo.

10. Dermal papilla cell properties are lost in two dimensional cell culture systems

One key step in generating HF-like structures from cultured human DP cells was a change from culturing the DP cells in monolayers to 3-D culture systems, which allowed the formation of spheroids [132]. Similar to Kang et al., Yen et al. reported the formation of keratinocyte-DP hybrid spheroidal microtissues comprising DPs surrounded by epidermal cells by coculturing DP and epidermal cells on a surface that is nonadherent for keratinocytes (EVAL), thereby forcing the epidermal cells to attach to DP cells. These aggregates eventually merged to larger microtissues that detached and formed spheroids capable of inducing hair in vivo [135]. As reported recently, human DP cells lose their innate characteristics in response to an altered gene expression almost immediately upon seeding the cells in a conventional tissue culture dish [95]. In this study, Higgins et al. compared the expression profile of native human DPs and DP cells cultured in monolayers for several passages. Dramatic changes in the transcriptome were seen as soon as DP cells were put in culture, while passaging the cells did not induce prominent changes any more. The changes involved almost 4000 transcripts, many of them involved in HF development. Importantly, by growing human DP cells in hanging drop cultures and thereby creating a microenvironment allowing self-aggregation, spheroids of roughly 3000 cells mimicking physiological DPs formed, which partially rescued the transcriptional changes caused by monolayer cultures and thus the hair-inductive properties (Fig. 4). These human DP spheroids induced human HFs in in vivo reconstitution assays, as mentioned above.

Miao et al. [136] created DP spheroids from cultured DPs on a Matrigel scaffold. As reported by Higgins et al. [95], the DP aggregates restored the transcriptome of normal DPs, which is lost in 2-D culture systems.

Similarly, Ohyama et al. performed a microarray from freshly isolated, microdissected human anagen DP cells and human DP cells cultured in conventional fibroblast medium [137], which also revealed significant differences in the expression pattern and loss of DP signature gene expression upon 2-D culture. This study reported 118 human DP signature genes, which partially overlapped with the DP genes reported from studies performed on pure flow-sorted mouse DP cells [50,138,139]. Analysis of the expression data confirmed upregulation of the WNT, BMP, and FGF signaling pathways in intact DPs. What is more, the authors developed a culture medium containing 6-bromoindirubin-39-oxime, recombinant BMP2, and basic FGF to stimulate these signaling pathways, which not only maintained expression of in situ DP signature genes in primarily cultured human DP cells but also restored DP signature gene expression in conventionally cultured DP cells. Importantly, DP-like cell aggregates cultured in this medium resulted in de novo HFs of human origin when placed between enzymatically separated afollicular murine sole epidermis and dermis and transplanted onto immunocompromised mice [137]. Although the authors did not report generation of new human HFs in vitro and DP cell growth in the so-called "DP activation culture (DPAC)" medium was only moderate [137], the newly developed culture medium provides an enormous advantage as DP cells can be cultured and rapidly expanded under conventional conditions before restoring DP gene signature and function with DPAC medium.

11. A hairy revolution for next-generation skin equivalents

Despite the advances in culturing DP cells in 3-D culture systems that allowed to multiply DP cells while maintaining the hair-inductive properties of DP cells, the studies described above reported de novo generation of human HF-like structures from cultured human epithelial and DC cells only upon transplantation onto immunocompromised mice but not in vitro.

A group of scientists from the Technical University in Berlin has finally managed to grow human hair entirely in the dish [140]. They isolated DP and epidermal cells and melanocytes from dissected human HFs and expanded them into multiple passages. DP cells were kept under low adherent culture conditions resulting in DP-like spheroids, which were subsequently coated with extracellular matrix protein components mimicking the basal membrane, and thus retained the hair-inductive properties. Thereafter, in cocultures, epidermal cells and melanocytes were forced to attach to the DP spheroids, thereby permitting further HF development. The resulting bioengineered microfollicles, which produced hair fibers [140], were inserted into SEs cultured under dynamic perfusion in a chip-based bioreactor (multiorgan chip, MOC) developed by the same group as reported at the European Society for Animal Cell Technology (ESACT) Meeting [141]. This microfluidic chip-based system extended the culture period of microfollicles inserted into SEs in the MOC.

These SEs with integrated, in vitro-generated human microfollicles cultured under dynamic perfusion take us one step closer to mimicking the complexity of the skin and provide a next-generation model system to study epithelial-mesenchymal interactions and many other aspects of HF biology and for the in vitro testing of hair growth modulating agents.

Considering that HF development involves a plethora of molecular and cellular events that are tightly controlled in a spatial and temporal manner, it is not surprising that bioengineered HFs are imperfect or incomplete structures. Nonetheless, these incomplete HF-like structures are valuable for researchers to analyze events in HF morphogenesis and to test various compounds, but what is more, they provide the basis for improved structures yet to be developed. These next-generation SEs that promote de novo HF generation are expected to promote normal skin function, wound healing, and skin appearance and can be used to study human HF morphogenesis and regeneration with cultured adult skin cells.

12. Conclusion

Grafting of hair-inductive dermal tissue is limited since only as many new HFs can be generated as DPs have been obtained from donor tissue. Similarly, characterization of human DPs has been unsatisfactory on the one hand because of the lack of efficient isolation and propagation methods of DP cells and on the other hand due to the loss of innate, hair-inductive characteristics when taken out of the context of their HF microenvironment. Recent advances in culturing DP cells in three-dimensional culture systems allowing the maintenance of the gene signature of DP cells in vitro and thus the hair-inductive properties of these specialized cells offer new opportunities to multiply hair-inductive DP cells isolated from the adult human skin and to grow HF-like structures in vitro. These minifollicles are still far from being optimal but are a promising evidence that by modulating culture conditions, hairy SEs and in vitro-generated human HFs for transplantation will be possible in near future.

For people suffering from alopecia or patients with insufficient HFs to provide DP cells, induced pluripotent stem cell (iPSC) technology in combination with defined protocols to differentiate these cells into DP-like cells may provide an additional cell source for in vitro generation of HFs. Notwithstanding, integration of HFs into bioengineered SEs by incorporating cultured dermal cells with hair-inductive properties will be a key step to recreating a functional skin resembling the in vivo situation. However, to mimic the full complexity of human skin, many other skin-derived cell types including endothelial or nerve cells or adipocytes need to be introduced as well.

Acknowledgments

We are grateful to Drs Kifayathullah Liakath-Ali and Ajay Mishra for providing the images of skin equivalents shown in Fig. 2. This work was supported by grants to B.M.L. from the Austrian Science Fund (FWF, V 525-B28) and the Federation of European Biochemical Societies (FEBS) Follow-up Research Fund.

Competing financial interests

The authors declare no competing financial interests.

Abbreviations

DC dermal condensate
DP dermal papilla
DS dermal sheath
HF hair follicle
EF ectopic follicle
HFOC human follicle organ culture
SC stem cell
SE skin equivalent
MOC multiorgan chip
DEC dermal-epidermal composite

References

[1] Hsu YC, Li L, Fuchs E. Emerging interactions between skin stem cells and their niches. Nat Med 2014;20(8):847–56. PubMed PMID: 25100530. Pubmed Central PMCID: 4358898.
[2] Botchkarev VA, Paus R. Molecular biology of hair morphogenesis: development and cycling. J Exp Zool B Mol Dev Evol 2003;298(1):164–80. PubMed PMID: 12949776.
[3] Schneider MR, Schmidt-Ullrich R, Paus R. The hair follicle as a dynamic miniorgan. Curr Biol 2009;19(3):R132–42. PubMed PMID: 19211055.
[4] Fuchs E. Scratching the surface of skin development. Nature 2007;445(7130):834–42. PubMed PMID: 17314969. Pubmed Central PMCID: 2405926. Epub 2007/02/23.eng.

[5] Sieber-Blum M, Grim M, Hu YF, Szeder V. Pluripotent neural crest stem cells in the adult hair follicle. Dev Dyn: Off Publ Am Assoc Anat 2004;231(2):258–69. PubMed PMID: 15366003.
[6] Christiano AM. Epithelial stem cells: stepping out of their niche. Cell 2004;118(5):-530–2. PubMed PMID: 15339656.
[7] Sennett R, Rendl M. Mesenchymal-epithelial interactions during hair follicle morphogenesis and cycling. Semin Cell Dev Biol 2012;23(8):917–27. PubMed PMID: 22960356. Pubmed Central PMCID: 3496047.
[8] Fuchs E, Nowak JA. Building epithelial tissues from skin stem cells. Cold Spring Harb Symp Quant Biol 2008;73:333–50. PubMed PMID: 19022769. Pubmed Central PMCID: 2693088.
[9] Alonso L, Fuchs E. Stem cells of the skin epithelium. Proc Natl Acad Sci U S A 2003;100 Suppl 1:11830–5. PubMed PMID: 12913119. Pubmed Central PMCID: 304094.
[10] Blanpain C, Fuchs E. Stem cell plasticity. Plasticity of epithelial stem cells in tissue regeneration. Science. 2014;344(6189):1242281. PubMed PMID: 24926024. Pubmed Central PMCID: 4523269.
[11] Arwert EN, Hoste E, Watt FM. Epithelial stem cells, wound healing and cancer. Nat Rev Cancer 2012;12(3):170–80. PubMed PMID: 22362215.
[12] Hsu YC, Pasolli HA, Fuchs E. Dynamics between stem cells, niche, and progeny in the hair follicle. Cell 2011;144(1):92–105. PubMed PMID: 21215372. Pubmed Central PMCID: 3050564.
[13] Fujiwara H, Ferreira M, Donati G, Marciano DK, Linton JM, Sato Y, et al. The basement membrane of hair follicle stem cells is a muscle cell niche. Cell 2011;144(4):577–89. PubMed PMID: 21335239. Pubmed Central PMCID: 3056115.
[14] Mesa KR, Rompolas P, Greco V. The dynamic duo: niche/stem cell interdependency. Stem Cell Rep 2015;4(6):961–6. PubMed PMID: 26028534. Pubmed Central PMCID: 4471832.
[15] Meyer KC, Klatte JE, Dinh HV, Harries MJ, Reithmayer K, Meyer W, et al. Evidence that the bulge region is a site of relative immune privilege in human hair follicles. Br J Dermatol 2008;159(5):1077–85. PubMed PMID: 18795933.
[16] Paus R, Bertolini M. The role of hair follicle immune privilege collapse in alopecia areata: status and perspectives. J Invest Dermatol Symp Proc 2013;16(1):S25–7. PubMed PMID: 24326544.
[17] Paus R, Foitzik K. In search of the "hair cycle clock": a guided tour. Differ Res Biol Divers 2004;72(9–10):489–511. PubMed PMID: 15617561.
[18] Philpott MP, Kealey T. Metabolic studies on isolated hair follicles: hair follicles engage in aerobic glycolysis and do not demonstrate the glucose fatty acid cycle. J Invest Dermatol 1991;96(6):875–9. PubMed PMID: 2045676.
[19] Hendrix S, Picker B, Liezmann C, Peters EM. Skin and hair follicle innervation in experimental models: a guide for the exact and reproducible evaluation of neuronal plasticity. Exp Dermatol 2008;17(3):214–27. PubMed PMID: 18261087.
[20] Nishimura EK. Melanocyte stem cells: a melanocyte reservoir in hair follicles for hair and skin pigmentation. Pigment Cell Melanoma Res 2011;24(3):401–10. PubMed PMID: 21466661.
[21] Paus R, Langan EA, Vidali S, Ramot Y, Andersen B. Neuroendocrinology of the hair follicle: principles and clinical perspectives. Trends Mol Med 2014;20(10):559–70. PubMed PMID: 25066729.
[22] Donati G, Proserpio V, Lichtenberger BM, Natsuga K, Sinclair R, Fujiwara H, et al. Epidermal Wnt/beta-catenin signaling regulates adipocyte differentiation via secretion

of adipogenic factors. Proc Natl Acad Sci U S A 2014;111(15):E1501–9. PubMed PMID: 24706781. Pubmed Central PMCID: 3992657.
[23] Paus R, van der Veen C, Eichmuller S, Kopp T, Hagen E, Muller-Rover S, et al. Generation and cyclic remodeling of the hair follicle immune system in mice. J Invest Dermatol 1998;111(1):7–18. PubMed PMID: 9665380.
[24] Botchkarev VA, Eichmuller S, Johansson O, Paus R. Hair cycle-dependent plasticity of skin and hair follicle innervation in normal murine skin. J Comp Neurol 1997;386(3):379–95. PubMed PMID: 9303424.
[25] Mecklenburg L, Tobin DJ, Muller-Rover S, Handjiski B, Wendt G, Peters EM, et al. Active hair growth (anagen) is associated with angiogenesis. J Invest Dermatol 2000;114(5):909–16. PubMed PMID: 10771470.
[26] Green H, Rheinwald JG, Sun TT. Properties of an epithelial cell type in culture: the epidermal keratinocyte and its dependence on products of the fibroblast. Prog Clin Biol Res 1977;17:493–500. PubMed PMID: 928463.
[27] Rheinwald JG, Green H. Epidermal growth factor and the multiplication of cultured human epidermal keratinocytes. Nature 1977;265(5593):421–4. PubMed PMID: 299924.
[28] Rheinwald JG, Green H. Serial cultivation of strains of human epidermal keratinocytes: the formation of keratinizing colonies from single cells. Cell 1975;6(3):331–43. PubMed PMID: 1052771.
[29] Rheinwald JG, Green H. Formation of a keratinizing epithelium in culture by a cloned cell line derived from a teratoma. Cell 1975;6(3):317–30. PubMed PMID: 1052770.
[30] Philpott MP, Green MR, Kealey T. Human hair growth in vitro. J Cell Sci 1990;97 (Pt 3):463–71. PubMed PMID: 1705941.
[31] Philpott MP, Green MR, Kealey T. Rat hair follicle growth in vitro. Br J Dermatol 1992;127(6):600–7. PubMed PMID: 1476919.
[32] Philpott MP, Sanders D, Kealey T. Cultured human hair follicles and growth factors. J Invest Dermatol 1995;104(5 Suppl):44S-5S. PubMed PMID: 7738396.
[33] Philpott MP, Sanders D, Westgate GE, Kealey T. Human hair growth in vitro: a model for the study of hair follicle biology. J Dermatol Sci 1994;7 Suppl:S55–72. PubMed PMID: 7999676.
[34] Philpott MP, Sanders DA, Kealey T. Whole hair follicle culture. Dermatol Clin 1996;14(4):595–607. PubMed PMID: 9238319.
[35] Philpott MP, Westgate GE, Kealey T. An in vitro model for the study of human hair growth. Ann N Y Acad Sci 1991;642:148–64. PubMed PMID: 1725579 [discussion 64–6].
[36] Balana ME, Charreau HE, Leiros GJ. Epidermal stem cells and skin tissue engineering in hair follicle regeneration. World J Stem Cells 2015;7(4):711–27. PubMed PMID: 26029343. Pubmed Central PMCID: 4444612.
[37] Schmidt-Ullrich R, Paus R. Molecular principles of hair follicle induction and morphogenesis. BioEssays: News Rev Mol Cell Dev Biolo 2005;27(3):247–61. PubMed PMID: 15714560.
[38] Veniaminova NA, Vagnozzi AN, Kopinke D, Do TT, Murtaugh LC, Maillard I, et al. Keratin 79 identifies a novel population of migratory epithelial cells that initiates hair canal morphogenesis and regeneration. Development 2013;140(24):4870–80. PubMed PMID: 24198274. Pubmed Central PMCID: 3848186.
[39] Botchkarev VA, Paus R. Molecular biology of hair morphogenesis: development and cycling. J Exp Zool B Mol Dev Evol 2003;298(1):164–80. PubMed PMID: 12949776 [Epub 2003/09/02.eng].
[40] Fuchs E, Horsley V. More than one way to skin. Genes Dev 2008;22(8):976–85. PubMed PMID: 18413712 [Epub 2008/04/17.eng].

[41] Rompolas P, Greco V. Stem cell dynamics in the hair follicle niche. Semin Cell Dev Biol. 2014;25–26:34–42. PubMed PMID: 24361866. Pubmed Central PMCID: 3988239.
[42] Alonso L, Fuchs E. The hair cycle. J Cell Sci 2006;119(Pt 3):391–3. PubMed PMID: 16443746 [Epub 2006/01/31.eng].
[43] Muller-Rover S, Handjiski B, van der Veen C, Eichmuller S, Foitzik K, McKay IA, et al. A comprehensive guide for the accurate classification of murine hair follicles in distinct hair cycle stages. J Invest Dermatol 2001;117(1):3–15. PubMed PMID: 11442744.
[44] Paus R, Cotsarelis G. The biology of hair follicles. N Engl J Med 1999;341(7):491–7. PubMed PMID: 10441606.
[45] Fuchs E. Skin stem cells: rising to the surface. J Cell Biol 2008;180(2):273–84. PubMed PMID: 18209104. Pubmed Central PMCID: 2213592.
[46] Grisanti L, Clavel C, Cai X, Rezza A, Tsai SY, Sennett R, et al. Tbx18 targets dermal condensates for labeling, isolation, and gene ablation during embryonic hair follicle formation. J Invest Dermatol 2013;133(2):344–53. PubMed PMID: 22992803. Pubmed Central PMCID: 3530628.
[47] Kaushal GS, Rognoni E, Lichtenberger BM, Driskell RR, Kretzschmar K, Hoste E, et al. Fate of prominin-1 expressing dermal papilla cells during homeostasis, wound healing and Wnt activation. J Invest Dermatol 2015;135(12):2926–34. PubMed PMID: 26288357. Pubmed Central PMCID: 4650270.
[48] Rezza A, Sennett R, Tanguy M, Clavel C, Rendl M. PDGF signalling in the dermis and in dermal condensates is dispensable for hair follicle induction and formation. Exp Dermatol 2015;24(6):468–70. PubMed PMID: 25708924. Pubmed Central PMCID: 4943754.
[49] Enshell-Seijffers D, Lindon C, Kashiwagi M, Morgan BA. Beta-catenin activity in the dermal papilla regulates morphogenesis and regeneration of hair. Dev Cell 2010;18(4):633–42. PubMed PMID: 20412777. Pubmed Central PMCID: 2893731.
[50] Rendl M, Lewis L, Fuchs E. Molecular dissection of mesenchymal-epithelial interactions in the hair follicle. PLoS Biol 2005;3(11):e331. PubMed PMID: 16162033. Pubmed Central PMCID: 1216328.
[51] Zhang Y, Tomann P, Andl T, Gallant NM, Huelsken J, Jerchow B, et al. Reciprocal requirements for EDA/EDAR/NF-kappaB and Wnt/beta-catenin signaling pathways in hair follicle induction. Dev Cell 2009;17(1):49–61. PubMed PMID: 19619491. Pubmed Central PMCID: 2859042.
[52] Huh SH, Narhi K, Lindfors PH, Haara O, Yang L, Ornitz DM, et al. Fgf20 governs formation of primary and secondary dermal condensations in developing hair follicles. Genes Dev 2013;27(4):450–8. PubMed PMID: 23431057. Pubmed Central PMCID: 3589561.
[53] Collins CA, Kretzschmar K, Watt FM. Reprogramming adult dermis to a neonatal state through epidermal activation of beta-catenin. Development 2011;138(23):5189–99. PubMed PMID: 22031549. Pubmed Central PMCID: 3210498.
[54] Karlsson L, Bondjers C, Betsholtz C. Roles for PDGF-A and sonic hedgehog in development of mesenchymal components of the hair follicle. Development 1999;126(12):2611–21. PubMed PMID: 10331973.
[55] Woo WM, Zhen HH, Oro AE. Shh maintains dermal papilla identity and hair morphogenesis via a Noggin-Shh regulatory loop. Genes Dev 2012;26(11):1235–46. PubMed PMID: 22661232. Pubmed Central PMCID: 3371411.
[56] Millar SE. Molecular mechanisms regulating hair follicle development. J Invest Dermatol 2002;118(2):216–25. PubMed PMID: 11841536.
[57] Lichtenberger BM, Mastrogiannaki M, Watt FM. Epidermal beta-catenin activation remodels the dermis via paracrine signalling to distinct fibroblast lineages. Nat Commun 2016 Feb 03;7:10537. PubMed PMID: 26837596. Pubmed Central PMCID: 4742837.

[58] Driskell RR, Lichtenberger BM, Hoste E, Kretzschmar K, Simons BD, Charalambous M, et al. Distinct fibroblast lineages determine dermal architecture in skin development and repair. Nature 2013;504(7479):277–81. PubMed PMID: 24336287. Pubmed Central PMCID: 3868929.

[59] Yang CC, Cotsarelis G. Review of hair follicle dermal cells. J Dermatol Sci 2010;57(1):-2–11. PubMed PMID: 20022473. Pubmed Central PMCID: 2818774.

[60] Leavitt T, Hu MS, Marshall CD, Barnes LA, Lorenz HP, Longaker MT. Scarless wound healing: finding the right cells and signals. Cell Tissue Res 2016;365(3):483–93. PubMed PMID: 27256396. Pubmed Central PMCID: 5010960.

[61] Billingham RE, Russell PS. Incomplete wound contracture and the phenomenon of hair neogenesis in rabbits' skin. Nature 1956;177(4513):791–2. PubMed PMID: 13321965.

[62] Breedis C. Regeneration of hair follicles and sebaceous glands from the epithelium of scars in the rabbit. Cancer Res 1954;14(8):575–9. PubMed PMID: 13199800.

[63] Ito M, Yang Z, Andl T, Cui C, Kim N, Millar SE, et al. Wnt-dependent de novo hair follicle regeneration in adult mouse skin after wounding. Nature 2007;447(7142):316–20. PubMed PMID: 17507982.

[64] Rognoni E, Gomez C, Pisco AO, Rawlins EL, Simons BD, Watt FM, et al. Inhibition of beta-catenin signalling in dermal fibroblasts enhances hair follicle regeneration during wound healing. Development 2016;143(14):2522–35. PubMed PMID: 27287810. Pubmed Central PMCID: 4958333.

[65] Silva-Vargas V, Lo Celso C, Giangreco A, Ofstad T, Prowse DM, Braun KM, et al. Beta-catenin and Hedgehog signal strength can specify number and location of hair follicles in adult epidermis without recruitment of bulge stem cells. Dev Cell 2005;9(1):121–31. PubMed PMID: 15992546.

[66] Jahoda CA. Induction of follicle formation and hair growth by vibrissa dermal papillae implanted into rat ear wounds: vibrissa-type fibres are specified. Development 1992;115(4):1103–9. PubMed PMID: 1451660.

[67] Jahoda CA, Horne KA, Oliver RF. Induction of hair growth by implantation of cultured dermal papilla cells. Nature. 1984 Oct 11–17;311(5986):560–2. PubMed PMID: 6482967.

[68] Oliver RF. The experimental induction of whisker growth in the hooded rat by implantation of dermal papillae. J Embryol Exp Morphol 1967;18(1):43–51. PubMed PMID: 6048979.

[69] Oliver RF. The induction of hair follicle formation in the adult hooded rat by vibrissa dermal papillae. J Embryol Exp Morphol 1970;23(1):219–36. PubMed PMID: 4926619.

[70] Oliver RF. Histological studies of whisker regeneration in the hooded rat. J Embryol Exp Morphol 1966;16(2):231–44. PubMed PMID: 5971987.

[71] Jahoda CA, Horne KA, Mauger A, Bard S, Sengel P. Cellular and extracellular involvement in the regeneration of the rat lower vibrissa follicle. Development 1992;114(4):887–97. PubMed PMID: 1618150.

[72] Kobayashi K, Nishimura E. Ectopic growth of mouse whiskers from implanted lengths of plucked vibrissa follicles. J Invest Dermatol 1989;92(2):278–82. PubMed PMID: 2918234.

[73] Reynolds AJ, Jahoda CA. Hair matrix germinative epidermal cells confer follicle-inducing capabilities on dermal sheath and high passage papilla cells. Development 1996;122(10):3085–94. PubMed PMID: 8898222.

[74] Jahoda CA, Oliver RF, Reynolds AJ, Forrester JC, Gillespie JW, Cserhalmi-Friedman PB, et al. Trans-species hair growth induction by human hair follicle dermal papillae. Exp Dermatol 2001;10(4):229–37. PubMed PMID: 11493311.

[75] Reynolds AJ, Lawrence C, Cserhalmi-Friedman PB, Christiano AM, Jahoda CA. Transgender induction of hair follicles. Nature 1999;402(6757):33–4. PubMed PMID: 10573414.
[76] Paus R, Ito N, Takigawa M, Ito T. The hair follicle and immune privilege. J Invest Dermatol Symp Proc 2003;8(2):188–94. PubMed PMID: 14582671.
[77] Rahmani W, Abbasi S, Hagner A, Raharjo E, Kumar R, Hotta A, et al. Hair follicle dermal stem cells regenerate the dermal sheath, repopulate the dermal papilla, and modulate hair type. Dev Cell 2014;31(5):543–58. PubMed PMID: 25465495.
[78] Chi W, Wu E, Morgan BA. Dermal papilla cell number specifies hair size, shape and cycling and its reduction causes follicular decline. Development 2013;140(8):1676–83. PubMed PMID: 23487317. Pubmed Central PMCID: 3621486.
[79] Zhou L, Xu M, Yang Y, Yang K, Wickett RR, Andl T, et al. Activation of beta-catenin signaling in CD133-positive dermal papilla cells drives postnatal hair growth. PLoS One 2016;11(7):e0160425. PubMed PMID: 27472062. Pubmed Central PMCID: 4966972.
[80] Zhou L, Yang K, Xu M, Andl T, Millar SE, Boyce S, et al. Activating beta-catenin signaling in CD133-positive dermal papilla cells increases hair inductivity. FEBS J 2016;283(15):2823–35. PubMed PMID: 27312243. Pubmed Central PMCID: 4975668.
[81] Rezza A, Wang Z, Sennett R, Qiao W, Wang D, Heitman N, et al. Signaling networks among stem cell precursors, transit-amplifying progenitors, and their niche in developing hair follicles. Cell Rep 2016;14(12):3001–18. PubMed PMID: 27009580. Pubmed Central PMCID: 4826467.
[82] Driskell RR, Jahoda CA, Chuong CM, Watt FM, Horsley V. Defining dermal adipose tissue. Exp Dermatol 2014;23(9):629–31. PubMed PMID: 24841073. Pubmed Central PMCID: 4282701.
[83] Tsai SY, Clavel C, Kim S, Ang YS, Grisanti L, Lee DF, et al. Oct4 and klf4 reprogram dermal papilla cells into induced pluripotent stem cells. Stem Cells 2010;28(2):221–8. PubMed PMID: 20014278.
[84] Higgins CA, Itoh M, Inoue K, Richardson GD, Jahoda CA, Christiano AM. Reprogramming of human hair follicle dermal papilla cells into induced pluripotent stem cells. J Invest Dermatol 2012;132(6):1725–7. PubMed PMID: 22336943.
[85] Garza LA, Yang CC, Zhao T, Blatt HB, Lee M, He H, et al. Bald scalp in men with androgenetic alopecia retains hair follicle stem cells but lacks CD200-rich and CD34-positive hair follicle progenitor cells. J Clin Invest 2011;121(2):613–22. PubMed PMID: 21206086. Pubmed Central PMCID: 3026732.
[86] Chew EG, Tan JH, Bahta AW, Ho BS, Liu X, Lim TC, et al. Differential expression between human dermal papilla cells from balding and non-balding scalps reveals new candidate genes for androgenetic alopecia. J Invest Dermatol 2016;136(8):1559–67. PubMed PMID: 27060448.
[87] Randall VA, Sundberg JP, Philpott MP. Animal and in vitro models for the study of hair follicles. J Invest Dermatol Symp Proc 2003;8(1):39–45. PubMed PMID: 12894993.
[88] Nakamura M, Schneider MR, Schmidt-Ullrich R, Paus R. Mutant laboratory mice with abnormalities in hair follicle morphogenesis, cycling, and/or structure: an update. J Dermatol Sci 2013;69(1):6–29. PubMed PMID: 23165165.
[89] Liakath-Ali K, Vancollie VE, Heath E, Smedley DP, Estabel J, Sunter D, et al. Novel skin phenotypes revealed by a genome-wide mouse reverse genetic screen. Nat Commun 2014;5:3540. PubMed PMID: 24721909. Pubmed Central PMCID: 3996542.
[90] Randall VA, Ebling FJ. Seasonal changes in human hair growth. Br J Dermatol 1991;124(2):146–51. PubMed PMID: 2003996.

[91] Randall VA, Thornton MJ, Hamada K, Redfern CP, Nutbrown M, Ebling FJ, et al. Androgens and the hair follicle. Cultured human dermal papilla cells as a model system. Ann N Y Acad Sci 1991;642:355–75. PubMed PMID: 1809092.

[92] Langan EA, Philpott MP, Kloepper JE, Paus R. Human hair follicle organ culture: theory, application and perspectives. Exp Dermatol 2015;24(12):903–11. PubMed PMID: 26284830.

[93] Jensen KB, Driskell RR, Watt FM. Assaying proliferation and differentiation capacity of stem cells using disaggregated adult mouse epidermis. Nat Protoc 2010;5(5):898–911. PubMed PMID: 20431535.

[94] Limat A, Hunziker T. Cultivation of keratinocytes from the outer root sheath of human hair follicles. Methods Mol Med 1996;2:21–31. PubMed PMID: 21359730.

[95] Higgins CA, Chen JC, Cerise JE, Jahoda CA, Christiano AM. Microenvironmental reprogramming by three-dimensional culture enables dermal papilla cells to induce de novo human hair-follicle growth. Proc Natl Acad Sci U S A 2013;110(49):19679–88. PubMed PMID: 24145441. Pubmed Central PMCID: 3856847.

[96] Tobin DJ, Colen SR, Bystryn JC. Isolation and long-term culture of human hair-follicle melanocytes. J Invest Dermatol 1995;104(1):86–9. PubMed PMID: 7528247.

[97] Zouboulis CC, Xia LQ, Detmar M, Bogdanoff B, Giannakopoulos G, Gollnick H, et al. Culture of human sebocytes and markers of sebocytic differentiation in vitro. Skin Pharmacol: Off J Skin Pharmacol Soc 1991;4(2):74–83. PubMed PMID: 1715175.

[98] Limat A, Hunziker T, Waelti ER, Inaebnit SP, Wiesmann U, Braathen LR. Soluble factors from human hair papilla cells and dermal fibroblasts dramatically increase the clonal growth of outer root sheath cells. Arch Dermatol Res 1993;285(4):205–10. PubMed PMID: 8342964.

[99] Fujie T, Katoh S, Oura H, Urano Y, Arase S. The chemotactic effect of a dermal papilla cell-derived factor on outer root sheath cells. J Dermatol Sci 2001;25(3):206–12. PubMed PMID: 11240268.

[100] Roh C, Tao Q, Lyle S. Dermal papilla-induced hair differentiation of adult epithelial stem cells from human skin. Physiol Genomics 2004;19(2):207–17. PubMed PMID: 15292489.

[101] Leiros GJ, Attorresi AI, Balana ME. Hair follicle stem cell differentiation is inhibited through cross-talk between Wnt/beta-catenin and androgen signalling in dermal papilla cells from patients with androgenetic alopecia. Br J Dermatol 2012;166(5):1035–42. PubMed PMID: 22283397.

[102] Wright KA, Nadire KB, Busto P, Tubo R, McPherson JM, Wentworth BM. Alternative delivery of keratinocytes using a polyurethane membrane and the implications for its use in the treatment of full-thickness burn injury. Burns: J Int Soc Burn Inj 1998;24(1):7–17. PubMed PMID: 9601584.

[103] Bidic SM, Dauwe PB, Heller J, Brown S, Rohrich RJ. Reconstructing large keloids with neodermis: a systematic review. Plast Reconstr Surg. 2012;129(2):380e-382e. PubMed PMID: 22286474.

[104] Barber C, Watt A, Pham C, Humphreys K, Penington A, Mutimer K, et al. Influence of bioengineered skin substitutes on diabetic foot ulcer and venous leg ulcer outcomes. J Wound Care 2008;17(12):517–27. PubMed PMID: 19052516.

[105] Limat A, Mauri D, Hunziker T. Successful treatment of chronic leg ulcers with epidermal equivalents generated from cultured autologous outer root sheath cells. J Invest Dermatol 1996;107(1):128–35. PubMed PMID: 8752851.

[106] El Ghalbzouri A, Commandeur S, Rietveld MH, Mulder AA, Willemze R. Replacement of animal-derived collagen matrix by human fibroblast-derived dermal matrix for

human skin equivalent products. Biomaterials 2009;30(1):71–8. PubMed PMID: 18838164.

[107] Schlabe J, Johnen C, Schwartlander R, Moser V, Hartmann B, Gerlach JC, et al. Isolation and culture of different epidermal and dermal cell types from human scalp suitable for the development of a therapeutical cell spray. Burns: J Int Soc Burn Inj 2008;34(3):376–84. PubMed PMID: 17869000.

[108] Liu P, Deng Z, Han S, Liu T, Wen N, Lu W, et al. Tissue-engineered skin containing mesenchymal stem cells improves burn wounds. Artif Organs 2008;32(12):925–31. PubMed PMID: 19133020.

[109] Li H, Chu Y, Zhang Z, Zhang G, Jiang L, Wu H, et al. Construction of bilayered tissue-engineered skin with human amniotic mesenchymal cells and human amniotic epithelial cells. Artif Organs 2012;36(10):911–9. PubMed PMID: 22607197.

[110] Mine S, Fortunel NO, Pageon H, Asselineau D. Aging alters functionally human dermal papillary fibroblasts but not reticular fibroblasts: a new view of skin morphogenesis and aging. PLoS One 2008;3(12):e4066. PubMed PMID: 19115004. Pubmed Central PMCID: 2605251.

[111] Shamis Y, Hewitt KJ, Carlson MW, Margvelashvilli M, Dong S, Kuo CK, et al. Fibroblasts derived from human embryonic stem cells direct development and repair of 3D human skin equivalents. Stem Cell Res Ther 2011;2(1):10. PubMed PMID: 21338517. Pubmed Central PMCID: 3092150.

[112] Higgins CA, Roger M, Hill R, Ali-Khan AS, Garlick J, Christiano AM, et al. Multifaceted role of hair follicle dermal cells in bioengineered skins. Br J Dermatol. 2016. PubMed PMID: 27679975.

[113] Black AF, Berthod F, L'Heureux N, Germain L, Auger FA. In vitro reconstruction of a human capillary-like network in a tissue-engineered skin equivalent. FASEB J: Off Publ Fed Am Soc Exp Biol 1998;12(13):1331–40. PubMed PMID: 9761776.

[114] Freiman A, Shandalov Y, Rozenfeld D, Shor E, Segal S, Ben-David D, et al. Adipose-derived endothelial and mesenchymal stem cells enhance vascular network formation on three-dimensional constructs in vitro. Stem Cell Res Ther 2016;7:5. PubMed PMID: 26753517. Pubmed Central PMCID: 4709933.

[115] Chan RK, Zamora DO, Wrice NL, Baer DG, Renz EM, Christy RJ, et al. Development of a vascularized skin construct using adipose-derived stem cells from debrided burned skin. Stem Cells Int 2012;2012:841203. PubMed PMID: 22848228. Pubmed Central PMCID: 3399490.

[116] Collawn SS, Banerjee NS, de la Torre J, Vasconez L, Chow LT. Adipose-derived stromal cells accelerate wound healing in an organotypic raft culture model. Ann Plast Surg 2012;68(5):501–4. PubMed PMID: 22510896. Pubmed Central PMCID: 3477580.

[117] Mori N, Morimoto Y, Takeuchi S. Skin integrated with perfusable vascular channels on a chip. Biomaterials 2017;116:48–56. PubMed PMID: 27914266.

[118] Bellas E, Seiberg M, Garlick J, Kaplan DL. In vitro 3D full-thickness skin-equivalent tissue model using silk and collagen biomaterials. Macromol Biosci 2012;12(12):1627–36. PubMed PMID: 23161763. Pubmed Central PMCID: 3724336.

[119] Frueh FS, Menger MD, Lindenblatt N, Giovanoli P, Laschke MW. Current and emerging vascularization strategies in skin tissue engineering. Crit Rev Biotechnol 2016 20:1–13. PubMed PMID: 27439727.

[120] Jimenez F, Garde C, Poblet E, Jimeno B, Ortiz J, Martinez ML, et al. A pilot clinical study of hair grafting in chronic leg ulcers Wound Repair Regener: Off Publ Wound Heal Soc Eur Tissue Repair Soc 2012;20(6):806–14. PubMed PMID: 23110506.

[121] Ansell DM, Kloepper JE, Thomason HA, Paus R, Hardman MJ. Exploring the "hair growth-wound healing connection": anagen phase promotes wound re-epithelialization. J Invest Dermatol 2011;131(2):518–28. PubMed PMID: 20927125.
[122] Blanpain C, Lowry WE, Geoghegan A, Polak L, Fuchs E. Self-renewal, multipotency, and the existence of two cell populations within an epithelial stem cell niche. Cell 2004;118(5):635–48. PubMed PMID: 15339667.
[123] Kamimura J, Lee D, Baden HP, Brissette J, Dotto GP. Primary mouse keratinocyte cultures contain hair follicle progenitor cells with multiple differentiation potential. J Invest Dermatol 1997;109(4):534–40. PubMed PMID: 9326386.
[124] Lichti U, Weinberg WC, Goodman L, Ledbetter S, Dooley T, Morgan D, et al. In vivo regulation of murine hair growth: insights from grafting defined cell populations onto nude mice. J Invest Dermatol 1993;101(1 Suppl):124S-129S. PubMed PMID: 8326145.
[125] Zheng Y, Du X, Wang W, Boucher M, Parimoo S, Stenn K. Organogenesis from dissociated cells: generation of mature cycling hair follicles from skin-derived cells. J Invest Dermatol 2005;124(5):867–76. PubMed PMID: 15854024.
[126] Ohyama M, Zheng Y, Paus R, Stenn KS. The mesenchymal component of hair follicle neogenesis: background, methods and molecular characterization. Exp Dermatol 2010;19(2):89–99. PubMed PMID: 19650868.
[127] Ehama R, Ishimatsu-Tsuji Y, Iriyama S, Ideta R, Soma T, Yano K, et al. Hair follicle regeneration using grafted rodent and human cells. J Invest Dermatol 2007;127(9):2106–15. PubMed PMID: 17429436.
[128] Chermnykh ES, Vorotelyak EA, Gnedeva KY, Moldaver MV, Yegorov YE, Vasiliev AV, et al. Dermal papilla cells induce keratinocyte tubulogenesis in culture. Histochem Cell Biol 2010;133(5):567–76. PubMed PMID: 20336308.
[129] Havlickova B, Biro T, Mescalchin A, Tschirschmann M, Mollenkopf H, Bettermann A, et al. A human folliculoid microsphere assay for exploring epithelial-mesenchymal interactions in the human hair follicle. J Invest Dermatol 2009;129(4):972–83. PubMed PMID: 18923448.
[130] Sriwiriyanont P, Lynch KA, Maier EA, Hahn JM, Supp DM, Boyce ST. Morphogenesis of chimeric hair follicles in engineered skin substitutes with human keratinocytes and murine dermal papilla cells. Exp Dermatol 2012;21(10):783–5. PubMed PMID: 23078401.
[131] Qiao J, Zawadzka A, Philips E, Turetsky A, Batchelor S, Peacock J, et al. Hair follicle neogenesis induced by cultured human scalp dermal papilla cells. Regen Med 2009;4(5):667–76. PubMed PMID: 19761392.
[132] Kang BM, Kwack MH, Kim MK, Kim JC, Sung YK. Sphere formation increases the ability of cultured human dermal papilla cells to induce hair follicles from mouse epidermal cells in a reconstitution assay. J Invest Dermatol 2012;132(1):237–9. PubMed PMID: 21850026.
[133] Thangapazham RL, Klover P, Wang JA, Zheng Y, Devine A, Li S, et al. Dissociated human dermal papilla cells induce hair follicle neogenesis in grafted dermal-epidermal composites. J Invest Dermatol 2014;134(2):538–40. PubMed PMID: 23924901. Pubmed Central PMCID: 3947143.
[134] Leiros GJ, Kusinsky AG, Drago H, Bossi S, Sturla F, Castellanos ML, et al. Dermal papilla cells improve the wound healing process and generate hair bud-like structures in grafted skin substitutes using hair follicle stem cells. Stem Cells Transl Med 2014;3(10):1209–19. PubMed PMID: 25161315. Pubmed Central PMCID: 4181392.
[135] Yen CM, Chan CC, Lin SJ. High-throughput reconstitution of epithelial-mesenchymal interaction in folliculoid microtissues by biomaterial-facilitated self-assembly of dis-

sociated heterotypic adult cells. Biomaterials 2010;31(15):4341–52. PubMed PMID: 20206989.

[136] Miao Y, Sun YB, Liu BC, Jiang JD, Hu ZQ. Controllable production of transplantable adult human high-passage dermal papilla spheroids using 3D matrigel culture. Tissue Eng A 2014;20(17–18):2329–38. PubMed PMID: 24528213. Pubmed Central PMCID: 4161057.

[137] Ohyama M, Kobayashi T, Sasaki T, Shimizu A, Amagai M. Restoration of the intrinsic properties of human dermal papilla in vitro. J Cell Sci 2012;125(Pt 17):4114–25. PubMed PMID: 22623722.

[138] Driskell RR, Clavel C, Rendl M, Watt FM. Hair follicle dermal papilla cells at a glance. J Cell Sci 2011;124(Pt 8):1179–82. PubMed PMID: 21444748. Pubmed Central PMCID: 3115771.

[139] Driskell RR, Giangreco A, Jensen KB, Mulder KW, Watt FM. Sox2-positive dermal papilla cells specify hair follicle type in mammalian epidermis. Development 2009;136(16):2815–23. PubMed PMID: 19605494. Pubmed Central PMCID: 2730408.

[140] Lindner G, Horland R, Wagner I, Atac B, Lauster R. De novo formation and ultrastructural characterization of a fiber-producing human hair follicle equivalent in vitro. J Biotechnol 2011;152(3):108–12. PubMed PMID: 21277344.

[141] Wagner I, Atac B, Lindner G, Horland R, Busek M, Sonntag F, et al. Skin and hair-on-a-chip: hair and skin assembly versus native skin maintenance in a chip-based perfusion system. BMC Proc 2013;7(6):93.

In vitro models to study cutaneous innervation mechanisms

Nicolas Lebonvallet, Christelle Le Gall-Ianotto, Jérémy Chéret, Raphaël Leschiera, Matthieu Talagas, Raphaële Le Garrec, Virginie Buhé, Killian L'Hérondelle, Olivier Gouin, Mehdi Sakka, Nicholas Boulais, Ulysse Pereira, Jean-Luc Carré, Laurent Misery
Laboratory of Neurosciences of Brest (EA4685), University of Brest, Brest, France

1. Introduction

Skin innervation includes a wide variety of sensory end organs, among which intraepidermal free nerve endings, dermal free nerve endings and dermal nerve endings associated with (sensory) corpuscules [1]. These nerve endings allow skin sensoriality and included multiple receptor types (e.g., thermoreceptors, mechanoreceptors, nociceptors, and pruriceptors) that act through sensory proteins such as those in the transient receptor potential (TRP) family [2]. Furthermore, neurotransmitter, growth factors, and cytokine secretion also occur, which mediate the control of skin functions and skin pathologies via the nervous system [3–5]. Conversely, the skin produces growth factors and neuromediators that guide neuronal growth within the skin [5–8]. The density of innervation, the nerve regrowth, the maintenance of innervation, the survival of neurons, and the formation of dendritic trees are all influenced by a complex balance among growth factors, neurotrophins, and semaphorins [6,9,10]. To grow in the extracellular matrix, neurons must produce matrix proteases. The cellular mechanisms that govern neurocutaneous interactions in this communication are dynamic and multilateral [7,11]. Furthermore, nerve endings and skin cells share the same receptors, mediators, and anatomical connections, as well as many of the same interactions. Hence, they constitute the neuro-immuno-cutaneous system (NICS) [12].

The interactions between the neurons and skin are increasingly apparent [12,13]. Because these interactions play very important roles in physiological and pathological conditions, there is a need to understand the mechanisms that govern the NICS at molecular, cellular, and tissue levels. For these reasons, in vitro models are required [14]. Over the last decade, new, interesting, and sophisticated in vitro models have been developed. However, there is a need to adapt these models to better understand the mechanisms underlying innervation and the interactions between the neurons and skin. These interactions are particularly difficult to study because the structure of skin innervation must be reconstructed in vitro. Indeed, nerve endings rapidly lose their connection with skin cells when the neurons are cut, which is induced by collecting skin samples. The study of neurons in vitro without skin

Skin Tissue Models. https://doi.org/10.1016/B978-0-12-810545-0.00013-9

cells is feasible but is not sufficient to understand all the mechanisms underlying cutaneous innervation. Thus, coculture and 3-D culture methods were developed. However, such studies are extremely limited for human neurons. Current models use primary cells and cell lines, but future studies will likely utilize stem cells that have been differentiated into sensory neurons to replace animal and human neurons and neuronal cell lines [15–18].

The aim of this review is to present and explain the importance of these different models for studying mechanisms of innervation and neurocutaneous interactions.

2. Models

In vitro models used to study cutaneous innervation and interactions in the NICS may be a monoculture or multicellular or three-dimensional (3-D). Monoculture models consist of one skin cell type (including neurons) cultured alone. Multicellular models comprise two or more skin cell types cultured together. 3-D models are cultured in three dimensions and involve one or more skin cell types. 3-D models that have characteristics similar to normal skin, as permitted by the reconstructed skin and skin explant (organotypic). For these models, primary cells, cell lines, and cells differentiated from stem cells or tissue may be used (Fig. 1).

2.1 Origin and preparation of cells

All types of skin cells can currently be extracted and cultured as primary cells from the human or animal skin, including keratinocytes, melanocytes, Langerhans cells, Merkel cells (MCs), fibroblasts, mast cells, endothelial cells, and adipocytes. Additionally, the skin element can be an explant, reconstructed skin, a dermal or epidermal equivalent, or extracellular matrix or an equivalent. Primary sensory neurons (PSNs) and Schwann cells are easily obtained from the dorsal root ganglia (DRGs) of animals. For neuronal or skin part, cell line may be used, but only a few cell lines have been evaluated for each cell type to study innervation mechanisms. For stem cells, differentiation, validation of the characteristics of the desired cellular type, and integration of the cells into a model are currently being assessed; although not described here, sensory neuron-like cells can be obtained.

2.1.1 Neurons

Classically, PSNs from DRGs (rats or mice), pigs, or rodent cell lines are used. For primary neuronal cells, the animals are euthanized and prepared. DRGs are extracted using a clamp along the spine and are then used as a complete explant [19], or they are mechanically and/or enzymatically dissociated to obtain single cells [20–22]. Dissociation results principally in a mixture of PSNs and Schwann cells. After extraction, sensory neurons are in a postmitotic state. For cell lines, different neurons or sensory neurons are used. ND7-23 and F-11 cells are sensory neuron-like cell lines [23,24]. They were obtained via cell fusion between the mouse neuroblastoma cell

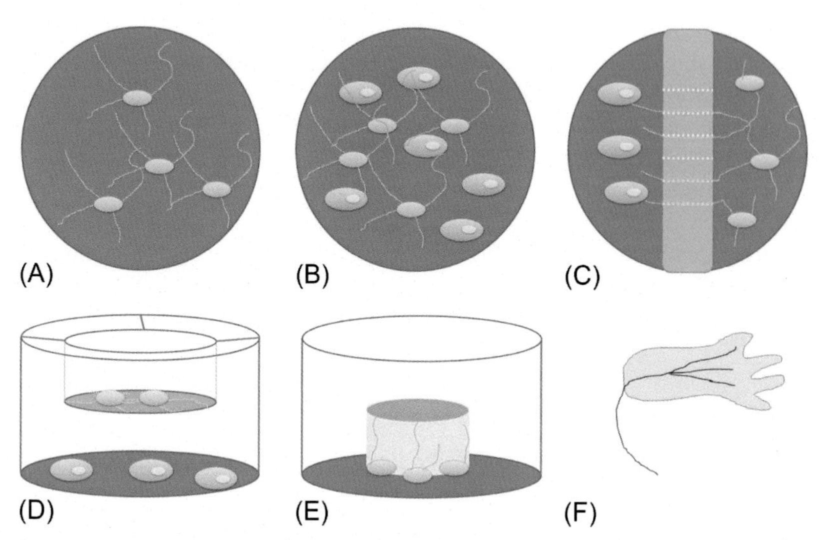

Fig. 1 Main types of models. The main types of models used for studying innervation mechanisms. (A) Monoculture model: one cell type, neurons or skin cells, is cultured alone. (B) Coculture model: two or more cell types are cultured together. (C) Coculture compartmentalized with microchannel: one or more cell types are cultured in distinct compartments. Neuronal extensions may move between compartments but not the cell bodies and slightly trophic factors. Contact between cells occurs via nerve endings. (D) Coculture compartmentalized with a Boyden channel: one or more cell types are cultured in distinct compartments. Neuronal extensions may move between compartments but not cell bodies and slightly trophic factors. Contact between cells is avoided. (E) 3-D coculture model: a skin explant or reconstructed skin are cultured with neurons. (F) Skin-nerve preparation: segments of skin and the attached neurons are extracted from an animal as a complete unit.

line N18Tg2 and rat neonatal, for ND7/23, or embryonic, for F-11, sensory DRG neurons; therefore, these cells contain the genetic material of both species. PC12 cells were derived from a rat pheochromocytoma that differentiated into cells with the characteristics of neurons following the addition of NGF [25]. The cell lines are cultured under growth conditions for multiplication using a classic commercial medium containing serum and other additives. Classically, F-11 cells are maintained in Ham's F12 (F12) containing 10% fetal bovine serum (FBS) and hypoxanthine, aminopterin, and thymidine (HAT). ND7/23 cells are cultured in DMEM with 10% FBS. PC12 cells are cultured in RPMI-1640 with 10% horse serum and 5% FBS or in DMEM/F12 with 10% FBS. After extraction or maintenance in growth conditions, the cells are seeded, and the medium is changed.

2.1.2 Skin cells

Primary skin cells of human or animal origin are extracted via the enzymatic and mechanical dissociation of dermal or epidermal tissue. The dermis and epidermis must first be separated via enzymatic digestion. Keratinocytes, melanocytes, MCs, and Langerhans cells are then obtained from the epidermis, and fibroblasts are obtained from the dermis. After the dissociation of these cells, it is equally possible to generate reconstructed skin or its equivalent made up of a unique epidermal sheet or dermal matrix and including all the other desired skin cells. The different techniques used to obtain cells and reconstructed skin samples will not be explained in this chapter. From the skin sample, it is equally possible to generate skin explants using punch biopsies or epidermal sheets after the dissociation of the dermis and epidermis [14,26]. For keratinocytes, HaCaT or A431 are the most commonly used cell line. For Merkel or Langerhans cells, there are no satisfactory cell lines to use in studying innervation mechanisms. For mast cells, the mouse raw264.7 or human U937 cell lines are used. For Langerhans cells, the mouse XS52 and XS106 cell lines are used.

2.2 Monoculture models

Monoculture models consist of one of the multiple cell types that make up the skin, including neurons. These cells are cultured alone (without other cell types) and are subjected to an controlled environment that permits the study of cutaneous innervation mechanisms, such as those related to the many factors produced by skin cells.

2.2.1 Neurons

To use neurons in this type of model, the medium may contain calcium, one or more types of neurotrophins and a specific complement, such as B27 or N2. Serum is usually avoided because its composition cannot be controlled. In the majority of cases, the support (culture plate) must be prepared with a coating that can be used alone or mixed, including polylysine, Matrigel, laminin, collagen, and fibronectin. For the extraction of primary neurons, the neurons are crushed. Consequently, in culture conditions, the neurites (cytoplasmic extensions from the body of a neuron, such as axons or dendrites) regenerate and may be studied. For sensory cell lines, serum deprivation and the addition of neurotrophins or cell-producing neurotrophins promote the extension of neurites for analysis.

The entire peripheral nerve explant may be used to understand innervation mechanisms. Tonge et al. presented different methods to extract, culture, and study amphibian and mammalian peripheral nerve explants, including those from mice [19]. For example, explanted mouse DRGs present robust axonal growth and can thus be studied. The nerves could be compartmentalized or cultured on Matrigel or collagen coating, and axonal outgrowth may be clearly observed from the cut ends of peripheral nerves upon the addition of trophic factors or drugs. Matrigel, collagen, or laminin were able to support neurite growth. However, collagen produced a lower growth of the axons than the Matrigel due to the composition. Results obtained with these models showed than NGF, NT-3, GDNF but not BDNF, or NT-4 increase the axonal growth in Matrigel. This growth is blocked by K252a (a kinase inhibitor) in NGF and NT3 conditions. The authors report also the presence of subpopulations of neurons expressing CGRP or phosphorylated heavy neurofilaments.

Generally, determining the effects of trophic or toxic factors on the regeneration of sensory neuron neurites is achieved via comparative measurements, with a control, of the length of neurites. These measurements are based on the total length of the neurons, including the branches. To preferentially identify complete phenotypes, types, or subtypes of neurons and/or the nerve fiber morphology of cells, immunocytochemistry (ICC) and patch-clamp analyses are performed.

Dissociated cells from DRG are a simple model to study cutaneous innervation mechanisms. Taneda et al. cultured DRG neurons in serum-free conditions with 10 ng/mL of NGF in DMEM/F12 and an N2 supplement [27]. The cells were seeded at 5000 cells per well in a 96-well plate coated with poly-D-lysine and laminin. After 24 or 48 h, the cells were fixed in 4% paraformaldehyde and analyzed using phase-contrast microscopy. The authors examined the effects of neurotrophin on NGF-induced neurite growth. In this model, they showed that the neurotrophin inhibited the NGF-induced neurite growth. In the absence of NGF, neurotrophin had no effect.

The cell lines represent an alternative interesting way to study cutaneous innervation. Lebonvallet et al. used a monoculture of PC12 cells differentiated into neurons using NGF to study neurite growth. After dissociation, PC12 cells were seeded at 5.10^4 cells per well in 12-well plates coated with poly-L-lysine (alternatively, plates may be coated with collagen I). The cells were incubated in growth medium (DMEM/F12 10% FCS) for 24 h. The medium was removed and replaced with DMEM/12 containing 10 ng/mL of hydrocortisone, 5 μg/mL of insulin, B27, and 10 ng/mL of NGF. Cells were maintained for 5 days alone or with a skin explant, but any other conditions with the addition of neurotrophins or active ingredients may be realized. After these 5 days, random live-cell imaging was performed for each condition, and images were analyzed to measure lengths. Results suggest that skin explant (representing a skin lesion) produce a mix of attractive growth factors.

2.2.2 Skin cells

Keratinocytes, fibroblasts, MCs, endothelial cells, immune cells, and other skin cell types may be used to study the molecular process involved in neuron guidance. However, neurons control skin functions, homeostasis, and wound-healing processes through the use of soluble neuronal growth factors and neuropeptides [3,13,28,29]. The addition of neuronal factors (neuropeptides, neurotransmitters, or neurotrophins) in skin cell cultures is obviously a potentially simple way to study the putative effects of neurons on these cells. Methods to culture keratinocytes are abundant but are generally based on low-calcium medium supplemented with epidermal growth factor and other additives, such as bovine pituitary gland extract. Commercial medium has been recommended for keratinocytes. Differentiated keratinocytes require high levels of calcium (between 1.5 and 2 mM).

After extraction, fibroblasts may be amplified and cultured through the use of, for example, DMEM/F12 with 10% FBS. Cheret et al. presented an interesting model for culturing fibroblasts [29]. Human fibroblasts were plated in individual wells coated with fibronectin and containing cloning rings in their middle (to mimic scratching), and after 2 days of culture, the cloning rings were removed, and fresh culture medium containing different neuropeptides (CGRP, VIP, substance P, and NGF) was added to the fibroblasts for either 24 or 48 h. A positive control for myofibroblast differentiation was performed in parallel by adding TGF-β1 to the culture medium. After 24 and 48 h, the culture medium was removed, and RNA was extracted to perform a real-time PCR analysis for α-SMA. The results revealed a significant increase of the expression of α-SMA in fibroblasts treated with the different neuropeptides in comparison with expression in the untreated sample, indicating that the different neuropeptides tested stimulated the differentiation of fibroblasts into myofibroblasts and, consequently, may participate at the wound-healing process.

Leon et al. used mast cells to determine their effects on innervation [30]. They used conditioned medium and identified multiple factors that are produced by mast cells. Peritoneal mast cells were obtained from rats at a purity of 90%. Conditioned medium was obtained from mast cells cultured in DMEM with 10% FCS for 24 h and was then applied to chicken embryonic DRGs for 24 h or 48 h. Leon et al. showed that neurite extension was promoted by the conditioned medium. This effect was altered when the medium was incubated with anti-NGF antibodies suggesting a role for NGF. Using ICC, ELISA, and PCR analyses, the authors confirmed that the NGF was produced by mast cells.

The roles of MCs in the NICS have been studied [31]. Factors associated with neuron survival are expressed by MCs. Because MCs represent a summary of the NICS by themselves, the addition of neurotransmitters or nerve growth factors to MC monocultures allows the effects of the peripheral nervous system (PNS) on these skin cells to be studied. Epidermal cells from piglet snout skin samples were dissociated via enzymatic digestion with trypsin. The cells were incubated with a magnetic microbead-conjugated antibody against CD56, MCs purified using magnetic cell sorting and seeded. Different media were used to promote cell survival, proliferation, and cytoplasmic extension. Neurotrophins such as NT3, BDNF, and NGF were unable to modify the MCs morphology. However, bFGF induced cytoplasmic extension outgrowth. Finally, to perform more long-term cultures, DMEM/F12 with 7% FCS and 7% horse serum was used, with the goal of identifying a factor that promotes MC survival.

2.3 Multicellular models

Cocultures of neurons and skin cells have recently been proposed. The cells of the skin and neurons can be cultured in a compartmentalized manner or mixed together in a monolayer or 3-D culture. It is possible to coculture neurons and all skin cell types or with a skin equivalent such as reconstructed skin, reconstructed epidermis, or a skin explant. Different methodologies, including with compartmentalization, of coculturing between DRG neurons and keratinocytes [32–34], fibroblasts [35], eosinophils/mast cells [36], Langerhans cells [37,38], and sweat glands [39] have been proposed. However, the medium must be adapted to permit the survival and functionality of all cell types in the culture [7,33,40]. Most models are flexible and may be adapted for all cell types and applications.

2.3.1 2-D compartmentalized models

Compartmentalized models are an interesting way because cellular body of neuron is separated of direct contact of skin cells to mimic physiology of the skin. Depending of experiment, contact between cells or only trophic factors are analyzed.

A simple model in two step permits to avoid direct contact between keratinocytes and cellular bodies of neurons using a temporary divider while at the beginning of culture [41]. Human primary keratinocytes and DRG neurons were seeded and cultured at the maximum levels in a petri dish with distinct compartments formed by a two-well silicon rubber chamber. The keratinocytes were maintained in Epilife 7S (with low calcium). DRGs were maintained in Epilife 7S with 1.8 mM calcium and 10% FBS. After 1 day, separation between the chambers was removed, and the medium was replaced with Epilife 7S with 1.8 mM calcium (thus, keratinocyte differentiation was initiated) for a maximum of 8 days. After removal of the compartmentalization, the neurites grew toward the keratinocytes and formed an organized structure in a cluster of cells, as observed via ICC using PGP9.5 and β-tubulin antibodies. The neurites, using ICC, showed a penetration into the keratinocyte cluster. These data suggested that keratinocytes produce neurotrophic factors and neurons possess the machinery to progress in a cluster of cells such as an epidermis. Blockage of NGF or its receptor showed no modification of the results suggesting an absence or low implication of NGF in this model. Furthermore, the authors described than Schwann cells (using vimentin stain) were present in the model. Schwann cells were described present near to neurons, up to the cluster of keratinocytes but do not penetrate into.

More complex, the model of Chateau et al. described cocultures between neurons and epithelial cells using specific tricompartmentalized chambers. The authors showed that MCs form neuroepithelial connections in vitro, and they used a homemade tricompartmentalized model separated by microchannels. Cells from the spinal cord were used in the first well, DRGs were seeded in the second well and a mixture of epidermal cells was seeded in the last well. The culture using a glial conditioned medium for 2 days was maintained for 15 days before analysis. After 15 days, many neurites could be observed. Contact between MCs and neurons was observed using ICC with cytokeratin 20 antibodies to stain MCs and PGP9.5 to stain neurons. The advantages of tricompartmentalization are that a third cell type may be studied. Results showed that it is possible to reconstruct neuroepithelial connexion in vitro, and this could be a synapsis.

It is also possible to realize a completely separate culture but with a communication between cells using the classical Boyden chamber or transwell. Ulmann et al. used the primary or sensory neuronal cell line ND7/23 and primary keratinocytes or the cell line A431 [7,33]. To coculture neurons and keratinocytes, they were seeded at 5000 cells/mL and 12,500 cells/mL, respectively, in MCDB medium in a 24-well plate. If compartmentalization was needed, the keratinocytes were seeded at 5×10^4 cells on Transwell cell culture inserts (0.4 μm pore size) to avoid contact but permit the diffusion of soluble factors. The calcium concentration was adjusted in a manner that was dependent of the experimental treatment. This model permits observation of the effects of keratinocytes on neurons, the effects of components of the medium, such calcium, on the cells and the identification of diffusible factors, such as NT3, BDNF, and NGF. Keratinocytes were analyzed via ICC using cytokeratin antibodies, and for neurons, PGP9.5 antibodies were used. The authors showed that the high level of Ca^{2+} was required for neurite extension, but not for survival of all neurons. Neurite expansion was also induced by NGF, BDNF, keratinocytes, or dehydroepiandrosterone (DHEA) but not by NT3, GDNF, pregnenolone, or progesterone. The keratinocytes induced neurite growth, which impacted essentially a subpopulation of neurons (cellular bodies at 17–25 μm) expressing SP and responding to capsaicin. This effect was diminished but not abolished by K252a suggesting the presence of a dependent and independent trk pathway.

As an alternative with previously described model, it exits possibility to provide a model using a Teflon chambers. Roggenkamp et al. proposed a model adapted to a Campenot chamber for studying compartmentalized neuron and keratinocyte cocultures [35,42,43]. A coculture between porcine DRG neurons and healthy or atopic skin cells was established. The cells were separated by a Teflon divider to prevent the blending of the media. In this model, cells can communicate via neurite extension through artificial microchannels or slowly via trophic factors. The DRG neurons were seeded in the middle compartment. After 1 day, keratinocytes or fibroblasts or neurotrophins and/or the associated neutralizing antibodies (GDNF and NGF) were seeded in the other well. In each compartment, the medium was changed every 2 days. The cells and overall effectiveness of the model were analyzed using ICC, PCR, and ELISA. The authors showed that NGF and GDNF induced neurite growth. The cumulative neurite length CGRP positive induced by NGF or GDNF was more important with NGF than GDNF suggesting that the neurotrophins impact different subpopulation of neurons. Healthy and atopic keratinocytes and fibroblasts expressed NGF and BDNF in distinct proportion and were able to induce neurite growth. The atopic keratinocyte produced more quantity of NGF. Kumamoto et al. developed an alternative to the Campenot chamber by adapting a commercial product, Axis(r), to study neurite growth [44,45]. The Axis chamber consisted of two distinct compartments separated by microgrooves. Human keratinocytes and/or mouse primary DRG neurons were seeded in the chamber, and the medium, which consisted of Epilife S7 and low level of calcium, was changed daily. For the analysis, the previously described method was used. The authors noted that there is neurite extension when keratinocyte are present. This growth was decreased when anti-BDNF antibody was used but not anti-NGF and anti-GDNF. Application of semaphorin 3A (a repulsive factor) in the medium reduced the neurite extension. The authors showed equally that microflow induced by a difference of quantity of medium in the two chambers induced neurite extension independently of NGF.

2.3.2 3-D models

Using a modified Boyden chamber, Tominaga developed a 3-D culture model using a type I collagen matrix [46,47]. Ice-cold collagen I or Matrigel was prepared and deposited in a 24-well Boyden chamber (0.4 μm pore). After gel formation, DRG neurons were seeded with 0.1 ng/mL of NGF. In the well under the Boyden chamber, medium containing 10 ng/mL of NGF was placed. After 48 h, the cells were fixed, and the number of tau fibers (revealed via ICC) crossing the membrane was determined. The expression of matrix metalloproteinase was evaluated via PCR and Western blot analyses, and their activities were determined using the zymogram method. In this model, only neurons were cultured in the upper matrix, and an attractive factor was added to the well, but we can imagine that any skin cell type could be placed under the Boyden chamber. The authors showed that neurites passed through the membrane dependently of NGF concentration in the well under the chamber. The neurite extension was limited by GM6001, an MMP inhibitor. More precisely, MMP-2, MMP8, and MT5-MMP were implicated. Furthermore, the expression levels of these MMPs were reduced with semaphorin 3A.

Currently, three models of reinnervated reconstructed skin and their derivatives have been published. In 2003, Gingras proposed a method for producing a reinnervated reconstruction of the skin [40]. A chitosan/collagen sponge was colonized by human fibroblasts and, when necessary, by human umbilical vein endothelial cells (HUVECs). Keratinocytes were added to the model opposite to the DRG neurons from mice. After preparing the sponge, fibroblasts and HUVECs were seeded on the sponge and maintained for 17 days in DMEM containing 10% FBS and 100 μg/mL of ascorbic acid. Next, mouse DRG neurons were deposited on top of the sponge. The medium was replaced with DMEM/F12 containing 10% SVF, 10 ng/mL of NGF, 0.4 μg/mL of hydrocortisone, 5 μg/mL of bovine insulin, and 100 μg/mL of ascorbic acid. After 7 days, the sponge was flipped over and reseeded with fibroblasts. After 3 days, the same quantity of keratinocytes was added. The culture was maintained for 4 days with immersion and then stopped or for 7 days with the immersion initiated on day 14 at the air-liquid interface. In this model, the neurite extensions were observed in the dermis reaching more than 770 μm at 14 days but not in the epidermis. The authors showed that NGF was necessary for neurite extension. The endothelial cells induced an increase of neurite growth, and the keratinocytes maintained the neurite formation. Roggenkamp developed a similar model using a polyester/polypropylene matrix and collagen gel inserts. Pig DRG neurons were placed in collagen I gel, and a portion was placed in the matrix. Subsequently, they were recovered by fibroblasts. After 24 h, keratinocytes were seeded. After 48 h, the model was cultured at the air-liquid interface for 12 days. The medium was refreshed daily with medium containing 10% FBS, 0.4 μg/mL of hydrocortisone, and 50 μg/mL of ascorbic acid. Using this model, the authors assessed epidermal homeostasis, pathological skin, and neurite growth. They showed that atopic keratinocytes induced an increase of epidermal thickness, neurite length, and concentration of CGRP. An alternative model was recently described by Martorina et al. that involves using human dermal microtissue, including fibroblasts, and epithelialization by HaCaT cells [48]. Neurons are very close to the epidermis. The authors showed the functionality of the models via calcium imaging using capsaicin as a cell activator.

At the interface between reconstructed skin and a skin explant, Nagase used epidermal sheets on collagen gels seeded with PC12 cells [26]. PC12 cells were seeded in collagen I gel in transwells in six-well plates. Hence, five epidermal sheets from footpad skin of adult Wistar rats were prepared via enzymatic dissociation with dispase, embedded in 0.5 mL of collagen, and placed in the upper well containing gel with PC12 cells. In this model, the authors principally observed that PC12 cells participated in the proliferation of MCs in the epidermis and maintained histological homeostasis of epidermal sheets, but other phenomena may also be assessed using this model.

The most complete model is the skin explant, but the environment is lesser controlled. Lebonvallet et al. developed a model of skin explants that were reinnervated with sensory neurons from rat DRGs [49–51]. An equivalent of one-fourth of the total DRGs from one rat was seeded in one well of an uncoated 12-well plate. After 1 or 2 h, the cells were exposed to DMEM/F12 containing 10 ng/mL of hydrocortisone, 5 μg/mL of insulin, and 25 ng/mL of NGF. Subsequently, a human skin explant 6 mm in diameter was placed at the center of the well and maintained for 10 days. The medium was changed after 2–3 days. After 10 days, many changes were observed. The authors showed that epidermal thickness, epidermal density, and innervation were higher when the skin explants were reinnervated. With this model, it is possible to perform electrophysiological analyses of neurons using topical applications to the epidermis, a histological analysis of reinnervation in the explant or of epidermal and dermal characteristics, and an analysis of neuronal growth around the explants. It is equally possible to perform omics analyses on the supernatant and the explants. In place of the normal skin, it is also possible to study pathological or wounded skin. Chéret et al. adapted the model to assess reinnervation during wound healing by making slight modifications and adding a wound in the center of skin samples [29].

Another interesting model uses a complete skin-nerve preparation that is collected from a mouse or rat and is maintained for survival [52,53]. The nerve and skin are microdissected, maintaining their unique organic structure, and placed in a bath specifically adapted to the chamber for the analysis of electric stimuli. This model is ideal for studying neuronal activity, for phenotyping the sensory nerves in the skin, and for making comparisons with the human skin using pharmacological treatments.

A comparison of the possible monoculture and multicellular models is shown in Table 1.

Table 1 Main features of monoculture and multicellular models to study cutaneous innervation

	Monolayer		**3-D models**
	Monoculture	**Multicellular**	
Cells	Neuronal/cutaneous Fibroblast, keratinocytes, Merkel cells, neurons	Neuronal and cutaneous, typically consisting of keratinocytes and neurons	Skin equivalent, pseudodermis or epidermis, skin explant or reconstructed epidermis, epidermal sheets, ex vivo
Source	Primary cells or cell lines		Primary cells, cell lines, or tissues
Level/scale	Mechanistic, cellular, physiological/ pathological	Mechanistic, cellular interaction, compartmented or not, communication or not, physiological/ pathological	Mechanistic, cellular interaction, *tissue*, physiological/pathological, *integrative*
Methods/ techniques	Cytotoxicity, migration and neurite growth, electrophysiology, biochemistry, molecular biology, labeling, imagery, video microscopy	Cytotoxicity, migration and neurite growth, electrophysiology, biochemistry, molecular biology, labeling, imagery, video microscopy	*Histology*, *topic*, cytotoxicity, migration and neurite growth, electrophysiology, biochemistry, molecular biology, labeling, imagery, video microscopy

3. Choice of model design and limitations

Using skin cells and neurons, different models have been developed to study innervation mechanisms and skin interactions. Monocultures are available but are not sufficient to understand and integrate all of the required parameters; thus, complementary cocultures and 3-D models provide very interesting approaches. The latter are more difficult to develop than a simple culture because the medium must be adapted for two cell types, but they offer different and more possibilities. All cells and elements may or may not be compartmentalized and in contact, and the model must be adapted depending on several parameters, such as the type of study performed and the level of interaction or cellular integration chosen. The goal of the project is important for the design of the model. The choice of the model will depend on [1] the method and scale of work (e.g., physiological, pathophysiological, or preclinical studies; cytological, histological, or functional studies; morphological, biochemical, or electrophysiological studies; and whether the culture is in a solution or uses a topical application or both) and [2] the focus of the work (e.g., cutaneous neurogenic inflammation, neurite growth, cellular tropism, reinnervation, epidermal homeostasis, signal integration and modality, wound healing, or pruritic pathologies, such as atopic dermatitis and psoriasis). Indeed, the constraints and results may be completely different depending on the model chosen, which is why early study choices are important.

3.1 Living materials (human, animal or cell line)

The choice of the living material and origin of the samples will depend on the study performed and the associated technical constraints. Cells or tissues are derived from humans or animals (commonly pigs, rats, and mice), primary cells (including stem cell) or cell lines. Skin element can be an explant, reconstructed skin, a dermal or epidermal equivalent, extracellular matrix or an equivalent, keratinocytes, and skin cells, such as MCs, fibroblasts, mast cells, Schwann cells, bone marrow cells, endothelial cells, or adipocytes. For neuronal equivalents, sensory neurons or sensory-like neurons are used.

A major difficulty associated with these models is the routine culturing of normal human sensory neurons, which is limited due to the limitations related to obtaining neurons from humans for ethical reasons. Consequently, sensory neurons from an animal origin, cell line or other source recently differentiated from human stem cells are used. Because it is very difficult to obtain human neurons, the principal bias in the coculture models could be the heterologous origin of the cells; skin cells usually have a human origin, whereas neurons are usually obtained from rodents. Unlike with immunologic research, there are no scientific data showing that the coculturing of neurons and skin cells of different origins modifies their properties in these models. The problems associated with heterologous models may be resolved by eliminating human cell types, but this would not elucidate the potentially different reactions between human and animal cells. However, a major advantage of heterologous models is that, for example, the mRNA may be differentially analyzed by using a specific animal probe, which would not be possible in a homologous model. A major advance in neurocutaneous interaction research is and will continue to be the use of neurons, especially sensory neurons, differentiated from human stem cells to replace nonhuman neurons [16,17,54–57].

The culture of primary neurons of an animal origin (especially rats or mice) is frequently performed but must be restrained, especially in the context of cosmetological research because it is forbidden by the European Union. Consequently, sensory neuron-like cell lines may be used in the models. PC12 cells are widely used to study the NGF receptor. However, major problems related to their similarity to primary cells may be present. For each cell type, not all potentialities of sensory neurons are reproduced. Rat PSNs present very good neurite growth after 1–3 days, whereas PC12 cells clearly show reduced performance; however, this difference may be due to the absence of Schwann cells with the PC12 cells [49]. Another difference is related to function. In fact, TRPV1 is less sensitive to the capsaicin on PC12 cells than on the PSN. On PC12 cells, TRPV1 is activated by a minimum of 80 μM of capsaicin, whereas less than 1 μM is sufficient to affect PSNs. In 2016, Yin et al. performed a transcriptome analysis comparing ND7/23 and F-11 cells and mouse DRG sensory neurons. ND7/23 and F-11 cells presented major differences in expression from one another and from the DRG neurons [58]. The principal explanation given by the authors is that the cell lines do not represent all of the sensory neuron subtypes present in and extracted from DRGs. Other studies investigating functionality in response to different agonists using calcium imagery of cell lines showed important differences in responses between cells depending on the agonist [59]. F-11 cells do not respond to a PAR-2 agonist, and ND7/23 cells do not respond to trypsin. Regarding neurite growth, Ulmann et al. showed that neurogenesis was completely suppressed in ND7/23 cells by the trk inhibitor K252a when compared with the response of primary neurons [60]. Concerning skin cells, few cell lines are available for each cell type. The results from these lines are not always identical to those from primary cells, but only limited data are available to support the observed effects on innervation mechanisms. For keratinocytes, human HaCaT or A431 is the most commonly used cell line. Ulmann et al. showed that A431 cells provoked a tendency for reduced neurite growth (approximately 25%) in rat PSNs when compared with the response in rat primary keratinocytes [7]. This difference may be due a reduced potential of A431 cells or to interspecies incompatibilities. In place of normal primary cells or cell lines, it is equally possible to use skin cells from a pathological individual. Roggenkamp et al. used a compartment model of primary sensory pig neurons and healthy or atopic keratinocytes or fibroblasts [35]. The team showed that atopic keratinocytes induced an increase in neurite length relative to results in healthy controls. However, this effect was not sufficient to demonstrate that pathological characteristics were conserved in all case in in vitro cultures. Such studies are needed. Similarly, cells that have been modified via transfection or KO/I may be introduced in cultures. These approaches are a more recently developing trend, and they are providing new and promising data.

In conclusion, for an optimal model, the use of PSNs from animals is currently recommended because they are simple to obtain and represent the full range of subtype diversity of neurons. However, for ethical reasons, this choice is not optimal; thus, it is necessary to develop lines from stem cells and validate the efficiency of sensory neurons. PC12 and ND7/23 cells may be an interesting alternative to study neurite growth. If the use of sacrificed animals is forbidden for cosmetic research, PC12, ND7/23, or F-11 cells may be used, with extreme caution taken in the interpretation of the results due to biases of the cell lines in terms of the function of the studied domain. Human, animal, or skin cell lines may be used to study innervation mechanisms. The limits of skin cell lines have not been studied.

3.2 Design and medium

The choice will depend on the scale of the analysis and whether it is necessary to integrate all parameters or if a specific feature is needed. In the case of the global approach, with all interactions and a histological analysis, a model based on a skin explant in coculture with neurons is optimal. It is basically possible to analyze [1] the protein, RNA, and tissue of the skin explant and [2] the protein, RNA, and electrophysiological profile/activity of the neurons. Furthermore, analysis of nerve fiber growth is possible both within and outside the skin explant. The supernatant may be collected and analyzed, with the disadvantage that the molecules produced by skin explants (keratinocytes and fibroblasts) and neurons are mixed. Reconstructed skin has similar characteristics except that all cells are not present in the model; consequently, all interactions are not present. However, a culture may be better maintained with reconstructed skin without the model degrading. Skin explants are more representative of lesional/inflamed skin, and the survival and maintenance of a skin explant for more than 10–14 days is affected by apoptosis and necrosis of the cells and tissue. Another difficulty related to reinnervated skin is the low reproducibility of the density of skin innervation between experiments, but low reproducibility is probably similar to that of reconstructed skins. A reference protocol is available to study and compare epidermal innervation density and may be adapted for these models [61]. However, performing a comparison between reinnervated skin and normal human skin currently lacks interest because densities are highly variable among experiments and among individuals and locations under in vivo conditions. However, with some additional development, reinnervated skin is very close to being used as a preclinical model. Skin-nerve preparations (dissection of the skin with attached nerves from the spinal cord) are also possible from rats and mice, which provide the advantage that the models are fully integrative; however, they are strictly limited to characterizations of neurons.

For a specific and more controlled approach, cocultures and monocultures are more adaptable techniques. Innervation mechanisms and interactions can be observed at the cellular level. Under monoculture conditions, transcriptomic, proteomic, and electrophysiological analyses are possible, as are measurements of cell morphology and neurite extension. Supernatants contain unique elements produced by cells and those added based on the protocol, and analyzing it is simple. In noncompartmented cocultures, two or more cell types (generally keratinocytes and neurons) are mixed, but the same methodology used for monocultures may be utilized. Tsutsumi et al. revealed the importance of the manner and order of seeding cells between neurons and keratinocytes [62]. They showed that different seed models resulted in different organizational structures and concluded that the method for seeding the cells is important for the interpretation of the results. Whether the cells communicate is of interest; consequently, responses between them may be studied, for example, trophic effect of keratinocytes on the growth of neurites of neurons or the proliferative effect of neurons on keratinocytes. Furthermore, the presence of additional cells, in comparison with monoculture models, may modify the responses of one or more cells to the studied parameters. As with reinnervated skin explants or reconstructed skin, the disadvantage associated with multicellular cultures is the presence in the supernatant of molecules produced/released by the mixture of all cells. This problem may be mitigated by compartmentalization with microgrooves between neurons and skin cells. Many models can be established using modified Boyden, Campenot, and Axis(r) chambers [35,43,44]. In these cases, diffusible factors (added or produced) may diffuse slowly between compartments, consequently, permitting the tropism of neurites, without the cellular body, into the target compartment. The uncontrolled diffusion of factors between compartments remains an unresolved problem and has been discussed [62]. Another difficulty with multicellular models is the compatibility of the medium for two cellular types and the seeding procedure. The medium must be adapted to permit the survival and functionality of all cells in the culture [7,33]. Even if the medium enables survival, it may also interfere with the culture, or an ingredient may be unnecessary. A simple example is calcium: neurons need calcium to correctly develop; however, high calcium concentrations provoke the differentiation of keratinocytes, which may or may not be desirable as the characteristics of keratinocytes are very different in basal or differentiated states. Ulmann et al., using a coculture model of neurons and keratinocytes or trophic factors, showed that under low-calcium conditions, axonal growth in sensory neurons was induced by NGF and BDNF but not by NT3 and GDNF, which is in contrast to the findings of another study that used high-calcium conditions [7,63]. Tsutsumi showed that there is no morphological interaction between keratinocytes and neurons in the presence of low calcium concentrations, unlike the results obtained using 1.8 mM calcium [62]. The team suggested that differentiated keratinocytes produce a trophic substance that affects neurons and that basal keratinocytes do not. B27, another component of medium used to culture or coculture neurons, does not appear to be necessary in cocultures of neurons and keratinocytes or of neurons and a skin equivalent. Using a coculture model including DRGs and human reconstructed skin, Gingras et al. showed that B27 supplementation was not necessary for the model; in contrast, it was necessary for culturing neurons alone [40]. A more recent model of rat DRG and human skin explant coculturing did not use B27 in the medium [49].

The choice of the model depends on the type of analyses needed to resolve the question. Different models are available for 2-D and 3-D cultures. In my opinion, a simple monoculture or 2-D model should first be implemented to avoid interfering parameters. In further experiments, more complex models may be used to permit a global and mostly histological approach.

4. Applications

These different models are very versatile, and most techniques are applicable to them. Furthermore, the 3-D models may be studied using histological techniques, which provide an interesting new dimension to studies of the skin. A compilation of the applications and associated models is presented in Table 2, but it is not exhaustive because many mixed combination and applications of the models presented here are possible.

Table 2 **Applications and associated models**

Application/theme of the studies	Models/cellular type
Neurite/axonal growth and cellular tropism, interaction/contact	3-D [26,40,64,65]
	Multicellular, compartmented, monolayer [7,35,43,44,66]
	Multicellular, monolayer [7,41,62,67,68]
	Monoculture [46,69]
	Monoculture, monolayer [21,29,70–72]
Cutaneous neurogenic inflammation	Multicellular, monolayer [32,73,74]
	Monoculture, monolayer [75]
Reinnervation	3-D [40,49,50]
	Monoculture [46]
Epidermal homeostasis	3-D [26,29,49,76]
Signal integration and modality	3-D/ex vivo [48,51–53]
	Monoculture, monolayer [77,78]
	Multicellular, monolayer [68]
Atopic dermatitis	3-D [76]
	Monoculture [46]
	Multicellular, compartmented, monolayer [35,43]
Wound healing	3-D [29]
	Monoculture [69]
	Monoculture, monolayer [29,70]
Pruritus	3-D [51]
	Monoculture [46]
UV irradiation	Multicellular, multilayer [79]

4.1 Neurite/axonal growth, cellular tropism and cellular communication

Cultures of sensory neurons are helpful to understand the effects of factors that are produced by epidermal or dermal cells and that participate in the innervation mechanisms of these cells. Neurons are dependent on particular neurotrophins for survival and guidance, such as NGF, BDNF, NT-3, and GDNF. Axon cone growth of neurons may be guided by specific factors. The growth of the axon cone results from attractive factors or neurotrophins, such as NGF and repulsive factors, such as semaphorins. Cultures and cocultures of sensory neurons are used to identify/study/screen substances (like active ingredients, growth factors, or other biological molecules) that are implicated in neuron growth, tropism, cone growth, and associated molecular mechanisms. Similarly, stressors or pro-wound-healing factors may be analyzed using a particular model that utilizes neurons cultured in the presence of slices of injured skin [70].

A model based on the coculture of F-11 cells and primary keratinocytes has been described [33]. The F-11 cell line has been conventionally used to mimic peptidergic nociceptive neurons. The morphological and functional characteristics of F-11 were evaluated first, followed by the influence of keratinocytes on axonal development or neuropeptides release. These neurons express markers of sensory neurons and are able to release neuropeptides after activation by capsaicin. Neuropeptide release was detected even at a low concentration of calcium, and axonal growth was not influenced by variations in calcium concentration. These properties were demonstrated when F-11 cells were cocultured with keratinocytes, but they had no significant influence on the development of axons or the release of neuropeptides. This coculture model, in which keratinocytes and neurons are maintained under low-calcium conditions, may be useful for studying the close contact between keratinocytes and sensory neurons.

Another model may be employed to study the role of skin in neuronal growth. In this model, keratinocytes are replaced with a whole-skin explant, and sensory neurons, with PC12 cells that have been differentiated into neurons. PC12 cells were plated, and a skin punch biopsy (skin explant) was placed in the center of the culture well. The cocultures were maintained in medium for 10 days. Every 2 days, neurite lengths of the PC12 cells were evaluated. Two conditions were compared: PC12 cells cultured alone and PC12 cells culture with the skin explant. The results indicate that the skin explants had a trophic effect on the neurites of PC12 cells. In this model, the skin explants may also be considered as lesional skin [71].

Contact between cells can be specifically evaluated, and the trophic effect of keratinocytes or fibroblasts on neurons has been demonstrated using compartmentalized models. Neurons and other cellular types are placed in separate compartments of a culture using microchannels and a separate chamber that permits contact between different cells via neurites [35,43,44] or in a classical Boyden chamber/transwell that does not permit cell contact but solely the circulation of soluble factors [7]. Microchannels allow neurons to pass into the other compartment only by projecting neurites because the channels are narrower than the bodies of the cells [43,66],. Based on electron microscopy and photomicroscopy results and using the compartment model or another model, contacts between sensory axons and cells of the epidermis have been described for keratinocytes, Langerhans cells, and MCs [38,41,66–68].

4.2 Reinnervation and epidermal homeostasis models

The first approaches to the study of the skin utilized fresh animal and human skin explants. Skin innervation and its effect on the epidermis were evaluated [3–5], including the potent participation of newly studied molecules on wound healing, such as neuroserpin [80], or the mechanism underlying skin sensitivity [81]. However, this approach is static, and not all conditions can be applied for ethical and technical reasons [14].

3-D coculture models are the most complete models and allow for a more comprehensive view of, for example, reinnervation or epidermal homeostasis. In 3-D in vitro skin models, these interactions are lost because of the lack of nerve endings [14]. In vitro 3-D skin models have been developed using skin explants [49], reconstructed skin [40], or epidermal sheets [26].

The model using reconstructed skin developed by Gingras is ideal to study reinnervation and axonal growth due to the lack of interference caused by previous ancient reinnervation in the case of explant. In this model, a collagen sponge populated with human endothelial cells and/or human fibroblasts was used. The authors used histological methods to demonstrate that the elongation was maximal within 14 days of culture. Thus, different modalities may be evaluated using this model, such as the influence of Schwann cells [64] or hair follicles [65].

A skin explant model is ideal for studying epidermal homeostasis because the epidermis is reinnervated, and all cells are present [49,82]. Previous human skin explant models have included a coculture of PSNs from rats. After 10 days of culture, the skin explant was analyzed. Using histological techniques, the epidermal thickness, cell density, and cell integrity of the innervated skin explant were found to be higher when the explants were reinnervated in comparison with changes in the noninnervated controls. The proliferation of epidermal cells was not modified, but apoptosis decreased significantly. Therefore, this innovative model using skin explants and neuronal cocultures allows improved skin integrity and could be useful to study interactions between the skin and the PNS. Furthermore, electrophysiological activity was assessed, and the role of the trophic effects of skin explants was confirmed [51,71]. Based on a model using reinnervated atopic reconstructed skin, Roggenkamp suggested that neurons strongly influence keratinocyte proliferation [76].

4.3 Wound-healing models

Based on the close interactions between the skin and the PNS, there is increasing evidence that cutaneous innervation is an important modulator of the normal wound-healing process. It is well established that neurotransmitters and nerve growth factors that are released into the skin have immunoregulatory roles and can exert mitogenic actions; they can also influence the functions of the different types of skin cells during the wound-healing process [28].

The study of wound healing using a 3-D model [29] was conducted using an adapted model of a reinnervated skin explant, as previously described [49]. A human skin wound-healing assay was designed with internal injuries on 6 mmØ human skin explants inflicted with a smaller 3 mmØ punch. Skin explants were incubated with a panel of molecules secreted by neurons or PSNs from rats. In a simple culture (without neurons), injured human skin explants were maintained in the same medium and supplemented with SP, CGRP, or VIP at different concentrations. Every 2 days, half of the culture was discarded and replaced with fresh culture medium containing neuropeptides. The results were compared with results from a coculture of neurons and wounded skin. Different culture times were evaluated, corresponding to different phases of wound healing. This study showed that the peripheral sensory nervous system plays a crucial role in this process by affecting the proliferation and remodeling phases of skin wound healing. Communication between skin cells and sensory neurons is mediated by neuropeptides, such as SP, CGRP, and VIP.

Another coculture type has been established between fibroblasts and neurons in a collagen sponge [69]. In this case, the role of neurons in the contractile potential of fibroblasts on a sponge was evaluated. PC12 cells were differentiated by the addition of NGF to the culture medium. The fibroblasts were obtained from rats (as were the PC12 cells). PC12 cells and fibroblasts were seeded on a 3-D collagen matrix. The differentiation of fibroblasts into myofibroblasts, as a marker of the progression of the wound-healing process, was evaluated in the presence or absence of PC12 cells, and the resulting contraction of the collagen matrix was considered. The results show that PC12 cells had a strong influence on fibroblast differentiation and the contractibility of the collagen matrix.

4.4 Models of cutaneous neurogenic inflammation

A coculture model of pig keratinocytes and pig sensory neurons can be used to understand/mimic more complex mechanisms, such as neurogenic inflammation skin tests, or for screening active ingredients and to understand their cellular functions [32,73,74]. In this model, epidermal keratinocytes and sensory neurons of piglets were extracted via enzymatic and mechanical dissociation. The cells were seeded in a 96-well plate and dissociated. The keratinocytes and neurons adhered to the support. After several days, the neurons had long extensions and formed contacts with neighboring cells. Immunolabeling showed that keratinocytes and neurons expressed SP and TRPV1. To evaluate cutaneous neurogenic inflammation, the cells were incubated with capsaicin, a TRPV1 agonist. The release of SP in the coculture was measured via ELISA. The assay revealed that the neuropeptide SP is released in response to capsaicin. This model allows for the evaluation of neurogenic inflammation based on the level of SP release into the supernatant via TRPV1 activation [32]. Using this model, the effect of tacrolimus as a regulator of neurogenic inflammation was validated, as was its mechanism of action [73]. It is also possible to assess the "soothing" effect of a factor or a sample by pretreating the cells. Thus, it has been shown that *sangre de drago* has a "soothing" effect on cutaneous neurogenic inflammation by reducing the release of SP after the addition of capsaicin if the product is incubated with the cells prior to exposure to capsaicin [74]. Similarly, Le Garrec et al. adapted a model using rat DRG neurons and human keratinocytes instead of pig cells to study ciguatoxins [34]. Using this model, the authors show that ciguatoxin (P-CTX-2) induced a release of SP and CGRP dependant of voltage-gated sodium channels. The ciguatoxin-induced sensory disturbances in ciguatera fish poisoning may involve the release of these neuropeptides. Moreover, the authors showed the influence of extracellular calcium on the release of neuropeptides elicited by P-CTX-2.

4.5 Pathological skin

Coculture models may be used to understand skin pathologies, such as atopic dermatitis, pruritus, or lesional skin resulting from UV irradiation. The model may be adapted to study pathological skin samples, for example, by replacing normal human keratinocytes, fibroblasts, or skin explants with the corresponding atopic cells/tissues. UV irradiation may also be used on cells to mimic naturally UV-damaged cutaneous cells (dermal, epidermal, and neuronal cells) [79]. Furthermore, using pathological skin or components of pathological skin, secondary treatments may be applied.

To study atopic dermatitis, Roggenkamp et al. [35] adapted a model using the compartmentalized chambers of Campenot et al. [43]. In this model, the authors used PSNs from pigs and human keratinocytes or fibroblasts from atopic patients. The cells were separated using a Teflon module that permitted the passage of neuron fibers via microchannels. In this model, the authors highlighted the roles of pathological keratinocytes and fibroblasts in neurite outgrowth from neurons. More recently, in a 3-D atopic reconstructed skin model, Roggenkamp confirmed the role of keratinocytes in neurite growth and showed that neurons increase the growth of keratinocytes in response to CGRP [76].

5. Conclusions

New models are available to study different features of neuron-skin interactions. These models may be characterized by monolayers or by 3-D, multicellular, or monoculture designs to study different parameters of physiological and pathological neurocutaneous interactions, such as axon growth and tropism, epidermal homeostasis, wound healing, atopic dermatitis, pruritus, and neurogenic inflammation. These models could also be helpful for preliminary investigations and the screening of active substances that could be used by cosmetic or pharmaceutical industries. Further advances could be made through the use of sensory neurons from stem cells to replace animal neurons. Also, the realization and validation of a reinnervated skin explant model for preclinical use would be a great advance.

Abbreviations

2-D/3-D two-dimensional/three-dimensional
A431 keratinocyte cell line
B27 neuronal medium supplement
BDNF brain-derived neurotrophic factor
bFGF basic fibroblast growth factor
Capsaicin TRPV1 agonist
CD56 cluster of differentiation 56
DMEM Dulbecco's modified Eagle's medium
DRG dorsal root ganglia
ELISA enzyme-linked immunosorbent assay
F11 sensory neuronal cell line. Rat-mouse hybridoma
F12 Ham's F12
FBS fetal bovine serum
GAP43 growth associated protein 43
GDNF glial-derived neurotrophic factor
HaCaT keratinocyte cell line
HAT hypoxanthine-aminopterin-thymidine
HUVEC human umbilical vein endothelial cell
ICC/IHC immunocytochemistry/immunohistochemistry
K252a Trk inhibitor
KO/I knock-out/knock-in
MC Merkel cells
N18TG2 mouse neuroblastoma

N2	neuronal medium supplement
ND7/23	sensory neuronal cell line. Rat-mouse hybridoma
NGF	nerve growth factor
NICS	neuro-immuno-cutaneous system
NT	neurotrophin
PAR	protease-activated receptor
PC12	pheochromocytoma cell line
PCR	polymerase chain reaction
PGP9.5	protein gene product 9.5
PSN	primary sensory neuron
raw264.7	mouse macrophage cell line
TGF	transforming growth factor
TRP	transient receptor potential
U937	human myeloid cell line
UV	ultraviolet
VIP	vasoactive intestinal peptide
XS52/106	Langerhans-like cell line
α-SMA	smooth muscle actin

References

[1] Hilliges M, Wang L, Johansson O. Ultrastructural evidence for nerve fibers within all vital layers of the human epidermis. J Invest Dermatol 1995;104(1):134–7.

[2] Boulais N, Misery L. The epidermis: a sensory tissue. Eur J Dermatol 2008;18(2):119–27.

[3] Hsieh S-T, Lin W-M, Chiang H-Y, Huang I-T, Ko M-H, Chang Y-C, et al. Skin innervation and its effects on the epidermis. J Biomed Sci 1997;4(5):264–8.

[4] Huang IT, Lin WM, Shun CT, Hsieh ST. Influence of cutaneous nerves on keratinocyte proliferation and epidermal thickness in mice. Neuroscience 1999;94(3):965–73.

[5] Tominaga M, Ogawa H, Takamori K. Decreased production of semaphorin 3A in the lesional skin of atopic dermatitis. Br J Dermatol 2008;158(4):842–4.

[6] Albers KM, Davis BM. The skin as a neurotrophic organ. Neuroscientist 2007;13:371–82.

[7] Ulmann L, Rodeau J-L, Danoux L, Contet-Audonneau J-L, Pauly G, Schlichter R. Trophic effects of keratinocytes on the axonal development of sensory neurons in a coculture model. Eur J Neurosci 2007;26(1):113–25.

[8] Tominaga M, Tengara S, Kamo A, Ogawa H, Takamori K. Psoralen-ultraviolet a therapy alters epidermal Sema3A and NGF levels and modulates epidermal innervation in atopic dermatitis. J Dermatol Sci 2009;55(1):40–6.

[9] Montaño JA, Pérez-Piñera P, García-Suárez O, Cobo J, Vega JA. Development and neuronal dependence of cutaneous sensory nerve formations: lessons from neurotrophins. Microsc Res Tech 2010;73(5):513–29.

[10] Montano JA, Perez-Pinera P, Garcia-Suarez O, Cobo J, Vega JA. Development and neuronal dependence of cutaneous sensory nerve formations: lessons from neurotrophins. Microsc Res Tech 2010;73(5):513–29.

[11] Gouin O, Lebonvallet N, L'Herondelle K, Le Gall-Ianotto C, Buhé V, Plée-Gautier E, et al. Self-maintenance of neurogenic inflammation contributes to a vicious cycle in skin. Exp Dermatol 2015;24(10):723–6.

[12] Misery L. Skin, immunity and the nervous system. Br J Dermatol 1997;137(6):843–50.

[13] Roosterman D, Goerge T, Schneider SW, Bunnett NW, Steinhoff M. Neuronal control of skin function: the skin as a neuroimmunoendocrine organ. Physiol Rev 2006;86(4):1309–79.
[14] Lebonvallet N, Jeanmaire C, Danoux L, Sibille P, Pauly G, Misery L. The evolution and use of skin explants: potential and limitations for dermatological research. Eur J Dermatol 2010;20(6):671–84.
[15] Fernandes KJ, Toma JG, Miller FD. Multipotent skin-derived precursors: adult neural crest-related precursors with therapeutic potential. Philos Trans R Soc Lond B Biol Sci 2008;363(1489):185–98.
[16] Lebonvallet N, Boulais N, Le Gall C, Chéret J, Pereira U, Mignen O, et al. Characterization of neurons from adult human skin-derived precursors in serum-free medium: a PCR array and immunocytological analysis. Exp Dermatol 2012;21(3):195–200.
[17] Lee G, Chambers SM, Tomishima MJ, Studer L. Derivation of neural crest cells from human pluripotent stem cells. Nat Protoc 2010;5(4):688–701.
[18] Reinhardt P, Glatza M, Hemmer K, Tsytsyura Y, Thiel CS, Höing S, et al. Derivation and expansion using only small molecules of human neural progenitors for neurodegenerative disease modeling. PLoS One 2013;8(3):e59252.
[19] Tonge D, Edstrom A, Ekstrom P. Use of explant cultures of peripheral nerves of adult vertebrates to study axonal regeneration in vitro. Prog Neurobiol 1998;54:459–80.
[20] Bowie D, Feltz P, Schlichter R. Subpopulations of neonatal rat sensory neurons express functional neurotransmitter receptors which elevate intracellular calcium. Neuroscience 1994;58(1):141–9.
[21] Lindsay RM, Shooter EM, Radeke MJ, Misko TP, Dechant G, Thoenen H, et al. Nerve growth factor regulates expression of the nerve growth factor receptor gene in adult sensory neurons. Eur J Neurosci 1990;2(5):389–96.
[22] Delree P, Leprince P, Schoenen J, Moonen G. Purification and culture of adult rat dorsal root ganglia neurons. J Neurosci Res 1989;23(2):198–206.
[23] Dunn PM, Coote PR, Wood JN, Burgess GM, Rang HP. Bradykinin evoked depolarization of a novel neuroblastoma x DRG neurone hybrid cell line (ND7/23). Brain Res 1991;545:80–6.
[24] Platika D, Boulos MH, Baizer L, Fishman MC. Neuronal traits of clonal cell lines derived by fusion of dorsal root ganglia neurons with neuroblastoma cells. Proc Natl Acad Sci U S A 1985;82:3499–503.
[25] Greene LA, Tischler AS. Establishment of a noradrenergic clonal line of rat adrenal pheochromocytoma cells which respond to nerve growth factor. Proc Natl Acad Sci U S A 1976;73:2424–8.
[26] Nagase K, Aoki S, Uchihashi K, Misago N, Shimohira-Yamasaki M, Toda S, et al. An organotypic culture system of Merkel cells using isolated epidermal sheets. Br J Dermatol 2009;161(6):1239–47.
[27] Taneda K, Tominaga M, Tengara S, Ogawa H, Takamori K. Neurotropin inhibits both capsaicin-induced substance P release and nerve growth factor-induced neurite outgrowth in cultured rat dorsal root ganglion neurones. Clin Exp Dermatol 2010;35(1):73–7.
[28] Chéret J, Lebonvallet N, Carré J-L, Misery L, Le Gall-Ianotto C. Role of neuropeptides, neurotrophins, and neurohormones in skin wound healing. Wound Repair Regen 2013;21(6):772–88.
[29] Chéret J, Lebonvallet N, Buhé V, Carre JL, Misery L, Le Gall-Ianotto C. Influence of sensory neuropeptides on human cutaneous wound healing process. J Dermatol Sci 2014;74(3):193–203.

[30] Leon A, Buriani A, Toso RD, Fabris M, Romanello S, Aloe L, et al. Mast cells synthesize, store, and release nerve growth factor. Proc Natl Acad Sci 1994;91(9):3739–43.

[31] Boulais N, Misery L. Merkel cells. J Am Acad Dermatol 2007;57:147–65.

[32] Pereira U, Boulais N, Lebonvallet N, Lefeuvre L, Gougerot A, Misery L. Development of an in vitro coculture of primary sensitive pig neurons and keratinocytes for the study of cutaneous neurogenic inflammation. Exp Dermatol 2010;19(10):931–5.

[33] Le Gall-Ianotto C, Andres E, Hurtado SP, Pereira U, Misery L. Characterization of the first coculture between human primary keratinocytes and the dorsal root ganglion-derived neuronal cell line F-11. Neuroscience 2012;210:47–57.

[34] Le Garrec R, L'hirondelle K, Le Gall-Ianotto C, Lebonvallet N, Leschiera R, Buhe V, et al. Release of neuropeptides from a neuro-cutaneous co-culture model: a novel in vitro model for studying sensory effects of ciguatoxins. Toxicon 2016;116:4–10.

[35] Roggenkamp D, Falkner S, Stäb F, Petersen M, Schmelz M, Neufang G. Atopic keratinocytes induce increased neurite outgrowth in a coculture model of porcine dorsal root ganglia neurons and human skin cells. J Invest Dermatol 2012;132(7):1892–900.

[36] Foster EL, Simpson EL, Fredrikson LJ, Lee JJ, Lee NA, Fryer AD, et al. Eosinophils increase neuron branching in human and murine skin and in vitro. PLoS One 2011;6(7):e22029.

[37] Torii H, Yan Z, Hosoi J, Granstein RD. Expression of neurotrophic factors and neuropeptide receptors by Langerhans cells and the Langerhans cell-like cell line XS52: further support for a functional relationship between Langerhans cells and epidermal nerves. J Invest Dermatol 1997;109(4):586–91.

[38] Gaudillere A, Misery L, Souchier C, Claudy A, Schmitt D. Intimate associations between PGP9.5-positive nerve fibres and Langerhans cells. Br J Dermatol 1996;135(2):343–4.

[39] Mehnert JM, Kisch T, Brandenburger M. Co-culture systems of human sweat gland derived stem cells and peripheral nerve cells: an in vitro approach for peripheral nerve regeneration. Cell Physiol Biochem 2014;34(4):1027–37.

[40] Gingras M, Bergeron J, Déry J, Durham HD, Berthod F. In vitro development of a tissue-engineered model of peripheral nerve regeneration to study neurite growth. FASEB J 2003;17(14):2124–6.

[41] Tsutsumi M, Nakatani M, Kumamoto J, Denda S, Denda M. In vitro formation of organized structure between keratinocytes and dorsal-root-ganglion cells. Exp Dermatol 2012;21(11):886–8.

[42] Campenot RB. Local control of neurite development by nerve growth factor. Proc Natl Acad Sci 1977;74(10):4516–9.

[43] Campenot RB, Lund K, Mok S-A. Production of compartmented cultures of rat sympathetic neurons. Nat Protoc 2009;4(12):1869–87.

[44] Kumamoto J, Nakatani M, Tsutsumi M, Goto M, Denda S, Takei K, et al. Coculture system of keratinocytes and dorsal-root-ganglion-derived cells for screening neurotrophic factors involved in guidance of neuronal axon growth in the skin. Exp Dermatol 2014;23(1):58–60.

[45] Kumamoto J, Kitahata H, Goto M, Nagayama M, Denda M. Effects of medium flow on axon growth with or without nerve growth factor. Biochem Biophys Res Commun 2015;465(1):26–9.

[46] Tominaga M, Kamo A, Tengara S, Ogawa H, Takamori K. In vitro model for penetration of sensory nerve fibres on a Matrigel basement membrane: implications for possible application to intractable pruritus. Br J Dermatol 2009;161(5):1028–37.

[47] Tominaga M, Tengara S, Kamo A, Ogawa H, Takamori K. Matrix metalloproteinase-8 is involved in dermal nerve growth: implications for possible application to pruritus from in vitro models. J Invest Dermatol 2011;131(10):2105–12.

[48] Martorina F, Casale C, Urciuolo F, Netti PA, Imparato G. In vitro activation of the neuro-transduction mechanism in sensitive organotypic human skin model. Biomaterials 2017;113:217–29.

[49] Lebonvallet N, Boulais N, Le Gall C, Pereira U, Gauché D, Gobin E, et al. Effects of the re-innervation of organotypic skin explants on the epidermis. Exp Dermatol 2012;21(2):156–8.

[50] Sevrain D, Le Grand Y, Buhé V, Jeanmaire C, Pauly G, Carré J-L, et al. Two-photon microscopy of dermal innervation in a human re-innervated model of skin. Exp Dermatol 2013;22(4):290–1.

[51] Lebonvallet N, Pennec J-P, Le Gall-Ianotto C, Chéret J, Jeanmaire C, Carré J-L, et al. Activation of primary sensory neurons by the topical application of capsaicin on the epidermis of a re-innervated organotypic human skin model. Exp Dermatol 2014;23(1):73–5.

[52] Zimmermann K, Hein A, Hager U, Kaczmarek JS, Turnquist BP, Clapham DE, et al. Phenotyping sensory nerve endings in vitro in the mouse. Nat Protoc 2009;4(2):174–96.

[53] Maurer K, Bostock H, Koltzenburg M. A rat in vitro model for the measurement of multiple excitability properties of cutaneous axons. Clin Neurophysiol 2007;118(11):2404–12.

[54] Chambers SM, Qi Y, Mica Y, Lee G, Zhang X-J, Niu L, et al. Combined small-molecule inhibition accelerates developmental timing and converts human pluripotent stem cells into nociceptors. Nat Biotechnol 2012;30(7):715–20.

[55] Gingras M, Champigny MF, Berthod F. Differentiation of human adult skin-derived neuronal precursors into mature neurons. J Cell Physiol 2007;210:498–506.

[56] Denham M, Hasegawa K, Menheniott T, Rollo B, Zhang D, Hough S, et al. Multipotent caudal neural progenitors derived from human pluripotent stem cells that give rise to lineages of the central and peripheral nervous system. Stem Cells 2015;33(6):1759–70.

[57] Guo X, Spradling S, Stancescu M, Lambert S, Hickman JJ. Derivation of sensory neurons and neural crest stem cells from human neural progenitor hNP1. Biomaterials 2013;34(18):4418–27.

[58] Yin K, Baillie GJ, Vetter I. Neuronal cell lines as model dorsal root ganglion neurons: a transcriptomic comparison. Mol Pain 2016;12.

[59] Vetter I, Lewis RJ. Characterization of endogenous calcium responses in neuronal cell lines. Biochem Pharmacol 2010;79(6):908–20.

[60] Ulmann L, Rodeau J-L, Danoux L, Contet-Audonneau J-L, Pauly G, Schlichter R. Dehydroepiandrosterone and neurotrophins favor axonal growth in a sensory neuron-keratinocyte coculture model. Neuroscience 2009;159(2):514–25.

[61] Ebenezer GJ, Hauer P, Gibbons C, McArthur JC, Polydefkis M. Assessment of epidermal nerve fibers: a new diagnostic and predictive tool for peripheral neuropathies. J Neuropathol Exp Neurol 2007;66(12):1059–73.

[62] Tsutsumi M, Goto M, Denda S, Denda M. Morphological and functional differences in coculture system of keratinocytes and dorsal-root-ganglion-derived cells depending on time of seeding. Exp Dermatol 2011;20(6):464–7.

[63] Lentz SI, Knudson CM, Korsmeyer SJ, Snider WD. Neurotrophins support the development of diverse sensory axon morphologies. J Neurosci 1999;19(3):1038–48.

[64] Blais M, Grenier M, Berthod F. Improvement of nerve regeneration in tissue-engineered skin enriched with Schwann cells. J Invest Dermatol 2009;129(12):2895–900.

[65] Gagnon V, Larouche D, Parenteau-Bareil R, Gingras M, Germain L, Berthod F. Hair follicles guide nerve migration in vitro and in vivo in tissue-engineered skin. J Invest Dermatol 2011;131(6):1375–8.

[66] Château Y, Dorange G, Clément J-F, Pennec J-P, Gobin E, Griscom L, et al. In vitro reconstruction of neuro-epidermal connections. J Invest Dermatol 2007;127(4):979–81.
[67] Shimohira-Yamasaki M, Toda S, Narisawa Y, Sugihara H. Merkel cell-nerve cell interaction undergoes formation of a synapse-like structure in a primary culture. Cell Struct Funct 2006;31(1):39–45.
[68] Klusch A, Ponce L, Gorzelanny C, Schäfer I, Schneider SW, Ringkamp M, et al. Coculture model of sensory neurites and keratinocytes to investigate functional interaction: chemical stimulation and atomic force microscope-transmitted mechanical stimulation combined with live-cell imaging. J Invest Dermatol 2013;133(5):1387–90.
[69] Fujiwara T, Kubo T, Kanazawa S, Shingaki K, Taniguchi M, Matsuzaki S, et al. Direct contact of fibroblasts with neuronal processes promotes differentiation to myofibroblasts and induces contraction of collagen matrix in vitro. Wound Repair Regen 2013;21(4):588–94.
[70] Taherzadeh O, Otto WR, Anand U, Nanchahal J, Anand P. Influence of human skin injury on regeneration of sensory neurons. Cell Tissue Res 2003;312(3):275–80.
[71] Lebonvallet N, Pennec J-P, Le Gall C, Pereira U, Boulais N, Cheret J, et al. Effect of human skin explants on the neurite growth of the PC12 cell line. Exp Dermatol 2013;22(3):224–5.
[72] Molliver DC, Wright DE, Leitner ML, Parsadanian AS, Doster K, Wen D, et al. IB4-binding DRG neurons switch from NGF to GDNF dependence in early postnatal life. Neuron 1997;19(4):849–61.
[73] Pereira U, Boulais N, Lebonvallet N, Pennec JP, Dorange G, Misery L. Mechanisms of the sensory effects of tacrolimus on the skin. Br J Dermatol 2010;163(1):70–7.
[74] Pereira U, Garcia-Le Gal C, Le Gal G, Boulais N, Lebonvallet N, Dorange G, et al. Effects of sangre de drago in an in vitro model of cutaneous neurogenic inflammation. Exp Dermatol 2010;19(9):796–9.
[75] Boulais N, Pereira U, Lebonvallet N, Gobin E, Dorange G, Rougier N, et al. Merkel cells as putative regulatory cells in skin disorders: an in vitro study. PLoS One 2009;4(8):e6528.
[76] Roggenkamp D, Köpnick S, Stäb F, Wenck H, Schmelz M, Neufang G. Epidermal nerve fibers modulate keratinocyte growth via neuropeptide signaling in an innervated skin model. J Invest Dermatol 2013;133(6):1620–8.
[77] Boulais N, Pennec J-P, Lebonvallet N, Pereira U, Rougier N, Dorange G, et al. Rat Merkel cells are mechanoreceptors and osmoreceptors. PloS One 2009;4(11):e7759.
[78] Hayes P, Meadows HJ, Gunthorpe MJ, Harries MH, Duckworth DM, Cairns W, et al. Cloning and functional expression of a human orthologue of rat vanilloid receptor-1. Pain 2000;88(2):205–15.
[79] Gruber JV, Holtz R. Examining communication between ultraviolet (UV)-damaged cutaneous nerve cells and epidermal keratinocytes in vitro. Toxicol Ind Health 2009;25(4–5):225–30.
[80] Chéret J, Lebonvallet N, Misery L, Le Gall-Ianotto C. Expression of neuroserpin, a selective inhibitor of tissue-type plasminogen activator in the human skin. Exp Dermatol 2012;21(9):710–1.
[81] Buhé V, Vié K, Guéré C, Natalizio A, Lhéritier C, Le Gall-Ianotto C, et al. Pathophysiological study of sensitive skin. Acta Derm Venereol 2016;96(3):314–8.
[82] Abels C. Intra-epidermal nerve fibers in human skin: back to the roots. Exp Dermatol 2014;23(4):232–3.

Skin in vitro models to study dermal white adipose tissue role in skin healing

14

Manuela E.L. Lago[*,†], *Mariana T. Cerqueira*[*,†], *Rogério P. Pirraco*[*,†], *Rui L. Reis*[*,†,‡], *Alexandra P. Marques*[*,†,‡]

[*]3B's Research Group—Biomaterials, Biodegradables and Biomimetics, Headquarters of the European Institute of Excellence on Tissue Engineering and Regenerative Medicine, University of Minho, Avepark, Barco, Guimarães, Portugal, [†]ICVS/3B's—PT Government Associate Laboratory, Braga/Guimarães, Portugal, [‡]The Discoveries Centre for Regenerative and Precision Medicine, Headquarters at University of Minho, Avepark, Barco, Guimarães, Portugal

1. Introduction

The notion of adipose tissue as an inert energy storage depot has long been foregone. The discovery of adipokines and their systemic effects has fueled the emergence of the concept of adipose tissue as an endocrine organ. Adipose tissue has been found to have key roles in regulating immunity and inflammation in both physiological and pathological settings. In humans, recent studies suggest that there are two distinct adipose layers associated with the skin. One is a subcutaneous adipose depot, and the other is the so-called dermal white adipose tissue that underlies the reticular dermis and encloses mature hair follicles. There is increasing evidence indicating a major role of adipose tissue in a variety of skin diseases. Concurrently, dermal white adipose tissue has been found to be essential in several cell-related mechanisms in dermal and epidermal tissues. Despite these facts, research aiming at characterizing the full role of this adipose tissue depot in skin tissue homeostasis and wound healing is scarce. This is partially due to the lack of adequate in vitro tools that allow an accurate recapitulation of what happens in the actual skin tissue. The development of novel in vitro models of the skin to study dermal white adipose tissue and its role on wound healing is therefore a need.

Advances in the biomaterials and cell biology fields are paving the way to the fabrication of in vitro constructs that mimic the structure and function of the native adipose tissue considering its individual components and its mechanical and biochemical properties. Such constructs will allow unveiling the inner workings of each component of adipose tissue, contributing to achieve functional adipose tissue models and to better understand its role in the skin.

This chapter overviews the main components of adipose tissue, what is currently known about their role in wound-healing events and hair formation and the state of the art in terms of available in vitro models. Furthermore, future trends are explored toward the fabrication of adequate in vitro models to study the role of dermal white adipose tissue.

Skin Tissue Models. https://doi.org/10.1016/B978-0-12-810545-0.00014-0

2. Adipose tissue

Adipose tissue has been commonly distinguished in two main types, white and brown adipose tissue [1,2]. More recently, beige adipose tissue was added to this distinction [3,4]. At different stages of development, different adipose types dominate the body. Newborn tissue consists mostly of brown adipose tissue to prevent hypothermia, while in adult humans, white adipose tissue dominates [2,3,5,6]. It is also known that mesenchymal precursor cells of neural crest or mesodermal origin give rise to different adipose depots. However, while it is known that brown adipocytes are derived from a myogenic factor 5+ (Myf5+) lineage, white adipose tissue adipocyte precursors origin has not been fully described [3,7,8]. Nonetheless, it is consensual that the depot location, along with morphological and functional cues, allows distinguishing each type of adipose tissue.

Adipose depots can be found mainly in subcutaneous and visceral locations; the major subcutaneous distribution of white adipose tissue includes superficial and deep abdominal and gluteal-femoral depots, while the visceral depots surround vital organs within the peritoneum and rib cage [9–12]. On the other hand, subcutaneous brown adipose tissue is mainly located around the shoulders, ribs, and axilla and visceral brown adipose tissue around perivascular, periviscus, and solid organs [9–11,13]. The distinction between white and brown adipose tissue also relates to their function. White adipose tissue has a major endocrine role controlling whole-body homeostasis, while brown adipose tissue is mainly thermoregulatory, being their components and respective interactions crucial for regulation of nonshivering events [1,3,14].

Adipose tissue is maintained from a pool of precursor cells, precommitted to differentiate into adipocytes. Self-replication and adipogenic differentiation of these preadipocytes are driven upon activation of the transcriptional factors nuclear hormone receptor peroxisome proliferator-activated receptor γ (PPARγ) and CCAAT/enhancer binding protein-α (C/EBPα) [15–17]. PPARγ is required during terminal embryonic adipogenesis, while C/EBPα is crucial for the dynamics of adipose tissue in adults [16,17]. Throughout life, the size and remodeling of the adipose tissue is dependent on the energy balance of the whole body. Expansion of the tissue occurs through the lipid enrichment of preexisting adipocytes, which increase in size (hypertrophy), or by the recruitment of preadipocytes from resident pools to increase of the number of adipocytes (hyperplasia) [16,18–22].

In a homeostatic state, diet-derived fatty acids synthesized in the liver and incorporated into water-soluble lipoproteins are uptaken by adipocytes through very-low-density lipoprotein (VLDL) receptors present in the cell membrane, and free fatty acids (FFAs) enter in the cell mediated by facilitated diffusion [23,24]. FFAs and cytokines/adipokines are the key regulatory elements for different processes mediated by adipose tissue. In the absence of glucose, adipocytes release fatty acids in the circulation to balance the carbohydrates required for cell metabolism [25]. Adipokines produced by adipocytes such as leptin and adiponectin also have important roles in different organs. Leptin receptor is highly expressed in the mediobasal hypothalamus [26–28], and its activation leads to repression of the orexigenic pathways and induction of anorexigenic pathways leading to food intake and energy expenditure [25].

This adipokine also regulates insulin sensitivity in the muscles and in the liver. In the first case, leptin acts directly on AMP-activated protein kinase (AMPK) reducing intramyocellular lipid levels and indirectly through central neural pathways [29]. In the liver, leptin inhibits proinsulin synthesis and secretion [25,30–32]. Similarly, adiponectin has an effect on insulin sensitivity by mediating the activation of AMPK in the skeletal muscle and liver. The expression of adiponectin receptor 1 and 2 regulated by the activation of AMPK leads to increased fatty acid oxidation and glucose uptake in the liver and inhibition of glycogenesis in the liver [33,34].

2.1 Adipose tissue cellular content

Depending on the location and physiological/pathological condition, adipose tissue has a different cellular composition [35]. Adipocytes are the main cellular component of adipose tissue, but the morphology of the adipocytes determines each type of adipose tissue varying from white to brown adipose tissue. Adipocytes in white adipose tissue are characterized by having a single lipid droplet for efficient energy storage, while adipocytes from brown adipose tissue present multiple lipid droplets and numerous mitochondria expressing UCP-1 [1–3,7,9]. Regarding beige adipocytes, the characteristics are shared with brown and white adipose tissue [9,36].

Besides adipocytes, there are other cell types interacting and coordinating adipose tissue functions (Fig. 1). Subcutaneous and visceral adipose depots are innervated, vascularized endocrine organs comprising precursor cells, fibroblasts, vascular cells, immune cells, and neurons, which altogether constitutes a stromal vascular fraction of adipose tissue [7,33].

The interaction between adipocytes and vascular cells such as endothelial cells and pericytes modulates adipose tissue angiogenesis and vascular remodeling, which are critical to accommodate increasing fat stores [7,33,37–39]. This is mainly mediated by a range of angiogenic factors (VEGF-A, bFGF, HGF, and leptin) and cytokines and adipokines (TNF-α, resistin, and A-FABP) produced by the adipocytes [40–42]. Other adipokines have also been implicated in the interaction between adipocytes and endothelial cells due to vasoactive properties. Leptin [43] and adiponectin [44,45] enhance the NO-dependent vasorelaxation through the activation of AMPK/protein kinase B (Akt) signaling cascade in the endothelial cells. Leptin also unleashes an NO-independent vasorelaxation mechanism promoted by the production of endothelium-derived hyperpolarizing factor (EDHF) [40,43].

Fibroblasts, the main element of adipose tissue associated to the production of extra components [33], interact with adipocytes through endocrine signaling [46]. Adipokines such as adiponectin and leptin are released by adipocytes and recognized by human dermal fibroblasts through their receptors adiponectin receptor 1 (AdipoR1) and receptor 2 (AdipoR2) and leptin receptor (Ob-Rb) [47]. It is known that fibroblasts respond to these signals by increasing hyaluronic acid and collagen production as a regulatory pathway to maintain the dermal homeostasis [46,47].

Resident adipose tissue macrophages (ATM) are critical cellular constituents of adipose tissue due to their role in systemic homeostasis such as in the control of the release of lipids into the bloodstream, through a balanced storage of the excess released

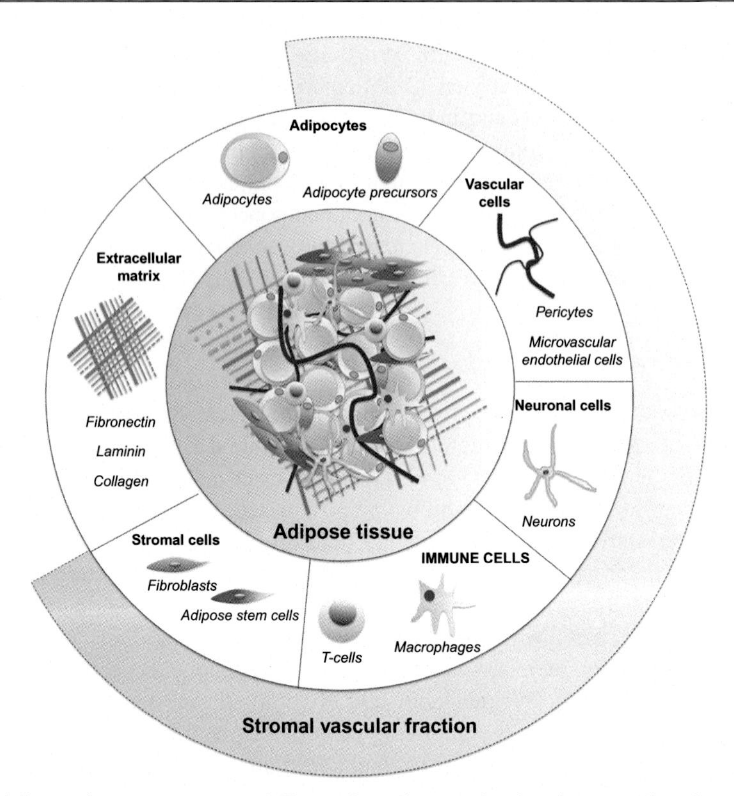

Fig. 1 Adipose tissue components. Adipose tissue is an endocrine, innervated, and vascularized organ with a collagen-, laminin-, and fibronectin-rich extracellular matrix. Mature adipocytes are the main component cellular of the adipose tissue. However, other cell types such as precursor cells, fibroblast, vascular cells, immune cells, and neurons, which altogether constitute a stromal vascular fraction, are also present.

by adipocytes [35,48]. Critically, they are also implicated in adipose tissue-related disorders, such as obesity and insulin resistance [48–50].

Adipose tissue T cells (ATT) commonly referred to T cells residing in adipose tissue are key mediators of metabolic and inflammatory states. Subpopulations of T cells differ according to regional fat depot and are associated with both systemic inflammation and insulin resistance [51]. However, the mechanisms of adaptive immune system surveillance in adipose tissue are still quite elusive. There are evidences that the immune response in adipose tissue is controlled by ATT via an antigen-driven process involving a fat-specific antigen that may be critical in maintaining a low inflammatory state in lean individuals [52]. But what protein or carbohydrate antigens are being recognized is yet to be discovered [53].

2.2 Adipose tissue extracellular matrix

Adipose tissue microenvironment is regulated not only through mono- and heterotypic cell-cell interactions [7] but also through the interaction of cells with the extracellular matrix (ECM), which also defines the mechanical features of each type of adipose tissue. In general, adipose tissue ECM is mainly composed of collagen, elastin, laminin, and sulfated glycosaminoglycans [7,54,55]. Some authors suggest that the composition varies with the species, type of fat depots, and differentiation process [56], but complete evidence that these variations are not dependent on protein characteristics, experimental conditions, and methodologies is yet to be clarified.

Adipose tissue ECM is as dynamic as the tissue itself in the sense that ECM remodeling is regulated by insulin, by the energy metabolism, and by mechanical forces that vary with the adipogenesis and the state of mature adipocytes. In fact, up to 6 days of differentiation, the expression of collagen IV, nidogen I (entactin), and various laminin complexes was upregulated in mouse 3T3-L1 preadipocyte cultures [57,58], while a decrease of collagens type I and type III was observed [56], marking a transition from a fibrillar to a laminar state. Moreover, after 12 days of differentiation, the values of collagens I and III returned to the levels of day 0, and the levels of collagens IV, V, and VI, although decreasing, remained above the initial ones [56,59]. The high and low expression of, respectively, collagens I, III, IV, VII, VIII, and X and collagens II and IX has been also confirmed in human adipose tissue in addition to adipocytes in culture [60].

2.3 Pathophysiology of adipose tissue

Adipose tissue malfunctioning has been associated to obesity, type 2 diabetes, and atherosclerosis. In obesity cases, an unbalanced diet and lifestyle leads to an increased size of adipose tissue depots, either by hypertrophy or hyperplasia of adipocytes, and fat deposition in other vital organs [16,25,33]. This in turn results in unbalanced adipokine secretome and to increased secretion of inflammatory cytokines [16,25,33]. According to several studies, abnormal levels of adipokines such as adiponectin [61], nesfatin [62,63], omentin [64], and resistin [65] result in disturbances in glucose tolerance and insulin action. Similar effects on insulin action have been noticed in lipidic metabolites or lipokines [66,67]. Concomitantly, hypertrophic adipocytes can present necrotic-like abnormalities and are responsible for a decrease in adipose tissue function [11,68]. This functional atrophy is accompanied by an enhanced released of pro-inflammatory factors, such as TNF, interleukin-1β (IL-1β), IL-6, IL-8, leptin, resistin, and monocyte chemoattractant protein-1 (MCP1), and a reduction in anti-inflammatory factors, such as IL-10 and adiponectin [11,16,33,69]. All of this has detrimental effects on insulin-related mechanisms and on glucose output, which ultimately may lead to type 2 diabetes. Furthermore, the increased release of pro-inflammatory adipokines promotes the recruitment of macrophages and T cells, while local hypoxia, due to a reduction of vasculature, activates hypoxia-inducible factor (HIF) 1α inducing adipose tissue fibrosis [11,19]. The ensuing lipolysis results in the release of FFAs that are taken by other organs, resulting in its accumulation and lipotoxicity [11,16,70], which in turn is bound to disturbed insulin action and glucose metabolism. Interestingly, contrary to hypertrophy, adipocyte hyperplasia may counterbalance the deleterious effects of hypertrophy since it results in metabolic correct adipose tissue with normal insulin-related parameters [11]. However, the mechanisms by which hyperplasic events may have beneficial effects over hypertrophic ones are not completely elucidated.

Therefore, a combination of factors triggered by obesity, namely, inflammation, is probably what leads to diabetes and atherosclerosis. Moreover, while the exact mechanisms leading to atherosclerosis in obese patients are elusive, its pronounced association with type 2 diabetes suggests that the underlying causes might be common [71].

3. Dermal white adipose tissue

Traditionally, abdominal subcutaneous white adipose tissue was considered for many years as a unique and the largest depot of skin-associated white adipose tissue. However, latest studies in mice revealed that, during development, a common precursor in dermal mesenchyme gives rise to both dermal fibroblasts anddermal white adipocytes [72]. Moreover, it was also demonstrated that the development of adipocytes in the skin is independent from subcutaneous white adipose tissue development [73]. Dermal white adipocyte precursors were also shown to be capable of regenerating skin adipocytes during hair cycle and after wounding [74,75], as discussed in the next sections. Based on this, a new nomenclature to refer to the skin-associated adipose tissue as dermal white adipose tissue was defined, which is expected to retrospectively change the view on many previous studies [76]. This new depot refers to the thick layer of dermal white adipocytes that encloses mature hair follicles.

3.1 Role in skin function

In humans, unlike mice, the absence of a continuous muscle layer separating dermal from subcutaneous adipose tissue impedes a clear anatomical distinction between them. However, several studies in mice have highlighted that these differences go beyond the depot location and that dermal white adipose tissue responds to several independent cues [73,74,77,78]. Among them, the dermal white adipose tissue depot appears to react to temperature [77] by increasing its thickness in cold or thinning out when in warm environments. Moreover, the increase in dermal white adipose tissue size has been also connected to defensive responses due to an abundant production of antimicrobial peptides by differentiating preadipocytes after bacteria exposure [78]. In accordance to that, the hypertrophy of dermal white adipose tissue was associated to increased susceptibility to and severity of skin infection [78].

3.2 Role in hair follicle cycle

In addition to a role in the overall skin function, dermal white adipose tissue depots display a high kinetic activity surrounding the hair follicles, varying with the hair cycle and then impacting their function and metabolism [74,79]. Hair formation relies on cyclic events that are organized in an anagen phase, where proliferative epithelial stem cells closely communicate with dermal papillae cells and hair shaft is produced, a catagen phase that comprises the regression of the hair follicle, and by a telogen phase, where cells are relatively quiescent [80,81]. During this last phase, signaling from adipocyte precursor cells activates hair follicle growth. This has been particularly evident in a bidirectional regulation through BMP [74] and WnT/β-catenin [79] signaling. In the anagen phase, keratinocytes produce proadipogenic factors downstream of Wnt/β-catenin signaling cascade, directly stimulating the expansion of dermal white adipose tissue depot size [79]. Dermal white adipose tissue adipocytes have shown to produce BMP2 during catagen/early telogen phase, causing follicles to be noncompliant to activation cues [82]. As this BMP signaling blocks anagen induction [83], it has been suggested that it is required for hair follicle stem cells quiescence [83]. The hair follicles are activated again, entering in anagen phase only when that signaling is finally reduced. While the involvement of BMP signaling in hair follicle cycling seems to be consensual, how it occurs still needs to be further investigated, as it appears that both hair follicle epithelial and dermal papillae cells are BMP-producing cells [84,85].

4. Dermal white adipose tissue action during wound healing

Another set of highly dynamic events where dermal white adipose tissue role has been more and more recognized is cutaneous wound healing (Fig. 2) [75]. This process is quite complex as it involves a coordinated action of multiple cell types and factors toward skin repair, as extensively reviewed elsewhere [86]. The main phases of skin wound healing include three different stages after injury: inflammation, proliferation, and remodeling [87]. In a very simplistic way, after the formation of a clot by platelet cells, peripheral blood mononuclear cells infiltrate the wounded area to clear debris. Subsequently, proliferation phase is characterized by keratinocyte migration from wound edges and by the migration of fibroblasts that generate a transitory ECM, which is then remodeled in the final stage of wound healing lasting several weeks [87]. Upon disruption of skin homeostasis by an insult/injury, the first alterations detected in the dermal white adipose tissue depot are a change in size, as a consequence of preadipocyte proliferation and repopulation by mature adipocytes [75], and in the secretome [75]. Consequently, dermal white adipose tissue affects the cutaneous healing cascade. A more evident effect over the proliferation phase of healing, particularly in keratinocyte and fibroblast migration, has been associated to a potential role of dermal white adipose tissue, respectively, in wound reepithelialization [88–93] and cutaneous fibrosis [94,95].

4.1 Action in reepithelialization

Factors such as leptin, adiponectin, and possibly TNFα play a pivotal role in keratinocyte proliferation/migration [88–90] during skin wound healing. It is known that leptin, predominantly produced and released by adipose tissue [96], is also actively produced by skin resident cells. Its secretion is triggered shortly after injury and prolonged throughout the various healing stages at the wound site [97]. Nonetheless, its specific role and its origin have just started to be unveiled. So far, studies on leptin-deficient obese mice (ob/ob) characterized by a severely delayed wound healing that is related with their mild diabetic phenotype revealed that systemic or topical application of leptin stimulates keratinocyte proliferation at the wound edges and consequently

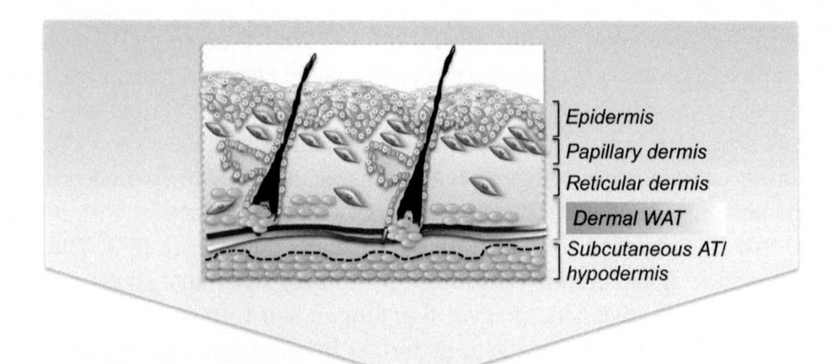

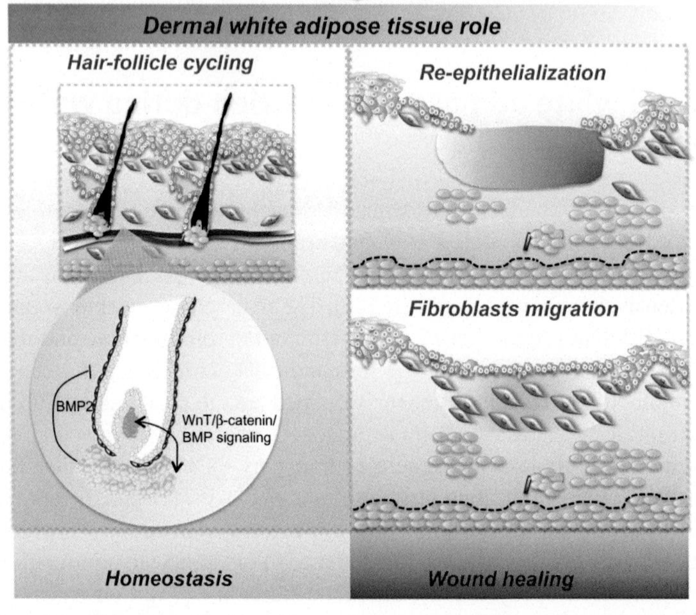

Fig. 2 Overview of the dermal *white* adipose tissue role in different skin-related events. It is known that dermal *white* adipose tissue is involved in the regulation of hair cycle function and metabolism contributing to skin homeostasis and in the reepithelialization and cutaneous fibrosis during wound healing.

reepithelialization of skin wounds [88]. Moreover, leptin topically administered to wild-type mice led to accelerated healing. The effect of dermal white adipose tissue on reepithelialization seems to be corroborated by studies focused on the role of adiponectin, exclusively secreted from adipose tissue. Low levels of this adipokine, either in adiponectin-deficient mice [89] or in diabetic ulcers [91], lead to a severely delayed reepithelialization and peripheral hyperkeratinosis in diabetic wounds. Accordingly, after administration of adiponectin, both systemically and topically, impaired healing was partially reversed and linked to an activation of ERK signaling in keratinocytes [89]. Considering that ERK pathways are required for keratinocyte proliferation and migration, these findings suggest that adiponectin positively regulates the reepithelialization process during cutaneous wound healing [89]. Adiponectin was also shown to promote upregulation of keratin gene transcripts within hours after injection in mice ear and dorsal excisional wounds [92]. In this study, the faster reepithelialization and the generally improved wound healing appeared to result from a downstream effect of adiponectin promoting cellular proliferation at the basal epithelial cell layer and pilosebaceous units.

Although there are no particular works demonstrating the involvement of dermal white adipose tissue, other factors besides the referred adipokines produced by white adipose tissue in general might also play a role in reepithelialization. This is the case of TNFα that was reported as a factor actively secreted by the subcutaneous depot of white adipose tissue [98]. This cytokine has been pointed out to be a common initiator of keratinocyte activation with downstream paracrine effects [99]. It is known that TNFα acts in an autocrine way directly in keratinocyte migration, with further paracrine effects on fibroblasts to produce FGF7 (KGF) [93]. This action in particular is related with fibroblasts migration and deposition of ECM, thus indirectly affecting keratinocyte motility and consequently reepithelialization. However, little is known about the expression/secretion of this molecule by dermal white adipocytes and its direct involvement in skin wound-healing mechanisms.

4.2 Dermal fibroblasts recruitment and wound matrix production

As mentioned above, dermal white adipose tissue appears to have also an active role in fibroblasts recruitment and proliferation, leading to a subsequent downstream effect on ECM production at the later remodeling phase. In fact, as soon as 5 days after wounding, resident adipocyte precursors were found to rapidly proliferate and differentiate, recruiting fibroblasts which then repopulate the wound bed along with the mature adipocytes [75]. Moreover, the importance of a mature adipogenic phenotype for an efficient cross talk with fibroblasts that promote migration is demonstrated. In lipoatrophic mice, when mature adipocytes are absent, an aberrant recruitment of fibroblasts to the wound site occurs. Activated fibroblasts with ER-TR7+ phenotype were detected in wound borders but not in wound bed, suggesting that they are just recruited there in the presence of mature adipocytes [75]. These observations are also related with another work that blocked adipogenesis using pharmacological antagonists of the master adipogenic factor PPARγ [75]. In vitro fibroblasts treated with increased concentrations of the antagonist remained unaltered. After injection of this antagonist in mice, a striking reduction of mature adipocytes was observed, but not of adipocyte precursors. Following that, despite normal reepithelialization, revascularization, and recruitment of macrophages and neutrophils during the wound healing, low migration of fibroblasts to wound site led to diminished ECM deposition and consequent reduced dermal reconstruction [75]. This showed that the effect in this migration is not due to the antagonist administration itself, but due to the absence of mature adipocytes. Despite this knowledge, the exact particularities of the communication between fibroblasts and dermal white adipocytes and their mechanistic action are still elusive. Additionally, the fact that the adipocytes maturation degree strongly influences the wound healing cascade might be related with the cytokine profile produced differently by these cell types in distinct differentiation degrees. This suggests a possible paracrine effect of mature components of dermal white adipose tissue over other skin cells.

Furthermore, it has been more and more recognized the involvement of adipocytes and their precursors in skin fibrosis. This appears as a consequence of fibrotic dermis deposition during wound healing by the activated myofibroblasts, which is normally associated with an atrophy of subcutaneous adipose tissue [94]. The origin of these myofibroblasts has been highly debatable due to the variety of models of fibrosis that have been used. However, recent findings using a lineage-tracing strategy to label mature adipocytes in a bleomycin-induced fibrosis model revealed that, in the dermal white adipose tissue, the adipocyte lipid vesicle marker perilipin colocalized with αSMA myofibroblast marker [94]. This was also accompanied by a morphological change from a round shape characteristic of adipocytic cells to a spindle-like shape. This data revealed that in dermal fibrosis some of the myofibroblasts in dermal fibrosis derive in fact from adipocytes. Additional studies also validated these observations, showing that the fibrosis-associated protein FIZZ1 was capable of inducing the transformation of cultured adipocytes into myofibroblasts expressing αSMA [95]. Moreover, the deletion of this profibrotic protein in transgenic mice significantly reduced the severity of skin fibrosis, highlighting the contribution of adipocyte-origin myofibroblasts in this phenomenon. Nonetheless, the origin of these myofibroblasts is not only restricted to mature adipocytes but also linked to adipocyte precursor cells. When stimulated in vitro by TGF-beta 1 and bleomycin, these cells are able to acquire a myofibroblast phenotype. Also, the induction or the inhibition of adipogenesis in mice had reversing effects in skin fibrosis, underlining the close correlation between adipogenesis and fibrosis. Taken together, these findings demonstrate that these adipocytic cells play a key role in fibrosis. However, further insights on the interactions of these cells as part of the dermal white adipose tissue depot with surrounding skin cells during fibrosis are still needed.

5. Models to study the role of dermal white adipose tissue

5.1 Animal models

Several animal models have been used to study particular functional and mechanistic traits of skin homeostasis and healing. Rats, mice, and pigs have been traditionally the most used animals in such experimentations. In the case of the small rodents, their skin presents several important differences in comparison with the human skin. The rat and mice skin is loosely attached to the subcutis [100], and healing happens mainly through contraction [100]. In opposition, the skin of pig is the one that presents more similarities with the human skin in terms of structure and because it is firmly attached to the subcutis [100,101]. Critically, porcine skin wound healing is achieved by reepithelialization, such as in the human skin [101]. Notwithstanding the differences between these animal models, there are many more studies in rodents, and only some of these focuses on dermal white adipose tissue.

A-ZIP mice that completely lack white adipose tissue and early B-cell factor-null (*EBF*$^{-/-}$, a protein present in dermal white adipocytes) mice were compared to assess the role of adipocyte progenitors in skin homeostasis [74]. While A-ZIP mice did not show any skin changes, in *EBF*-null mice, hair follicle cycle was disrupted, and hair follicles remained in late catagen or telogen phases. These effects were correlated with the absence of adipogenic lineage cells, establishing a regulatory role of these cells, and consequently of dermal white adipose tissue, in epidermal stem cell activity and hair follicle cycling. Knockout mice for *Mpzl3*, a protein related with the *rough coat* phenotype, were used to determine the role of that protein in the skin. These mice presented a significant reduction of overall adipose tissue, but in particular dermal white adipose tissue [102]. These changes were accompanied by increased dermal thickness, sebaceous gland hypertrophy, and a tendency to develop inflammatory skin diseases, which reinforces the regulatory role of dermal white adipose tissue both at the physiological and pathological levels. This is also confirmed with other transgenic models that present defects in dermal white adipose tissue, *Fatp4*-null mice [103] and *Dgat2*-null mice [104] showed disturbances in adipose tissue content that were accompanied by abnormalities in dermal and/or epidermal function. In the first case, disruption of *Fatp4* gene, involved in fatty acid transport, generated an altered fatty acid composition of epidermal ceramides that originated hyperproliferative hyperkeratosis with decreased epidermal barrier function. Defects in epidermal barrier function were also found in *Dgat2*-null mice that lack DGAT2 enzyme. This enzyme is involved in triglyceride synthesis, and mouse lacking the enzyme was found to be lipopenic, which had a clear effect on skin barrier function due to thinner epidermis, orthohyperkeratosis of the *stratum corneum*, and flattened dermal-epidermal junction. Similarly, APOC1 mice produced to study the role of APOC1 protein in lipoprotein metabolism and the lack of dermal white fat presented a set of cutaneous abnormalities, such as epidermal hyperplasia and hyperkeratosis and sebaceous gland atrophy [105]. Mouse models where cutaneous fibrosis was induced show consistent atrophy of the adipose component of the skin [94,106–109], which is in line to what is reported in human subjects with the same disease. A study using bleomycin-challenged mice to induce skin fibrosis demonstrated that the loss of dermal white adipose tissue precedes skin fibrotic lesions [94]. Critically, the authors showed that adiponectin-positive dermal white progenitors or adipogenic cells gave rise to dermal myofibroblasts, clearly suggesting a huge role of dermal white adipose tissue in the disease.

While transgenic mouse models are versatile and very useful to uncover particular molecular pathways, their contribution to reveal the role of dermal white adipose tissue in humans is somewhat limited by the significant morphological differences between human and rat/mouse skin and by the differing skin wound-healing mechanisms. On the other hand, although having the greatest similarities to humans, including the hypodermis adipose chambers and its lipid content [110], pig models are expensive to maintain and require specialized facilities that are not accessible for most groups. Furthermore, there are variations in the biological response of skin tissue according to the breed of pigs [111]. In alternative, humanized rodent models have long been proposed to overcome the functional differences between small rodent skin and human skin [112,113]. Nonetheless, although the human xenografts maintain some functionality, the wound-healing process is obviously severely affected by the fact that the recipient animals must be immunocompromised [114]. All these facts only illustrate the need to develop adequate in vitro models to study dermal white adipose tissue function that are cheap and easy to set up in any lab with cell culture capabilities and that can effectively bridge fundamental research with clinical research.

5.2 In vitro models: A tissue engineering perspective

Tissue engineering is a relatively new field that appeared in the context of severe shortage of donor material to replace injured tissues and organs. It is based on a multidisciplinary approach involving engineering, biology, and medical sciences to create or engineer tissue analogues [115]. Classically, Tissue Engineering involves the use of cells, an ECM surrogate, or scaffold, and growth factors that under the right conditions form tissue or organ analogues. Additionally, given the potential of tissue-engineered constructs to mimic native tissues, they have also been used as 3-D in vitro models with superior reliability and reproducibility as compared with typical 2-D cultures [116].

Based on the concept of tissue engineering, the pursuit for reliable 3-D adipose tissue models has been following different routes. Regarding the type of cells used, 3T3-L1, a murine preadipocyte cell line, is one of the most common cell source used to study various aspects of adipocyte biology including adipogenesis. In fact, a third of works in the last 5 years reporting its use are related to adipogenesis and obesity-related characteristics [117]. The main advantages of cell lines are that they are easier to culture, have fewer associated costs, and deliver homogenous response to experimental conditions. However, they are a not good option to extrapolate the in vitro results to in vivo conditions [117] and therefore depending on the questions to be answered might not be the best ones to generate in vitro models. Others such as primary cells (mature adipocytes) and adult stem cells have been explored as well. Primary adipocytes have the advantage of being obtained from multiple sources thus allowing the study of age- and gender-dependent adipogenic. However, limited proliferation is a significant hindrance in the use of primary adipocytes for the generation of reproducible models. The higher proliferation capacity of adult stem cells and the potential to differentiate into the adipogenic lineage have been the basis for the greater number of approaches that have been exploring these cells to obtain 3-D in vitro adipose tissue models [117].

In what concerns the 3-D structure in which cells are organized in the mimicking of the native tissue, scaffold-free approaches [118] have been recently introduced in the field taking advantage of the high degree of similarity between cell-produced ECM and tissues' native ECM. Nonetheless, the classical perspective of growing/differentiating cells within synthetic- or natural-origin scaffolds has been highly explored to recreate some features of the native adipose tissue (Fig. 3). Independently of the type of cells used, an overview of in vitro strategies to obtain 3-D models of adipose tissue and the main outcomes achieved so far is provided next.

5.2.1 Cell-sheet based models

Scaffold-free 3-D models, such as spheroids, have been proposed as 3-D tissue models with enhanced value to address and study cellular mechanisms in vitro [119,120].

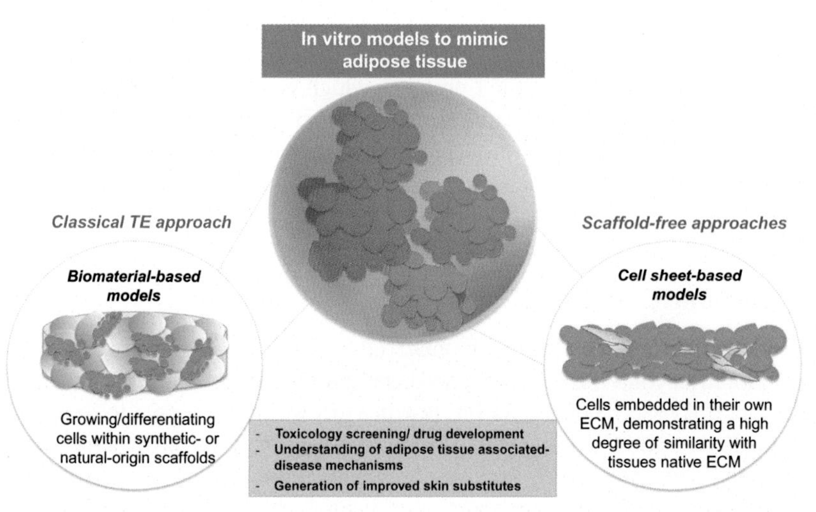

Fig. 3 In vitro models of adipose tissue. Both the classical tissue engineering approach of growing/differentiating cells within synthetic- or natural-origin scaffolds and the scaffold-free strategy have been explored to recreate some features of the native adipose tissue aiming to achieve in vitro 3-D models of adipose tissue useful not only for toxicology screening and drug development but also to better understand associated disease mechanisms and ultimately to lead the generation of improved skin substitutes.

Despite this, one of the disadvantages of these models is the residual or absent ECM that is known to play a key role in tissue physiology and pathophysiology. Cell sheet engineering, which relies on the recovery of cells as sheets retaining cell-cell and cell-ECM junctions, has been posed as an impressive alternative to overcome that issue [118]. In fact, by stacking different cell sheets, it is possible to create 3-D constructs composed of cells and their self-produced ECM components that therefore mimic in a highly accurate manner native tissue microenvironment. Adipose tissue microenvironments based on this technology were recreated by inducing human adipose-derived stem cells to produce ECM and then to differentiate toward the adipogenic lineage [121–124]. One of the works provided a deep understanding of the adipogenic differentiation process by analyzing the effect of the culture conditions over the differentiation of the generated cell sheets [124]. Cell sheets with different degrees of differentiation were obtained, and 3-D constructs with diverse secretomes were created by assembling the cell sheets in varied combinations. This strategy will be of benefit to mimic different adipose tissue microenvironments according to (patho)physiological features, thus allowing the study of mechanisms of interest in a more accurate mode. In other works, adipose cell sheets were stacked to engineer adipose substitutes that engage in adipokine secretion and lipolysis after β-adrenergic receptor stimulation, two major native adipose tissue functions [122]. The same strategy was used to produce adipose tissue substitutes with a surface area superior to 33 cm^2 where adipocytes were embedded in an ECM composed of fibronectin, collagens I and IV, as in native adipose tissue, and adipokine production was comparable with native adipose tissue [123]. Importantly, similar constructs were combined with endothelial cells to create prevascularized adipose tissue-like analogues [123]. These were stable for at least 11 weeks and were able to produce significant amounts of leptin, PAI-1, and angiopoietin-1 proteins, which was reflected in a superior ability to support capillary-like structure formation. The strategy described in these last three works [122–124] aims at producing fully functional adipose tissue analogues for tissue replacement but which can be of great value for toxicology screening and drug development.

5.2.2 Biomaterial-based models

As mentioned before, the classical biomaterial-based tissue engineering concept has been explored beyond the tissue replacement and repair/regeneration context in the development of tissue models. So far, few in vitro works were able to develop reliable models that allow distinguishing or representing features of the adipose tissue. Additionally, these are mainly focusing on white adipose tissue potentially because there are still many uncertainties about what distinguishes the different adipose tissue depots.

A wide range of materials (both synthetic and of natural origin including decellularized matrices) and culture conditions (differentiation or differentiation plus maintenance culture media, number of cells, and dynamic environment) have been tested to recreate in vitro adipose tissue. The achieved degree of similarity with the native adipose tissue has been mainly focusing on the level of adipogenic differentiation achieved within the different 3-D structures, the stability of the obtained construct, and on their vascularization (Table 1).

Like for many other tissue engineering strategies, the use of ECM-derived proteins/peptides or decellularized tissue has been a highly explored approach in the construction of adipose tissue-like analogues. While this rationale has demonstrated successful achievements in different areas [125,126], it is somehow questionable concerning adipose tissue engineering since, as mentioned above, adipose tissue ECM composition greatly reflects a balance between collagens I and IV, depending on its maturation/differentiation stage. Nonetheless, collagen I [127] and collagen I-hyaluronic acid [128] sponges, collagen and alginate microspheres [129], and gelatin-hyaluronic acid gels [130,131] are among the different materials used as platforms to differentiate adipose-derived stem cells into the adipogenic lineage. Although cells were able to differentiate, in a direct comparison with human platelet-poor plasma (PPP), alginate and fibrin gel, a collagen sponge did not perform and the PPP and alginate systems [127]. Interestingly, other works have also shown enhanced differentiation of 3T3-L1 preadipocytes in alginate modified with either early embryonic laminin and hyaluronic acid [132] or laminin [133]. Peptides present in ECM proteins and known to directly interact with integrins such as αVβ5 and α1β1 have been also incorporated in poly(ethylene glycol) (PEG) hydrogels [134] and poly(amidoamine) foams [135] to promote adipogenic differentiation. However, either the condition with RGDs was not compared with the nonfunctionalized one, or the lack of striking differences prevents concluding about the relevance of this approach in the generation of adipose tissue-like environments. Alternatively, the decellularization of human adipose tissue

Table 1 **Overview of the strategies followed to attain biomaterial-based adipose tissue-like microenvironments**

Similarity with the native adipose tissue	Biomaterial	Cell types and number	Main achievements	References
ECM-derived proteins	Platelet-poor plasma (PPP), alginate, fibrin gel, and collagen I sponge	Human adipose-derived stem cells (5×10^6 cells/mL of polymer solution)	Differentiation was lower in collagen sponges than in PPP and alginate systems	[127]
	Gelatin-hyaluronic acid	Porcine adipose-derived stem cells (5×10^4 cells/scaffold)	Cryogels with higher percentage of gelatin were further considered for adipogenic differentiation due to higher similarities with the matrix	[131]
	Alginate, laminin, and hyaluronic acid (HA-L-Alg)	Murine preadipocyte cell line 3T3-L1 ($\sim3\times10^5$ cells)	Enhanced adipogenic differentiation when alginate is modified with either early embryonic laminin and hyaluronic acid or laminin	[132,133]
	Laminin-alginate	Murine preadipocyte cell line 3T3-L1 (5×10^4 cells/mL of beads)		
Functionalization with peptides derived from ECM proteins	Poly(ethylene glycol) functionalized with RGD	Human adipose-derived stem cells (2×10^6–2×10^7 cells/mL of hydrogel)	The degree of differentiation varied with the peptide conformation and consequently associated to initial cell adhesion number	[134]
	Poly(amidoamine) oligomer microporous foam (OPAAF) functionalized with RGD	Murine preadipocyte cell line 3T3-L1 (2×10^6 cells/scaffold)	OPAAF scaffolds support adipogenic differentiation, but comparison with RGD-free structures is not performed	[135]

Continued

Table 1 **Continued**

Similarity with the native adipose tissue	Biomaterial	Cell types and number	Main achievements	References
Decellularized adipose tissue	Polyacrylamide functionalized with decellularized human lipoaspirate	Human adipose-derived stem cells (0.5×10^4 cells/cm^2)	The presence of the biochemical cues from the native adipose tissue favors adipogenic differentiation	[136,137,139]
	Methacrylated chondroitin sulfate in combination with milled decellularized adipose tissue	Human adipose-derived stem cells (2.5×10^5 and 5.0×10^5 cells/scaffold)		
	Decellularized tissue	Human adipose-derived stem cells (1×10^6 cells/scaffold)		
Mechanical properties	Collagen and hyaluronic acid	Murine preadipocyte cell line 3T3-L1 (1×10^6 cell/mL)	Scaffolds with increasing percentages of hyaluronic acid reaching a Young's modulus of approximately 6.7 kPa showed higher expression of the adipogenic differentiation marker adipsin	[128]
	Hyaluronic acid and gelatin	Human adipose-derived stem cells (2.5×10^4 cells/cm^2 of hydrogel)	Hydrogels with a Young's modulus close to the one of native tissue (3.5 kPa) supported adipogenic differentiation on their surface	[130]
	Human decellularized adipose tissue	Human adipose-derived stem cells (1×10^6 cells/scaffold)	Higher adipogenic differentiation was attained in the scaffolds with a Young's modulus of 3.67 kPa, which seemed to add to the biochemical cues of the decellularized adipose tissue forming the 3-D structures	[138]

Stability and vascularization	Laminin-coated silk scaffolds	Endothelial cells (8×10^5 cells/scaffold) cocultured with human adipose-derived stem cells (4×10^5 cells/scaffold) triggered into the adipogenic differentiation	At late time points in a dynamic culture, endothelial cells were organized into continuous capillary-like structures within a stable adipogenic construct	[143]
	Silk scaffolds	Liquefied adipose tissue containing unilocular adipocytes and endothelial cells	Under dynamic conditions of a dynamic culture, a vascularized tissue-like microenvironment containing physiologically relevant mature unilocular adipocytes responsive to an acute inflammatory stimulus was attained	[144]
Dermal white adipose tissue mimicking	Collagen type I	Human mature adipocytes ($\sim 1.07 \times 10^5$ cells per hydrogel)	Full-skin equivalent containing the underlying subcutaneous adipose tissue able to release adipocyte-specific factors such as adipsin, visfatin, leptin, and PAI-1 in levels comparable with those of control cultures of mature adipocytes	[145]

has been explored as a way to preserve in a higher degree the composition and the structure of the native ECM. In fact, the biochemical cues of the decellularized ECM have been shown to contribute to an inductive microenvironment for the adipogenic differentiation not only if used to functionalize the surface of polyacrylamide gels [136] and within photo-cross-linkable methacrylated chondroitin sulfate as a bioactive matrix [137] but also as a part of the whole decellularized 3-D structure [136,138,139].

Despite being known that the biomechanics of the 3-D support structure adds into the role of the biochemical factors in stimulating the adipogenic differentiation, it is not clear if the decellularized adipose tissue actually preserves this function. In fact, decellularized adipose tissue ECM was solubilized and used to prepare microporous and bead foams under conditions that, together with the concentration of soluble ECM, affected the mechanical properties of the generated 3-D structure. Moreover, higher adipogenic differentiation was observed in the microporous foam with a Young's modulus of 3.67 ± 0.38 kPa [138], which is the 3–4 kPa range measured for human adipose tissue [140,141]. Similar Young's modulus (3.5 kPa) was achieved for glutaraldehyde cross-linked hyaluronic acid/gelatin and human platelet-poor plasma gels, but adipogenic differentiation was not carried out in the materials but on their surface [130]. Interestingly, higher expression of adipogenic differentiation markers was observed with scaffolds composed of collagen and with increasing percentages of hyaluronic acid that reached the Young's modulus of approximately 6.7 kPa [128].

To be reliable and functional, in vitro adipose model tissue-like structures have to possess appropriate stability. From the works that have addressed the stability of adipose tissue-like constructs generated in vitro up to 12 weeks after implantation in different animal models [129,131,138,142], some suggested that this stability is in great part provided by their vascularization. This has been also demonstrated in vitro in two works that aimed developing physiologically relevant and long-term models of human adipose tissue. Human adipose-derived stem cells cultured in adipogenic differentiation medium for 7 days were seeded onto a 3-D silk porous scaffold precultured with microvascular endothelial cells, and the cocultured structure was further kept in static conditions and then transferred into spinner flasks for up to 6 months [143]. The dynamic culture allowed not only achieving the required stability but also the organization of the endothelial cells into continuous capillary-like structures. In a subsequent work [144], liquefied adipose tissue was used to repopulate the silk scaffolds in order to better represent the complexity of native adipose tissue and in particular the physiologically relevant mature unilocular adipocytes. The adipocyte functionality in vitro and the applicability of the generated model to respond to an acute inflammatory stimulus (TNFα) was demonstrated.

Despite this exciting work that provides the most advanced proof of concept of a biomaterial-based white adipose tissue model to study associated disease mechanisms and drug responses, the specificities of the dermal white adipose tissue are not addressed. Recently, a full-skin equivalent containing the underlying subcutaneous adipose tissue was reported [145]. The culture conditions were optimized to allow a release of adipocyte-specific factors such as adipsin, visfatin, leptin, and PAI-1 in levels comparable with those of control cultures of mature adipocytes.

6. Future trends

Adipose tissue moved from being understood as a simple storage of lipids and energy to being a complex endocrine organ that regulates many different processes dependent on their depot location. In particular, dermal white adipose tissue complexity is finally being unveiled, and its role has been described to be of major importance in both skin homeostasis, namely, in hair follicle cycle signaling, and in skin wound healing. So far, taken together the data herein described, one can extrapolate which adipokines, generally detected in white adipose tissue, have shown promising skin wound-healing effects, but its secretion by dermal white adipose tissue is yet to be investigated. Developmental cues that are currently under investigation will have a striking effect on the design of new approaches to act as models to explore the therapeutic role of this depot under the wound-healing context. Moreover, these findings will also impact wound management and treatment strategies by supporting the development of improved skin substitutes aimed to regenerate and restore full tissue functionality.

The majority of adipose tissue models described up to date take advantage of tissue engineering approaches; nonetheless, they are still in their infancy regarding the complexity of the niche. In this sense, there is the need for generating models with higher complexity taking advantage of cutting-edge tools including advanced material processing, incorporation of factors, and different cell players with unprecedented precision. Moreover, in order to understand its action, there is the urgent requirement of addressing specific questions regarding how this depot is regulated. It is critical to understand that the ECM of this depot, in addition to unveiled factors and cellular roles, also reflects its (patho)physiological state being therefore key in the construction of accurate and reliable models of great interest to understand adipose tissue-associated diseases and to study new therapeutics.

Abbreviations

AdipoR1 adiponectin receptor 1
AdipoR2 adiponectin receptor 2
Akt protein kinase B
Alg alginate
AMPK AMP-activated protein kinase
AT adipose tissue
ATM adipose tissue macrophages
ATT adipose tissue T cells
C/EBPα CCAAT/enhancer binding protein-α
$EBF^{-/-}$ early B-cell factor-null
ECM extracellular matrix
EDHF endothelium-derived hyperpolarizing factor
FFA free fatty acids
HA hyaluronic acid
HIF hypoxia-inducible factor
IL interleukin

MCP1 monocyte chemoattractant protein 1
Myf5+ myogenic factor 5+
Ob/ob leptin-deficient obese mice
Ob-Rb leptin receptor
OPAAF poly(amidoamine) oligomer microporous foam
PEG poly(ethylene glycol)
PPARγ peroxisome proliferator-activated receptor γ
PPP platelet-poor plasma
VLDL very-low-density lipoprotein
WAT white adipose tissue

References

[1] Cannon B, Hedin A, Nedergaard J. Exclusive occurrence of thermogenin antigen in brown adipose tissue. FEBS Lett 1982;150:129–32.
[2] Dawkins MJ, Scopes JW. Non-shivering thermogenesis and brown adipose tissue in the human new-born infant. Nature 1965;206:201–2.
[3] Rosenwald M, Perdikari A, Rulicke T, Wolfrum C. Bi-directional interconversion of brite and white adipocytes. Nat Cell Biol 2013;15:659–67.
[4] Wu J, et al. Beige adipocytes are a distinct type of thermogenic fat cell in mouse and human. Cell 2012;150:366–76.
[5] Lean ME, James WP, Jennings G, Trayhurn P. Brown adipose tissue uncoupling protein content in human infants, children and adults. Clin Sci (Lond) 1986;71:291–7.
[6] Frontini A, Cinti S. Distribution and development of brown adipocytes in the murine and human adipose organ. Cell Metab 2010;11:253–6.
[7] Pope BD, Warren CR, Parker KK, Cowan CA. Microenvironmental control of adipocyte fate and function. Trends Cell Biol 2016;26:745–55.
[8] Seale P, et al. PRDM16 controls a brown fat/skeletal muscle switch. Nature 2008;454:961–7.
[9] Kwok KH, Lam KS, Xu A. Heterogeneity of white adipose tissue: molecular basis and clinical implications. Exp Mol Med 2016;48:e215.
[10] Sharma RC, Kramsch DM, Lee PL, Colletti P, Jiao Q. Quantitation and localization of regional body fat distribution—a comparison between magnetic resonance imaging and somatometry. Obes Res 1996;4:167–78.
[11] Choe SS, Huh JY, Hwang IJ, Kim JI, Kim JB. Adipose tissue remodeling: its role in energy metabolism and metabolic disorders. Front Endocrinol 2016;7:30.
[12] Gesta S, Tseng Y-H, Kahn CR. Developmental origin of fat: tracking obesity to its source. Cell 2007;131:242–56.
[13] Sacks H, Symonds ME. Anatomical locations of human brown adipose tissue: functional relevance and implications in obesity and type 2 diabetes. Diabetes 2013;62:1783–90.
[14] Lin CS, Klingenberg M. Isolation of the uncoupling protein from brown adipose tissue mitochondria. FEBS Lett 1980;113:299–303.
[15] Hyvönen MT, Spalding KL. Maintenance of white adipose tissue in man. Int J Biochem Cell Biol 2014;56:123–32.
[16] Kusminski CM, Bickel PE, Scherer PE. Targeting adipose tissue in the treatment of obesity-associated diabetes. Nat Rev Drug Discov 2016;15:639–60.
[17] Wang QA, et al. Distinct regulatory mechanisms governing embryonic versus adult adipocyte maturation. Nat Cell Biol 2015;17:1099–111.

[18] Rutkowski JM, Stern JH, Scherer PE. The cell biology of fat expansion. J Cell Biol 2015;208:501–12.
[19] Halberg N, et al. Hypoxia-inducible factor 1alpha induces fibrosis and insulin resistance in white adipose tissue. Mol Cell Biol 2009;29:4467–83.
[20] Zeve D, Tang W, Graff J. Fighting fat with fat: the expanding field of adipose stem cells. Cell Stem Cell 2009;5:472–81.
[21] Kubota N, et al. PPAR gamma mediates high-fat diet-induced adipocyte hypertrophy and insulin resistance. Mol Cell 1999;4:597–609.
[22] Lee Y-H, Petkova AP, Mottillo EP, Granneman JG. In vivo identification of bipotential adipocyte progenitors recruited by beta3-adrenoceptor activation and high-fat feeding. Cell Metab 2012;15:480–91.
[23] Manteiga S, Choi K, Jayaraman A, Lee K. Systems biology of adipose tissue metabolism: regulation of growth, signaling and inflammation. Wiley Interdiscip Rev Syst Biol Med 2013;5:425–47.
[24] Kamp F, Hamilton JA. How fatty acids of different chain length enter and leave cells by free diffusion. Prostaglandins Leukot Essent Fatty Acids 2006;75:149–59.
[25] Rosen ED, Spiegelman BM. Adipocytes as regulators of energy balance and glucose homeostasis. Nature 2006;444:847–53.
[26] Schwartz MW, Seeley RJ, Campfield LA, Burn P, Baskin DG. Identification of targets of leptin action in rat hypothalamus. J Clin Invest 1996;98:1101–6.
[27] Bjorbaek C, Kahn BB. Leptin signaling in the central nervous system and the periphery. Recent Prog Horm Res 2004;59:305–31.
[28] Fei H, et al. Anatomic localization of alternatively spliced leptin receptors (Ob-R) in mouse brain and other tissues. Proc Natl Acad Sci U S A 1997;94:7001–5.
[29] Minokoshi Y, et al. Leptin stimulates fatty-acid oxidation by activating AMP-activated protein kinase. Nature 2002;415:339–43.
[30] Kamohara S, Burcelin R, Halaas JL, Friedman JM, Charron MJ. Acute stimulation of glucose metabolism in mice by leptin treatment. Nature 1997;389:374–7.
[31] Kieffer TJ, Habener JF. The adipoinsular axis: effects of leptin on pancreatic beta-cells. Am J Physiol Endocrinol Metab 2000;278:E1–14.
[32] Covey SD, et al. The pancreatic beta cell is a key site for mediating the effects of leptin on glucose homeostasis. Cell Metab 2006;4:291–302.
[33] Ouchi N, Parker JL, Lugus JJ, Walsh K. Adipokines in inflammation and metabolic disease. Nat Rev Immunol 2011;11:85–97.
[34] Yamauchi T, et al. Cloning of adiponectin receptors that mediate antidiabetic metabolic effects. Nature 2003;423:762–9.
[35] Garg SK, Delaney C, Shi H, Yung R. Changes in adipose tissue macrophage and T cell during aging. Crit Rev Immunol 2014;34:1–14.
[36] Harms M, Seale P. Brown and beige fat: development, function and therapeutic potential. Nat Med 2013;19:1252–63.
[37] Cao Y. Angiogenesis and vascular functions in modulation of obesity, adipose metabolism, and insulin sensitivity. Cell Metab 2013;18:478–89.
[38] Rupnick MA, et al. Adipose tissue mass can be regulated through the vasculature. Proc Natl Acad Sci U S A 2002;99:10730–5.
[39] Sung H-K, et al. Adipose vascular endothelial growth factor regulates metabolic homeostasis through angiogenesis. Cell Metab 2013;17:61–72.
[40] Gu P, Xu A. Interplay between adipose tissue and blood vessels in obesity and vascular dysfunction. Rev Endocr Metab Disord 2013;14:49–58.
[41] Sun K, Kusminski CM, Scherer PE. Adipose tissue remodeling and obesity. J Clin Invest 2011;121:2094–101.

[42] Rehman J, et al. Secretion of angiogenic and antiapoptotic factors by human adipose stromal cells. Circulation 2004;109:1292–8.

[43] Procopio C, et al. Leptin-stimulated endothelial nitric-oxide synthase via an adenosine 5′-monophosphate-activated protein kinase/Akt signaling pathway is attenuated by interaction with C-reactive protein. Endocrinology 2009;150:3584–93.

[44] Li FYL, Cheng KKY, Lam KSL, Vanhoutte PM, Xu A. Cross-talk between adipose tissue and vasculature: role of adiponectin. Acta Physiol (Oxf) 2011;203(167–180).

[45] Cheng KKY, et al. Adiponectin-induced endothelial nitric oxide synthase activation and nitric oxide production are mediated by APPL1 in endothelial cells. Diabetes 2007;56:1387–94.

[46] Ezure T, Amano S. Negative regulation of dermal fibroblasts by enlarged adipocytes through release of free fatty acids. J Invest Dermatol 2011;131:2004–9.

[47] Ezure T, Amano S. Adiponectin and leptin up-regulate extracellular matrix production by dermal fibroblasts. Biofactors 2007;31:229–36.

[48] Boutens L, Stienstra R. Adipose tissue macrophages: going off track during obesity. Diabetologia 2016;59:879–94.

[49] Morris DL, Singer K, Lumeng CN. Adipose tissue macrophages: phenotypic plasticity and diversity in lean and obese states. Curr Opin Clin Nutr Metab Care 2011;14:341–6.

[50] Lumeng CN, et al. Aging is associated with an increase in T cells and inflammatory macrophages in visceral adipose tissue. J Immunol 2011;187:6208–16.

[51] McLaughlin T, et al. T-cell profile in adipose tissue is associated with insulin resistance and systemic inflammation in humans. Arterioscler Thromb Vasc Biol 2014;34: 2637–43.

[52] Mathis D, Shoelson SE. Immunometabolism: an emerging frontier. Nat Rev Immunol 2011;11:81.

[53] Ferrante AWJ. The immune cells in adipose tissue. Diabetes Obes Metab 2013;15(Suppl 3):34–8.

[54] Choi JS, et al. Decellularized extracellular matrix derived from human adipose tissue as a potential scaffold for allograft tissue engineering. J Biomed Mater Res A 2011;97:292–9.

[55] Young DA, Ibrahim DO, Hu D, Christman KL. Injectable hydrogel scaffold from decellularized human lipoaspirate. Acta Biomater 2011;7:1040–9.

[56] Mariman EC, Wang P. Adipocyte extracellular matrix composition, dynamics and role in obesity. Cell Mol Life Sci 2010;67:1277–92.

[57] Aratani Y, Kitagawa Y. Enhanced synthesis and secretion of type IV collagen and entactin during adipose conversion of 3T3-L1 cells and production of unorthodox laminin complex. J Biol Chem 1988;263:16163–9.

[58] Molina H, et al. Temporal profiling of the adipocyte proteome during differentiation using a five-plex SILAC based strategy. J Proteome Res 2009;8:48–58.

[59] Nakajima I, Muroya S, Tanabe R, Chikuni K. Extracellular matrix development during differentiation into adipocytes with a unique increase in type V and VI collagen. Biol Cell 2002;94:197–203.

[60] Spencer M, et al. Adipose tissue extracellular matrix and vascular abnormalities in obesity and insulin resistance. J Clin Endocrinol Metab 2011;96:E1990–8.

[61] Yamauchi T, et al. The fat-derived hormone adiponectin reverses insulin resistance associated with both lipoatrophy and obesity. Nat Med 2001;7:941–6.

[62] Ramanjaneya M, et al. Identification of nesfatin-1 in human and murine adipose tissue: a novel depot-specific adipokine with increased levels in obesity. Endocrinology 2010;151:3169–80.

[63] Zhang Z, et al. Increased plasma levels of nesfatin-1 in patients with newly diagnosed type 2 diabetes mellitus. Exp Clin Endocrinol Diabetes 2012;120:91–5.

[64] Yang R-Z, et al. Identification of omentin as a novel depot-specific adipokine in human adipose tissue: possible role in modulating insulin action. Am J Physiol Endocrinol Metab 2006;290:E1253–61.

[65] Steppan CM, et al. The hormone resistin links obesity to diabetes. Nature 2001;409:307–12.

[66] Cao H, et al. Identification of a lipokine, a lipid hormone linking adipose tissue to systemic metabolism. Cell 2008;134:933–44.

[67] Murdolo G, et al. Lipokines and oxysterols: novel adipose-derived lipid hormones linking adipose dysfunction and insulin resistance. Free Radic Biol Med 2013;65:811–20.

[68] Strissel KJ, et al. Adipocyte death, adipose tissue remodeling, and obesity complications. Diabetes 2007;56:2910–8.

[69] Skurk T, Alberti-Huber C, Herder C, Hauner H. Relationship between adipocyte size and adipokine expression and secretion. J Clin Endocrinol Metab 2007;92:1023–33.

[70] Wueest S, Rapold RA, Rytka JM, Schoenle EJ, Konrad D. Basal lipolysis, not the degree of insulin resistance, differentiates large from small isolated adipocytes in high-fat fed mice. Diabetologia 2009;52:541–6.

[71] Moreno-Viedma V, et al. Common dysregulated pathways in obese adipose tissue and atherosclerosis. Cardiovasc Diabetol 2016;15:120.

[72] Driskell RR, et al. Distinct fibroblast lineages determine dermal architecture in skin development and repair. Nature 2013;504:277–81.

[73] Wojciechowicz K, Gledhill K, Ambler CA, Manning C, Jahoda CA. Development of the mouse dermal fat layer is linked to hair follicle development, occurs independently of subcutaneous fat, and is marked by restricted early expression of FABP4. J Invest Dermatol 2013;133:1417.

[74] Festa E, et al. Adipocyte lineage cells contribute to the skin stem cell niche to drive hair cycling. Cell 2011;146:761–71.

[75] Schmidt BA, Horsley V. Intradermal adipocytes mediate fibroblast recruitment during skin wound healing. Development 2013;140:1517–27.

[76] Driskell RR, Jahoda CA, Chuong CM, Watt FM, Horsley V. Defining dermal adipose tissue. Exp Dermatol 2014;23:629–31.

[77] Kasza I, et al. Syndecan-1 is required to maintain intradermal fat and prevent cold stress. PLoS Genet 2014;10:e1004514.

[78] Zhang Lj, et al. Dermal adipocytes protect against invasive *Staphylococcus aureus* skin infection. Science 2015;347:67–71.

[79] Donati G, et al. Epidermal Wnt/β-catenin signaling regulates adipocyte differentiation via secretion of adipogenic factors. Proc Natl Acad Sci U S A 2014;111:E1501–9.

[80] Blanpain C, Lowry WE, Geoghegan A, Polak L, Fuchs E. Self-renewal, multipotency, and the existence of two cell populations within an epithelial stem cell niche. Cell 2004;118:635–48.

[81] Jahoda CAB, Horne KA, Oliver RF. Induction of hair growth by implantation of cultured dermal papilla cells. Nature 1984;311:560–2.

[82] Plikus MV, et al. Cyclic dermal BMP signalling regulates stem cell activation during hair regeneration. Nature 2008;451:340–4.

[83] Kandyba E, et al. Competitive balance of intrabulge BMP/Wnt signaling reveals a robust gene network ruling stem cell homeostasis and cyclic activation. Proc Natl Acad Sci U S A 2013;110:1351–6.
[84] Rivera-Gonzalez G, Shook B, Horsley V. Adipocytes in skin health and disease. Cold Spring Harb Perspect Med 2014;4:.
[85] Rendl M, Polak L, Fuchs E. BMP signaling in dermal papilla cells is required for their hair follicle-inductive properties. Genes Dev 2008;22:543–57.
[86] Gurtner GC, Werner S, Barrandon Y, Longaker MT. Wound repair and regeneration. Nature 2008;453:314–21.
[87] Cerqueira M, Reis R, Marques A. Wound healing microenvironmental cues: from tissue analogs to skin regeneration. Curr Tissue Eng 2013;2:145–53.
[88] Frank S, Stallmeyer B, Kämpfer H, Kolb N, Pfeilschifter J. Leptin enhances wound re-epithelialization and constitutes a direct function of leptin in skin repair. J Clin Invest 2000;106:501–9.
[89] Shibata S, et al. Adiponectin regulates cutaneous wound healing by promoting keratinocyte proliferation and migration via the ERK signaling pathway. J Immunol 2012;189:3231–41.
[90] Sogabe Y, Abe M, Yokoyama Y, Ishikawa O. Basic fibroblast growth factor stimulates human keratinocyte motility by Rac activation. Wound Repair Regen 2006;14:457–62.
[91] Kawai K, et al. Effects of adiponectin on growth and differentiation of human keratinocytes-implication of impaired wound healing in diabetes. Biochem Biophys Res Commun 2008;374:269–73.
[92] Salathia NS, Shi J, Zhang J, Glynne RJ. An in vivo screen of secreted proteins identifies adiponectin as a regulator of murine cutaneous wound healing. J Invest Dermatol 2013;133:812–21.
[93] Werner S, et al. Large induction of keratinocyte growth factor expression in the dermis during wound healing. Proc Natl Acad Sci U S A 1992;89:6896–900.
[94] Marangoni RG, et al. Myofibroblasts in murine cutaneous fibrosis originate from adiponectin-positive intradermal progenitors. Arthritis Rheumatol 2015;67:1062–73.
[95] Martins V, et al. FIZZ1-induced Myofibroblast transdifferentiation from adipocytes and its potential role in dermal fibrosis and lipoatrophy. Am J Pathol 2015;185:2768–76.
[96] Zhang Y, et al. Positional cloning of the mouse obese gene and its human homologue. Nature 1994;372:425–32.
[97] Murad A, et al. Leptin is an autocrine/paracrine regulator of wound healing. FASEB J 2003;17:1895–7.
[98] Gealekman O, et al. Depot-specific differences and insufficient subcutaneous adipose tissue angiogenesis in human obesity. Circulation 2011;123:186–94.
[99] Komine M, et al. Inflammatory versus proliferative processes epidermis. Tumor necrosis factor α induces K6b keratin synthesis through a transcriptional complex containing NFκB and C/EBPβ. J Biol Chem 2000;275:32077–88.
[100] Davidson JM. Animal models for wound repair. Arch Dermatol Res 1998;290:S1–11.
[101] Sullivan TP, Eaglstein WH, Davis SC, Mertz P. The pig as a model for human wound healing. Wound Repair Regen 2001;9:66–76.
[102] Leiva AG, et al. Loss of Mpzl3 function causes various skin abnormalities and greatly reduced adipose depots. J Invest Dermatol 2014;134:1817–27.
[103] Herrmann T, et al. Mice with targeted disruption of the fatty acid transport protein 4 (Fatp 4, Slc27a4) gene show features of lethal restrictive dermopathy. J Cell Biol 2003;161:1105–15.
[104] Stone SJ, et al. Lipopenia and skin barrier abnormalities in DGAT2-deficient mice. J Biol Chem 2004;279:11767–76.

[105] Jong MC, et al. Hyperlipidemia and cutaneous abnormalities in transgenic mice overexpressing human apolipoprotein C1. J Clin Invest 1998;101:145–52.
[106] Wei J, et al. Canonical Wnt signaling induces skin fibrosis and subcutaneous lipoatrophy: a novel mouse model for scleroderma? Arthritis Rheum 2011;63:1707–17.
[107] Servettaz A, et al. Selective oxidation of DNA topoisomerase 1 induces systemic sclerosis in the mouse. J Immunol 2009;182:5855–64.
[108] Gerber EE, et al. Integrin-modulating therapy prevents fibrosis and autoimmunity in mouse models of scleroderma. Nature 2013;503:126–30.
[109] Sonnylal S, et al. Postnatal induction of transforming growth factor β signaling in fibroblasts of mice recapitulates clinical, histologic, and biochemical features of scleroderma. Arthritis Rheum 2007;56:334–44.
[110] Zhu KQ, Carrougher GJ, Gibran NS, Isik FF, Engrav LH. Review of the female Duroc/Yorkshire pig model of human fibroproliferative scarring. Wound Repair Regen 2007;15:S32–9.
[111] Ignacio G, El-Amin I, Mendenhall V. In: Albanna MZ, Holmes JH IV, editors. Skin tissue engineering regenerative medicine. London: Academic Press; 2016. p. 387–400. https://doi.org/10.1016/B978-0-12-801654-1.00019-X.
[112] Kischer CW, Pindur J, Shetlar MR, Shetlar CL. Implants of hypertrophic scars and keloids into the nude (athymic) mouse: viability and morphology. J Trauma 1989;29:672–7.
[113] Alrobaiea SM, Ding J, Ma Z, Tredget EE. A novel nude mouse model of hypertrophic scarring using scratched full thickness human skin grafts. Adv Wound Care 2016;5:299–313.
[114] Wong VW, Sorkin M, Glotzbach JP, Longaker MT, Gurtner GC. Surgical approaches to create murine models of human wound healing. J Biomed Biotechnol 2011;2011:1–8.
[115] Langer R, Vacanti JP. Tissue engineering. Science 1993;260:920–6.
[116] Edmondson R, Broglie JJ, Adcock AF, Yang L. Three-dimensional cell culture systems and their applications in drug discovery and cell-based biosensors. Assay Drug Dev Technol 2014;12:207–18.
[117] Ruiz-Ojeda FJ, Ruperez AI, Gomez-Llorente C, Gil A, Aguilera CM. Cell models and their application for studying adipogenic differentiation in relation to obesity: a review. Int J Mol Sci 2016;17:.
[118] Yang J, et al. Cell sheet engineering: recreating tissues without biodegradable scaffolds. Biomaterials 2005;26:6415–22.
[119] Xu X, Farach-Carson MC, Jia X. Three-dimensional in vitro tumor models for cancer research and drug evaluation. Biotechnol Adv 2014;32:1256–68.
[120] Kelm JM, et al. A novel concept for scaffold-free vessel tissue engineering: self-assembly of microtissue building blocks. J Biotechnol 2010;148:46–55.
[121] Vallee M, Cote JF, Fradette J. Adipose-tissue engineering: taking advantage of the properties of human adipose-derived stem/stromal cells. Pathol Biol 2009;57:309–17.
[122] Vermette M, et al. Production of a new tissue-engineered adipose substitute from human adipose-derived stromal cells. Biomaterials 2007;28:2850–60.
[123] Aubin K, et al. Characterization of in vitro engineered human adipose tissues: relevant adipokine secretion and impact of TNF-alpha. PLoS One 2015;10:e0137612.
[124] Lago MEL, et al. Modulation of the secretory potential of in vitro adipose tissue microenvironments. Eur Cells Mater 2017;33:729.
[125] Cheng CW, Solorio LD, Alsberg E. Decellularized tissue and cell-derived extracellular matrices as scaffolds for orthopaedic tissue engineering. Biotechnol Adv 2014;32:462–84.
[126] Gomes S, Leonor IB, Mano JF, Reis RL, Kaplan DL. Natural and genetically engineered proteins for tissue engineering. Prog Polym Sci 2012;37:1–17.

[127] Girandon L, Kregar-Velikonja N, Bozikov K, Barlic A. In vitro models for adipose tissue engineering with adipose-derived stem cells using different scaffolds of natural origin. Folia Biol 2011;57:47–56.

[128] Davidenko N, Campbell JJ, Thian ES, Watson CJ, Cameron RE. Collagen-hyaluronic acid scaffolds for adipose tissue engineering. Acta Biomater 2010;6:3957–68.

[129] Yao R, Zhang R, Lin F, Luan J. Injectable cell/hydrogel microspheres induce the formation of fat lobule-like microtissues and vascularized adipose tissue regeneration. Biofabrication 2012;4:45003.

[130] Korurer E, Kenar H, Doger E, Karaoz E. Production of a composite hyaluronic acid/gelatin blood plasma gel for hydrogel-based adipose tissue engineering applications. J Biomed Mater Res A 2014;102:2220–9.

[131] Chang KH, Liao HT, Chen JP. Preparation and characterization of gelatin/hyaluronic acid cryogels for adipose tissue engineering: in vitro and in vivo studies. Acta Biomater 2013;9:9012–26.

[132] Chen YS, Chen YY, Hsueh YS, Tai HC, Lin FH. Modifying alginate with early embryonic extracellular matrix, laminin and hyaluronic acid for adipose tissue engineering. J Biomed Mater Res A 2015;104(3):669–77.

[133] Hsueh Y-S, et al. Laminin-alginate beads as preadipocyte carriers to enhance adipogenesis in vitro and in vivo. Tissue Eng Part A 2017;23(5–6):185–94.

[134] Clevenger TN, et al. Vitronectin-based, biomimetic encapsulating hydrogel scaffolds support adipogenesis of adipose stem cells. Tissue Eng Part A 2016;22:597–609.

[135] Rossi E, et al. Biologically and mechanically driven design of an RGD-mimetic macroporous foam for adipose tissue engineering applications. Biomaterials 2016;104:65–77.

[136] Young DA, Choi YS, Engler AJ, Christman KL. Stimulation of adipogenesis of adult adipose-derived stem cells using substrates that mimic the stiffness of adipose tissue. Biomaterials 2013;34:8581–8.

[137] Brown CFC, et al. Effect of decellularized adipose tissue particle size and cell density on adipose-derived stem cell proliferation and adipogenic differentiation in composite methacrylated chondroitin sulphate hydrogels. Biomed Mater 2015;10:45010.

[138] Yu C, et al. Porous decellularized adipose tissue foams for soft tissue regeneration. Biomaterials 2013;34:3290–302.

[139] Flynn LE. The use of decellularized adipose tissue to provide an inductive microenvironment for the adipogenic differentiation of human adipose-derived stem cells. Biomaterials 2010;31:4715–24.

[140] Samani A, Zubovits J, Plewes D. Elastic moduli of normal and pathological human breast tissues: an inversion-technique-based investigation of 169 samples. Phys Med Biol 2007;52:1565–76.

[141] Alkhouli N, et al. The mechanical properties of human adipose tissues and their relationships to the structure and composition of the extracellular matrix. Am J Physiol Endocrinol Metab 2013;305:E1427–35.

[142] Wittmann K, et al. Development of volume-stable adipose tissue constructs using polycaprolactone-based polyurethane scaffolds and fibrin hydrogels. J Tissue Eng Regen Med 2016;10:E409–18.

[143] Bellas E, Marra KG, Kaplan DL. Sustainable three-dimensional tissue model of human adipose tissue. Tissue Eng Part C Methods 2013;19:745–54.

[144] Abbott RD, et al. The use of silk as a scaffold for mature, sustainable unilocular adipose 3D tissue engineered systems. Adv Healthc Mater 2016;5:1667–77.

[145] Huber B, et al. Integration of mature adipocytes to build-up a functional three-layered full-skin equivalent. Tissue Eng Part C Methods 2016;22:756–64.

Immunocompetent human in vitro skin models

Victoria Hutter, Stewart B. Kirton, David Y.S. Chau
Research Centre in Topical Drug Delivery and Toxicology, Department of Pharmacy, Pharmacology and Postgraduate Medicine, School of Life and Medical Sciences, University of Hertfordshire, Hatfield, United Kingdom

1. Introduction

The skin is the largest organ of the body and provides the main barrier between the internal and external environment. As well as providing protection to the body against chemical and physical insults and microbiological invasion, it also contributes to both temperature and water regulation by acting as a barrier [1]. The barrier function of the skin is critical and, if disrupted, results in nonspecific innate and specific adaptive immune responses as part of the body's defense mechanism. In addition to eliciting immune responses to external insults, the skin participates in local "sterile" immunity including allergic and autoimmune responses and tumor immunity [2].

The skin is a structurally complex organ that consists of three separate and anatomically unique layers with each stratum having a distinct composition, characteristic, and function. The epidermis, which consists of the stratum corneum and viable epidermis, forms the outermost layer of innate skin. The stratum corneum, which presents the greatest barrier to permeation, is a 10–25 μm-thick layer that possesses a unique arrangement commonly referred to as a "brick and mortar" construct: the bricks being relatively nonpermeable protein-rich corneocytes and the mortar consisting of an orderly configured lipid-rich medium, that is, cholesterol, triglycerides, and ceramides [3,4]. Essentially, the function of this layer is to control the absorption of substances into the skin and also for maintenance of fluid homeostasis. The other layers of the epidermis include the stratum granulosum, stratum spinosum, and stratum basale within which a dynamic activity takes place whereby cells from the stratum basale undergo differentiation and move upward to form the stratum corneum, a process known as cornification [5]. The viable epidermis is additionally populated with a number of immune cells and melanocytes. Located beneath the epidermis is the dermis layer that is a 3–5 mm-thick stratum composed of collagen, reticulin, and elastin woven within mucopolysaccharide network [3]. The dermis layer also houses a variety of skin appendages that include apocrine glands, blood vessels, hair follicles, and aspects of the lymphatic system. The innermost and often the thickest layer of the skin structure is known as the hypodermis. This layer is essentially populated with adipocytes that specialize in accumulating fats and hence provide thermoregulation. The authors recommend the comprehensive and detailed review of the skin and its structure by Monteiro-Riviere [6].

Skin Tissue Models. https://doi.org/10.1016/B978-0-12-810545-0.00015-2

2. Immunological overview of the skin

The immunologic response of the skin can be attributed to a variety of distinct aspects within its gross structure and often described in context of a conceptual framework known as skin-associated lymphoid tissues (SALT), which mediates and coordinates the response between the different layers [7]. At a rudimentary level, the epidermis houses a number of immunocompetent cells and biological response modifiers including keratinocytes, Langerhans cells (LCs), melanocytes, and epidermotrophic lymphocytes. In contrast, the dermis is mostly populated with fibroblasts and dendritic cells (DCs) but work together with the epidermal immunocompetent cells to bring about the appropriate immune response post stimuli [8,9]. Importantly, all pathological diseases and/or responses have, to some extent, an underlying immune component [10].

3. Cells in epidermal immunity

The stratified, squamous epidermis is not only made up of more than 90% keratinocytes that provide the barrier integrity required for protection but also responsible for the production of cytokines that regulate the skin immune response cascade. One of the most important families of cytokines in the skin is the interleukin-1 (IL-1) family that is composed of a group of 11 cytokines that are synthesized not only by keratinocytes but also fibroblasts and macrophages. These work by modulating key responsive effects such as the activation of T and B lymphocytes, increase the expression of adhesion factors on endothelial cells (i.e., the intercellular adhesion molecule, ICAM-1), and affect the activity of the hypothalamus during an inflammatory insult [11,12]. Keratinocytes in the skin are able to produce a number of interleukins, colony-stimulating factors, growth factors, chemotactic factors, and immunosuppressive factors, which have a myriad of effects on keratinocytes and immunologic cells in the skin [13]. Keratinocyte production of IL-1 and TNF-α is well established and is known to activate immunologic cells in the skin including lymphocytes, macrophages, and DCs [14].

LCs are a specific subset of DCs that are associated with the epidermis. They are antigen-presenting cells that originate from the bone marrow and populate the epidermis—being most prominent in the stratum spinosum but absent in the stratum corneum. Identified by their unique racket-shaped Birbeck granules and expression of CD1a and CD207 surface marker proteins, LCs are members of the major histocompatibility complex class II (MHC II) molecules and adopt a "sentinel" role by being directly responsible for the initiation of both innate and adaptive immune responses to skin-relevant antigens [15]. LCs are antigen-presenting cells with one of their main roles being to bind and process antigens and migrate to the lymph nodes to present the antigens to naive T cells.

Melanocytes are DCs located in the stratum basale layer of the epidermis and epidermal appendages and have a close anatomical relationship with keratinocytes [16].

Although intensively documented for their modulation of skin pigmentation, melanocytes have also been suggested to contribute to immunologic responses within the skin via the secretion of cytokines including IL-3, IL-6, and TNFα complexes [17]. Melanocytes have also been shown to express MHC II molecules, which are important for initiating immune responses. However, recent studies have highlighted that the evidence is not, as yet, fully conclusive of the exact role melanocytes play in skin immunology [18–20].

4. Cells in dermal immunity

Two distinct populations of DCs are located within the dermis, namely, dermal dendritic cells (DDCs) and plasmacytoid dendritic cells (pDCs). The populations are characterized by the production of different cytokines and may initiate different inflammatory pathways after activation although it is unlikely that these roles are exclusive. The DCs in the dermal layer possess more motility than the LCs in the epidermis and are able to migrate within the collagen fibril structure of the dermis [2]. In contrast with LCs, the DDCs lack Birbeck granules and have higher expression of CD1a but show near absent expression of CD207 and CD36 [21].

Macrophages can be found localized within the dermis and have a high degree of motility and phagocytic capacity. While they are present in much fewer numbers in the dermis in comparison with DCs, they are capable of presenting antigens to T cells and activating immune responses through cell surface signaling and cytokine secretion [2]. Circulating monocytes are immature macrophages, which can be recruited to the skin in response to infection or injury. They are often regarded as guardians within the immunologic cascade and are attracted to a compromised wound site through the localization of growth factors/cytokines/chemokines that are released by platelets and other cells following a barrier insult [22]. These monocytes then differentiate accordingly into macrophages that are thereafter focused on the phagocytosis of the foreign bodies and associated damaged tissues at the site of injury. Furthermore, it has been documented that LCs may also bring about an immune response by phagocytosis as a result of stimulation by immunoglobulins IgE and IgG and thereby contribute to the overall regulatory mechanism [23].

Other innate immune cells including neutrophils, eosinophils, basophils, and mast cells may also be recruited to the skin following activation. The first cells recruited by DC or macrophage activation are typically neutrophils that primarily phagocytose pathogens. Eosinophils, basophils, and mast cells are associated with allergic and antiparasitic responses in the skin [2]. Mast cells in the skin are located in close proximity to T lymphocytes (TCs) and are well-known for their participation in a number of immune and inflammatory processes [24,25].

The majority of adaptive immunologic responses in the skin are determined by TCs. There is almost double the number of TCs present in the skin (approximately 20 million) in comparison with those found in the blood [26]. Within the skin, TCs are predominantly located in the blood vessels and the lymphatic system within the dermis. A smaller number of TCs are located in the epidermis for T-cell immune surveillance and mainly express the α/β T antigen receptors: $CD2^+$, $CD3^+$, $CD4^+$ inducer, and $CD8^+$ suppressor lymphocytes. These specialist cells may migrate from the vessels into the dermis by adhering to the expressed ICAM-1 in response to the presence of antigens or a compromised skin barrier. Cytotoxic TCs express CD8+ and kill cells that express specific antigens presented on a cell by human leukocyte antigen I. Helper TCs produce cytokines and promaturation signals that enable effector cells to mediate an attack and steer the response. Regulatory TCs suppress autoimmune responses and help to resolve inflammation [2]. An overview of the immune functions present within the stratified layers of the native skin in response to inflammation can be seen in Fig. 1.

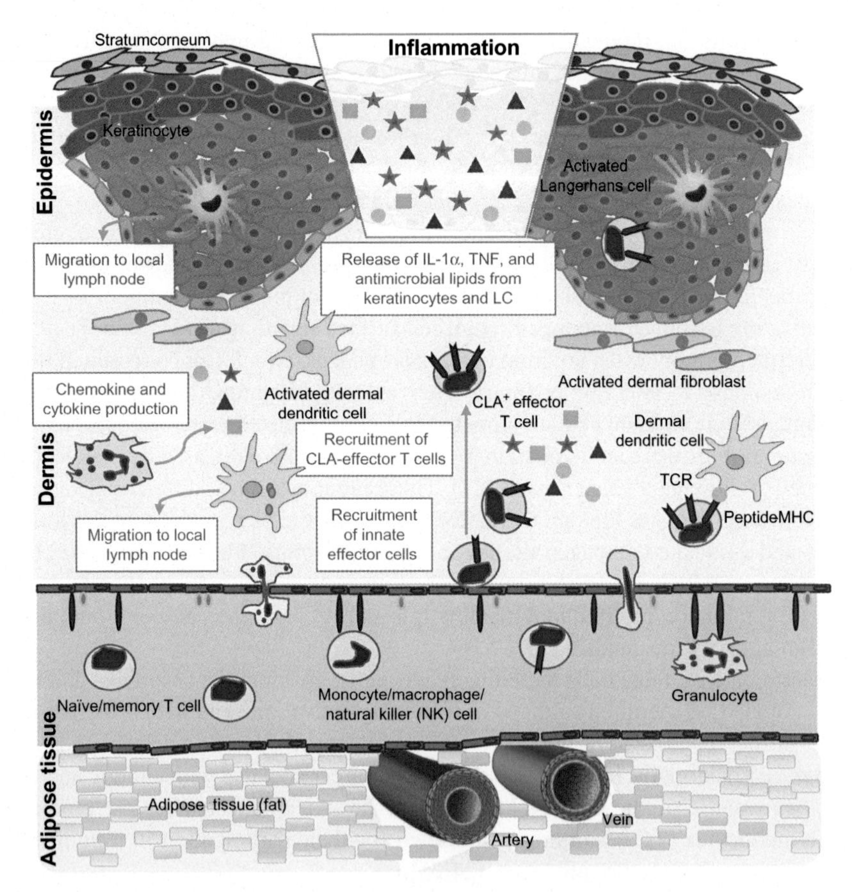

Fig. 1 A schematic representation of the mechanisms of immune functions in response to inflammation from a stimulus.
Modified from Stamatas GN, Morello AP, Mays DA. Early inflammatory processes in the skin. Curr Mol Med 2013;13(8):1250–69.

5. Justification for skin models

With the ever-increasing exposure to a variety of new chemicals found in personal, health-care, and environmental products, there is a critical need for the detailed characterization of these compounds in line with current legislative regulations and protocols. Similar requirements for safety testing apply for novel excipients and permeation enhancers. Although toxicity is often considered as the primary objective of an assessment, the ability of the compound to elicit an adverse dermal allergenic or immunogenic response also needs to be considered due to the ease of exposure/route of entry [27]. In addition, as the immune system plays an important role in the development and pathogenesis of most skin diseases, there is a requirement for more robust in vivo-like in vitro models when developing novel therapies [28]. Historically, test methods such as the guinea pig maximization test [29], the Buehler test [30], and the local lymph node assays [31,32] were exploited. However, the majority of these models suffer from questionable human biological relevance and severe ethical constraints. With an underlying abetment of the 3Rs for animal use in experiments (i.e., reduction, replacement, and refinement), the implementation of the seventh amendment to the EU Cosmetics Directive, and the updated Helskinki ECHA/NI/17/01 legislative guideline within the Registration, Evaluation, Authorization and Restriction of Chemicals (REACH) 1907/2006 regulation, more physiological accurate and biologically responsive models of these tissues/organs are required [27]. Moreover, these pseudotissues and pseudoorgans can additionally or, alternatively, be used as models to elucidate disease pathways/mechanisms/progression to further medical applications and therapeutic understanding.

6. In vitro cell culture modeling

Tissue or cell culture is a generalized term used to describe the process of removing cells, tissue, or organs from an animal (or plant) and subsequently grow them in a sustained artificial environment that mimics the native in vivo setting. In simplistic terminology, this environment consists of a suitable vessel containing a liquid feed—termed culture medium—and being incubated at the appropriate temperature and atmosphere. Originally devised by Ross Harrison in 1907 as a technique to establish long-term nerve cell culture, the field has now rapidly progressed over the years with significant technological advancements and development of a number of high-impact commercialized products and protocols [33].

Cells directly dissociated from parental organ/tissue and cultured thereafter are often referred to as "primary cell cultures" and regarded as the "gold standard" as they essentially maintain the same characteristics as the native tissue. However, sourcing of these cells often possesses difficulties due to ethical constraints, and the cells, themselves, only offer a limited working life span due to their innate characteristic of senescence, that is, biological aging and/or the Hayflick limit [34]. An alternative option is to use established or immortalized cell "lines" that are derived from a native cancerous state or generated using viral vectors, hybridoma technology, or overexpression of regulatory proteins such as telomerase and the P450 family of enzymes. However, a key disadvantage of using cell lines is that they tend to deviate genetically and, more importantly, phenotypically compared with their tissue of origin [35]. In contrast, their availability, pseudocharacteristic likeness, and ease of use offer researchers a valuable complement to existing in vivo experiments by allowing for a more controlled manipulation of cellular functions and processes.

Tissue engineering is a term coined to describe an interdisciplinary field that applies engineering and life science principles toward the development of biological substitutes that restore, maintain, or improve tissue function [36]. Ideologically, the field aimed to develop and create transplantable tissues/organs to replace those undergoing natural aging, having diseases, or subjected to traumatic injury [37]. However, current demand and advances in novel technologies have caused a bias during model development toward toxicity assessment and biomedical applications as these, in vitro, cell culture models can ultimately provide a simple, fast, and cost-effective tool alternative and also help minimize the exploitation of animals during the drug development pathway. Although a number of human skin model equivalents exist and have been exploited for skin irritation studies, they have not been able to accurately mimic the underlying immune component present in most disease states leading to failure of many investigational drugs at the preclinical phase or during clinical trials [28]. As such, only a handful of these can be considered to be truly "immunocompetent" and hence suitable for performing skin sensitization studies [38,39].

7. Epidermal cell culture models: Keratinocyte monocultures

Single-cell keratinocyte cultures are the most simple in vitro cell culture models available that mimic the epidermal layer of the skin. These may be constructed from human cell lines or from human primary keratinocyte sources (Table 1). The immortalized human keratinocyte cell line, HaCaT, is one of the best characterized and widely used cell line models that maintains full epidermal differentiation and proliferation capacity up to 140 passages [46,47]. Additionally, there has been extensive research regarding the culture of human primary keratinocytes over the past 20 years, and there are now many standardized isolation and propagation protocols established for keratinocyte skin models and many which are available commercially [27,48,49]. Reconstituted human epidermal equivalent (RHE) models are designed to represent the epidermis and are constructed from primary normal human keratinocytes cultured on a collagen matrix at an air-liquid interface. The primary normal human epidermal keratinocyte (NHEK) source varies between models but typically originates from neonatal foreskin or adult mammary tissue samples. Cells are differentiated at an air-liquid interface producing 8–12 layers of human keratinocytes that mimic a pseudo stratum corneum and functional epidermal keratinocyte layers. While the viable keratinocyte functional layers are representative of human skin, the barrier properties of RHE models possess significantly enhanced permeability properties in comparison with the human skin

Table 1 Keratinocyte-only skin-based models

Skin model	Specifications	Readout parameter	Limitations	References
EDI-Co Skin	The epidermis only, primary keratinocytes	Histology, cytotoxicity assessment assay (MTT/MTS), IL-1α and IL-8 release	Pseudo 2-D/3-D microenvironment, no immune cells	www.edigmbh.de
EpiDerm	Reconstituted human epidermal equivalent (RHE) based on normal human-derived epidermal keratinocytes (NHEK) on collagen-coated TCP	Histology; cytotoxicity assessment assay (MTT/MTS); IL-1α, PGE2, and LDH release; sodium fluorescein permeability	Pseudo 2-D/3-D microenvironment, no immune cells	www.mattek.com
Epidermal Skin Test 1000 (EST1000)	Reconstructed epidermal model from primary human keratinocytes	Histology, cytotoxicity assessment assay (MTT/MTS), expression markers: KI-67, CK 1/10/5/14, transglutaminase, collagen IV, involucrin, β1 integrin	Pseudo 2-D/3-D microenvironment, no immune cells	[40]
EpiSkin (L'Oreal: Invitroskin)	RHE based on adult human keratinocytes on a collagen substrate	Histology, cytotoxicity assessment assay (MTT/MTS), IL-1α release	No immune cells	[41-43]
LabCyte EPI-MODEL	Keratinocytes embedded within agarose gel	Cytotoxicity assessment assay (MTT/MTS)	No immune cells	[44,45]
Prediskin (Transkin)	Donor sample and/or RHE	Histology, cytotoxicity assessment assay (MTT/MTS)	No immune cells	www.biopredic.com
Reconstituted human epidermis (RHE-EPI/001)	Human primary keratinocytes seeded on a polycarbonate filter	Histology, cytotoxicity assessment assay (MTT/MTS)	No immune cells	www.straticell.com

Modified from Grindon C, Combes R, Cronin MT, Roberts DW, Garrod JF. Integrated decision-tree testing strategies for skin corrosion and irritation with respect to the requirements of the EU REACH legislation. Altern Lab Anim 2008;36(Suppl. 1):65–74.

in vivo [50]. Nevertheless, RHE models are widely used for research purposes including assessment of skin irritation, corrosion, hydration, and toxicological screening (phototoxicity and genotoxicity) and the assessment of drug candidates. Typical readouts of these models include cell histology, cell viability (cell mitochondrial activity MTT assay and LDH release profile), barrier integrity (TEER/TEWL measurements and sodium fluorescein permeability), and cytokine release, for example, IL-1α, IL-6, IL-8, and TNF α.

8. Epidermal and dermal cell culture models: Keratinocyte and fibroblast co-cultures

More recently, full-thickness in vitro human skin models or living skin equivalents (LSE) have been developed that are designed to mimic both the dermis and epidermis layers (Table 2). The lower dermal compartment is composed of a collagen matrix containing primary normal human dermal fibroblasts (NHDF) that supports the growth of an upper epidermal compartment of layers of primary NHEK cells cultured at an air-liquid interface. These models benefit from a more in vivo-like morphology and growth characteristics including a pseudo basement membrane structure that demonstrates keratin production and appropriate cell junctions [51,52]. Importantly, the coculture setup also allows for interplay of cellular signals between the fibroblasts and keratinocytes in the model. These models are used for similar purposes as the keratinocyte and RHE models described above including skin responses, dermal toxicity, and drug screening with similar readouts based on observational histology, basic cell health assays, barrier integrity, and cytokine release [53,54].

9. Epidermal and dermal cell culture models: Co-cultures systems incorporating melanocytes

Primary melanocytes have also been incorporated into coculture models alongside keratinocytes and/or fibroblasts (Table 3). These models are utilized particularly by the cosmetics industry to assess the impact of skin-lightening agents and formulations on skin pigmentation [56]. While melanocytes are DCs within the epidermis, these cells are not considered to be the key immunologic cells within the skin and have limited relevance outside of the cosmetics industry [57].

10. Epidermal and dermal cell culture models: Diseased state models

Several in vitro cell culture models of skin diseases including psoriasis, keloid scars, dystrophic epidermolysis, and squamous cell carcinoma have been developed (Table 4). However, the majority of these are based on primary keratinocytes sourced

Table 2 **Keratinocyte and fibroblast skin-based models**

Skin model	Specifications	Readout parameter	Limitations	References
EpiDermFT	NKEK and NHF	Histology; cytotoxicity assessment assay (MTT/MTS); IL-1α, PGE2, and LDH release; sodium fluorescein permeability	Pseudo 2-D/3-D microenvironment, no immune cells	www.mattek.com
EpiSkin RealSkin		Histology, cytotoxicity assessment assay (MTT/MTS), IL-1α release	No immune cells	www.episkin.com
IGB human skin model (Fraunhofer EU 00F36047)	Keratinocytes and fibroblasts in collagen matrix	Histology, cytotoxicity assessment assay (MTT/MTS), IL-1α release	No immune cells	www.tissue-factory.com
Phenion FTSM	Primary keratinocytes and fibroblasts seeded onto novel 3-D collagen scaffold	Histology, TUNEL cell viability, IL-6 and IL-8 release	No immune cells	[85]
StrataTesT human 3-D skin	Proprietary NIKS human keratinocytes. NIKS cells and normal human fibroblasts. Twenty-four well-plate format	Histology, cytotoxicity assessment assay (MTT/MTS), IL-1α release, membrane integrity based on changes in transepithelial electrical resistance (TEER) measurement	No immune cells	www.stratatechcorp.com
Vitrolife-Skin RHE	Keratinocytes and fibroblast coculture on collagen sponges. A-L interface	Histology, cytotoxicity assessment assay (MTT/MTS), LDH release, IL-2 release	No immune cells	www.gunze.co.jp; [86]

Modified from Grindon C, Combes R, Cronin MT, Roberts DW, Garrod JF. Integrated decision-tree testing strategies for skin corrosion and irritation with respect to the requirements of the EU REACH legislation. Altern Lab Anim 2008;36(Suppl. 1):65–74.

Table 3 **Keratinocyte and fibroblast skin-based models incorporating melanocytes**

Skin model	Specifications	Readout parameter	Limitations	References
iPSC-based models	Induced pluripotent stem cells as fibroblasts, keratinocytes, melanocytes	Histology, cytotoxicity assessment assay (MTT/MTS), cytokine release and biomarker profiling, melanin content	Pseudo 2-D/3-D microenvironment due to culture technique, early development, ethical constraints, limited immunologic aspects, limited barrier properties	[55]
MelanoDerm	Primary normal human melanocytes (NHM) and NHEK cells seeded on a collagen matrix	Histology, melanin content assay	Pseudo 2-D/3-D microenvironment due to culture technique, ethical constraints, limited immunologic aspects	www.mattek.com
Reconstituted human pigmented epidermis (RHE-MEL/001)	Primary human melanocytes and keratinocytes seeded on a polycarbonate filter	Histology, cytotoxicity assessment assay (MTT/MTS), cytokine release and biomarker profiling, melanin content	Pseudo 2-D/3-D microenvironment due to culture technique, limited immunologic aspects, limited barrier properties	www.straticell.com
SkinEthic RHPE	Primary human melanocytes and keratinocytes seeded on a polycarbonate filter	Histology, cytotoxicity assessment assay (MTT/MTS), cytokine release and biomarker profiling, melanin content	Pseudo 2-D/3-D microenvironment due to culture technique, early development, ethical constraints, limited immunologic aspects, limited barrier properties	www.skinEthic.com

Table 4 Disease state models

Skin model	Specifications	Readout parameter	Limitations	References
Keloid epidermis (fibroblasts)	Indirect coculture of keloid epidermis with keloid fibroblasts	Histology, cytotoxicity assessment assay (MTT/MTS), cytokine release and biomarker profiling	Pseudo 2-D/3-D microenvironment, no immune cells	[58,59]
3-D hypertrophic scar SE	Reconstituted epidermis of keratinocytes on a matrix containing fibroblasts	Histology, cytotoxicity assessment assay (MTT/MTS), cytokine release and biomarker profiling	No immune cells	[59,60]
Recessive dystrophic epidermolysis bullosa 3-D SE	Immortalized patient-derived keratinocytes and fibroblasts in deepidermalized porcine dermis	Histology, cytotoxicity assessment assay (MTT/MTS), cytokine release and biomarker profiling	No immune cells	[61,62]
Melanoma	Human malignant melanoma cells (A375), NHEK and NHDF cells	Histology, cytotoxicity assessment assay (MTT/MTS), cytokine release and biomarker profiling	No immune cells	www.mattek.com/products/melanoma/
3-D psoriasis SE	Normal human primary epidermal keratinocytes and human primary psoriatic fibroblasts harvested from psoriatic lesions	Histology, cytotoxicity assessment assay (MTT/MTS), cytokine release and biomarker profiling, validated drug panel screening	No immune cells	www.mattek.com; [63,64]
Squamous cell carcinoma SE	Commercially available and/or patient-derived cell lines (A375, WM35, SK-mel-28, SSC-12B2), human keratinocytes and fibroblasts, dermal matrix	Histology, cytotoxicity assessment assay (MTT/MTS), cytokine release and biomarker profiling, validated drug panel screening	No immune cells	[65,66]

Modified from Bergers LI, Reijnders CM, van den Broek LJ, Spiekstra SW, de Gruijl TD, Weijers EM, et al. Immune-competent human skin disease models. Drug Discov Today 2016;21(9):1479–88.

from patients with the disease, cultured alone, or in association with fibroblasts [28]. While this provides a good model with respect to understanding the differences in keratinocyte function between normal and healthy cells, keratinocytes only play a minor role in many immunologic skin disorders. For example, psoriasis is a chronic inflammatory skin disease that alongside epidermal hyperplasia and abnormal keratinocyte differentiation is also associated with abnormal circulation of polymorphonuclear leukocytes, infiltration of activated TCs and DCs, and altered levels of cytokines [67]. Gaining a better understanding of immune skin responses (e.g., sensitization, irritation, and inflammation) and immune skin diseases (e.g., eczema, dermatitis, and psoriasis) in vitro will require the use of more complex models that incorporate key innate or adaptive immune responses represented in the skin in vivo.

11. Advanced epidermal and dermal cell culture models

Although the dermal models mentioned above have contributed significantly to the field of skin research, a number of limitations exist that prevent their continual and/or translational application. These defining factors include a range of interrelated properties and center on (i) the expressed phenotypic differences when cultured in a 2-D environment compared with a 3-D; (ii) the implementation of an air-liquid interface during culture; (iii) the native source and pedigree of cells, that is, species and cell type; (iv) the limited presence and/or localization of skin-specific appendages, for example, hair follicles, sweat pores, and sebaceous glands; (v) the lack of complementary immunocompetent cell subpopulations; and (vi) the identification/detection/quantification of the (required) readout signal from the model following a test stimuli.

With the ongoing desire to establish a unified skin cell model that could be exploited by the multiarray of dermal researcher, technological advances have led to the step-by-step standardization of a human dermal equivalent. The initial characterization of a suitable cell model requires confirmation that a "valid" skin equivalent exists despite the origin of the cells and their in vitro culture microenvironment. As such, an extensive inventory of a variety of products and techniques has been developed over the past two decades that includes novel (3-D) biomaterials, surface modification techniques (e.g., charge, topography, and chemical moieties), incorporated perfusion/microfluidic loops, external stimuli, and growth factor supplementation—resulting in the generation of a human stratified skin equivalent [68]. This "trueness" factor is often determined through morphological characterization, correct profiling/expression of cell surface markers, and the presence of a functional barrier integrity through TEER/TEWL measurements. Following this justification, assessment of the detrimental effects of novel compounds tends to be based on a combination of visual observations of barrier integrity through simple histology and feedback from induced cytotoxic effects using (LD_{50}) viability assays and/or end point assays such as the MTT/MTS mitochondria activity assay, redox reaction, and/or LDH release profile.

However, the most crucial aspect of these models is the limitation in the incorporation of a population of immune cells to truly represent the in vivo (adaptive and innate immunologic) environment of the skin. The ability of a substance to cause damage—either through corrosion, irritation, or sensitization—when applied to the skin is determined as part of the assessment criteria for the topical testing of new chemicals and/or novel therapeutics. Corrosion can be regarded as a severe irreversible process leading to cell death (i.e., necrosis), whereas irritation initiates a complex cascade within the inflammatory pathway that causes changes to epithelial and endothelial cells within the tissues. Therefore, it is now considered fundamental that the next-generation immunocompetent skin models should encompass these immunologically responsive cells within a complex construct of stratified skin layers. This representation of native skin based on keratinocytes- and fibroblasts-populated layers would therefore allow immune cells to localize into and/or migrate out from the 3-D microenvironment as would be in an in vivo setting following the appropriate stimuli or insult.

12. In silico epidermal and dermal cell culture models

Given the complex nature and specialist skills required to ascertain permeation and sensitization data in skin models, it is unsurprising that the scientific community has also investigated the possibilities of predicting these parameters via the use of computational models. Accurate and robust in silico predictions of skin permeation have the potential to significantly reduce the time and costs of research projects and as such are of general interest to researchers. The emergence of ever-increasingly sophisticated predictive models was arguably catalyzed by the publication of a database of 97 permeation coefficients for 94 compounds by GL Flynn in 1990 [69]. It was the first time that data of this type had been compiled in a single repository, and it allowed Flynn to propose a simple model for skin permeation related to the molecular water and lipophilicity of the compound.

Flynn's dataset was rapidly adopted and exploited by the scientific community, which quickly led to the development of a number of quantitative structure-property relationship models, which attempted to predict, not just explain, the permeability coefficients of small molecules. The first of these, established by Potts and Guy [70], showed a quantitative linear relationship between permeation, lipophilicity, and molecular weight (Eq. 1) that was able to accurately explain 67% of the variance seen in the Flynn dataset:

$$\log K_{\mathrm{p}} = 0.71 \log K_{ow} - 0.0061 MW - 6.3 \qquad (1)$$

where $\log K_{\mathrm{p}}$ is the permeation coefficient, $\log K_{\mathrm{ow}}$ is the octan-1-ol-water partition coefficient, and MW is the molecular weight of a compound.

By quantifying Flynn's initial algorithmic ideas, Potts and Guy demonstrated that the physicochemical properties of a compound could be linked to its permeability through the skin using multiple linear regression analysis techniques. However, the correlation between predicted and experimental values for this model was relatively low. This led to investigations into other properties that could possibly be influencing permeation through the skin. Using a subset of 60 compounds from the Flynn dataset (deliberately excluding outliers such as the hydrocortisones), Barratt [71] showed that a robust and predictive QSPR model could be achieved by building a QSPR incorporating molecular volume, the octanol-water partition coefficient, and the melting point of a compound (Eq. 2):

$$\log K_p = 0.82 \log K_{ow} - 0.0093 MW - 0.039 MPt - 2.36 \tag{2}$$

where $\log K_p$ is the permeation coefficient, $\log K_{ow}$ is the octan-1-ol-water partition coefficient, MV is the molecular volume, and MPt is the melting point of the compound.

In the same year, Potts and Guy refined their initial model [72] and from a subset of the Flynn dataset that contained only 37 nonelectrolyte compounds generated a QSPR model that also included a term for molecular volume and replaced their original partition coefficient term, with descriptors related to the number of hydrogen bond donor and receptor activity of the molecule (Eq. 3):

$$\log K_p = 0.0256 MV - 1.72 \sum \alpha_2{}^H - 3.93 \sum \beta_2{}^H - 4.85 \tag{3}$$

where $\log K_p$ is the permeation coefficient, MV is the molecular volume, $\sum \alpha_2{}^H$ represents the total hydrogen bond donor activity, and $\sum \beta_2{}^H$ accounts for the total hydrogen bond acceptor activity.

A number of incremental QSPR models of percutaneous absorption followed, but it was not long before attention turned toward the limitations of the Flynn dataset for constructing predictive models and the limitations of linear modeling techniques in general given the multiplicity of routes that molecules can access to permeate through the skin (i.e., transcellular, inter/paracellular, and transappendageal transport). As Flynn had collated the information for the initial database from 15 different sources, the values lacked internal consistency because they had been collected in a variety of experimental manners. By critically evaluating the experimental methodologies used to obtain permeation coefficients and combining datasets from Kirchner [73], Degim [74], and Johnson [75], among others, Mass and Cronin [76] were able to produce a new dataset of 116 molecules that showed an improvement in predictive ability of a QSPR ($r^2 = .87$) for skin permeation using only the original descriptors of molecular weight and $\log P$ initially identified by Potts and Guy (Eq. 4):

$$\log K_p = 0.74 \log P - 0.091 MW - 2.39 \tag{4}$$

where $\log K_p$ is the permeation coefficient and MW is the molecular weight of a compound.

High-quality datasets such as this have since been employed by scientists to build and validate nonlinear computational models, such as artificial neural networks (see, e.g., [77,78]), designed to enhance predictions by accounting for the complex and competing mechanisms of permeation across the skin. Success in being able to accurately predict permeation has also led to the development of models that are able to provide information on potential skin sensitization, such as Pred-Skin [79,80], which is able to give accurate predictions of sensitization for compounds based on their molecular structure.

13. In vitro immunocompetent skin models

Only a relatively few validated, in vitro, cell culture models exist that explicitly incorporate immune-responsive cells as a functional component of the entirety (Table 5). Although diseased state skin models exist and considered as immunologic competent, this "immunology" is stand-alone and physiologically biased toward the projected prognosis. As such, these models tend to be exploited solely in therapeutic developmental studies for focused diseased state—the reader is directed to an excellent review article by Bergers and colleagues [28].

Table 5 Immunocompetent in vitro skin models

Skin model	Specifications	Readout parameter	Limitations	References
Nott-Sheff 3-D K-F-DC	Coculture of dendritic cells, within an agarose gel, alongside primary fibroblasts and keratinocytes	Histology, cytotoxicity assessment assay (MTT/MTS), cytokine release and biomarker profiling, DC migration	Pseudo 2-D/3-D microenvironment, early development, ethics and cell procurement limitations, no T cells	[27]
Skin equivalent with T cells	Decellularized deepidermized dermis seeded with human primary keratinocytes cocultured with activated T cells obtained from human donors allogeneic to the skin	Histology, cell viability, TUNEL staining, cytokine release, PCR, flow cytometry	Pseudo 2-D/3-D microenvironment, early development, limited fibroblast interaction	[59]
Skin equivalent with Langerhans cells	Air-liquid interface-differentiated epidermis integrated with MUTZ-3 cells on fibroblast-populated collagen I hydrogel	Histology, cytotoxicity assessment assay (MTT/MTS), cytokine release and biomarker profiling, DC plasticity and migration	Pseudo 2-D/3-D microenvironment, early development, no T cells	[81]

Although the culture of a single dermal-derived immune cell population (e.g., TCs, macrophages, DCs, and monocytes) may provide an immunologic response readout signal following stimulation, recent evidence suggests that for a complex organ, such as the skin, it is the interaction between different cell populations in a suitable culture environment that exerts the resultant outcome [82–84]. Accordingly, it is known that keratinocytes account for 90% of the epidermis and secrete cytokines and interleukins capable of modulating and directing the immune cells accordingly. Similarly, sedate and activated fibroblasts also participate in this immunologic pathway and coordinate the required response to the stimuli. As such, it can be considered that, in the case of the skin, it is the interplay between keratinocytes, fibroblasts, and a population of DCs to be the simplest representation of an immunocompetent skin equivalent. This hypothesis was taken to practice by Chau and colleagues who developed a human 3-D immunocompetent skin model based on primary keratinocytes, fibroblasts, and DCs differentiated from donor blood-derived monocytes [27]. In response to sensitization, the model allowed dendritic cell migration and demonstrated an upregulation of CD86 and HLA-DR. However, IL-1α, IL-6, and IL-8 were not upregulated in response to DNCB or SDS stimulation and put into question the reliability and/or indicative soluble marker for validation. Similarly, Kosten and colleagues developed a "skin equivalent" model that incorporated LCs alongside human keratinocytes and fibroblasts [81], whereas van de Bogaard and colleagues' skin equivalent model exploited T cells on a keratinocyte-loaded decellularized dermis [59]. Although both teams demonstrated positive cross talk between the cells and a degree of cytokine/chemokine interaction, they report that the mechanisms controlling general inflammation are different to the native environment and so are at an early stage of development. Taken together, the innate immunologic response in the skin is a highly complex and controlled environment that is orchestrated by a variety of cell types and requires further investigation to fully understand and/or mimic using an in vitro system.

14. Conclusion

It can be seen that the number of human skin equivalent skin models has significantly expanded over the past few years. Originally developed as layer-specific monoculture of skin cells on a flat culture substrate, these next-generation advanced dermal models incorporate a number of factors that drive the formation of the required in vivo tissue architecture. It has been found that not only do the physical and mechanical architecture (i.e., 3-D, biomaterial, topography, and perfusion) of the culture substrate contribute to the correct phenotypic expression of the cells but also there is a need to ensure that correct cell-to-cell signaling and interaction occur through the incorporation of the many different cell types naturally present in native skin. Although computational modeling and in silico programs can provide additional insight to the field of skin research, a fully functional in vitro model of immuno-responsive skin is still at developmental infancy. Fortunately, with the advances in tissue engineering, material science, and cell biology, the future adaptation of these models in both their construction and biological readout parameters will continue to further the understanding of skin disease, drug delivery, and in vitro toxicology assessment.

Conflict of interests

The authors have no conflict of interests.

References

[1] Aulton ME. Aulton's pharmaceutics: the design and manufacture of medicines. Edinburgh: Churchill Livingstone; 2007.

[2] Richmond JM, Harris JE. Immunology and skin in health and disease. Cold Spring Harb Perspect Med 2014;4(12):a015339.

[3] Benson HAE, Watkins AC. Topical and transdermal principles and practice. New Jersey: John Wiley and Sons, Inc.; 2012.

[4] Elias PM, Menon GK. Structural and lipid biochemical correlates of the epidermal permeability barrier. Adv Lipid Res 1991;24:1–26.

[5] Hadgraft J. Skin, the final frontier. Int J Pharm 2001;224(1–2):1–18.

[6] Monterio-Riviere NA. Structure and function of skin. In: Monterio-Riviere NA, editor. Toxicology of the skin. New York: Informa Healthcare; 2010.

[7] Streilein JW. Skin-associated lymphoid tissues (SALT): origins and functions. J Invest Dermatol 1983;80(Suppl):12s–6s.

[8] Haniffa M, Gunawan M, Jardine L. 2015. Human skin dendritic cells in health and disease. J Dermatol Sci 2015 Feb;77(2):85–92.

[9] Salmon JK, Armstrong CA, Ansel JC. The skin as an immune organ. West J Med 1994;160(2):146–52.

[10] Pasparakis M, Haase I, Nestle FO. Mechanisms regulating skin immunity and inflammation. Nat Rev Immunol 2014;14(5):289–301.

[11] Dinarello CA. A clinical perspective of IL-1β as the gatekeeper of inflammation. Eur J Immunol 2011;41(5):1203–17.

[12] Flacher V, Sparber F, Tripp CH, Romani N, Stoitzner P. Targeting of epidermal Langerhans cells with antigenic proteins: attempts to harness their properties for immunotherapy. Cancer Immunol Immunother 2009;58(7):1137–47.

[13] Schmitt D. Immune functions of the human skin. Models of in vitro studies using Langerhans cells. Cell Biol Toxicol 1999;15(1):41–5.

[14] Kupper TS. The activated keratinocyte: a model for inducible cytokine production by non-bone marrow-derived cells in cutaneous inflammatory and immune responses. J Invest Dermatol 1990;94(6 Suppl):146S–50S.

[15] Mizumoto N, Takashima A. CD1a and langerin: acting as more than Langerhans cell markers. J Clin Invest 2004;113(5):658–60.

[16] Ramata-Stunda A, Boroduskis M, Vorobjeva V, Ancans J. Cell and tissue culture-based in vitro test systems for evaluation of natural skin care product ingredients. Environ Exp Biol 2013;11:159–77.

[17] Ericsson AD, Dora J, Cao S. Melanocytes: morphological basis for an exteroceptive sensory system for monitoring ultraviolet radiation. Physiol Chem Phys Med NMR 2003;35(1):27–42.

[18] Dell'anna ML, Picardo M. A review and a new hypothesis for non-immunological pathogenetic mechanisms in vitiligo. Pigment Cell Res 2006;19(5):406–11.

[19] Tokura Y, Fujiyama T, Ikeya S, Tatsuno K, Aoshima M, Kasuya A, et al. Biochemical, cytological, and immunological mechanisms of rhododendrol-induced leukoderma. J Dermatol Sci 2015;77(3):146–9.

[20] Xie H, Zhou F, Liu L, Zhu G, Li Q, Li C, et al. Vitiligo: how do oxidative stress-induced autoantigens trigger autoimmunity? J Dermatol Sci 2016;81(1):3–9.
[21] Bueno V, Sant'Anna OA, Lord JM. Ageing and myeloid-derived suppressor cells: possible involvement in immunosenescence and age-related disease. Age (Dordr) 2014;36(6):9729.
[22] Beanes SR, Dang C, Soo C, Ting K. Skin repair and scar formation: the central role of TGF-beta. Expert Rev Mol Med 2003;5(8):1–22.
[23] Bruynzeel-Koomen C, van der Donk EM, Bruynzeel PL, Capron M, de Gast GC, Mudde GC. Associated expression of CD1 antigen and fc receptor for IgE on epidermal Langerhans cells from patients with atopic dermatitis. Clin Exp Immunol 1988;74(1):137–42.
[24] Metcalfe DD, Baram D, Mekori YA. Mast cells. Physiol Rev 1997;77(4):1033–79.
[25] Theoharides TC, Alysandratos K-D, Angelidou A, Delivanis D-A, Sismanopoulos N, Zhang B, et al. Mast cells and inflammation. Biochim Biophys Acta 2012;1822(1):21–33.
[26] Clark RA. Skin resident T cells: the ups and downs of on site immunity. J Invest Dermatol 2010;130(2):362–70.
[27] Chau DYS, Johnson C, MacNeil S, Haycock JW, Ghaemmaghami AM. The development of a 3D immunocompetent model of human skin. Biofabrication 2013;5(3):035011.
[28] Bergers LI, Reijnders CM, van den Broek LJ, Spiekstra SW, de Gruijl TD, Weijers EM, et al. Immune-competent human skin disease models. Drug Discov Today 2016;21(9):1479–88.
[29] Magnusson B, Kligman AM. The identification of contact allergens by animal assay. The guinea pig maximization test. J Invest Dermatol 1969;52(3):268–76.
[30] Robinson MK, Nusair TL, Fletcher ER, Ritz HL. A review of the Buehler guinea pig skin sensitization test and its use in a risk assessment process for human skin sensitization. Toxicology 1990;61(2):91–107.
[31] Kimber I, Pichowski JS, Betts CJ, Cumberbatch M, Basketter DA, Dearman RJ. Alternative approaches to the identification and characterization of chemical allergens. Toxicol In Vitro 2001;15(4–5):307–12.
[32] Kimber I, Hilton J, Dearman RJ, Gerberick GF, Ryan CA, Basketter DA, et al. An international evaluation of the murine local lymph node assay and comparison of modified procedures. Toxicology 1995;103(1):63–73.
[33] Ambrose CT. An amended history of tissue culture: concerning Harrison, burrows, mall, and carrel. J Med Biogr 2016;https://doi.org/10.1177/0967772016685033.
[34] Khokhlov AN. Does aging need its own program, or is the program of development quite sufficient for it? Stationary cell cultures as a tool to search for anti-aging factors. Curr Aging Sci 2013;6(1):14–20.
[35] Geraghty RJ, Capes-Davis A, Davis JM, Downward J, Freshney RI, Knezevic I, et al. Guidelines for the use of cell lines in biomedical research. Br J Cancer 2014;111(6):1021–46.
[36] Langer R, Vacanti JP. Tissue engineering. Science 1993;260(5110):920–6.
[37] Berthiaume F, Maguire TJ, Yarmush ML. Tissue engineering and regenerative medicine: history, progress, and challenges. Annu Rev Chem Biomol Eng 2011;2:403–30.
[38] Johansson H, Albrekt AS, Borrebaeck CA, Lindstedt M. The GARD assay for assessment of chemical skin sensitizers. Toxicol In Vitro 2013;27(3):1163–9.
[39] Reuter H, Spieker J, Gerlach S, Engels U, Pape W, Kolbe L, et al. In vitro detection of contact allergens: development of an optimized protocol using human peripheral blood monocyte-derived dendritic cells. Toxicol In Vitro 2011;25(1):315–23.
[40] Hoffmann J, Heisler E, Karpinski S, Losse J, Thomas D, Siefken W, et al. Epidermal-skin-test 1,000 (EST-1,000)—a new reconstructed epidermis for in vitro skin corrosivity testing. Toxicol In Vitro 2005;19(7):925–9.

[41] Botham PA. The validation of in vitro methods for skin irritation. Toxicol Lett 2004;149:387–90.
[42] Spielmann H, Liebsch M, Botham PA, Fentem JH, Roguet R, Cotovio J, et al. The ECVAM international validation study on in vitro tests for acute skin irritation: report on the validity of the EPISKIN and EpiDerm assays and on the skin integrity function test. Altern Lab Anim 2007;35(6):559–601.
[43] Zuang V, Balls M, Botham PA, Coquette A, Corsini E, Curren RD, et al. Follow-up to the ECVAM prevalidation study on in vitro tests for acute skin irritation. The European Centre for the Validation of alternative methods skin irritation task force report 2. Altern Lab Anim 2002;30(1):109–29.
[44] Katoh M, Hamajima F, Ogasawara T, Hata K. Assessment of human epidermal model LabCyte EPI-MODEL for in vitro skin irritation testing according to European Centre for the Validation of alternative methods (ECVAM)-validated protocol. J Toxicol Sci 2009;34(3):327–34.
[45] Katoh M, Hamajima F, Ogasawara T, Hata K. Assessment of the human epidermal model LabCyte EPI-MODEL for in vitro skin corrosion testing according to the OECD test guideline 431. J Toxicol Sci 2010;35(3):411–7.
[46] Boukamp P, Petrussevska RT, Breitkreutz D, Hornung J, Markham A, Fusenig NE. Normal keratinization in a spontaneously immortalized aneuploid human keratinocyte cell line. J Cell Biol 1988;106(3):761–71.
[47] Boukamp P, Popp S, Altmeyer S, Hülsen A, Fasching C, Cremer T, et al. Sustained nontumorigenic phenotype correlates with a largely stable chromosome content during long-term culture of the human keratinocyte line HaCaT. Genes Chromosomes Cancer 1997;19(4):201–14.
[48] Li L, Fukunaga-Kalabis M, Herlyn M. Establishing human skin grafts in mice as model for melanoma progression. Methods Mol Biol 2015;13:1–10.
[49] van de Sandt J, Roguet R, Cohen C, Esdaile D, Ponec M, Corsini E, et al. The use of human keratinocytes and human skin models for predicting skin irritation. Altern Lab Anim 1999;27(5):723–43.
[50] Kandárová H, Hayden P, Klausner M, Kubilus J, Kearney P, Sheasgreen J. In vitro skin irritation testing: improving the sensitivity of the EpiDerm skin irritation test protocol. Altern Lab Anim 2009;37(6):671–89.
[51] Chioni AM, Grose R. Organotypic modelling as a means of investigating epithelial-stromal interactions during tumourigenesis. Fibrogenesis Tissue Repair 2008;1(1):8.
[52] Pruniéras M, Delescluse C, Regnier M. The culture of skin. A review of theories and experimental methods. J Invest Dermatol 1976;67(1):58–65.
[53] Hakelius M, Saiepour D, Göransson H, Rubin K, Gerdin B, Nowinski D. Differential gene regulation in fibroblasts in co-culture with keratinocytes and head and neck SCC cells. Anticancer Res 2015;35(6):3253–65.
[54] Nolte SV, Xu W, Rennekampff HO, Rodemann HP. Diversity of fibroblasts—a review on implications for skin tissue engineering. Cells Tissues Organs 2007;187(3):165–76.
[55] Gledhill K, Guo Z, Umegaki-Arao N, Higgins CA, Itoh M, Christiano AM. Melanin transfer in human 3D skin equivalents generated exclusively from induced pluripotent stem cells. PLoS One 2015;10(8):e0136713.
[56] Ryu JH, Seok JK, An SM, Baek JH, Koh JS, Boo YC. A study of the human skin-whitening effects of resveratryl triacetate. Arch Dermatol Res 2015;307(3):239–47.
[57] Duval C, Cohen C, Chagnoleau C, Flouret V, Bourreau E, Bernerd F. Key regulatory role of dermal fibroblasts in pigmentation as demonstrated using a reconstructed skin model: impact of photo-aging. PLoS One 2014;9(12):e114182.

[58] Moulin VJ. Reconstitution of skin fibrosis development using a tissue engineering approach. Methods Mol Biol 2013;961:287–303.
[59] van den Bogaard EH, Tjabringa GS, Joosten I, Vonk-Bergers M, van Rijssen E, Tijssen HJ, et al. Crosstalk between keratinocytes and T cells in a 3D microenvironment: a model to study inflammatory skin diseases. J Invest Dermatol 2014;134(3):719–27.
[60] van den Broek LJ, Niessen FB, Scheper RJ, Gibbs S. Development, validation and testing of a human tissue engineered hypertrophic scar model. ALTEX 2012;29(4):389–402.
[61] Cogan J, Weinstein J, Wang X, Hou Y, Martin S, South AP, et al. Aminoglycosides restore full-length type VII collagen by overcoming premature termination codons: therapeutic implications for dystrophic epidermolysis bullosa. Mol Ther 2014;22(10):1741–52.
[62] El Ghalbzouri A, Jonkman M, Kempenaar J, Ponec M. Recessive epidermolysis bullosa simplex phenotype reproduced in vitro: ablation of keratin 14 is partially compensated by keratin 17. Am J Pathol 2003;163(5):1771–9.
[63] Jean J, Lapointe M, Soucy J, Pouliot R. Development of an in vitro psoriatic skin model by tissue engineering. J Dermatol Sci 2009;53(1):19–25.
[64] Tjabringa G, Bergers M, van Rens D, de Boer R, Lamme E, Schalkwijk J. Development and validation of human psoriatic skin equivalents. Am J Pathol 2008;173(3):815–23.
[65] Commandeur S, de Gruijl FR, Willemze R, Tensen CP, El Ghalbzouri A. An in vitro three-dimensional model of primary human cutaneous squamous cell carcinoma. Exp Dermatol 2009;18(10):849–56.
[66] Linde N, Gutschalk CM, Hoffmann C, Yilmaz D, Mueller MM. Integrating macrophages into organotypic co-cultures: a 3D in vitro model to study tumor-associated macrophages. PLoS One 2012;7(7):e40058.
[67] Korthuis RJ, Unthank JL. Experimental models to investigate inflammatory processes in chronic venous insufficiency. Microcirculation 2000;7(6 Pt 2):S13–22.
[68] Elliott NT, Yuan F. A microfluidic system for investigation of extravascular transport and cellular uptake of drugs in tumors. Biotechnol Bioeng 2012;109(5):1326–35.
[69] Gerrity TR, Henry CJ. Principles of route-to-route extrapolation for risk assessment. New York: Elsevier Science Ltd.; 1991.
[70] Potts RO, Guy RH. Predicting skin permeability. Pharm Res 1992;9(5):663–9.
[71] Barratt MD. Quantitative structure-activity relationships for skin permeability. Toxicol In Vitro 1995;9(1):27–37.
[72] Potts RO, Guy RH. A predictive algorithm for skin permeability: the effects of molecular size and hydrogen bond activity. Pharm Res 1995;12(11):1628–33.
[73] Kirchner LA, Moody R, Doyle E, Bose R, Jeffery J, Chu I. The prediction of skin permeability data by using physicochemical data. Altern Lab Anim 1997;25:359–70.
[74] Degim IT, Pugh WJ, Hadgraft J. Skin permeability data: anomalous results. Int J Pharm 1998;170(1):129–33.
[75] Johnson ME, Blankschtein D, Langer R. Permeation of steroids through human skin. J Pharm Sci 1995;84(9):1144–6.
[76] Moss GP, Cronin MTD. Quantitative structure-permeability relationships for percutaneous absorption: re-analysis of steroid data. Int J Pharm 2002;238(1):105–9.
[77] Değím T, Hadgraft J, İlbasmiş S, Özkan Y. Prediction of skin penetration using artificial neural network (ANN) modeling. J Pharm Sci 2003;92(3):656–64.
[78] Lim CW, Fujiwara S, Yamashita F, Hashida M. Prediction of human skin permeability using a combination of molecular orbital calculations and artificial neural network. Biol Pharm Bull 2002;25(3):361–6.
[79] Alves VM, Capuzzi SJ, Muratov EN, Braga RC, Thornton TE, Fourches D, et al. QSAR models of human data can enrich or replace LLNA testing for human skin sensitization. Green Chem 2016;18(24):6501–15.

[80] Braga RC, Alves VM, Muratov EN, Strickland J, Kleinstreuer N, Trospsha A, et al. Pred-skin: a fast and reliable web application to assess skin sensitization effect of chemicals. J Chem Inf Model 2017;57(5):1013–7.
[81] Kosten IJ, Buskermolen JK, Spiekstra SW, de Gruijl TD, Gibbs S. Gingiva equivalents secrete negligible amounts of key chemokines involved in langerhans cell migration compared to skin equivalents. J Immunol Res 2015;627125.
[82] Chung E, Choi H, Lim JE. Development of skin inflammation test model by co-culture of reconstituted 3D skin and RAW264.7 cells. Tissue Eng Regen Med 2014;11:87–92.
[83] De Wever B, Sandrine K, Pascal D. Human skin models for research applications in pharmacology and toxicology: introducing NativeSkin®, the "missing link" bridging cell culture and/or reconstructed skin models and human clinical testing. Appl In Vitro Toxicol 2015;1(1):26–32.
[84] Hewitt NJ, Edwards RJ, Fritsche E, Goebel C, Aeby P, Scheel J, et al. Use of human in vitro skin models for accurate and ethical risk assessment: metabolic considerations. Toxicol Sci 2013;133(2):209–17.
[85] Zöller NN, Kippenberger S, Thaçi D, Mewes K, Spiegel M, Sättler A, et al. Evaluation of beneficial and adverse effects of glucocorticoids on a newly developed full-thickness skin model. Toxicol In Vitro 2008;22(3):747–59.
[86] Uchino T, Tokunaga H, Onodera H, Ando M. Effect of squalene monohydroperoxide on cytotoxicity and cytokine release in a three-dimensional human skin model and human epidermal keratinocytes. Biol Pharm Bull 2002;25(5):605–10.

Further reading

[1] Grindon C, Combes R, Cronin MT, Roberts DW, Garrod JF. Integrated decision-tree testing strategies for skin corrosion and irritation with respect to the requirements of the EU REACH legislation. Altern Lab Anim 2008;36(Suppl. 1):65–74.
[2] Stamatas GN, Morello AP, Mays DA. Early inflammatory processes in the skin. Curr Mol Med 2013;13(8):1250–69.
[3] Young B, Heath JW. Wheater's functional histology. 4th ed. London: Churchill Livingstone; 2000. p. 162. ISBN 0-443-05612-9.
[4] Carlson MW, Alt-Holland A, Egles C, Garlick JA. Three-dimensional tissue models of normal and diseased skin. Curr Protoc Cell Biol 2008. Chapter 19:Unit 19.9.

Section E

Emerging technologies for the development of 3D skin models

Additive manufacturing in the development of 3D skin tissues

Kelsey N. Retting, Deborah G. Nguyen
Organovo Inc., San Diego, CA, United States

1 Introduction

Additive manufacturing is the process of building three-dimensional (3D) structures layer by layer using a computer-controlled system or computer-aided design (CAD). Three-dimensional printing is a form of additive manufacturing that offers both geometric and material flexibility to allow a high degree of customization in many applications. A 3D printing device uses materials ranging from plastics, resins, alloys, metals, polymers, and ceramics to create objects through precise deposition in the *x*, *y*, and *z* dimensions [1]. In recent years, 3D printing has gone from a specialized niche technology to a widely applicable approach, allowing both engineers and the public the ability to produce customized objects from cars to medical devices to machine parts. In medicine, 3D bioprinting has allowed physicians to produce skull implants, orthodontic devices, hearing aids, and external orthopedic devices that are tailored to a patient's unique dimensions [2]. Progress in this arena has been so rapid that the FDA has had to issue "leapfrog" draft guidance, even as it continues to try and understand the impact of the technology on developing new devices [3].

On the heels of the accelerated use of 3D printing in general, 3D bioprinting has also begun to emerge as a transformational technology. Human organs are not random collections of cells, but instead contain cells arranged in specific locations in three dimensions. This spatial context enables cells to form specific contacts with each other and deliver local chemical cues, leading to activation of certain pathways and inhibition of others. Bioprinting incorporates the deposition of cells and/or cell-containing materials (referred to as "bioink") into an automated approach with the potential to mimic native biologically accurate structure and function of tissues and organs. Much like traditional additive manufacturing, the bioink is deposited in specific locations in three-dimensional space layer by layer, with the resulting structure demonstrating the sort of spatial arrangement more typical of human tissues. The flexibility of this tissue-engineering platform allows bioprinted tissues to be produced in various forms including tubes, patches, sheets, and organoids and can generate compartmentalized architecture within the *x*, *y*, and *z* axes to create laminar and planar patterning. The utilization of bioprinting is a disruptive technology in the field of biomedical research with implications for advancing drug development and personalized medicine by creating tissue models for preclinical in vitro use and restoring and replacing damaged and diseased tissue. In this chapter, we will briefly review the different bioprinting approaches, outline the unique structural requirements of an engineered skin tissue, and then describe efforts to employ bioprinting for the fabrication of the human skin.

Skin Tissue Models. https://doi.org/10.1016/B978-0-12-810545-0.00016-4

2 Bioprinting: Deposition approaches and biomaterials

2.1 Deposition approaches

The desire to create complex 3D tissues that accurately model the biological responses of native human tissue has driven the development of several fundamentally different deposition technologies, each with different biomaterial compatibility (Table 1 and Fig. 1; modified review article table below). Because the process needs to enable fabrication of a wide variety of complex structures, the 3D bioprinting system must be capable of spatially directing a wide variety of cell types, biomaterials, and bioreagents to the print surface with a high level of control, all while preserving cell viability. The fabrication time and overall cytocompatibility of the engineered tissue will depend on the size and complexity of the structure geometry, choice of cellular input, and biomaterials.

2.1.1 Extrusion-based bioprinting

Perhaps the most widely utilized approach in 3D printing is extrusion-based printing, which produces a continuous deposition of material from a cassette or syringe cartridge through an orifice under pressure that is dispensed through pneumatic, piston-driven, or screw-driven force [4]. The volume of material dispensed is directly proportional to the area displaced. A major advantage of bioprinting with an extrusion-based approach is the ability to utilize high-viscosity hydrogels including highly cellular bioinks and cellular aggregates. Material can be deposited as aggregates including cylindrical filaments or spheres [5,6]. Incorporation of multiple print heads can further enhance versatility by allowing for combinations of materials to be bioprinted simultaneously. The minimum feature size of this methodology can vary depending upon the cell-containing biomaterials and on the cell type(s) included. An advantage of this methodology is that it can generate larger feature sizes often around 100–500 μm, which can allow for scalability and rapid printing of both large volumes and replicate structures [6]. The resolution, however, may be lower than other printing approaches and may limit finer cell patterning.

2.1.2 Droplet-based bioprinting

To achieve finer feature sizes like thin lines, small compartments, or single cell layers, a variety of methods that involve the creation and deposition of droplets at defined positions in three-dimensional space have been developed. Inkjet bioprinting involves the ejection of droplets containing either cells or biomaterials from a reservoir in response to a pressure pulse generated by thermal or piezoelectric forces. Thermal approaches electrically heat the print head to quickly create a small bubble that forces droplets from the nozzle. In contrast, piezoelectric approaches generate vibrations or acoustic waves inside the printer head to break the bioink into droplets that are then ejected [7,8]. Both inkjet methods generate droplets in the picoliter to microliter range, enabling the creation of very fine patterns. Microsolenoid valve bioprinting utilizes an electromechanical valve to generate droplets. The voltage pulse directs current through a coil

Table 1 Summary of bioprinting approaches

Category	Technology	Description	Basic dispense types	Resolution	Volume dispensed	Cell density	Viscosity	References
Extrusion-based	Positive displacement/ extrusion	Deposition of cellular material with or without hydrogel by extrusion through an orifice	Cylinders, sheets, lines, spheres, dots	>50 μm	Microliter-milliliter	Low-high, up to 100% cellular	High	[4-6,37,58,59,63]
Droplet-based	Inkjet	Ejection of small volumes of material in a pressurized reservoir following a thermal or piezoelectric pulse	Droplets, liquid overlays	> 25 μm	Picoliter-microliter	Low	Low	[7,9,32]
	Microvalve/ microsolenoid valve	Controlled deposition of material in pressurized reservoir through electronically controlled valve	Droplets, thin films, liquid overlays	>2 mm	Microliter	Low	Low	[7,10,23,25,31]
	Laser-driven transfer	Generation of small droplets by a focused laser on a surface containing the print material	Droplets	1–125 μm	Femtoliter-nanoliter	High, up to 10^8 cells/mL	Low-medium	[12,13,15,33,34]

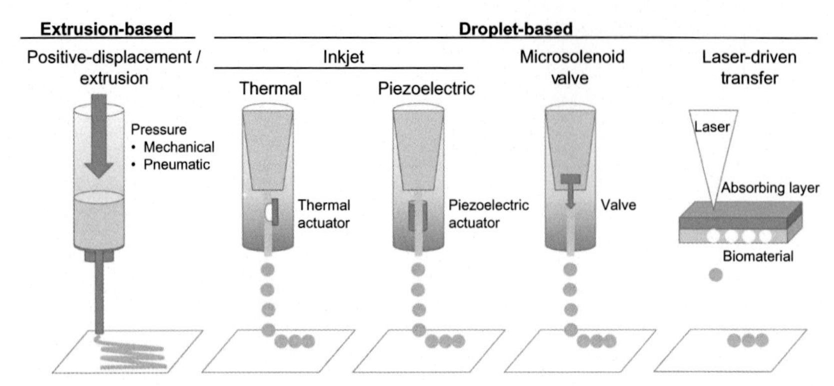

Fig. 1 Diagram of bioprinting approaches.

and in turn creates a coincident magnetic field that opens the solenoid valve to release bioink from a pressurized chamber [7]. This approach can employ a lower range of pneumatic pressure compared with inkjet bioprinting, therefore reducing the amount of shear stress applied to cellular material and potentially reducing cell damage. When compared with inkjet, microsolenoid valves generate larger cell-containing droplets of 10 μL or more and can also be utilized in a spraying mode to provide relatively thin layers of cellular or other biomaterials. In general, the droplet-based methods can generate relatively high-resolution structures from 20 to 100 μm, although the resolution of microvalve dispensing may be slightly lower. Both approaches are limited by the low range of viscosity and low cell density required to avoid shear stress or cell clogging. In addition, cell suspensions may encounter cell sedimentation and aggregation, which can alter the cell density in the dispensed droplets and reduce tissue to tissue reproducibility [9,10]. In some cases, droplet-based printing approaches rely on fast gelation rates to polymerize or stabilize material following deposition, which may limit the types of materials that can be used [11].

Laser-based bioprinting, laser-assisted bioprinting (LABP), or laser-induced forward transfer (LIFT) utilizes laser pulses to dispense droplets. The laser is focused on a "donor slide" or ribbon, comprising a surface with an energy absorbing substrate, such as gold, and a bioink layer. Energy from the focused laser is transferred from the absorbing layer to the bioink, inducing high gas pressure and evaporation that propels a droplet toward a print surface or "collector slide." The collector slide can contain hydrogel, which cells can be printed onto or embedded into. The donor and collector slides can be moved relative to each other to create patterning in the xy axis and then repeated to produce layers and 3D patterns. The size of the droplet can be adjusted by altering the pulse energy, the gap distance, the size of the focused laser spot, and the thickness of the absorbing substrate [12,13]. Laser bioprinting creates smaller droplets than inkjet approaches (picoliter to femtoliter range), which enables very high- resolution patterning both in 2D and in 3D [14,15]. In addition, because it is nozzle-free, this technique can accommodate higher cell densities (up to 10^8 cells/mL) and hydrogel viscosities [16]. Despite these advantages, the heat generated from laser energy and the prolonged fabrication time compared with other methods can damage cells and reduce viability, thus limiting its utility for some applications.

2.2 Biomaterials

2.2.1 Cellular materials

Current in vitro skin models incorporate a variety of cell types including primary cells and cell lines to recapitulate both epidermal and dermal layers. Common skin cell lines utilized in fabrication approaches include the immortalized human keratinocyte cell line HaCaT, the human foreskin fibroblast cell line HFF-1, and the murine NIH3T3 fibroblast cell line [17,18]. Although cell lines minimize variability, differences in proliferation and differentiation rates, metabolism, and species-specific physiology can be seen in comparison with the intact human skin [19,20]. Primary human keratinocytes and dermal fibroblasts may offer the most accurate representation of native cell function. However, it is important to consider the impact of variability in proliferative capacity and matrix production from donor to donor, as well as the differences between neonatal and adult cells. For disease modeling, primary cells derived from tissue biopsies or from iPS cells of diseased patients can be substituted for normal cells [21,22]. Additional aspects such as immunogenicity must be considered if the tissue is to be used for therapeutic applications such as tissue grafts. Because they can be expanded after removal from the body, autologous sources (i.e., the patient's own cells) of both keratinocytes and dermal fibroblasts are a viable option for clinical use. The high proliferative capacity and multipotency of stem cells make them an attractive alternative for skin engraftment approaches due to their lack of significant immunogenicity [23].

2.2.2 Biomaterials

To build architecturally defined tissues, bioprinting platforms use bioinks with a variety of cellular densities and matrix components. Reconstructing native tissue requires an interplay of cell-cell and cell-matrix interactions in appropriate cellular density, spatial patterning, and matrix composition to maintain tissue integrity, viability, and function. The relative cellularity of a bioink formulation can range from 0% to 100% [24]. Acellular materials or hydrogels can be applied for structural support or constitute space-saving regions in which there are no cells; hydrogels have also been used as sacrificial materials to create void spaces in vascular structures prior to tissue maturation [25-27]. High density cellular bioinks can harness the natural cell-cell interactions stimulated by close proximity to promote organized architecture based on physiological and biological properties [5,28-30]. Modifications to the bioink can be used to create architectural features (both large and small) and microenvironments and may incorporate specialized ECM to augment or inhibit intracellular interactions and differentiation. By combining cells and matrix and modifying the relative ratio of these components, both hypercellular and hypocellular tissues can be readily fabricated with high reproducibility. Engineered in vitro skin models incorporate a variety of biomaterials to recapitulate the architecture of the native skin, including both natural polymers (i.e., collagen [10,23,31-34], hyaluronic acid [35,36], fibrin [23,32,36], gelatin [37], and alginate [12,37]) and synthetic polymers (i.e., polyethylene gly*col* (PEG) [9,38], polyethylene glycol diacrylate (PEGDA) [36], polylactic acid (PLA) [39], polyglycolic acid (PGA) [40], poly(ε-caprolactone) (PCL) [41], and poly (lactic-co-glycolic acid)

(PLGA) [42]). The stiffness or viscosity of the hydrogel matrix can be tuned via chemical manipulation, temperature, cross-linking density, and polymer density [11]. Regardless of whether it is natural or synthetic, the matrix material must be noncytotoxic and compatible with the bioprinting approach; the selection of appropriate materials must take into account viscosity, gelation time, rheological properties, and sheer stress in order to balance the capability of the fabrication platform with the need for a viable tissue with the desired architecture [43,44].

3 Architectural considerations for skin tissue engineering

The native human skin exhibits a complex layered architecture consisting of epidermal, dermal, and hypodermal layers. The epidermis, or outermost layer, is composed mainly of keratinocytes that are organized into a highly stratified architecture distinguished by both morphology and differentiation state. The basal keratinocytes that form the proliferative base layer align atop a basement membrane and can be distinguished by their columnar morphology and expression of cytokeratins 5 and 14 (CK5 and CK14). As the basal cells differentiate, they migrate upward, become more flattened, and express cytokeratins 1 and 10 (CK1 and CK10), forming the spinous layer. These spinous keratinocytes continue to differentiate into a distinct granular layer, in which granules are present in the cells with expression of key markers involucrin and filaggrin. Cells in the granular layer then terminally differentiate, creating a waxy and lipid-rich layer known as the stratum corneum that serves as a protective barrier. Between the epidermal and dermal layer is the dermal-epidermal junction (DEJ), composed of basement membrane components including type IV collagen, laminin, and heparan sulfate proteoglycans. This basement membrane serves to anchor the epidermis to the dermis and provides cues to the basal keratinocytes that perpetuate their function; autoimmune targeting of basement membrane components leads to debilitating blistering conditions [45,46].

Supporting the epidermal layer is the dermis, composed mainly of dermal fibroblasts and connective tissue or extracellular matrix (ECM). The dermal layer is hypocellular compared with the epidermis with a complex mixture of ECM including collagens, elastin, proteoglycans, and glycosaminoglycans. The dermal layer can be subdivided into an upper papillary layer containing thinner collagen fibers and a lower reticular dermal layer composed of thicker collagen fibers and fewer fibroblasts. Below the dermal layer is the hypodermal layer, comprising adipose and connective tissue and containing blood and lymphatic vessels. Specialized cell types can be found in all layers of skin tissue including innate and adaptive immune cells and melanocytes. The native skin is also populated with vasculature and nerve networks as well as appendages including hair follicles, sebaceous glands, and sweat glands [46]. When engineering skin tissue, researchers must consider what architectural features are critical to include for their target application and then determine the appropriate fabrication methodology necessary to achieve those features (Fig. 2).

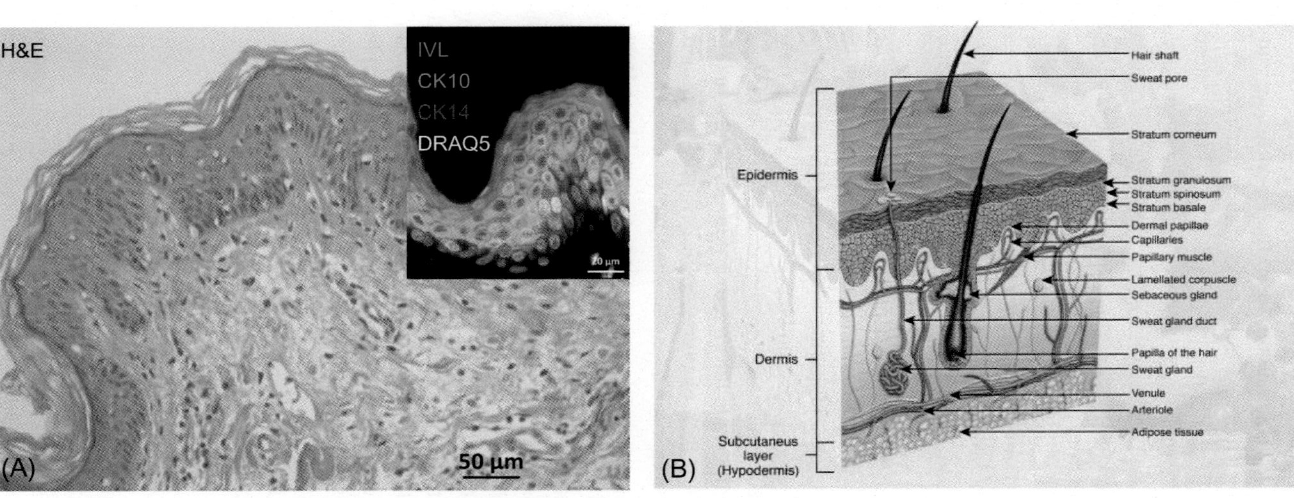

Fig. 2 Histology of native skin. (A) Hematoxylin and eosin staining (H&E) of bilayered architecture showing disinct epidermis (upper layer) and dermis (lower layer). Inset shows immunofluorescent staining of epidermis with differentiated and stratified keratinocyte layers: basal layer stained red for cytokeratin 14 (CK14), spinous layer stained green for cytokeratin 10 (CK10), spinous/granular layer stained blue for involucrin (IVL), and white cell nuclei (DRAQ5). (B) Schematic of skin tissue reproduced with permission from [50].

3.1 Manually fabricated 3D skin tissues

Manually fabricated in vitro skin models are commonly used for routine predictive assessment in the cosmetic and chemical industries, with end points including skin irritation, corrosion, absorption, and sensitization. Epidermal skin tissues can be produced in the laboratory by manually seeding a monolayer of keratinocytes from a cell suspension and subsequently differentiating the cells at an air-liquid interface to create an epidermal tissue layer. Epidermal tissue models containing keratinocytes alone can recapitulate many functional aspects of skin tissue; however, a full-thickness skin provides an advantage by allowing signaling cross talk between dermal fibroblasts and epidermal keratinocytes, which has been shown to improve epidermal thickness and organization [47,48]. A full-thickness skin tissue can be achieved in a similar fashion by manually seeding the epidermal keratinocytes onto a previously seeded and cultured mixture of dermal fibroblasts and collagen matrix [49,50].

Despite their widespread use, manually fabricated layered skin tissue models lack some complexity in relation to native architecture, which may reduce correlation with in vivo data [17,50,51]. In addition, reproducibility of manually fabricated skin tissues may be lower than desired. Alternative methods can be employed to achieve three-dimensional tissue structures including seeding of cells onto porous microcarrier scaffolds, spheroids/cell aggregates, and bioprinting [5,52,53]. Tissue-engineering approaches utilizing cell-seeded porous biomaterial scaffolds, such as static surface seeding and cell injection, can enable the creation of larger structures [54]. However, seeded scaffolds suffer from a lack of precision of cell placement and nonuniform cell distribution that can result in insufficient intracellular interactions and mechanical properties, impeding subsequent development of the native-like cellular density required for normal skin function. Spheroids and aggregates can provide higher cellular density and can be used as building blocks to create spatially arranged clusters by utilizing inherent properties of cell-cell and cell-matrix adhesion [5,30]. Spheroids can be formed by seeding cells on to nonadhesive surfaces or by incubating a suspension as a hanging drop [55]. When allowed to fuse in micromolds containing recesses, spheroids can also be used to fabricate simple shapes [56]. By virtue of their simple fabrication approaches and low cellular input requirements, spheroids can be adapted to high-throughput applications. It may be difficult, however, to control the uniformity of spheroid size and to achieve the desired microarchitecture and reproducible cell patterning associated with native tissue. Furthermore, the air-liquid interface required for native skin development and function may not be compatible with in vitro fabrication methodologies that require continuous cell submersion in culture media.

3.2 Bioprinted 3D skin tissues

When compared with both manual seeding and other 3D fabrication approaches, bioprinting confers many advantages. Automated processes like bioprinting often lead to improved reproducibility when compared with manual approaches. Manual full- thickness 3D skin models often require a minimum of 3–4 weeks to obtain a layered skin-like structure. Bioprinting can be employed to overlay sheets of cells simultaneously to create complex architecture from the beginning, thus enabling a shorter time period for maturation [49]. A key limitation of both cell-seeded scaffolds and spheroids is the ability to control deposition with microscale x, y, and z dimensional precision. Using a flexible bioprinting approach, engineers can incorporate specialized cell populations in precise controlled architecture into either full-thickness skin or individual layers such as epidermal tissue or dermal tissue alone to better model complex effects. Deposition of specialized cell types like follicular dermal papillae, sebocytes, or cancer cells within specific layers or 3D compartments within the tissue could allow for the formation of hair follicles and sebaceous glands and even melanomas for compound efficacy and disease modeling, as well as implantation where appropriate. Finally, another advantage of an automated bioprinting approach is scalability; larger bioprinted skin structures can be fabricated as implants or engraftments to reconstruct the 3D morphology of a wound, while miniaturized tissues can be produced to enable high-throughput applications for in vitro compound screening purposes in multiwell formats [24,57]. Bioprinted skin models therefore have the potential to fill unmet needs in many applications of skin research including cosmetic, chemical, pharmaceutical, wound healing, and disease modeling. Researchers have begun to utilize a variety of bioprinting approaches to achieve these goals (Table 2).

3.2.1 Extrusion-based bioprinted skin

Extrusion-based bioprinting has been successfully utilized to develop reproducible, spatially controlled architecture in vitro for multiple tissue types including liver and kidney, thus enabling biochemical, genetic, and histological interrogation following exposure to modulators of interest for toxicology and disease modeling [58-62]. Researchers at Organovo, Inc. recently demonstrated a novel scaffold-free extrusion-based bioprinted skin model using the NovoGen Bioprinter Platform. In this study, tissues were fabricated with normal primary human dermal fibroblasts and primary keratinocytes to produce tissue that exhibited distinct layered epidermal and dermal compartments with stratified epidermal architecture and collagen production in the dermal layer. The model was developed in a multiwell format for screening applications and exhibited an injury response to known toxicants SDS or Triton X upon treatment, with reduced viability and demonstrated spikes in IL1a production within the time frame of exposure [63].

Extrusion-based printing approaches have also recently been used to explore fabrication of skin appendages. Research by Liu et al. utilized a pneumatic 3D bioprinting system with two cell-laden bioinks composed of murine embryonic epithelial progenitor cells, gelatin, and cross-linked alginate in a layered crisscross pattern to generate square-pore structures. These structures, which also contained a milieu of ECM components including bone morphogenetic protein-4 (BMP-4) and planter dermis to mimic an inductive niche, were subsequently cultured in vitro to induce sweat gland morphogenesis [37]. Normal architecture of a sweat gland is a bilayered tissue consisting of a hollow center lined by an inner layer of secretory (luminal) cells and an outer layer of myoepithelial cells surrounded by a basement membrane. Glands are formed by invaginations of CK5/CK14 positive cells. As the tubular structure extends, cells of the inner layer differentiate into CK8/CK18 luminal cells, while cells of the outer layer remain CK5/CK14 positive [64]. In this study, cell aggregation and

Table 2 **Examples of bioprinted skin**

Technology	Biomaterials	Cell types	Bioink cell density	Resolution
Extrusion-based	nd	Primary human dermal fibroblasts, primary human epidermal keratinocytes	>50 million cells/mL	100 μm
	Alginate, gelatin	Murine embryonic epithelial progenitor cells	nd	300–400 μm
Droplet-based, microvalve	Collagen	Primary adult human dermal fibroblasts, primary adult human epidermal keratinocytes	1 million cells/mL	200–900 μm (interdispensing distance), nanoliter range droplets, 1–3 psi, valve open time 450–600 μs
	Collagen	HFF-1, HaCaT	0.5–5 million cells/mL	500 μm (interdispensing distance), nanoliter range droplets, 1.4–2.7 psi; valve open time 750 μs
	Collagen, fibrinogen, thrombin	Amniotic fluid-derived stem cells (AFS) and mesenchymal stem cells (MSC)	16.6 million cells/mL	nd
Droplet-based, inkjet	PEG-based bioink	Primary human dermal fibroblasts (HDFa), primary human epidermal keratinocytes	9 million cells/mL	Dosing distance 0.05–0.90 mm, valve open time 200–250 μs, air pressure 0.25–1.25 bar, feed rate 10 mm/s, thickness 0.01–0.05 mm
	Collagen, fibrinogen, thrombin	Human microvascular endothelial cells (HMVEC), neonatal human dermal fibroblasts (NHDF), neonatal human epidermal keratinocytes (NHEK)	2 million cells/mL	nd
Droplet-based; laser driven transfer	Alginate, blood plasma, Matrigel	NIH3T3, HaCaT	33–66 million cells/mL	80–140 μm droplet, 70 μm line width and 200 μm line spacing, laser fluences 3–6 J/cm^2, 50 μm of cell-containing bioink transferred
	Collagen, Matriderm	NIH3T3, HaCaT	33 million cells/mL	0.1–1 nL transferred volume, laser fluences 1–2 J/cm^2, 60 μm of cell-containing collagen transferred
	Collagen, Matriderm	NIH3T3, HaCaT	41 million cells/mL	60 μm of cell-containing collagen transferred

nd, not discussed; n/a, not applicable.

Aspect of skin bioprinted	Dermal approach	Epidermal approach	Key results achieved	Tissue structure, dimensions	References
Dermis + epidermis	Fibroblast layers	Keratinocyte layers	Proliferation observed by PCNA staining 14 days post print	$4 \times 4\,mm^2$	[63]
Gland structure	n/a	n/a	Viability observed by fluorescence 5 days post print	$20 \times 20 \times 5$ mm crisscross blocks and square pores	[37]
Dermis + epidermis	Alternating printed layers of fibroblast suspension and collagen	Alternating printed layers of keratinocyte suspension and collagen	85%–95% viable by calcein AM staining 3 h after dispensing	$10 \times 10\,mm^2$, 10 layers	[10]
Dermis + epidermis	Alternating printed layers of HFF-1 suspension and collagen	Alternating printed layers of HaCaT suspension and collagen	84%–98% viable by propidium iodide staining 7 days post print	$6 \times 6 \times 1.2$ mm, 16 layers	[31]
Dermis	Alternating layers of AFS or MSC in fibrinogen/collagen and thrombin	n/a	Wound healing enabled, GFP-tagged cells not observed after day 7	2.0×2.0 cm, 5 layers	[23]
Dermis	Alternating layers of fibroblast suspension and bioink topped with seven circular bioink layers	Manually deposited keratinocyte suspension	Dermal compartment viable up to 7 weeks by MTT	6×5.5 mm base, 15 layers; 2.7 mm radius cylinder, seven layers	[9]
HMVEC microvessel network	Manually deposited fibroblasts in collagen with fibrinogen overlay	Manually deposited keratinocytes in collagen	PKH26 staining visible 6 days post print, wound healing enabled	$2.5 \times 5\,cm \times$ 1.2 mm thickness with HMVEC grid pattern embedded	[32]
Dermis + epidermis	n/a	n/a	$90\% \pm 10\%$ cell viability immediately post transfer	9.6×9.6 mm 2D chessboard pattern	[12]
Dermis + epidermis	20 layers of NIH3T3 cells embedded within collagen	10–20 layers of HaCaT cells embedded within collagen	Proliferation observed by Ki-67 staining 10 days post print	$10 \times 10 \times 2$ mm	[33]
Dermis + epidermis	20 layers of NIH3T3 cells embedded within collagen	20 layers of HaCaT cells embedded within collagen	Ki-67 staining observed in suprabasal layers 11 days post engraftment	2.3×2.3 cm	[34]

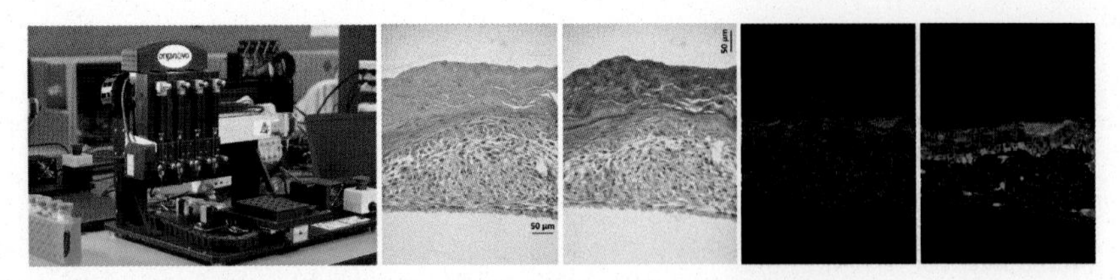

Fig. 3 Examples of skin tissue fabricated by extrusion-based bioprinting approaches. (A) Organovo NovoGen Bioprinter Platform. (B) Histological analysis showed bilayered architecture with distinct epidermal and dermal compartments and stratified epidermal keratinocytes (H&E). Trichrome staining (TCM) demonstrated collagen production *(blue)* in the dermal compartment. Immunofluorescent staining showed a distinct layer of cytokeratin 10 (CK10) positive keratinocytes *(red)* and blue nuclei (DAPI), indicating development of a spinous layer within the epidermis. Further immunofluorsencent analysis confirmed the presence of cytokeratin 5 (CK5)-positive basal keratinocytes *(green)* with distinct columnar morphology beneath a layer of flattened, differentiated spinous/granular involucrin (IVL)- positive keratinocytes *(red)* with blue nuclei (DAPI).

differentiation were observed within the pore-like structure with epidermal marker expression (CK5 and CK14) transitioning to luminal epithelial marker expression (CK18 and CK19), suggesting successful movement toward a sweat gland phenotype. These data demonstrate that a 3D bioprinted microenvironment can direct formation of skin appendages including glandular morphogenesis and may also pave the way toward formation of sebaceous glands and hair follicles (Fig. 3).

3.2.2 Droplet-based bioprinted skin

Several methods utilizing droplet-based approaches to bioprint multilayered 3D skin tissues composed of dermal fibroblasts and keratinocytes have been reported [9,10,23,31,32]. In two recent microvalve studies, skin tissue was fabricated by first dispensing a nebulized mist of cross-linking agent followed by collagen droplets to form a basal gel layer [10,31]. The process was repeated layer by layer to increase structure thickness. During the process, fibroblasts or keratinocytes were dispensed dropwise onto partially cross-linked hydrogel to embed within selected layers, producing spatially distinctive cell layers separated by collagen to represent the dermis and epidermis, respectively. The authors demonstrated that these engineered skin tissues composed of collagen-encapsulated primary adult epidermal keratinocytes or dermal fibroblasts maintained distinct layers visible by confocal microscopy after 4 days of culture [10]. One advantage described for the microvalve approach is that it permits noncontact droplet dispensing onto nonplanar surfaces such as polydimethylsiloxane (PDMS) molds, which may be critical for future fabrication of tailored skin grafts. Expanding on this method, V. Lee et al. performed a follow-up study increasing the number of cell layers printed (13 layers in total) in the tissue with HFF-1 and HaCaT cell lines and extending culture conditions to include a 14-day air-liquid interface. Distinct epidermal and dermal layers were again maintained. It was also observed that the epidermal layer thickened over time suggesting that the tissue was differentiating, although keratinocyte stratification was incomplete [31]. Both studies reported high viability (>85%) associated with the microvalve approach. This may have been due to their use of a relatively large-diameter nozzle (150 μm) and low pneumatic pressure (1–3 psi) to minimize the cell damage associated with inkjet and laser-based approaches. Despite these benefits, the authors cited some limitations of the approach including cell clogging at higher cell suspension concentrations (fibroblasts $>3.0 \times 10^6$ cells/mL) and reduced viability at low concentrations.

A combination of automated and manual approaches was utilized by Rimann et al. to produce a full-thickness model [9]. First, a microvalve droplet approach was employed to produce dermal equivalents in a predefined pattern with seven alternating layers each of chemically defined UV cross-linked PEG-based bioink and primary human dermal fibroblasts in suspension. The printing design also included deposition of seven circular acellular bioink layers on top of the dermal structure to form a cylinder shape used for collection of the manually added primary human keratinocyte suspension. The dermal equivalents alone cultured in the absence of keratinocytes maintained viability for up to 7 weeks, highlighting that bioprinted tissues can serve as long-term culture models. Upon addition of the keratinocytes, constructs were subsequently cultured for 14 days at an air-liquid interface. Histological evaluation did not show mature epidermal stratification and interestingly showed migration of the epidermal keratinocytes into the dermis, especially into dermal equivalents with short maturation times. This suggests that proper maturation and composition of the dermal layer may be important to achieve correct epidermal architecture. Although proper epidermal stratification and fibroblast spacing were not fully achieved, collectively, these microvalve studies demonstrate promising advances in achieving layered architecture in vitro.

Droplet-based bioprinted skin tissues have also been designed for therapeutic applications like wound healing and burn injuries. Recent work by Yanez et al. utilized a modified inkjet printer to deposit human microvascular endothelial cells (HMVEC) together with a manually mixed combination of collagen, neonatal human dermal fibroblast cells, and neonatal human epidermal keratinocytes to create a full- thickness skin graft. In this study, a dermal layer of fibroblast and collagen mixture was manually deposited followed by a layer of fibrinogen. A grid-like pattern of HMVEC in thrombin was then inkjet printed to create a microvasculature network on top of the dermal layer. Finally, an epidermal layer containing a mixture of keratinocytes in collagen was manually added to create a tissue with integrated printed vasculature embedded within a bilayered skin graft. When implanted into the back of athymic nude mice, tissue retained viability, epidermal and dermal layers, and promoted neovasculogenesis [32]. Droplet-based methodologies have also been incorporated into an in situ bioprinting approach for wound repair in vivo. Researchers dispensed alternating layers of cell-laden fibrin-collagen hydrogels (containing green fluorescent protein (GFP)-labeled amniotic fluid-derived stem (AFS) cells) and thrombin directly into a murine full-thickness wound [23]. When followed histologically over time, the GFP-labeled cells did not permanently integrate into surrounding tissue, but were believed to guide tissue repair through secretion of growth factors. The tissue covered 100% of the wound and enabled accelerated wound closure and increased neovasuclarization compared with hydrogel alone, suggesting that noncontact droplet-based bioprinting may be a viable therapeutic approach for wound healing (Fig. 4).

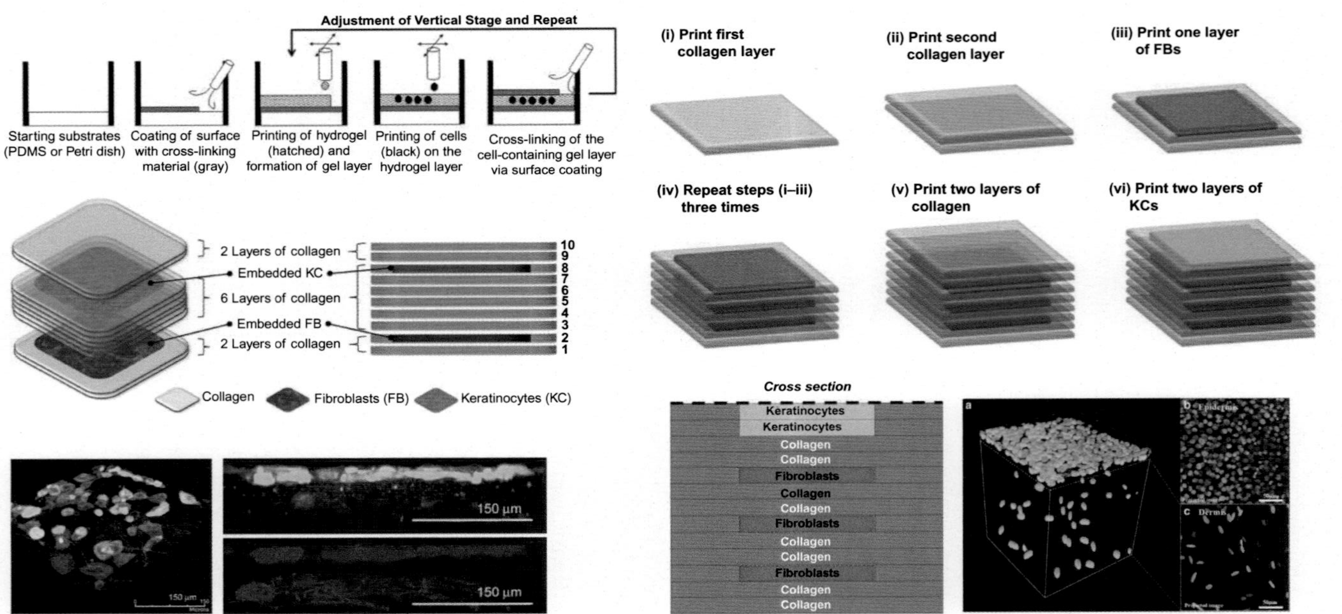

Fig. 4 Examples of skin tissue fabricated by droplet-based bioprinting approaches. (A) Schematic of deposition approach with multilayered skin cells and collagen. (B) Confocal microscopy images of printed skin after 4 days in submerged culture conditions with keratin-containing keratinocyte layer *(green)* and b-tubulin-containing fibroblasts *(red)*. (C) Confocal microscopy images of printed skin after 7 days in submerged culture conditions showed layered architecture with distinct epidermal and dermal compartments. Live cell nuclei are stained green with distinct rounded keratinocyte and elongated fibroblast morphologies.
Reproduced with permission from Lee W, Debasitis JC, Lee VK, Lee JH, Fischer K, Edminster K, et al. Multi-layered culture of human skin fibroblasts and keratinocytes through three-dimensional freeform fabrication. Biomaterials 2009;30(8):1587–95. Lee V, Singh G, Trasatti JP, Bjornsson C, Xu X, Tran TN, et al. Design and fabrication of human skin by three-dimensional bioprinting. Tissue Eng Part C Methods 2014;20(6):473–84.

Laser-based bioprinting of skin tissues has been demonstrated by a group from the Hannover Medical School and Laser Zentrum Hannover [12,33,34]. Their initial work utilized a donor slide with labeled murine NIH3T3 fibroblasts or human HaCaT keratinocytes encapsulated within a hydrogel of alginate and blood plasma printed onto a collector slide containing Matrigel. They produced 2D patterning in the form of a chessboard with alternating green- and blue-labeled cells, highlighting the ability of the bioprinter to produce specific predefined patterns with high resolution and without major changes in cell viability and proliferation [12]. Follow-up work demonstrated the production of 3D patterned structures by overlaying seven alternating layers of red- and green-labeled keratinocytes. Having verified both xy and z patterning capability, this method was then used to produce more complex bilayered skin tissue architecture by dispensing 20 layers of NIH3T3 cells embedded within neutralized collagen gel onto a Matriderm collector slide followed by 10–20 layers of HaCaT cells embedded within the same gel matrix on top, with droplet volumes ranging from 0.1 to 1 nL. [33,34]. The authors show histologically that the individual layers of the tissue do not intermix over the 10-day culture period. Furthermore, the cells are viable and form intercellular adhesions between epidermal cells and a basal lamina between epidermal and dermal layers. Although the tissues exhibit features of the normal skin, histological analysis showed that collagen remnants from the printing process still detected within the epidermis and keratinocytes that were not stratified into distinct differentiated layers, suggesting that the embedding of keratinocytes into a matrix may impair epidermal tissue formation. Interestingly, the bioprinted skin tissue morphology was improved when engrafted into a dorsal skinfold chamber in nude mice; bioprinted skin cultured in vivo had improved organization within the epidermis and vascular invasion into the Matriderm from the wound bed [34]. Side-by-side comparison of constructs cultured in vitro and in vivo demonstrates the importance of the surrounding microenvironment to modulate tissue differentiation and further supports the potential for bioprinted skin tissues in therapeutic applications (Fig. 5).

4 Summary and future directions

Three-dimensional bioprinting allows for the reproducible fabrication of spatially ordered and complex multicellular structures with controlled microenvironments, thus providing more native biophysical and biochemical cues to enhance cellular cross talk and tissue differentiation. Recent progress in bioprinting 3D skin tissues provides proof of concept that this approach can produce viable tissues composed of distinct dermal and epidermal layers. The variety of methodologies employed successfully to date highlights the versatility of these automated computer-controlled design platforms and the broad compatibility of the skin with the use of automated manufacturing approaches. However, most approaches to date have yielded tissues that lack both the cell density of native tissue and a fully mature stratified epidermis. To go beyond these basic features, complex tissue designs will require incorporation of additional specialized cell types such as immune cells, melanocytes, and dermal papilla cells and microcompartment appendages like sebaceous glands, eccrine glands, and

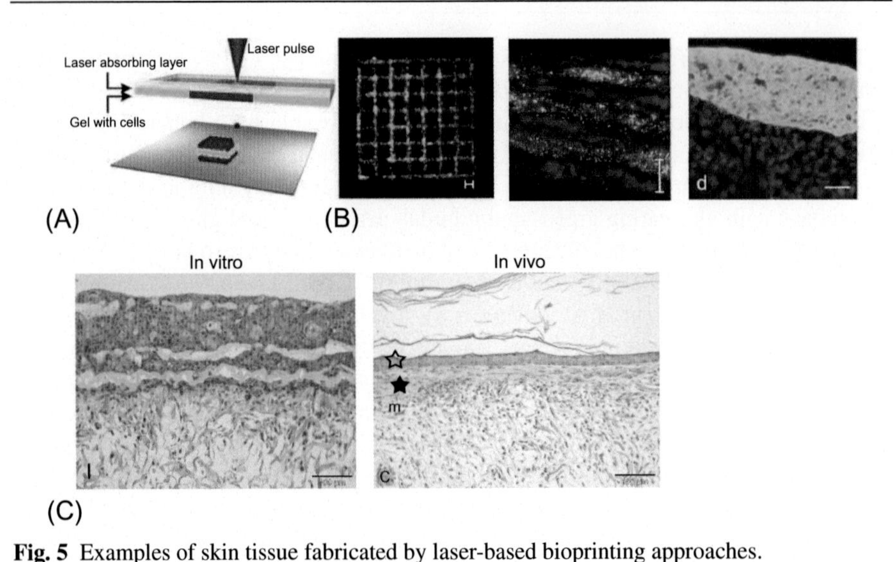

Fig. 5 Examples of skin tissue fabricated by laser-based bioprinting approaches.
(A) Schematic of laser deposition approach. (B) Fluorescent microscopy demonstrated 2D micropatterning of fluorescently labeled fibroblasts *(green)* and keratinocytes *(red, left panel)*, alternating red- and green-labeled keratinocyte layers *(center panel)*, and distinct epidermal and dermal layering of full-thickness skin tissue with fibroblasts stained in red *(pan-reticular fibroblast)*, and keratinocytes are stained in green *(cytokeratin 14, right panel)*. (C) Masson's trichrome staining of cells *(red)* and collagen *(green)* showed differences in morphology between bioprinted skin tissues cultured in vitro or in vivo.
Reproduced with permission from Koch L, Deiwick A, Schlie S, Michael S, Gruene M, Coger V, et al. Skin tissue generation by laser cell printing. Biotechnol Bioeng 2012;109(7):1855–63. Michael S, Sorg H, Peck CT, Koch L, Deiwick A, Chichkov B, et al. Tissue engineered skin substitutes created by laser-assisted bioprinting form skin-like structures in the dorsal skin fold chamber in mice. PLoS One 2013;8(3):e57741.

hair follicles. Incorporation of perfusable vasculature may also be necessary for larger tissues including skin grafts and for providing oxygen and nutrients to support the long-term viability of smaller full-thickness tissues. In most research to date, each investigator has applied their singular bioprinting technique of interest to generate 3D skin tissues. Continuing advancements in each specific deposition method and in the biomaterials available may improve printing resolution and tissue architecture to more closely mimic native skin. However, integration of these fabrication platforms will likely be key to increasing capability; the method that works best to create the layered dermal and epidermal compartments may not be effective for creating planar compartments containing additional features such as sweat glands, and the biomaterials currently compatible with keratinocytes and dermal fibroblasts may not support more specialized cell types. Despite the progress made on fabrication, many studies still fail to verify the functional attributes of their engineered skin equivalents. Postfabrication maintenance of complex engineered tissue requires tight regulation of environmental elements such as media and physical stress and strain to support viability, induce differentiation, and preserve the skin-like attributes for utilization in vitro. Bioprinted skin tissues remain to be validated for use in assessment of irritation, corrosion, absorption, and sensitization of topical chemicals. In addition, these engineered tissues have not yet been used as in vitro disease models, an area where novel culture systems are sorely needed [65,66]. Future studies will be required to define a framework of appropriate standards to validate the use of bioprinted models in these areas, thus contributing to the principle of the 3Rs (replacement, reduction, and refinement) in order to decrease reliance on animal research and to ultimately provide enhanced predictability of human response. While the field is young, advances in bioprinting in vitro skin tissues have the potential to transform disease modeling, drug development, and personalized medicine.

Abbreviations

3D three-dimensional
CAD computer-aided design
iPSC induced pluripotent stem cell
ECM extracellular matrix
CK cytokeratin

Glossary

Bioink Cells and/or cell-containing biomaterials used in the fabrication of tissue.
Biomaterials Substances compatible with living cells that can provide structural support during the fabrication or long-term culture of bioprinted 3D tissues.
Bioprinting An automated, spatially controlled 3D printing approach used to create ex vivo tissues and organs through the deposition of cell-containing "bioink."
Dermis Layer of tissue supporting the epidermal layer composed mainly of dermal fibroblasts and connective tissue.
Droplet-based bioprinting Bioprinting methods that involve the creation and deposition of droplets at defined positions in three-dimensional space.
Epidermis Outermost layer of the skin composed mainly of keratinocytes organized into a highly stratified architecture distinguished by morphology and differentiation state.
Extracellular matrix (ECM) Complex mixture of extracellular molecules surrounding cells that provides structural support and biochemical cues.
Extrusion bioprinting Bioprinting approach where material is deposited from a cassette or syringe cartridge through an orifice under pressure by pneumatic, piston-driven, or screw-driven force.
Inkjet bioprinting Droplet-based bioprinting approach involving the ejection of droplets containing either cells or biomaterials from a reservoir in response to a pressure pulse generated by thermal or piezoelectric forces.
Laser bioprinting Droplet-based bioprinting approach utilizing laser pulses to create and dispense droplets onto a receiving surface.
Microsolenoid valve bioprinting Droplet-based bioprinting approach that uses the opening and closing of an electromechanical valve to generate droplets.

References

[1] Melchels FPW, Domingos MAN, Klein TJ, Malda J, Bartolo PJ, Hutmacher DW. Additive manufacturing of tissues and organs. Prog Polym Sci 2012;37(8):1079–104.
[2] Dodziuk H. Applications of 3D printing in healthcare. Pol J Cardio-thorac Surg 2016;13(3):283–93.
[3] U.S. Food and Drug Administration. Technical considerations for additive manufactured devices: draft guidance for Industry and Food and Drug Administration Staff; 2016 [cited 2016]. Available from: http://www.fda.gov/downloads/MedicalDevices/DeviceRegulationandGuidance/GuidanceDocuments/UCM499809.pdf.
[4] Panwar A, Tan LP. Current status of bioinks for micro-extrusion-based 3D bioprinting. Molecules (Basel, Switzerland) 2016;21(6).
[5] Norotte C, Marga F, Niklason L, Forgacs G. Scaffold-free vascular tissue engineering using bioprinting. Biomaterials 2009;30(30):5910–7.
[6] Jakab K, Norotte C, Marga F, Murphy K, Vunjak-Novakovic G, Forgacs G. Tissue engineering by self-assembly and bio-printing of living cells. Biofabrication 2010;2(2):022001.
[7] Gudapati H, Dey M, Ozbolat I. A comprehensive review on droplet-based bioprinting: past, present and future. Biomaterials 2016;102:20–42.
[8] Vijayavenkataraman S, Lu WF, Fuh JY. 3D bioprinting of skin: A state-of-the-art review on modelling, materials, and processes. Biofabrication 2016;8(3):032001.
[9] Rimann M, Bono E, Annaheim H, Bleisch M, Graf-Hausner U. Standardized 3D bioprinting of soft tissue models with human primary cells. J Lab Automat 2016;21(4):496–509.
[10] Lee W, Debasitis JC, Lee VK, Lee JH, Fischer K, Edminster K, et al. Multi-layered culture of human skin fibroblasts and keratinocytes through three-dimensional freeform fabrication. Biomaterials 2009;30(8):1587–95.
[11] Skardal A, Atala A. Biomaterials for integration with 3D bioprinting. Ann Biomed Eng 2015;43(3):730–46.
[12] Koch L, Kuhn S, Sorg H, Gruene M, Schlie S, Gaebel R, et al. Laser printing of skin cells and human stem cells. Tissue Eng Part C Methods 2010;16(5):847–54.
[13] Guillotin B, Souquet A, Catros S, Duocastella M, Pippenger B, Bellance S, et al. Laser assisted bioprinting of engineered tissue with high cell density and microscale organization. Biomaterials 2010;31(28):7250–6.
[14] Barron JA, Krizman DB, Ringeisen BR. Laser printing of single cells: statistical analysis, cell viability, and stress. Ann Biomed Eng 2005;33(2):121–30.
[15] Gruene M, Pflaum M, Hess C, Diamantouros S, Schlie S, Deiwick A, et al. Laser printing of three-dimensional multicellular arrays for studies of cell-cell and cell-environment interactions. Tissue Eng Part C Methods 2011;17(10):973–82.
[16] Ringeisen BR, Othon CM, Barron JA, Young D, Spargo BJ. Jet-based methods to print living cells. Biotechnol J 2006;1(9):930–48.
[17] Zhang Z, Michniak-Kohn BB. Tissue engineered human skin equivalents. Pharmaceutics 2012;4(1):26–41.
[18] Brohem CA, da Silva Cardeal LB, Tiago M, Soengas MS, de Moraes Barros SB, Maria-Engler SS. Artificial skin in perspective: concepts and applications. Pigment Cell Melanoma Res 2011;24(1):35–50.
[19] Hewitt NJ, Edwards RJ, Fritsche E, Goebel C, Aeby P, Scheel J, et al. Use of human in vitro skin models for accurate and ethical risk assessment: metabolic considerations. Toxicol Sci 2013;133(2):209–17.

[20] Khorshid FA. Comparative study of keloid formation in humans and laboratory animals. Med Sci Monitor: Int Med J Exp Clin Res 2005;11(7):Br212–9.
[21] Barker CL, McHale MT, Gillies AK, Waller J, Pearce DM, Osborne J, et al. The development and characterization of an in vitro model of psoriasis. J Investig Dermatol.123(5):892–901.
[22] Itoh M, Kiuru M, Cairo MS, Christiano AM. Generation of keratinocytes from normal and recessive dystrophic epidermolysis bullosa-induced pluripotent stem cells. Proc Natl Acad Sci 2011;108(21):8797–802.
[23] Skardal A, Mack D, Kapetanovic E, Atala A, Jackson JD, Yoo J, et al. Bioprinted amniotic fluid-derived stem cells accelerate healing of large skin wounds. Stem Cells Transl Med 2012;1(11):792–802.
[24] Roskos K, Stuiver I, Pentoney S, Presnell S. Chapter 24—Bioprinting: an industrial perspective. In: Atala A, Yoo JJ, editors. Essentials of 3D biofabrication and translation. Boston: Academic Press; 2015. p. 395–411.
[25] Lee W, Lee V, Polio S, Keegan P, Lee JH, Fischer K, et al. On-demand three-dimensional freeform fabrication of multi-layered hydrogel scaffold with fluidic channels. Biotechnol Bioeng 2010;105(6):1178–86.
[26] Golden AP, Tien J. Fabrication of microfluidic hydrogels using molded gelatin as a sacrificial element. Lab Chip 2007;7(6):720–5.
[27] Richards D, Jia J, Yost M, Markwald R, Mei Y. 3D bioprinting for vascularized tissue fabrication. Ann Biomed Eng 2017;45(1):132–47. https://doi.org/10.1007/s10439-016-1653-z.
[28] KHATIWALA C, LAW R, SHEPHERD B, DORFMAN S, CSETE M. 3D cell bioprinting for regenerative medicine research and therapies. Gene Ther Regulat 2012;07(01):1230004.
[29] Mironov V, Visconti RP, Kasyanov V, Forgacs G, Drake CJ, Markwald RR. Organ printing: tissue spheroids as building blocks. Biomaterials 2009;30(12):2164–74.
[30] Jakab K, Neagu A, Mironov V, Markwald RR, Forgacs G. Engineering biological structures of prescribed shape using self-assembling multicellular systems. Proc Natl Acad Sci U S A 2004;101(9):2864–9.
[31] Lee V, Singh G, Trasatti JP, Bjornsson C, Xu X, Tran TN, et al. Design and fabrication of human skin by three-dimensional bioprinting. Tissue Eng Part C Methods 2014;20(6):473–84.
[32] Yanez M, Rincon J, Dones A, De Maria C, Gonzales R, Boland T. In vivo assessment of printed microvasculature in a bilayer skin graft to treat full-thickness wounds. Tissue Eng A 2015;21(1–2):224–33.
[33] Koch L, Deiwick A, Schlie S, Michael S, Gruene M, Coger V, et al. Skin tissue generation by laser cell printing. Biotechnol Bioeng 2012;109(7):1855–63.
[34] Michael S, Sorg H, Peck CT, Koch L, Deiwick A, Chichkov B, et al. Tissue engineered skin substitutes created by laser-assisted bioprinting form skin-like structures in the dorsal skin fold chamber in mice. PLoS One 2013;8(3):e57741.
[35] Zhang H, Yang K, Liu G, Zhu S, Yin R, Zhang W. 3D bioprinting of multi-biomaterial/crosslinked bioink for skin tissue engineering. In: Front Bioeng Biotechnol conference abstract: 10th World Biomaterials Congress; 2016. https://doi.org/10.3389/conf.FBIOE.2016.01.02354.
[36] Rey Legidos J, Cubo Mateo N, Montero Simon A, Garcia Diez M, Velasco Bayón D, Jorcano Noval J. Development of a hyaluronic acid/plasma-derived fibrin hydrogel for the optimization of dermo-epidermal autologous equivalents. Front Bioeng Biotechnol conference abstract: 10th World Biomaterials Congress. doi:10.3389/conf.FBIOE.2016.01.00743.

[37] Liu N, Huang S, Yao B, Xie J, Wu X, Fu X. 3D bioprinting matrices with controlled pore structure and release function guide in vitro self-organization of sweat gland. Sci Rep 2016;6:34410.

[38] Tsao C-T, Leung M, Chang JY-F, Zhang M. A simple material model to generate epidermal and dermal layers in vitro for skin regeneration. J Mater Chem B Mater Biol Med 2014;2(32):5256–64.

[39] Garric X, Guillaume O, Dabboue H, Vert M, Moles JP. Potential of a PLA-PEO-PLA-based scaffold for skin tissue engineering: in vitro evaluation. J Biomater Sci Polym Ed 2012;23(13):1687–700.

[40] Park KE, Kang HK, Lee SJ, Min BM, Park WH. Biomimetic nanofibrous scaffolds: preparation and characterization of PGA/chitin blend nanofibers. Biomacromolecules 2006;7(2):635–43.

[41] Lee JM, Chae T, Sheikh FA, Ju HW, Moon BM, Park HJ, et al. Three dimensional poly(epsilon-caprolactone) and silk fibroin nanocomposite fibrous matrix for artificial dermis. Mater Sci Eng C Mater Biol Appl 2016;68:758–67.

[42] Sadeghi AR, Nokhasteh S, Molavi AM, Khorsand-Ghayeni M, Naderi-Meshkin H, Mahdizadeh A. Surface modification of electrospun PLGA scaffold with collagen for bioengineered skin substitutes. Mater Sci Eng C Mater Biol Appl 2016;66:130–7.

[43] Murphy SV, Skardal A, Atala A. Evaluation of hydrogels for bio-printing applications. J Biomed Mater Res A 2013;101A(1):272–84.

[44] Malda J, Visser J, Melchels FP, Jüngst T, Hennink WE, Dhert WJA, et al. 25th anniversary article: engineering hydrogels for biofabrication. Adv Mater 2013;25(36):5011–28.

[45] Chan LS. Human skin basement membrane in health and in autoimmune diseases. Front Biosci: J Virt Lib 1997;2:d343–52.

[46] Kolarsick PAJ, Kolarsick MA, Goodwin C. Anatomy and physiology of the skin. J Dermatol Nurses' Assoc 2011;3(4):203–13.

[47] Andriani F, Margulis A, Lin N, Griffey S, Garlick JA. Analysis of microenvironmental factors contributing to basement membrane assembly and normalized epidermal phenotype. J Investig Dermatol 2003;120(6):923–31.

[48] Boehnke K, Mirancea N, Pavesio A, Fusenig NE, Boukamp P, Stark HJ. Effects of fibroblasts and microenvironment on epidermal regeneration and tissue function in long-term skin equivalents. Eur J Cell Biol 2007;86(11–12):731–46.

[49] Carlson MW, Alt-Holland A, Egles C, Garlick JA. Three-dimensional tissue models of normal and diseased skin. In: Bonifacino JS, et al., editors. Current protocols in cell biology/editorial board. Hoboken, NJ: John Wiley & Sons; 2008. [CHAPTER:Unit-19.9].

[50] Mathes SH, Ruffner H, Graf-Hausner U. The use of skin models in drug development. Adv Drug Del Rev 2014;69(70):81–102.

[51] Gordon S, Daneshian M, Bouwstra J, Caloni F, Constant S, Davies DE, et al. Non-animal models of epithelial barriers (skin, intestine and lung) in research, industrial applications and regulatory toxicology. ALTEX 2015;32(4):327–78.

[52] Li B, Wang X, Wang Y, Gou W, Yuan X, Peng J, et al. Past, present, and future of microcarrier-based tissue engineering. J Orthopaed Transl 2015;3(2):51–7.

[53] Loh QL, Choong C. Three-dimensional scaffolds for tissue engineering applications: role of porosity and pore size. Tissue Eng B Rev 2013;19(6):485–502.

[54] Thevenot P, Nair A, Dey J, Yang J, Tang L. Method to analyze three-dimensional cell distribution and infiltration in degradable scaffolds. Tissue Eng Part C Methods 2008;14(4):319–31.

[55] Foty R. A simple hanging drop cell culture protocol for generation of 3D spheroids. J Visual Exp: JoVE 2011;51:2720.

[56] Rago AP, Dean DM, Morgan JR. Controlling cell position in complex heterotypic 3D microtissues by tissue fusion. Biotechnol Bioeng 2009;102(4):1231–41.
[57] Murphy SV, Atala A. 3D bioprinting of tissues and organs. Nat Biotech 2014;32(8):773–85.
[58] Nguyen DG, Funk J, Robbins JB, Crogan-Grundy C, Presnell SC, Singer T, et al. Bioprinted 3D primary liver tissues allow assessment of organ-level response to clinical drug induced toxicity in vitro. PLoS One 2016;11(7):e0158674.
[59] Norona LM, Nguyen DG, Gerber DA, Presnell SC, LeCluyse EL. Editor's highlight: modeling compound-induced fibrogenesis in vitro using three-dimensional bioprinted human liver tissues. Toxicol Sci: Off J Soc Toxicol 2016;154(2):354–67.
[60] Homan KA, Kolesky DB, Skylar-Scott MA, Herrmann J, Obuobi H, Moisan A, et al. Bioprinting of 3D convoluted renal proximal tubules on perfusable chips. Sci Rep 2016;6:34845.
[61] Nguyen DLG, King, Shelby Marie, Presnell, Sharon C, Inventor engineered renal tissues, arrays thereof, and methods of making the same patent 20160097039; 2016.
[62] NDG IJT, Grundy CC, Smith TR, Retting KN, King SM, Presnell SC. In: Bioprinted human tissues for toxicology and disease modeling. World preclinical congress, June, Boston, MA; 2016.
[63] Nguyen DG, Grundy CC, Smith TR, Murphy TJ, Norona LM, Retting KN, et al. In: Bioprinted three-dimensional (3D) human tissues for toxicology and disease modeling. Keystone symposia, modern phenotypic drug discovery: defining the path forward, Big Sky, MO, USA; 2016.
[64] Lu C, Fuchs E. Sweat gland progenitors in development, homeostasis, and wound repair. Cold Spring Harbor Perspect Med 2014;4(2):a015222.
[65] Vanderburgh J, Sterling JA, Guelcher SA. 3D printing of tissue engineered constructs for in vitro modeling of disease progression and drug screening. Ann Biomed Eng 2016;1–16.
[66] Bergers LI, Reijnders CM, van den Broek LJ, Spiekstra SW, de Gruijl TD, Weijers EM, et al. Immune-competent human skin disease models. Drug Discov Today 2016;21(9):1479–88.

Induced pluripotent stem cells to generate skin tissue models

Olga Kashpur, Avi Smith*, Nailia Mukhamedshina*, Jeremy Baskin*, Yulia Shamis*, Kyle Hewitt†, Behzad Gerami-Naini*, Jonathan A. Garlick**

*Department of Diagnostic Sciences, School of Dental Medicine, Tufts University, Boston, MA, United States, †School of Stem Cell and Regenerative Medicine, University of Wisconsin Madison, Madison, WI, United States

1. Introduction

The fabrication and application of 3-D skin tissues using iPSC-derived cells lie at the interface of the fields of tissue engineering and stem cell biology [1,2]. Tissue engineering approaches to generate skin tissues use three-dimensional (3-D) extracellular matrices (ECM) as scaffolds that provide cells with a physical framework to create the spatial organization required to support cellular functions that occur in tissues [1,3,4]. For many years, 3-D skin tissues have been constructed with adult stem cells that differentiate to display in vivo-like tissue architecture and function [5]. These 3-D skin-like tissues have been derived from cells differentiated from multipotent stem cells, capable of giving rise to multiple cell types, or from unipotent stem cells that can differentiate into a single cell type [6,7]. Induced pluripotent stem cells (iPSCs) now offer a novel source of stem cells that can be guided to specific cell lineages to yield progenitor cells that can be differentiated to a broad spectrum of cell types with functions well suited to skin tissue fabrication [1,8,9].

Skin-like tissues are composed of a stratified epithelium that is formed by differentiated keratinocytes and dermis that contains fibroblasts that generate ECM and growth factors that guide epithelial growth and differentiation [1,10–12]. Three-dimensional skin tissues have been used to perform assays for tissue penetration, metabolism, phototoxicity, and irritation to predict the absorption, metabolism, and toxicity (ADME/Tox) features of test compounds. In addition, these tissues have been customized to study mechanisms of skin disease and acute and chronic skin processes such as wound healing, UV irradiation response, fibrosis, and inflammation. Fabricating skin tissues using iPSC-derived cells offers new opportunities to advance all of these experimental and drug screening outcomes by potentially eliminating barriers that currently limit skin tissue standardization and reproducibility needed for these applications.

Currently, potential new drugs are primarily screened using conventional, 2-D monolayer culture assays that do not display architectural and functional features of the human skin and are not highly predictive of drug responses in vivo [10]. The use of iPSC-derived cells has the potential to generate more reproducible, patient-specific skin tissues than are currently available to improve such preclinical safety and toxicity testing and drug development using a more "personalized" or patient-specific

Skin Tissue Models. https://doi.org/10.1016/B978-0-12-810545-0.00017-6

approach [1,13–16]. This will require the development of approaches to screen and identify iPSC-derived cells that are best suited for the fabrication of 3-D skin tissues. In addition, these tissues can be used to determine the developmental capacity and cellular functions of iPSC-differentiated cells when incorporated into the in vivo-like tissue microenvironment of skin-like tissues and to enable ways to better assess their cellular behavior, safety, and efficacy before the future therapeutic use of iPSC-derived cells in humans.

This review will present an overview of the relevance of iPSC reprogramming and specific functional cell types for the generation of skin-like tissue models. We will first describe the features of cell reprogramming and the subsequent differentiation of iPSCs into cells able to fabricate these tissues. We will then discuss approaches through which iPSC-based technologies may be broadly applied for the fabrication of skin tissues and how they may help overcome barriers currently limiting the translational use of skin tissues. A schematic overview of reprogramming of primary cells to iPSCs and their sequential differentiation and incorporation into 3-D tissues is shown in Fig. 1.

2. Reprogramming to iPSCs and differentiation into cell types needed for 3-D skin tissue fabrication

The breakthrough discovery of somatic cell reprogramming to create induced pluripotent stem cells (iPSCs) and decade of progress that has followed has

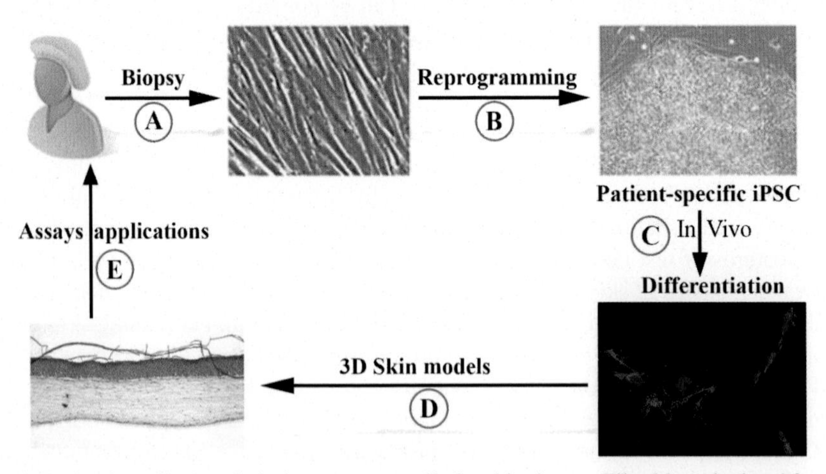

Fig. 1 Derivation of induced pluripotent stem cells for skin tissues. Fibroblasts isolated from patients (A) are reprogrammed to a pluripotent state in monolayer culture (B). Following reprogramming, iPSCs can be maintained indefinitely in culture and differentiated when needed into mature cell types that can be patient- or disease-specific (C). These differentiated cells can be incorporated into 3-D skin models (D) and used for a variety of assays and applications for drug discovery or studying skin disease pathways (E).

opened dramatic new opportunities to transform human health [1,7,17]. This transformative technology was first demonstrated by Yamanaka et al., who showed that mouse iPSCs could be generated by delivering four transcription factors *(Oct4/Klf4/Sox2/cMyc, OKSM)* into skin-derived fibroblasts that reprogrammed them to a pluripotent state [18]. This was followed by the reprogramming of human iPSCs using the same transcription factors [19]. These cells demonstrated many features similar to human embryonic stem cells (hESCs) including cellular morphology, growth kinetics, unlimited expansion potential [20], telomerase activity, global patterns of gene expression, and capacity to differentiate into cells from all germ layers [7,21]. Since their discovery a decade ago, tremendous progress has been made applying iPSC technology. Patient-specific, iPSC-derived cells are now being used for modeling both monogenic and sporadic disease types. In combination with 3-D organoid technology, iPSC-derived cells have recently been used in large-scale, drug screening applications for diseases such as Alzheimer's disease, amyotrophic lateral sclerosis, and fragile X syndrome and in regenerative medicine in an ongoing clinical trial using iPSC-derived retinal pigmented epithelial cells [17]. Importantly, iPSCs offer an alternative source of cells for 3-D tissue fabrication in addition to the primary, patient-derived cells typically used. Following their reprogramming from patient-derived somatic cells, iPSCs can be maintained indefinitely under defined culture conditions and subsequently differentiated into lineage- and patient-specific cells with a wide spectrum of cellular phenotypes. As a result, the controlled differentiation of functional cell types from iPSCs can establish a replenishing cell source for skin tissue fabrication.

To harness the potential of pluripotent stem cells for skin tissue fabrication, it is necessary to predictably and reproducibly control the differentiation of iPSCs to specific cell lineages and cell types. The directed differentiation of iPSCs recapitulates lineage fate decisions that occur during human embryonic development in vivo [8,22–24]. One important variable in establishing iPSC differentiation is the substrate or cell feeder layers that provides essential cell-cell or cell-matrix contact needed to direct iPSC-derived cells to specified lineages [25,26]. Another factor directing pluripotent stem cell differentiation is the presence of soluble growth factors [27], such as WNT, Nodal, and BMP, which are dynamically coordinated during specification to endodermal, ectodermal, or mesenchymal cell lineage fates. Outcomes of exposure of pluripotent stem cells to these signaling mediators are temporally controlled. For example, BMP signaling during the early stage of differentiation induces ectodermal specification but inhibits neuronal differentiation during later stages of differentiation where it promotes selection of definitive ectoderm or epidermal lineages needed for skin tissue fabrication [28]. By providing specific growth substrates and a well-defined soluble growth milieu, it is possible to provide a controlled environment that can be tuned to modulate gene expression profiles specifically needed for the differentiation of iPSCs to a skin cell type of interest.

Methods that rely on temporally controlled growth substrates and well-defined growth factor combinations have been developed to differentiate iPSCs into multiple cell types that can be incorporated into 3-D skin tissues. Deriving these cell types, including keratinocytes, fibroblasts, melanocytes, endothelial cells, macrophages, and folliculogenic epithelial stem cells, is as follows:

(A) Keratinocytes—iPSC-derived keratinocytes have been shown to support the differentiation of surface epithelial cells that possess long-term self-renewing properties in 3-D skin tissue models. These studies have shown that both the ECM substrate and the growth media used are critical for this process. Early reports showed that BMP4 effectively directs the differentiation of mouse ESCs into epidermal cells when undifferentiated ESCs were grown on ECM deposited by human neonatal fibroblasts [29,30]. Human ESCs have been differentiated into keratinocytes using BMP4 and ascorbic acid [31] or in growth media containing all-trans retinoic acid, BMP4, and FAD medium containing insulin, adenine, hydrocortisone, cholera toxin, and EGF and were supported by a fibroblast feeder layer [32]. In a report comparing different growth surfaces, keratinocytes were derived from iPSCs using supplementation of retinoic acid, BMP4, and cultured sequentially on vitronectin, fibronectin, and type I collagen, which increased differentiation efficiency. iPSCs have been differentiated to keratinocyte lineages by culturing them in the presence of BMP4, vitamin A, and all-trans retinoic acid on a decellularized human dermal fibroblast-derived ECM and through the preferential adherence to type IV collagen, which allowed selection based on the expression of keratinocyte markers, keratin 14 and p63 [33]. Treatment of iPSC-derived keratinocytes with Y-27632 and EGF stimulated proliferation and enabled derivation of sufficient number of cells to produce 3-D skin tissues [34]. Long-term renewal of human iPSC-derived keratinocytes was achieved by differentiating keratinocytes from teratomas generated from iPSCs. These human iPSC teratoma-derived keratinocytes were used to fabricate 3-D skin tissues that were engrafted into mice and successfully persisted for 12 weeks [35].

(B) Fibroblasts—As fibroblasts are known to regulate cell-cell cross talk and ECM production needed for skin tissue growth and differentiation [36,37], they are a critical cell type for the fabrication of 3-D skin tissues. Fibroblasts have been successfully generated from hESCs, which has been informative for their generation from iPSCs. hESCs have been differentiated into fibroblasts from embryoid bodies (EBs) embedded in type I collagen that expressed fibroblast markers such as vimentin and αSMA and displayed typical fibroblast phenotypes, migratory properties, and gel contraction [38, 39]. Similarly, fibroblasts have also been generated using EBs derived from iPSCs by using ascorbic acid and TGFβ2 supplementation [40]. iPSCs have been differentiated into fibroblast and epithelial lineages from the same source, suggesting an ectomesenchymal origin for these cell types [11]. The step-by-step differentiation of fibroblasts from iPSCs has been shown to occur over the course of 28 days as fibroblastic cells emerge from iPSCs that show a similar phenotype to fibroblasts differentiated from pluripotent hESCs (Fig. 2).

(C) Melanocytes—Melanocytes are melanin-producing cells that protect the skin from UV irradiation and are responsible for skin pigmentation. They were first generated from hESCs and were successfully grafted when incorporated into human 3-D skin tissue equivalents [41]. Melanocytes have been generated from both hESCs and hiPSCs using BMP4 to recapitulate developmental stages of differentiation from neural crest-like cells and have been differentiated from mouse and human iPSCs using Wnt3a, SCF, and ET-3 [42,43]. iPSC-derived melanocytes have been shown to transfer melanin to iPSC-derived keratinocytes which confirmed their functionality and provided evidence of an all iPSC-derived epidermal-melanin unit in the skin [44]. Several disease models have attempted to recapitulate pigmentation defects using human iPSC-derived melanocytes, including Hermansky-Pudlak and Chédiak-Higashi syndromes [45].

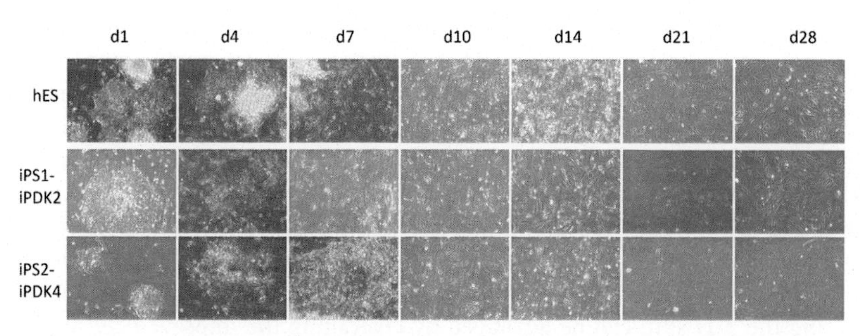

Fig. 2 iPSCs show emergence of fibroblasts during sequential stages of differentiation. Human embryonic stem cells (hESCs) and iPSCs were differentiated in parallel, using identical differentiation procedures, and monitored for cell morphology at various stages of differentiation. Representative images show the morphology of hESCs and two human foreskin-derived iPSC lines at days 1, 4, 7, 10, 14, 21, and 28 after initiation of differentiation. Fibroblast differentiation was first seen at the periphery of iPSC colonies (days 1 and 4). Cells soon acquired features of elongated stellate cells (day 10) that followed a similar time course to hESC differentiation. Cell lines showed characteristic fibroblast morphology at days 21 and 28 after iPSC differentiation [11].

(D) Macrophages and endothelial cells—Evaluating how inflammatory responses and angiogenesis may contribute to altered wound healing and a range of skin disorders could be achieved by incorporating macrophages and endothelial cells into 3-D skin tissues. Currently, human iPSCs have been differentiated into cells in monocyte/macrophage lineages [46,47] that were proved to be MYB-independent, tissue resident macrophages related to cell types such as microglia, Kupffer cells, and Langerhans cells and were not related to hematopoietic stem cell origin, which are known to be MYB-dependent (i.e., skin macrophages) [48]. Endothelial cells have also been derived from iPSCs [49–52]. Using these cells to create blood-vessel-like structures in 3-D skin tissues would model skin disorders and could enable more efficient in vivo grafting of 3-D tissues.

(E) Skin appendages—Folliculogenic epithelial stem cells (EpSCs) are cells that can generate keratinocytes and all hair follicle cell lineages and are defined by the presence of surface markers CD200 and ITGA6. Recently, EpSCs have been differentiated from iPSCs [53]. It has also been shown that a functional “integumentary organ system” can be generated after in vivo transplantation of iPSC-derived EBs that were embedded in collagen gels. This approach demonstrated the feasibility of efficiently generating epithelium and skin appendage cells, including hair follicles and sebaceous glands, from iPSCs that will be beneficial for the construction of 3-D skin-like tissues [54].

Since stromal and epithelial compartments of the skin are interdependent through their paracrine growth factor signaling [37], iPSCs offer an advantage by allowing 3-D skin tissues to be constructed from multiple, isotypic cell types that can be derived from patient-specific iPSC clones. Further refinement of reprogramming to iPSCs and differentiation to multiple skin-specific cell types will improve our understanding of key regulatory signals needed to generate iPSC-derived cells with optimal functional properties that can further improve the morphology and function of 3-D skin tissues.

Reprogramming somatic cells to a pluripotent state opens new opportunities for improvements in the fabrication of skin-like tissues. This will occur by developing approaches through which iPSC-derived cells can be tested to determine if they offer functional improvements when compared with the parental skin cells from which these iPSCs were originally reprogrammed [12, 20]. In order to be utilized for 3-D tissue fabrication, fibroblasts derived from iPSCs need to produce the ECM proteins and growth factors that are required to support epithelial growth and differentiation and maintain their functional stability throughout the tissue fabrication process. The potential uses of iPSC-derived fibroblasts for engineering the 3-D skin can be advanced by leveraging the improved cellular functions seen when fibroblasts have been differentiated from iPSCs and compared with the primary fibroblasts from which they were initially reprogrammed. For example, it has been shown that iPSC-derived fibroblasts that were initially reprogrammed from senescent fibroblasts from elderly patients showed augmented cellular function that included evasion of cell senescence, elongation of telomeres, and enhanced mitochondrial function [55]. This suggests that iPSC reprogramming may partially reverse the "biological clock" of the somatic cells from which they were reprogrammed.

Next, fibroblasts derived from iPSCs have been shown to produce greater amounts of ECM proteins than the parental cells from which they were derived, which may improve tissue fabrication when incorporated into skin tissues. These studies were performed in a skin tissue model that simulates the features of dermis when grown in the absence of keratinocytes [12]. In this tissue model, dermal fibroblasts produce and assemble an endogenous, human stroma-like tissue composed of collagen and fibronectin. Such dermis-like connective tissue structures have been defined as "self-assembled" ECM tissues [56]. When these tissues are grown in the presence of ascorbic acid, iPSC-derived fibroblasts have been shown to have elevated ECM production and deposition, when compared with the parental fibroblasts from which they were reprogrammed [12]. The analysis of the composition of these stromal tissues has revealed that those containing iPSC-derived fibroblasts produced greater amounts of type III collagen when compared with tissues derived from their parental fibroblasts, indicating their similarities to provisional matrix seen during the early stages of wound repair. This demonstrates that iPSC-derived fibroblasts may represent a replenishing source of potent fibroblasts that may be useful to study a range of connective tissue disorders of the skin, such as chronic wounds and fibrosis. Importantly, functional differences between iPSC-derived fibroblasts and fibroblasts from which they were reprogrammed were only fully revealed when these cells were incorporated into the complex, in vivo-like 3-D tissue environments that more closely mimic human dermis.

Another existing challenge currently facing skin tissue fabrication using adult-derived primary cells is the incomplete repertoire of structural proteins that are present in the dermal compartment of these tissues. For example, the absence of mature elastin and specific basement membrane proteins in current skin tissue models using adult-derived cells limits their ability to mimic chronic skin changes resulting from aging or UV exposure [36]. It may be possible to generate iPSC-derived tissue scaffolds that generate more complex dermal tissue constructs that more faithfully represent the stromal compartment of the human skin. This may be possible, as it has been shown that iPSC-derived fibroblasts produce and self-organize stromal tissue structures, known as "self-assembled" connective tissues that mimic the morphology and functional features of their in vivo counterparts [12]. Producing such broad repertoire of stable ECM components that have been shown to be assembled into a dermis-like tissue by iPSC-derived fibroblasts would be a helpful advance in developing sustainable skin tissues that would also be critical to developing tissue models that will mimic chronic conditions of the skin.

Together, this demonstrates that fibroblasts differentiated from iPSCs can acquire an augmented biological potency when compared with their parental fibroblasts as characterized by their increased production and assembly of ECM, which are functional features important for application of these cells in 3-D skin tissues [56]. This suggests that iPSC-derived cells will require rigorous phenotypic analysis in a variety of skin tissue models to determine if they have acquired functional properties that shift fibroblasts to an enhanced functional state, which can be leveraged to optimize skin tissue fabrication. A schematic of iPSC reprogramming and differentiation processes and their application in 3-D skin tissues is shown in Fig. 3.

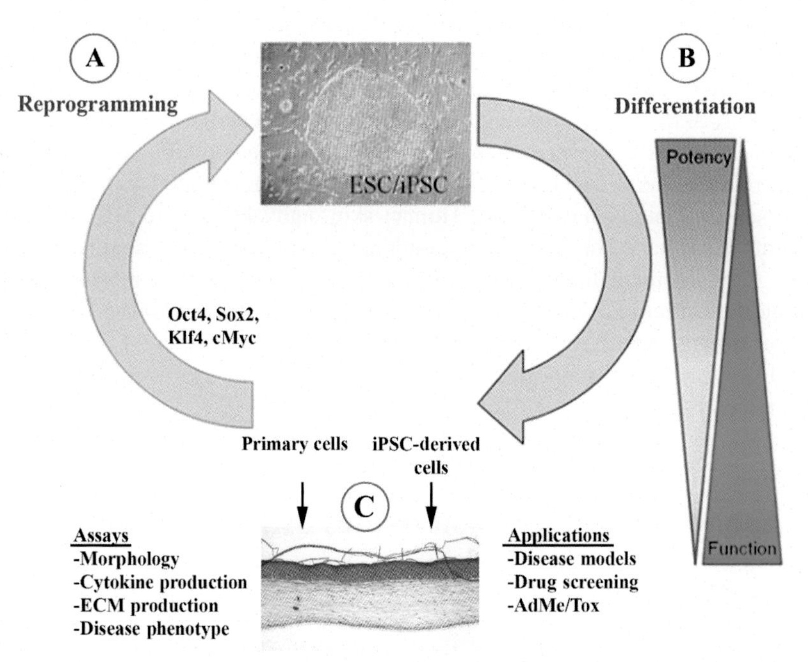

Fig. 3 Generation of skin tissue models from iPSC-derived skin cells and potential applications. (A) iPSCs are generated by the transfer of transcription factors to adult somatic cells to become pluripotent resulting in reprogramming to an embryonic-like state. (B) iPSCs are differentiated to cells with functional and phenotypic properties of skin cells such as fibroblasts and keratinocytes, which (C) can then be characterized and introduced into 3-D skin tissues. These tissues can be used for applications and assays listed above.

3. Achieving tissue complexity by incorporating multiple iPSC-derived cell types into 3-D skin tissues

The implementation of iPSC-based approaches for tissue engineering is an exciting and rapidly evolving field that can be applied to 3-D skin tissue fabrication. Levenberg et al. first demonstrated that ectodermal and mesenchymal cells differentiated from hESCs could self-organize into tissues displaying 3-D epithelial structures that could integrate into the host upon transplantation [57]. Subsequent studies have demonstrated that both epithelial cells [29] and stromal fibroblasts [11,12,24] generated from iPSCs could be incorporated into 3-D skin tissues that support the generation of a fully differentiated epithelium [29,40]. iPSCs derived from skin fibroblasts have been differentiated into stable and functional fibroblast lineages and incorporated into 3-D skin models that support morphological development of foreskin-derived keratinocytes [11] by producing soluble factors that stimulate epithelial growth and repair in 3-D skin tissues [12,58] (Fig. 4). Thus, tissues constructed with fibroblasts derived from iPSCs demonstrate 3-D tissue organization and functional features typical of the human skin. Three-dimensional human skin equivalents (HSEs) have been generated with iPSC-derived fibroblasts and iPSC-derived keratinocytes [40] or with primary human dermal fibroblasts that were combined with iPSC-derived keratinocytes [32]. More recently, human epidermal equivalents (HEEs) were generated from hESC- and iPSC-derived keratinocytes and subjected to sequential high-to-low humidity environments to generate a functional epidermal permeability barrier that will be essential for more precise skin disease modeling or drug development and testing [33].

Greater cellular complexity has been achieved by constructing 3-D skin tissues with fibroblasts, keratinocytes, and melanocytes that demonstrated melanosomes and the presence of pigmentation [44]. Human skin equivalents were also constructed with human primary fibroblasts and keratinocytes and micropatterned vascular networks. Alginate microchannels were created in the dermis of these tissues using 3-D printing that structured iPSC-derived endothelial cells, which allowed formation of vascular structures with robust endothelial barrier function. These 3-D human vascularized skin constructs supported neovascularization when they were engrafted into SCID mice [49].

4. Applications of 3-D skin tissues using iPSC-derived cells

An important opportunity in the development of 3-D skin tissues will be to generate personalized, tissue "surrogates" of their in vivo counterparts [3,14]. This will require the development of patient-derived cell lines that will be reprogrammed to iPSCs and that stably express their functional phenotypes after incorporation into 3-D tissues. The reprogramming of iPSCs from the dermis and epithelium of patients with skin diseases may enable the differentiation of iPSC-derived cells that could be used to fabricate disease-specific, skin tissue models that can recapitulate key stages of skin disease pathogenesis. Recent studies have shown that disease phenotypes of

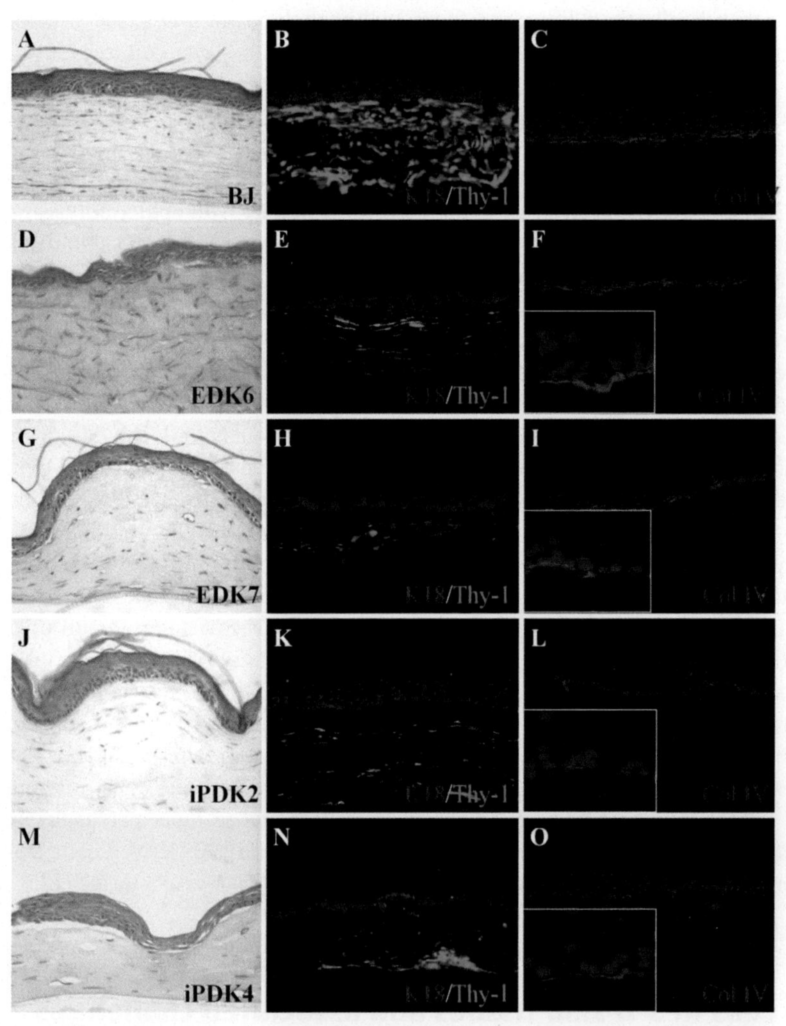

Fig. 4 iPSC-derived fibroblasts support the development of a fully differentiated epithelium when incorporated into the stroma of 3-D skin tissues. Fibroblasts from two independent differentiations of hESCs and iPSCs were incorporated into a collagen gel, seeded with foreskin keratinocytes, and grown at an air-liquid interface. H&E staining of tissues reveals that BJ (A), hESC-derived EDK6 (D), hESC-derived EDK7 (G), iPSC-derived iPDK2 (J), and iPSC-derived iPDK4 (M) cells had an elongated morphology in the collagen gel and supported the growth and differentiation of keratinocytes. Immunostaining of these tissues shows the expression of fibroblast marker Thy-1 *(green)* in the stromal compartment of all tissues and K18 *(red)* within ESC- and iPSC-derived cells (B, E, H, K, and N). Type IV collagen *(red)*, a basement membrane component, is localized at the region between the stromal and epidermal compartment, illustrating the initiation of basement membrane assembly in these skin tissues (C, F, I, L, and O) [11].

iPSC-derived cells show improved functional phenotypes in 3-D tissue structures in skin tissues rather when compared with conventional monolayer cultures [59–62]. To date, iPSC-derived keratinocytes have been generated from patients with recessive dystrophic epidermolysis bullosa (RDEB) harboring *COL7A1* mutation that causes chronic blistering. These RDEB patient-specific iPSC-derived keratinocytes were used to generate 3-D skin equivalents and formed all epidermal layers successfully [32]. Revertant iPSC-derived keratinocytes from patients with epidermolysis bullosa containing a mutation in the *COL17A1* gene have been used in a 3-D skin tissue and demonstrate their promise as a screening platform to develop new therapies to treat patients with this disorder [63]. Attempts to model melanocyte disorders such as neurofibromatosis type 1 have used pluripotent stem-cell-derived cells, but these have not yet been adapted for 3-D skin tissue models [64]. Similarly, melanocyte disorders including albinism, Griscelli syndrome, and vitiligo would benefit from developing iPSC-derived cells incorporated into 3-D skin tissue models [65]. In addition, a fetal rat model of myelomeningocele (MMC), a neural tube defect characterized by multiple central and peripheral nervous system defects and skin defects, has been used to test 3-D human skin equivalents as a mean of fetal tissue modeling. In this study, keratinocytes were differentiated from iPSCs that were reprogrammed from amniotic fluid of fetuses with twin-twin transfusion syndrome and trisomy 21. These iPSC-derived keratinocytes were then used to construct skin-like tissues leading to full or partial skin defect when implanted into a fetal rat MMC model [34].

Such application of iPSC-based technologies can facilitate the use of these skin disease models for precision medicine in ways that will lay the groundwork for personalized drug screening targeted to treating skin disease progression [1,2]. Through continued refinement of such disease-specific skin models using cells differentiated from patient-specific iPSCs, these skin tissues will lead to more personalized drug screening strategies that will enable a shift in the use of iPSC-derived cells from the current standard "disease in a dish" to more advanced "disease in a tissue" [2,7,16].

5. Future development and applications of iPSC-derived cells in 3-D skin tissues and associated challenges

Despite their potential, the application of cells derived from disease- and patient-specific iPSCs faces barriers that currently limit their implementation in 3-D skin tissues [14].

The application of iPSC-derived cells and tissues for regenerative therapies to treat human diseases will require the development of reliable methods to evaluate and ensure their safety and efficacy before application to humans [15,18,66,67]. For example, cells differentiated from iPSCs go through discrete developmental stages as early mesenchymal progenitors and emerge committed to specific mesenchymal lineage fates [68]. It has been previously shown that mesenchymal stem cells (MSCs) differentiated from iPSCs demonstrate differentiation to fat, cartilage, and the bone but do not support epithelial development of the skin in 3-D tissues [11]. Since the morphology of MSCs and fibroblasts appears identical in 2-D culture and since flow cytometry cannot definitely identify them as fibroblasts, it is only possible to confirm their fibroblast identity by confirming that they can support epithelial growth and differentiation in 3-D tissue context. Thus, 3-D skin tissues can be used to better understand the biological potency and functional phenotype of iPSC-derived cell lineages in a 3-D skin-like tissue context. By monitoring the capacity of iPSC-derived skin cells to contribute to normal skin morphogenesis, these tissues provide a platform to fully characterize the functional properties of iPSC-derived cells in ways that cannot be accomplished using conventional 2-D monolayer cultures. Importantly, incorporation of iPSC-derived cells in their in vivo-like tissue context will advance the study of developmental processes in the skin and can elucidate gene regulatory networks and signaling pathways that can modulate tissue organization and maturation. Skin tissues therefore offer a significant improvement when compared with 2-D cultures as a functional readout of the differentiation state of iPSC-derived cells in ways that will further determine optimal cues needed to guide their morphogenesis.

It has been demonstrated that reprogramming iPSCs from primary cells may result in the accumulation of genetic mutations and aberrant karyotypes due to viral integrations or following prolonged culture that could alter their phenotype or differentiation potential in 3-D tissues [67,69]. To overcome the problem of viral integration, several nonviral reprogramming methods have been applied to iPSC reprogramming. These include delivering a plasmid vector [70], using episomal plasmids as “minicircle vectors” [71,72], expressing OKSM mRNA [73], directly delivering OKSM proteins [7], or inducing expression of OKSM factors with microRNA [74] or with small molecules [21,69,75]. Additionally, nonintegrating viral methods such as Sendai virus have been developed. When reprogramming methods were compared, Sendai virus was shown to be the most efficient nonintegration method to reprogram somatic cells to iPSCs [76]. Whole-genome profiles of DNA methylation of iPSC clones have shown significant variability following reprogramming to iPSC, including retention of somatic cell memory and aberrant DNA methylation in these clones [77]. This suggests that future evaluations of iPSC-derived cells will require the development of reliable methods to assess their phenotypic maturation and functional outcomes before they can be incorporated into skin tissue models.

Changes in the epigenetic profile occur upon iPSC reprogramming and subsequent differentiation [77–79]. The stepwise differentiation of specific cell types from iPSC is characterized by a progressive restriction of their differentiation potential and an increase in lineage-specific gene expression. Such differentiation is associated with changes in nuclear arrangement, including remodeling of chromatin structure, histone modifications, and DNA methylation that guide expression of genes that determine cell lineage fate and function [78]. For example, iPSC differentiation is accompanied by increased DNA methylation of silenced chromatin and decreased methylation at sites of active gene transcription [77–79]. Profiling these changes can provide valuable information as to changes in the differentiation status and lineage stability of iPS cell populations differentiated to skin cell lineages. In order to generate stable cell phenotypes that will be most optimal for use of iPSC-derived skin tissues, it will be essential to develop sophisticated profiling tools that can rapidly identify epigenetic marks that can more accurately define lineage fate and functional properties of cells derived from iPSCs.

Epigenetic changes and establishment of a methylation signature early in the process of differentiation from a pluripotent cell are critical in defining the lineage identity and cell phenotype that will determine the properties of 3-D skin-like tissues. It is known that the tissue microenvironment plays a critical role in the epigenetic regulation of gene expression in skin tissues [80]. This indicates that the 3-D tissue microenvironment can modify DNA methylation in ways that make iPSC-derived cells more in vivo-like when incorporated into complex skin tissue context. As it is known that methylation changes that occur during reprogramming of iPSCs can be transmitted to cells differentiated from them at a relatively high frequency [81], it will be important to elucidate the underlying epigenetic mechanisms that occur upon reprogramming or subsequent differentiation from the pluripotent state. Further defining the differentiation state of iPSC-derived cells that display distinctive epigenetic features will serve as a fingerprint to identify optimal cells for skin tissue fabrication. For example, evaluating methylation status of iPSC-derived cells will be useful in determining whether a reprogrammed cell can differentiate into specific functional lineages. Methylation status may also help to identify iPSC clones that are incompletely reprogrammed and thus cannot be used to generate skin cells useful for 3-D tissue fabrication.

In addition, iPSC-derived cells are known to harbor a residual DNA methylation signature related to their cell of origin [77]. This "epigenetic memory" is thought to predispose iPSC clones to differentiate along lineages related to the cell type from which the iPSCs were initially derived, thus restricting differentiation to alternative cell fates [77,82]. This suggests that an epigenetic memory of the tissue of origin may impact the phenotypic stability of iPSC-derived cells used in tissue fabrication. There is still considerable debate over the extent of epigenetic changes associated with reprogramming [78,81,82] as whole-genome profiles of DNA methylation at single-base resolution show significant reprogramming variability [83]. Differentiation to generate stable cell lineages remains challenging due to these epigenetic changes [14,84]. Through continued refinement of such mechanisms that are currently limiting the stable and reproducible differentiation of skin cell lineages from iPSCs, we can further understand molecular and epigenetic changes critical to derivation of specified functional cell types best suited for skin tissue fabrication. Importantly, using 3-D tissues to assess cell functionality will markedly improve upon existing methods to determine the utility of iPSC-derived cells in 3-D tissues. The maturation status of cells differentiated in vitro from iPSCs is difficult to determine and will require even greater stringency to define biologically meaningful functions of these cells in skin disease models. Future findings that will further define the specific cell and tissue types that display distinctive phenotypic functional and epigenetic features will help to identify optimal cells for skin tissue fabrication.

Gene editing technologies, such as CRISPR/Cas9 [85–89], hold future promise to modify specific gene signatures in iPSC-derived skin cells either by correcting disease-causing mutations or through modification of expression of genes that may lead to an improved phenotype for tissue fabrication or for applications important for skin disease modeling. Optimizing skin disease-specific or customized 3-D tissue models will benefit from incorporating iPSC-derived cells that have undergone gene editing, which can provide novel ADME/Tox readouts. Combining CRISPR gene editing of iPSCs with skin tissue fabrication can improve disease-specific, skin-like tissues, when multiple isogenic cell types can be derived that all contains the same gene modification that will be valuable for developing customized skin disease phenotypes in 3-D skin tissues. The future utility of CRISPR/Cas9 to gene edit iPSCs or iPSC-derived cells needed for skin tissue fabrication may be complicated by the need to modify multiple genes needed to revert skin disease phenotypes. Currently, a maximum of nine genomic loci have been simultaneously targeted in a particular cell using CRISPR/Cas9. The use of CRISPR/Cas9 to modify iPSCs for skin tissue fabrication will also require considering possible off-target effects, which may cause unintended changes in expression of genes that will need to be addressed through refinements of Cas9 delivery [90–95].

An important consideration in the construction of skin tissues is the reproducibility and standardization of tissue fabrication. Tissue standardization using self-renewing and disease-specific cells currently limits the widespread deployment of commercially viable 3-D skin tissue platforms for drug screening [1,15,96]. While it will be advantageous to develop more tissue complexity by incorporating multiple cell types, this will require that these primary cells can be sourced in ways that consistently mimic the cells that they were initially derived from. iPSCs can serve as a replenishing, isogenic source of multiple cell types found in the human skin that are needed for skin tissue fabrication. Producing such isogenic cell types from a common source should advance the development and standardization of 3-D tissue fabrication for preclinical drug evaluation [15,16].

Differentiating well-standardized cells from iPSCs that can be used reproducibly for 3-D tissue fabrication has proved to be challenging. Reprogramming is a stochastic process, and individual reprogrammed iPSC clones may show great variability between them [7,21,75]. As a result, iPSC reprogramming and differentiation may not always establish a uniform source of skin cells that would improve upon existing heterogeneous sources of primary cells derived from adult skin. It is therefore important to establish criteria to assess this cellular heterogeneity in order to identify which iPSC clones can be used for differentiating skin cells for tissue fabrication. This will need to include analysis of molecular and functional benchmarks that are acquired by epithelial and stromal cells differentiated from iPSCs that reflect their capacity to support skin development in 3-D tissue models. As seen in Fig. 5, molecular benchmarks can be characterized by markers of sequential ectodermal specification (keratin 18) and by their subsequent keratinocyte commitment (p63 and keratin 14) and terminal differentiation of iPSC-derived epithelial cells (Fig. 5). In addition, iPSC-derived fibroblasts can be characterized by markers of specification (Thy-1 and PDGFR-β) and commitment (type I/III collagen and fibronectin) to fibroblast lineage (Fig. 5). Functional benchmarks for stromal fibroblasts differentiated from iPSC may include production of growth factors (HGF, KGF, GM-CSF, and IL1-B) and CD surface markers that can be detected in monolayer culture and indicate that iPSC-derived fibroblasts are likely to support skin tissue fabrication. Additional benchmarks for iPSC-derived fibroblast function that are important for 3-D skin tissue fabrication include the capacity to self-assemble an ECM stroma [12] and for that stroma to produce supportive growth factors needed for epithelial tissue maturation. Fabricating skin

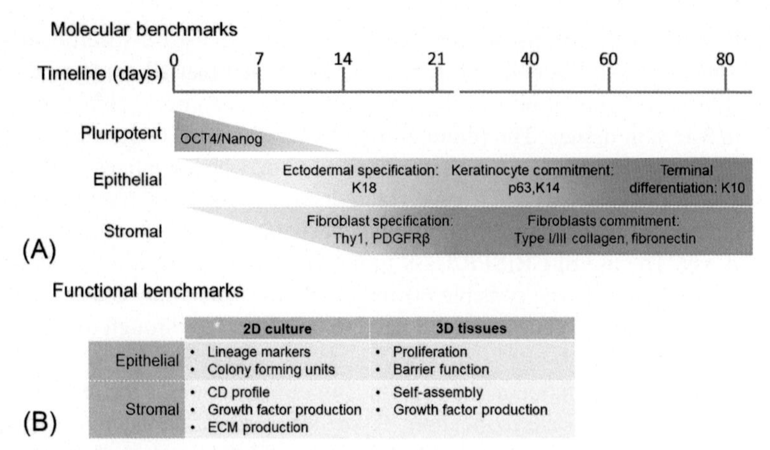

Fig. 5 Criteria to identify optimal iPSC-derived skin cells for tissue fabrication. Analysis of molecular benchmarks (A) will indicate the status of epithelial cells and fibroblasts through sequential stages of differentiation, specification, lineage commitment, and terminal differentiation following differentiation from iPSCs. Functional benchmarks (B) used to characterize epithelial and stromal cells differentiated from iPSCs are needed to show that these iPSC-derived cells have acquired the capacity to support skin development in 3-D tissue models. Benchmarks for stromal fibroblasts in 2-D culture include growth factor production and CD surface marker profiling, and 3-D skin tissues include the capacity to self-assemble an ECM stroma and production of supportive growth factors. iPSC-derived epithelial cell function can be characterized by monitoring expression of keratin protein lineage markers and clonogenic potential in 2-D culture and proliferation and barrier function in 3-D skin tissue.

tissues requires incorporation of keratinocyte progenitors that are enriched for cells that can proliferate extensively to generate fully stratified epithelium [29,40,97–100]. As a result, it is important to identify iPSC-derived cells that are enriched in highly replicative progenitor cells that will be required for successful 3-D tissue fabrication. The developmental stage of progenitor cells differentiated from iPSCs is difficult to assess, as these cells are often heterogeneous in nature [68]. As a result, it will be important to develop strategies to identify keratinocyte progenitors with the elevated proliferative cell fraction needed for the growth, differentiation, and morphological development of skin-like tissues. Optimal keratinocyte progenitors derived from iPSCs can be identified by determining the colony-forming efficiencies of these cells to best predict their utility when incorporated into skin tissues (Fig. 5B). Establishing such standardized functional [15,17] benchmarks for the proliferation of iPSC-derived skin cells should also improve cell source uniformity needed for skin tissue fabrication. This is likely to improve the qualities of iPSC-derived skin cells when compared with heterogeneous, primary adult-derived cells currently used for skin tissue fabrication.

The development of skin tissues from iPSCs will provide a replenishing isogenic cell source that can ideally obviate the need for sourcing cells from different primary tissue donors to improve reproducibility and scalability of these tissues when applied to higher throughput, tissue-based platforms. In this light, developing reliable and reproducible sources of iPSC-derived cells for skin tissue fabrication will be an important advancement over currently available monolayer tissue cultures, as their improved predictive value will greatly streamline preclinical decisions on the safety and efficacy of new drugs. Developing and optimizing iPSC-derived cells to generate disease-specific 3-D tissues will hopefully provide a renewable cell source that can produce all needed cell types in 3-D tissues and have the potential to revolutionize future scale-up and commercialization of disease- and patient-specific tissues for drug screening. In this way, iPSCs can streamline screening of drugs designed to improve treatment of skin diseases, so they can safely and efficiently enter human clinical trials. Optimally developed 3-D skin tissues can mimic complex, organ-level functions and can identify unexpected or untoward side effects that may fail to be detected using conventional 2-D monolayer cultures commonly used for preclinical drug screening [15,16,101]. As a result, skin tissue models may enable more efficient, cost-effective ways of predicting drug failures before embarking on human clinical trials. In the context of drug discovery, iPSC-derived skin tissues are likely to pave the way for novel preclinical drug screening models that may reveal molecular mechanisms of action, drug toxicities, and new biomarkers [2,10,15]. This holds potential to dramatically improve our ability to identify effective and safe candidate compounds at an early stage of drug development that will move the field of preclinical drug screening scale-up and commercialization to better prioritize the selection of new drugs.

The goal of developing a more streamlined approach to drug screening is also being advanced by the development of human "organs-on-chips," which are microfluidic cell culture devices harboring interconnected tissues to measure in vitro drug responses within and across different 3-D tissue types [3]. By incorporating 3-D skin tissues into this on-chip design using iPSC-derived cells, skin tissues can be generated and integrated with other tissues fabricated from these iPSCs to create screening systems that can reconstitute "whole-body" physiology and disease conditions [102]. Using iPSC-derived cells in these formats for drug screening will provide information that is more highly predictive of multiple human tissue responses and would shorten the preclinical drug development timeline, save animal lives, reduce failure rates, inform regulatory decision-making, and accelerate the development of new and more effective therapeutics to treat skin disease [15,16].

6. Conclusions

In conclusion, iPSCs offer a source of multiple patient-specific cell types necessary to fabricate 3-D skin tissues, including fibroblasts, keratinocytes, melanocytes, endothelial cells, macrophages, and folliculogenic epithelial stem cells that give rise to skin appendages. Three-dimensional skin tissue models harboring iPSC-derived cells will enable better understanding of skin biology and differentiation processes into cells needed for skin fabrication. Future incorporation of other important skin iPSC-derived cell types, such as adipocytes and sensory neurons, will allow to create improved 3-D skin tissues models. The use of increasingly complex tissues in disease modeling and drug testing will be fostered by incorporating 3-D skin tissues with multiple other organs in an organs-on-chips design. Even though challenges associated with 3-D skin tissues incorporating iPSC-derived cells that need to be resolved remain, these tissues hold great promise for drug testing and regenerative therapies. Complex 3-D skin tissues constructed with multiple iPSC-derived cells offer reliable, highly standardized, predictable models of human diseases and conditions that will allow to test efficacy and safety of drugs. Three-dimensional skin tissues constructed with iPSC-derived, patient-specific, isogenic cell types in combination with gene editing strategies will provide products to satisfy the needs of regenerative medicine at a large scale.

References

[1] Collins FS. Reengineering translational science: the time is right. Sci Transl Med 2011;3(90):90cm17.
[2] Daley GQ. Stem cells: roadmap to the clinic. J Clin Invest 2010;120(1):8–10.
[3] Bhatia SN, Ingber DE. Microfluidic organs-on-chips. Nat Biotechnol 2014;32(8):760–72.
[4] Lysaght MJ, Jaklenec A, Deweerd E. Great expectations: private sector activity in tissue engineering, regenerative medicine, and stem cell therapeutics. Tissue Eng A 2008;14(2):305–15.
[5] Carlson MW, Alt-Holland A, Egles C, Garlick JA. Three-dimensional tissue models of normal and diseased skin. Curr Protoc Cell Biol 2008. Chapter 19:Unit 19 9.
[6] Robey PG, Bianco P. The use of adult stem cells in rebuilding the human face. J Am Dent Assoc 2006;137(7):961–72.
[7] Gonzalez F, Boue S, Belmonte JCI. Methods for making induced pluripotent stem cells: reprogramming a la carte. Nat Rev Genet 2011;12(4):231–42.
[8] Yu J, Vodyanik MA, Smuga-Otto K, Antosiewicz-Bourget J, Frane JL, Tian S, et al. Induced pluripotent stem cell lines derived from human somatic cells. Science 2007;318(5858):1917–20.
[9] Nishikawa S, Goldstein RA, Nierras CR. The promise of human induced pluripotent stem cells for research and therapy. Nat Rev Mol Cell Biol 2008;9(9):725–9.
[10] Egles C, Garlick JA, Shamis Y. Three-dimensional human tissue models of wounded skin. Methods Mol Biol 2010;585:345–59.
[11] Hewitt KJ, Shamis Y, Hayman RB, Margvelashvili M, Dong S, Carlson MW, et al. Epigenetic and phenotypic profile of fibroblasts derived from induced pluripotent stem cells. PLoS One 2011;6(2):e17128.
[12] Shamis Y, Hewitt KJ, Bear SE, Alt-Holland A, Qari H, Margvelashvilli M, et al. iPSC-derived fibroblasts demonstrate augmented production and assembly of extracellular matrix proteins. In Vitro Cell Dev Biol Anim 2012;48(2):112–22.
[13] Park IH, Arora N, Huo H, Maherali N, Ahfeldt T, Shimamura A, et al. Disease-specific induced pluripotent stem cells. Cell 2008;134(5):877–86.
[14] Gunaseeli I, Doss MX, Antzelevitch C, Hescheler J, Sachinidis A. Induced pluripotent stem cells as a model for accelerated patient- and disease-specific drug discovery. Curr Med Chem 2010;17(8):759–66.
[15] Inoue H, Yamanaka S. The use of induced pluripotent stem cells in drug development. Clin Pharmacol Ther 2011;89(5):655–61.
[16] Rubin LL. Stem cells and drug discovery: the beginning of a new era? Cell 2008;132(4):549–52.
[17] Shi Y, Inoue H, JC W, Yamanaka S. Induced pluripotent stem cell technology: a decade of progress. Nat Rev Drug Discov 2017;16(2):115–30.

[18] Takahashi K, Yamanaka S. Induction of pluripotent stem cells from mouse embryonic and adult fibroblast cultures by defined factors. Cell 2006;126(4):663–76.
[19] Takahashi K, Tanabe K, Ohnuki M, Narita M, Ichisaka T, Tomoda K, et al. Induction of pluripotent stem cells from adult human fibroblasts by defined factors. Cell 2007;131(5):861–72.
[20] Lapasset L, Milhavet O, Prieur A, Besnard E, Babled A, Ait-Hamou N, et al. Rejuvenating senescent and centenarian human cells by reprogramming through the pluripotent state. Genes Dev 2011;25(21):2248–53.
[21] Huangfu D, Maehr R, Guo W, Eijkelenboom A, Snitow M, Chen AE, et al. Induction of pluripotent stem cells by defined factors is greatly improved by small-molecule compounds. Nat Biotechnol 2008;26(7):795–7.
[22] Lian QZ, Zhang YL, Zhang JQ, Zhang HK, XG W, Zhang Y, et al. Functional mesenchymal stem cells derived from human induced pluripotent stem cells attenuate limb ischemia in mice. Circulation 2010;121(9):1113–23.
[23] Villa-Diaz LG, Brown SE, Liu Y, Ross AM, Lahann J, Parent JM, et al. Derivation of mesenchymal stem cells from human induced pluripotent stem cells cultured on synthetic substrates. Stem Cells 2012;30(6):1174–81.
[24] Hewitt KJ, Shamis Y, Knight E, Smith A, Maione A, Alt-Holland A, et al. PDGFRbeta expression and function in fibroblasts derived from pluripotent cells is linked to DNA demethylation. J Cell Sci 2012;125(Pt 9):2276–87.
[25] Lutolf MP, Gilbert PM, Blau HM. Designing materials to direct stem-cell fate. Nature 2009;462(7272):433–41.
[26] Discher DE, Mooney DJ, Zandstra PW. Growth factors, matrices, and forces combine and control stem cells. Science 2009;324(5935):1673–7.
[27] Schuldiner M, Yanuka O, Itskovitz-Eldor J, Melton DA, Benvenisty N. Effects of eight growth factors on the differentiation of cells derived from human embryonic stem cells. Proc Natl Acad Sci U S A 2000;97(21):11307–12.
[28] Aberdam D, Gambaro K, Medawar A, Aberdam E, Rostagno P, de la Forest Divonne S, et al. Embryonic stem cells as a cellular model for neuroectodermal commitment and skin formation. C R Biol 2007;330(6–7):479–84.
[29] Aberdam D. Derivation of keratinocyte progenitor cells and skin formation from embryonic stem cells. Int J Dev Biol 2004;48(2–3):203–6.
[30] Coraux C, Hilmi C, Rouleau M, Spadafora A, Hinnrasky J, Ortonne JP, et al. Reconstituted skin from murine embryonic stem cells. Curr Biol 2003;13(10):849–53.
[31] Guenou H, Nissan X, Larcher F, Feteira J, Lemaitre G, Saidani M, et al. Human embryonic stem-cell derivatives for full reconstruction of the pluristratified epidermis: a preclinical study. Lancet 2009;374(9703):1745–53.
[32] Itoh M, Kiuru M, Cairo MS, Christiano AM. Generation of keratinocytes from normal and recessive dystrophic epidermolysis bullosa-induced pluripotent stem cells. Proc Natl Acad Sci U S A 2011;108(21):8797–802.
[33] Petrova A, Celli A, Jacquet L, Dafou D, Crumrine D, Hupe M, et al. 3D in vitro model of a functional epidermal permeability barrier from human embryonic stem cells and induced pluripotent stem cells. Stem Cell Rep 2014;2(5):675–89.
[34] Kajiwara K, Tanemoto T, Wada S, Karibe J, Ihara N, Ikemoto Y, et al. Fetal therapy model of myelomeningocele with three-dimensional skin using amniotic fluid cell-derived induced pluripotent stem cells. Stem Cell Rep 2017;8(6):1701–13.
[35] Garcia M, Quintana-Bustamante O, Segovia JC, Bueren J, Martinez-Santamaria L, Guerrero-Aspizua S, et al. Long-term skin regeneration in xenografts from iPSC teratoma-derived human keratinocytes. Exp Dermatol 2016;25(9):736–8.

[36] Smola H, Stark HJ, Thiekotter G, Mirancea N, Krieg T, Fusenig NE. Dynamics of basement membrane formation by keratinocyte-fibroblast interactions in organotypic skin culture. Exp Cell Res 1998;239(2):399–410.
[37] Maas-Szabowski N, Stark HJ, Fusenig NE. Keratinocyte growth regulation in defined organotypic cultures through IL-1-induced keratinocyte growth factor expression in resting fibroblasts. J Investig Dermatol 2000;114(6):1075–84.
[38] Togo S, Sato T, Sugiura H, Wang X, Basma H, Nelson A, et al. Differentiation of embryonic stem cells into fibroblast-like cells in three-dimensional type I collagen gel cultures. In Vitro Cell Dev Biol Anim 2011;47(2):114–24.
[39] Sato T, Liu X, Basma H, Togo S, Sugiura H, Nelson A, et al. IL-4 induces differentiation of human embryonic stem cells into fibrogenic fibroblast-like cells. J Allergy Clin Immunol 2011;127(6):1595–603 [e9].
[40] Itoh M, Umegaki-Arao N, Guo Z, Liu L, Higgins CA, Christiano AM. Generation of 3D skin equivalents fully reconstituted from human induced pluripotent stem cells (iPSCs). PLoS One 2013;8(10):e77673.
[41] Fang D, Leishear K, Nguyen TK, Finko R, Cai K, Fukunaga M, et al. Defining the conditions for the generation of melanocytes from human embryonic stem cells. Stem Cells 2006;24(7):1668–77.
[42] Ohta S, Imaizumi Y, Okada Y, Akamatsu W, Kuwahara R, Ohyama M, et al. Generation of human melanocytes from induced pluripotent stem cells. PLoS One 2011;6(1):e16182.
[43] Yang R, Jiang M, Kumar SM, Xu T, Wang F, Xiang L, et al. Generation of melanocytes from induced pluripotent stem cells. J Investig Dermatol 2011;131(12):2458–66.
[44] Gledhill K, Guo Z, Umegaki-Arao N, Higgins CA, Itoh M, Christiano AM. Melanin transfer in human 3D skin equivalents generated exclusively from induced pluripotent stem cells. PLoS One 2015;10(8):e0136713.
[45] Mica Y, Lee G, Chambers SM, Tomishima MJ, Studer L. Modeling neural crest induction, melanocyte specification, and disease-related pigmentation defects in hESCs and patient-specific iPSCs. Cell Rep 2013;3(4):1140–52.
[46] Karlsson KR, Cowley S, Martinez FO, Shaw M, Minger SL, James W. Homogeneous monocytes and macrophages from human embryonic stem cells following coculture-free differentiation in M-CSF and IL-3. Exp Hematol 2008;36(9):1167–75.
[47] van Wilgenburg B, Browne C, Vowles J, Cowley SA. Efficient, long term production of monocyte-derived macrophages from human pluripotent stem cells under partly-defined and fully-defined conditions. PLoS One 2013;8(8):e71098.
[48] Buchrieser J, James W, Moore MD. Human induced pluripotent stem cell-derived macrophages share ontogeny with MYB-independent tissue-resident macrophages. Stem Cell Rep 2017;8(2):334–45.
[49] Abaci HE, Guo Z, Coffman A, Gillette B, Lee WH, Sia SK, et al. Human skin constructs with spatially controlled vasculature using primary and iPSC-derived endothelial cells. Adv Healthc Mater 2016;5(14):1800–7.
[50] Kim KL, Song SH, Choi KS, Suh W. Cooperation of endothelial and smooth muscle cells derived from human induced pluripotent stem cells enhances neovascularization in dermal wounds. Tissue Eng A 2013;19(21–22):2478–85.
[51] Rufaihah AJ, Huang NF, Jame S, Lee JC, Nguyen HN, Byers B, et al. Endothelial cells derived from human iPSCS increase capillary density and improve perfusion in a mouse model of peripheral arterial disease. Arterioscler Thromb Vasc Biol 2011;31(11):e72–9.
[52] Lian X, Bao X, Al-Ahmad A, Liu J, Wu Y, Dong W, et al. Efficient differentiation of human pluripotent stem cells to endothelial progenitors via small-molecule activation of WNT signaling. Stem Cell Rep 2014;3(5):804–16.

[53] Yang R, Zheng Y, Burrows M, Liu S, Wei Z, Nace A, et al. Generation of folliculogenic human epithelial stem cells from induced pluripotent stem cells. Nat Commun 2014;5:3071.

[54] Takagi R, Ishimaru J, Sugawara A, Toyoshima K, Ishida K, Ogawa M, et al. Bioengineering a 3D integumentary organ system from iPS cells using an in vivo transplantation model. Sci Adv 2016;2(4).

[55] Suhr ST, Chang EA, Tjong J, Alcasid N, Perkins GA, Goissis MD, et al. Mitochondrial rejuvenation after induced pluripotency. PLoS One 2010;5(11):e14095.

[56] Pouyani T, Ronfard V, Scott PG, Dodd CM, Ahmed A, Gallo RL, et al. De novo synthesis of human dermis in vitro in the absence of a three-dimensional scaffold. In Vitro Cell Dev Biol Anim 2009;45(8):430–41.

[57] Zoldan J, Levenberg S. Engineering three-dimensional tissue structures using stem cells. Methods Enzymol 2006;420:381–91.

[58] Shamis Y, Hewitt KJ, Carlson MW, Margvelashvilli M, Dong S, Kuo CK, et al. Fibroblasts derived from human embryonic stem cells direct development and repair of 3D human skin equivalents. Stem Cell Res Ther 2011;2(1):10.

[59] Huch M, Koo BK. Modeling mouse and human development using organoid cultures. Development 2015;142(18):3113–25.

[60] Lancaster MA, Knoblich JA. Organogenesis in a dish: modeling development and disease using organoid technologies. Science 2014;345(6194).

[61] Eiraku M, Takata N, Ishibashi H, Kawada M, Sakakura E, Okuda S, et al. Self-organizing optic-cup morphogenesis in three-dimensional culture. Nature 2011;472(7341):51–6.

[62] Schwank G, Koo BK, Sasselli V, Dekkers JF, Heo I, Demircan T, et al. Functional repair of CFTR by CRISPR/Cas9 in intestinal stem cell organoids of cystic fibrosis patients. Cell Stem Cell 2013;13(6):653–8.

[63] Umegaki-Arao N, Pasmooij AM, Itoh M, Cerise JE, Guo Z, Levy B, et al. Induced pluripotent stem cells from human revertant keratinocytes for the treatment of epidermolysis bullosa. Sci Transl Med 2014;6(264):264ra164.

[64] Allouche J, Bellon N, Saidani M, Stanchina-Chatrousse L, Masson Y, Patwardhan A, et al. In vitro modeling of hyperpigmentation associated to neurofibromatosis type 1 using melanocytes derived from human embryonic stem cells. Proc Natl Acad Sci U S A 2015;112(29):9034–9.

[65] Nissan X, Lemaitre G, Peschanski M, Baldeschi C. Coloring skin with pluripotent stem cells. Cell Cycle 2011;10(23):3985–6.

[66] Lund RJ, Narva E, Lahesmaa R. Genetic and epigenetic stability of human pluripotent stem cells. Nat Rev Genet 2012;13(10):732–44.

[67] Carpenter MK, Frey-Vasconcells J, Rao MS. Developing safe therapies from human pluripotent stem cells. Nat Biotechnol 2009;27(7):606–13.

[68] Hynes K, Menicanin D, Mrozik K, Gronthos S, Bartold PM. Generation of functional mesenchymal stem cells from different induced pluripotent stem cell lines. Stem Cells Dev 2014;23(10):1084–96.

[69] Huangfu D, Osafune K, Maehr R, Guo W, Eijkelenboom A, Chen S, et al. Induction of pluripotent stem cells from primary human fibroblasts with only Oct4 and Sox2. Nat Biotechnol 2008;26(11):1269–75.

[70] Okita K, Matsumura Y, Sato Y, Okada A, Morizane A, Okamoto S, et al. A more efficient method to generate integration-free human iPS cells. Nat Methods 2011;8(5):409–12.

[71] Narsinh KH, Jia FJ, Robbins RC, Kay MA, Longaker MT, Wu JC. Generation of adult human induced pluripotent stem cells using nonviral minicircle DNA vectors. Nat Protoc 2011;6(1):78–88.

[72] JY Y, KJ H, Smuga-Otto K, Tian SL, Stewart R, Slukvin II, et al. Human induced pluripotent stem cells free of vector and transgene sequences. Science 2009;324(5928):797–801.
[73] Warren L, Manos PD, Ahfeldt T, Loh YH, Li H, Lau F, et al. Highly efficient reprogramming to pluripotency and directed differentiation of human cells with synthetic modified mRNA. Cell Stem Cell 2010;7(5):618–30.
[74] Lin SL, Chang DC, Lin CH, Ying SY, Leu D, Wu DT. Regulation of somatic cell reprogramming through inducible mir-302 expression. Nucleic Acids Res 2011;39(3):1054–65.
[75] Zhu S, Li W, Zhou H, Wei W, Ambasudhan R, Lin T, et al. Reprogramming of human primary somatic cells by OCT4 and chemical compounds. Cell Stem Cell 2010;7(6):651–5.
[76] Schlaeger TM, Daheron L, Brickler TR, Entwisle S, Chan K, Cianci A, et al. A comparison of non-integrating reprogramming methods. Nat Biotechnol 2015;33(1):58–230.
[77] Kim K, Doi A, Wen B, Ng K, Zhao R, Cahan P, et al. Epigenetic memory in induced pluripotent stem cells. Nature 2010;467(7313):285–90.
[78] Apostolou E, Hochedlinger K. Chromatin dynamics during cellular reprogramming. Nature 2013;502(7472):462–71.
[79] Bock C, Beerman I, Lien WH, Smith ZD, HC G, Boyle P, et al. DNA methylation dynamics during in vivo differentiation of blood and skin stem cells. Mol Cell 2012;47(4):633–47.
[80] DesRochers TM, Shamis Y, Alt-Holland A, Kudo Y, Takata T, Wang GW, et al. The 3D tissue microenvironment modulates DNA methylation and E-cadherin expression in squamous cell carcinoma. Epigenetics 2012;7(1):34–46.
[81] Hewitt KJ, Garlick JA. Cellular reprogramming to reset epigenetic signatures. Mol Asp Med 2013;34(4):841–8.
[82] Mandegar MA, Huebsch N, Frolov EB, Shin E, Truong A, Olvera MP, et al. CRISPR interference efficiently induces specific and reversible gene silencing in human iPSCs. Cell Stem Cell 2016;18(4):541–53.
[83] Hatzistergos KE, Blum A, Ince T, Grichnik JM, Hare JM. What is the oncologic risk of stem cell treatment for heart disease? Circ Res 2011;108(11):1300–3.
[84] Lister R, Pelizzola M, Kida YS, Hawkins RD, Nery JR, Hon G, et al. Hotspots of aberrant epigenomic reprogramming in human induced pluripotent stem cells. Nature 2011;471(7336):68–73.
[85] Garneau JE, Dupuis ME, Villion M, Romero DA, Barrangou R, Boyaval P, et al. The CRISPR/Cas bacterial immune system cleaves bacteriophage and plasmid DNA. Nature 2010;468(7320):67–71.
[86] Horvath P, Barrangou R. CRISPR/Cas, the immune system of bacteria and archaea. Science 2010;327(5962):167–70.
[87] Deveau H, Garneau JE, Moineau S. CRISPR/Cas system and its role in phage-bacteria interactions. Annu Rev Microbiol 2010;64:475–93.
[88] Cong L, Ran FA, Cox D, Lin S, Barretto R, Habib N, et al. Multiplex genome engineering using CRISPR/Cas systems. Science 2013;339(6121):819–23.
[89] Mali P, Yang L, Esvelt KM, Aach J, Guell M, DiCarlo JE, et al. RNA-guided human genome engineering via Cas9. Science 2013;339(6121):823–6.
[90] Cho SW, Kim S, Kim Y, Kweon J, Kim HS, Bae S, et al. Analysis of off-target effects of CRISPR/Cas-derived RNA-guided endonucleases and nickases. Genome Res 2014;24(1):132–41.
[91] Frock RL, JZ H, Meyers RM, Ho YJ, Kii E, Alt FW. Genome-wide detection of DNA double-stranded breaks induced by engineered nucleases. Nat Biotechnol 2015; 33(2):179–86.

[92] Fu YF, Foden JA, Khayter C, Maeder ML, Reyon D, Joung JK, et al. High-frequency off-target mutagenesis induced by CRISPR-Cas nucleases in human cells. Nat Biotechnol 2013;31(9):822.
[93] Hsu PD, Scott DA, Weinstein JA, Ran FA, Konermann S, Agarwala V, et al. DNA targeting specificity of RNA-guided Cas9 nucleases. Nat Biotechnol 2013;31(9):827.
[94] Lin YN, Cradick TJ, Brown MT, Deshmukh H, Bao G. CRISPR/Cas9 systems have off-target activity with insertions or deletions between target DNA and guide RNA sequences. Mol Ther 2014;22:S94–95.
[95] Pattanayak V, Lin S, Guilinger JP, Ma EB, Doudna JA, Liu DR. High-throughput profiling of off-target DNA cleavage reveals RNA-programmed Cas9 nuclease specificity. Nat Biotechnol 2013;31(9):839.
[96] Fabre KM, Livingston C, Tagle DA. Organs-on-chips (microphysiological systems): tools to expedite efficacy and toxicity testing in human tissue. Exp Biol Med (Maywood) 2014;239(9):1073–7.
[97] Kolodka TM, Garlick JA, Taichman LB. Evidence for keratinocyte stem cells in vitro: long term engraftment and persistence of transgene expression from retrovirus-transduced keratinocytes. Proc Natl Acad Sci U S A 1998;95(8):4356–61.
[98] Andriani F, Margulis A, Lin N, Griffey S, Garlick JA. Analysis of microenvironmental factors contributing to basement membrane assembly and normalized epidermal phenotype. J Investig Dermatol 2003;120(6):923–31.
[99] Segal N, Andriani F, Pfeiffer L, Kamath P, Lin N, Satyamurthy K, et al. The basement membrane microenvironment directs the normalization and survival of bioengineered human skin equivalents. Matrix Biol 2008;27(3):163–70.
[100] Carlson M, Faria K, Shamis Y, Leman J, Ronfard V, Garlick J. Epidermal stem cells are preserved during commercial-scale manufacture of a bilayered, living cellular construct (Apligraf(R)). Tissue Eng A 2011;17(3–4):487–93.
[101] Wikswo JP. The relevance and potential roles of microphysiological systems in biology and medicine. Exp Biol Med (Maywood) 2014;239(9):1061–72.
[102] van den Broek LJ, Bergers L, Reijnders CMA, Gibbs S. Progress and future prospectives in skin-on-chip development with emphasis on the use of different cell types and technical challenges. Stem Cell Rev 2017;13(3):418–29.

Requirements of skin tissue models for high-throughput screening

18

Stephanie H. Mathes, Christian N. Parker†*
*ICBT, Tissue Engineering, Zurich University of Applied Sciences, Waedenswil, Switzerland,
†Novartis Institutes of Biomolecular Research, Basel, Switzerland

1. Outline

This chapter will focus on the use of in vitro skin models for drug screening rather than at the automated production of the tissue constructs. The authors provide an overview concerning existing methods and approaches and crystallize the most demanding requirements for integration of human in vitro skin models in high-throughput screening (HTS) processes.

2. Introduction

When considering the use of human skin models for HTS purposes, two essential questions arise:

(1) What is driving the demand for the use of such in vitro human skin equivalents?
(2) What are the requirements and obstacles preventing the routine implementation of HTS of skin equivalents today?

The aim of this article is to answer these questions and to identify actions that have to be taken to enable HTS of in vitro skin models. The article will also discuss the factors required in order to achieve relevant and robust data that are representative of the in vivo situation being studied. The automated cultivation or the large-scale production of human skin models is not discussed here; the reader is referred to previous reviews on this subject [1-4]. Systems and processes have already been defined for the manufacture of cell- and matrix-based organotypic models and for models representing a disease phenotype [5]. There are companies that offer human skin models; these companies make use of robotic systems to guarantee a high and standardized quality of their products (e.g., EpiSkin, Mattek, and CellSystems). However, when it comes to the use of such skin models in the cosmetic or pharmaceutical industry or even in academia, the process steps (drug application, cultivation, and analysis) are mainly conducted manually, with only limited use of robotics. Another important issue of such models is to support scientific investigation of factors influencing skin biology [6].

These uses of in vitro skin models address the first question, that is, what is the impetus that drives the need for the development of such models. In particular, driven

Skin Tissue Models. https://doi.org/10.1016/B978-0-12-810545-0.00018-8

by the demands of regulatory authorities and customers, the cosmetic industry has supported the establishment of human skin models to serve as predictive tools for safety assessment studies [7,8]. Ultimately, once such in vitro skin models have been validated and showed to faithfully reflect the action of compounds on the skin, then such a model could be used to screen the 10,000 compounds that have been profiled in the Tox21 screening program (https://www.epa.gov/chemical-research/toxicity-testing-21st-century-tox21). Such a screening process would allow the correlation of compound structures and activity on a large scale, helping to improve in vitro toxicity testing and prediction.

In the cosmetic industry, reconstructed human epidermal models are routinely used nowadays to assess the risk of skin irritancy and corrosion by assessing relatively simple end points based on the metabolic status of the skin model and cytokine release [9-11]. However, as the number of new compounds to be assessed with such methods rarely exceeds 50 compounds per experiment, adaptation of the process for truly HTS has not been necessary. Besides evaluating any risk a compound may exhibit, more and more interest is currently arising concerning profiling and high-content analysis in order to scientifically support the product claims [12]. In the pharmaceutical industry, however, there is still a need for novel dermatologically acting drugs, and appropriate models have to be designed to enable target and hit validation in an HTS setup. Skin diseases such as psoriasis, atopic dermatitis, decubitus ulcer, eczema, acne, vitiligo, skin cancer, and abnormal scar formation after trauma remain illnesses requiring novel treatments [13-15]. However, the development of models suitable for greater understanding of such disorders and the validation of potential drug targets will require further assay development and validation.

Currently, there is a broad spectrum of models and ways to create 3-D skin replicas, which represent different levels of biological complexity [16,17]. The first and the most thoroughly characterized models consist of only one cell type (reconstructed human epidermis made of keratinocytes, RHE). Such models are routinely used for risk assessment of compounds [18]. Even so, more sophisticated models are needed when investigating subjects such as the efficacy, pharmacodynamics (PD), or pharmacokinetics (PK) of treatments for a particular disease. Such models may require the incorporation of multiple cell types (fibroblasts [5,19,20], melanocytes [21], stem cells [22,23], Langerhans cells [24], vascular structures [25], and adipocytes [26,27]). However, such models are more challenging to standardize due to the complexity of their creation and the readouts used to monitor activity, leading to greater assay variability. Parameters that contribute to increased variability of the phenotypic outcome of an in vitro created skin model can be divided into three clusters: (I) cell types, (II) epigenetics, and (III) cultivation condition (artificial environment). (I) Primary human cell lines are usually isolated and pooled from different donors for the creation of skin equivalent models. Differences exist between individuals, but even cells from one individual vary, for example, different subtypes of dermal fibroblasts that contribute to different responses in assays such as hypertrophic scar formation [28]. (II) Epigenetic modifiers can regulate epidermal differentiation processes, and diseases such as psoriasis and cancer exhibit shifts in these epigenetic mechanisms, resulting in variations of proliferation and differentiation [29]. (III) Cultivation of cells in the artificial environment of a cell culture plate, submerged with a mixture of undefined components like fetal bovine serum to achieve the phenotype of interest, has a crucial impact, and small changes can alter expression patterns significantly [30]. Consequently, to address the basic requirement for stable in vitro tissue equivalents, the establishment of an unlimited cell source is critical. Reproducible sources of cells can be obtained by using human telomerase reverse transcriptase (hTERT) immortalized cells [31] or redifferentiated iPS cells [32-35]. In the case of hTERT immortalization, the overexpression of telomerase is not sufficient to circumvent senescence [36]. The usage of pluripotent stem cells as an appropriate source leads to the concern that epigenetic differences may have been created that will result in diversity of the organotypic skin models being generated. In order to obtain some insight into this issue, different methylation patterns of the gene encoding trichohyalin (an intermediate filament-associated protein associated with keratin intermediate of the granular layer of the epidermis), which is important for barrier maturation of the *stratum corneum*, were investigated regarding the phenotypic outcome of the barrier properties. The results supported the assumption that methylation patterns of this specific gene do not effect the differentiation of the in vitro tissue model [35,37]. In order to ensure that epigenetic factors for other critical genes are not modulating the function of the skin model, further studies have to be performed to pave the way for iPS cell-based skin models for routine use.

3. Technical aspects

3.1 Automated platform for HTS lead discovery

Automation of cell-based assays for high-throughput lead discovery is well established, and various vendors of liquid-handling platforms offer solutions for numerous applications [38] (Table 1).

While there are systems that automate the procedure of conventional cell culture and passaging of cells in flasks, it is much more common to manually culture cells. Cell-based models are often developed and miniaturized to fit the ANSI/SLAS format of microplates. Assays that are established in this format can be transferred to robotic liquid-handling platforms available on the market. Automated pipetting in microplates from 1 well to 1536 wells is standard and, depending on the type of culture or assay, can be performed not only by individual 8- or 12-channel pipettes but also by the use of multiple 96- or 384-channel pipetting heads. Depending on the type of assay, samples can be prepared by the liquid-handling arms present in the automated work station or off-line by hardware specifically designed for sample dispensing (such as the Labcyte Echo dispenser or Tecan D300e Digital Dispenser). Such automated setups can be used in a broad range of different assay types, for example, viability assays or dose-response experiments, and in many different research areas, such as target and lead discovery or ADME. The assay itself is then performed on the same workstation including all liquid handling and robotic handling of the microplates to dedicated

Table 1 Vendors and systems of automated liquid-handling platforms experienced with cell and tissue cultures

Vendor	System	Mammalian cell culture and HTS analysis	3-D tissues	Transwell systems	Handling of skin cells
Beckman Coulter, Inc.	Biomek FX[P] Biomek NX[P]	Yes	Yes	Not specified	Not specified, 2-D cell cultures and related assays feasible
CyBio AG	CyBi-Drop 3D	Multiplexed viability, cytotoxicity, and apoptosis	Phenotypic screening [39] Spheroid liver model [40] Hanging drop array [41]	No	Not specified, 2-D cell cultures and related assays feasible
Hamilton	Microlab STAR pipetting workstation	Cell counting, confluence, and viability; genomics and proteomics	Not specified	Standard and Transwell plates	Production of skin cell experimental models [42]
Perkin-Elmer Inc.	cell::explorer	Yes	3-D spheriods	Not specified	Not specified, 2-D cell cultures and related assays feasible
Sartorius Stedim Biotech	SelecT and CompacT SelecT	Cell culture only	2-D cell culture	Not specified	Not specified, 2-D cell cultures and related assays feasible
Tecan	Freedom EVO Fluent	Media exchange, cell plating, incubation, and assay preparation. Cell-based assays, cell counting, bright-field-based cell imaging	Scaffold-based or scaffold-free cultures	Intestine models in Transwell	Not specified, 2-D cell cultures and skin models in Transwell system should be feasible

positions for specific functions, such as shaking, incubation, plate washing, and finally insertion of the plate into the reader for data generation. The seamless integration of dedicated devices required for an assay and full automation are essential for HTSs, unless a workstation approach is taken where each step in the process is performed by a stand-alone workstation. Both approaches lead to increased productivity, leaving additional time for laboratory personnel to perform other tasks, while the robotic system is performing the routine assays.

As shown in Table 1, there are multiple vendors around offering liquid-handling platforms for cell culture (including barcoding of the cell culture laboratory hardware, barcode readers, microplate centrifuges, plate readers and imagers, plate transport devices, shakers, plate washers, safety and biocontainment enclosures, temperature-controlled storage of liquids, and tube handling devices for capping/decapping).The cultivation and basic monitoring of the cells on robotic platforms are currently feasible, even for complex 3-D cultures. In particular, the use of spheroids has been described and used for multiple HTS approaches [43,44]. For such organoid 3-D models, there are examples where automation, data generation, and analysis have been shown to be possible, but this has not been widely applied for skin models. The only company that claims the production and handling of skin models on their platform is Hamilton [42]. Although rather designed for the production of standardized epithelial models than for the screening process, an adaptation might be conceivable. The liquid-handling platform Freedom EVO was used in the past to process another type of epithelium (intestine) and may be adaptable to skin models.

The formats for creating and cultivating of epithelial barrier models, which are used on automated systems, are typically grown in microplates: in a matrix (e.g., Matrigel), in hanging drops (e.g., GravityPLUS), or in round-bottom low-attachment plates (e.g., Corning) where the cells build spheroids or most commonly in Transwell plates where cells grow at the air-liquid interface on a membrane on the insert plate. The format that is chosen is highly dependent on the experimental question to investigate the interaction between the organism and the environment.

One feature of using skin models in the context of drug application and drug testing is that test compound can be lipophilic and yet has to be applied directly onto the dry *stratum corneum*. Such application of highly lipophilic fluids using automated liquid-handling systems requires optimization of the liquid-handling parameters to guarantee accurate dispensing of defined volumes evenly over the whole surface of the tissue microstructure.

3.2 Laboratory hardware adaptation

As stated above, truly HTS requires an automated assay platform, and so, the formats of the organotypic tissue models have to be adapted accordingly. As already mentioned, a commonly used system to grow human skin models in vitro is the Transwell system [5,45,46]. This system enables the cultivation of the cell culture at an air-liquid interface, which is important to trigger the differentiation of the keratinocytes and the maturation of the *stratum corneum* barrier. Transwell systems are available from different suppliers in formats ranging from 6 to 96 wells within the standard SBS plate dimensions (Table 2).

Table 2 List of common vendors selling Transwell plates

Company	Web-based information
Merck Millipore	http://www.merckmillipore.com/CH/de/product/Millicell-Zellkulturplatten-mit-24--96-Wells,MM_NF-C10584?ReferrerURL=https%3A%2F%2Fwww.google.ch%2F&bd=1
Corning Inc.	https://www.corning.com/worldwide/en/products/life-sciences/products/permeable-supports/transwell-snapwell-netwell-falcon-permeable-supports.html
Neuro Probe Inc.	http://www.neuroprobe.com/product/chemo_tx/
Greiner Bio-one	https://shop.gbo.com/en/germany/articles/catalogue/article-groups/0110_0110_0090/
Essen BioScience	http://www.essenbioscience.com/en/products/reagents-consumables/incucyte-clearview-96-well-cell-migration-plate/

HTS Transwell plates for robotic manipulation exist in a 96-well format with apical and basolateral access ports that allow for automated filling and sampling [47,48]. Assuming that the appropriate protocol for a suitable in vitro skin tissue exists, this setup allows for compound application and harvesting from the basolateral medium reservoir. The tissue constructs have to be removed from the Transwell insert to facilitate lysis or preparation for histological analysis, steps that have proved difficult to automate. Application of lysis buffer alone is not sufficient for full disruption of in vitro-derived skin models, and so, additional mechanical manipulation is required, again making full automation of the process challenging. The harvesting of tissue from Transwell inserts for standard immunohistological analysis is not trivial, especially in a HTS setup. However, customized solutions can be installed that enable the harvesting of the tissues for fixation and embedding [49] .

Alongside standard Transwell systems, human skin microtissues are also available from InSphero AG (Switzerland). Complementing a portfolio of human-derived in vitro 3-D model systems, skin microtissues were developed for applications that require larger throughput (e.g., dose- and time-dependent substance evaluation). Skin microtissues can be obtained with and without the keratinocyte layer, which allows direct comparison to the extent to which the epidermal barrier might impact tissue response. In contrast to Transwell skin models, which are used to test topically administered compounds (including final formulations or lipophilic actives), only soluble substances can be tested with skin microtissues. The tissues are grown to a size of 350–450 μm with a relative standard deviation below 10% and are composed of primary human dermal fibroblasts (dermal microtissues) and keratinocytes (epidermal microtissues). They are maintained in serum-free, defined medium conditions with an in vitro lifetime of at least 14 days. Similar to conventional 2-D cell culture, the application of microtissues in a 96-well format allows a multiparametric analysis of test compounds in a dose-dependent manner in an assay volume of 50–70 μL. The spherical skin microtissues show the expression of dermal matrix and basal lamina proteins and cytokeratin and involucrin occurrence at distinct layers in the epidermis (Fig. 1). Further functional characterization of the tissues

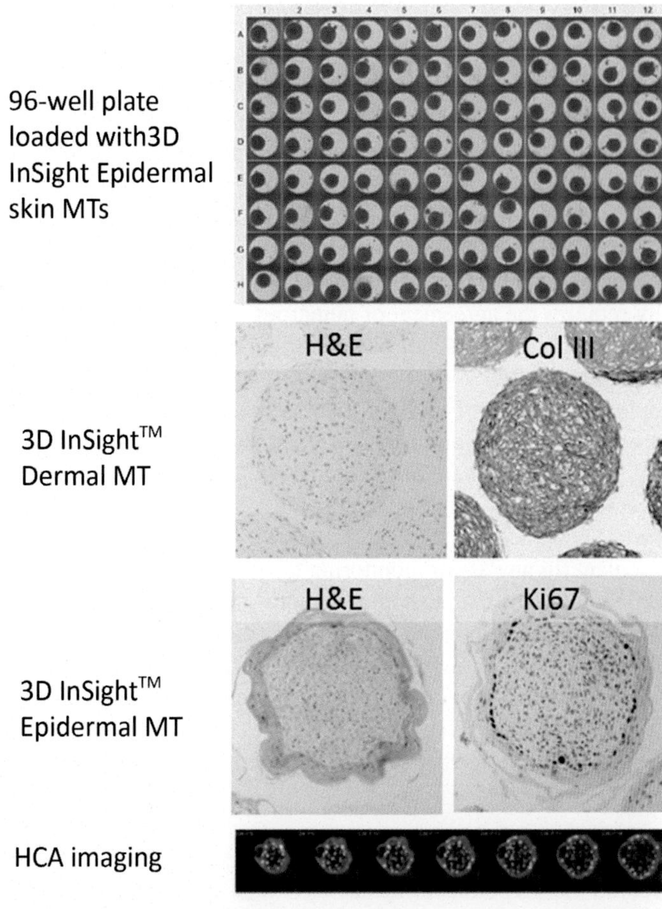

Fig. 1 3D InSight Skin MTs loaded on the GravityTRAP microtissue platform acquired with Cell3iMager plate scanner imaging device (Dainippon Screen) (top). Histological analysis of dermal only microtissues (3D InSight Dermal MT) on H&E and collagen type 3 and skin microtissues (3D InSight Epidermal MT), H&E and Ki67. HCA confocal imaging of 3D InSight skin MT with Opera Phenix (PerkinElmer) at 40× confocal, plane 8–14 μm at distance 2 μm (bottom) (provided by InSphero AG, Switzerland).

with regard to epidermal barrier quality, metabolic competence, and response to different test substances shows the similarity of standard skin tissue equivalents to the native human skin. As skin microtissues are particularly developed for HTS, assay reproducibility and intraplate, interplate, and interbatch variations are basic parameters evaluated at the very beginning of model establishment, in contrast to most Transwell-based tissue systems. Multiple common end points have been established for use with microtissue models that apply also for the skin microtissues, such as viability (CellTiter-Glo, Promega Corp.), histology, efficient RNA extraction, barrier function assays, high-content analysis, or genetic engineering of distinct cell populations within the model (personal communication, InSphero AG).

4. End points, read-out systems, and assay analysis

4.1 Requirements regarding target validation

Screening using phenotypic assays has been recognized as the major source of truly novel pharmaceutical treatments and an important tool for the identification of novel drug targets [50]. One of the major challenges (but also the greatest opportunity) is the identification of the efficacy target of any novel treatment [51]. Methods for target identification span the gamut from genetic to biochemical profiling methods, with no single method being applicable to all assay formats [52]. Compelling target identification usually requires the agreement of numerous combinations of these methods in order to provide supporting evidence from multiple orthogonal assay methods [53] or as exemplified in Ref. [17].

For example, in the case of vitiligo, multiple models were created, and different studies were conducted to try and reveal the molecular cause of the disease [54,55]. Unfortunately, these studies were not coordinated, leading to only a fragmented picture of the overall situation. A full understanding can only be acquired if all data are put into context, taking account of the differences of data acquisition [56].

The requirement for any assay, monitoring the phenotype of interest, to be suitable for target deconvolution, is that it should be robust, that is, the assay should be reproducible and specific. If the target of a congeneric series of compounds is being sought, then a series of related compounds showing a clearly structure-activity relationship (SAR), with active and inactive controls being present in the series, are of great help. Knowledge of the SAR for a given compound series is also helpful to identify sites on the compound suitable for labeling and attachment of affinity capture moieties [52].

4.2 Assay systems

HTS has become the predominant means by which novel drug leads are identified [57]. While there has been some controversy concerning the relative importance of target-based versus phenotypic-based drug discovery programs, in either case, screening remains the most efficient means by which as unbiased a set of treatment types as possible can be surveyed in order to identify the most promising lead.

The requirement for any assay system is that it accurately and reproducibly replicates an aspect of biology, in a quantifiable manner, and that the assay can be scaled up to allow large-scale testing of potential treatments. For pharmaceutical research, the assay system should reflect, as closely as possible, the biology of the human subject and preferentially the patient for whom the treatment is being sought. These requirements have been summarized by researchers from Pfizer as the "phenotypic screening" rule of three [58] that states that phenotypic assays should be disease-relevant, should respond to the endogenous stimuli (or none if that is appropriate), and should monitor a readout relevant for the disease. Such requirements have driven two trends in pharmaceutical assay development in recent years. First, the establishment of assay systems using primary cells or cells is derived from iPS cells [59,60]. The second trend that has developed in the last few years has been the development of 3-D organotypic models that hope to recapitulate the three-dimensional aspects of the organ being studied [61,62]. Not only models of skin disease have to be able to replicate the 3-D structure of the organ with its temporal and structural structure, but also there is increasingly a recognition that the models should include a wide number of different cell types, from hair follicles to immune cells and melanocytes, as recently reviewed by Weitz and Ritchlin [63].

4.3 Standard analysis of in vitro skin models

Not only for skin tissue but also for all other three-dimensional tissue constructs, the spatial distributions of target markers and the morphology of these markers are of interest besides information concerning the distribution and metabolism of an applied compound. During drug development, a deep understanding of the pathways leading to drug action is advantageous and also facilitates the identification of biomarkers; immunohistochemical (IHC) analysis represents the state-of-the-art method for such objectives. This method is well established and belongs to a routine procedure in pathology for evaluating human biopsies. But as immunohistological investigations require tissue fixation, dehydration using solvents, and demasking, these process steps may also disturb the applied substance and its impact leading to a false picture of its distribution and action. In cases where drug delivery processes are investigated, this may lead to biased conclusions. In order to adapt these histology methods to an automated high-content setting, the greatest challenge lies in the harvest of the tissues and the preparation of tissue sections. Further process steps of staining and automated image analysis and quantification are reasonably well addressed by a number of different image analysis products [64].

The methods and challenges facing image analysis of standard tissue culture systems have recently been reviewed [65]. Unfortunately, this article (and others in the field) fails to address the specific challenges faced by image analysis of in vitro skin models. Driven by the need for fast and reliable analysis of tissue biopsies in oncology, advances in imaging technologies have recently occurred. Shinde et al. have demonstrated the identification and quantification of two cancer markers using human skin biopsies. In an automated image analysis setup, they evaluated the outcome of an Aurora A inhibitor for mitotic and apoptotic indexes [66].

In addition to immunohistological characterization, the analysis of secreted cytokines or metabolites of the applied drug can be assessed in the cell culture medium by many, well-established quantitative bioanalytical and chemical methods (e.g., ELISA, HPLC, and GC-MS).

4.4 High-content screening

Image-based descriptions of cellular tissue or organism's phenotype (i.e., high-content screening (HCS)) have become an important approach for exploring biological systems [67,68], as it allows the collection of many features of a biological sample. This has been highlighted by the observation that the majority of first-class drugs were discovered by phenotypic screening, which is heavily reliant on image-based assays [69].

Image-based assays now have a role in all aspects of drug discovery and development, from helping to identify novel targets or mechanisms of action [70] and screening for novel treatments to including aspects of safety assessment [71].

Image-based assays for monitoring different aspects of skin biology often measure features such as cell migration [72,73] or proliferation [74] that can be measured using relatively simple 2-D models of skin cells. Not only single cultures but also a co-culture system of keratinocytes and melanocytes were adapted to be applicable in a high-throughput/HCS to screen 4000 compounds in respect to their ability to modulate pigmentation [75]. In order to visualize and evaluate certain tissue structures of 3-D models on histological sections, fluorescence microscopy is the most common method. In particular, confocal fluorescence microscopy is the state-of-the-art imaging technology when it comes to the visual analysis of 3-D tissue constructs. This technique allows the localization and intensity of fluorescent-labeled markers to be followed. Optical sectioning of the tissue layers is achieved by the combination of point illumination and point detection. Recently, Li et al. (2016) summarized existing 3-D high-throughput imaging and high-content imaging systems [76]. In order to overcome the limitations of excitation beam penetration depth, two-photon excitation fluorescence [77] represents a valuable improvement but will remain unsuitable for high-throughput applications due to the length of time needed for acquisition and data analysis.

Although routinely used, fluorescent-based imaging does have a number of limitations and constraints. Issues such as photobleaching, sensitivity, and autofluorescent properties of the tissue being investigated all serve to limit the sensitivity of this detection technique. Hence, the imaging methods needed to monitor more complex, 3-D models of the skin require more sophisticated imaging techniques (such as 3-D reconstruction of confocal images), which consequently makes quantification and measurement of the monitored phenotype more difficult. One such method for monitoring complex 3-D models is laser light sheet fluorescence microscopy that allows optical rather than physical sectioning of the specimen. As excitation and detection follow different paths and image acquisition does not require line scanning, this technology offers multiple advantages such as reduced time for image acquisition, low levels of photobleaching, and the decrease of phototoxic effects [78]. Many of the challenges facing the conversion of current 3-D image-based assays to a screening format have been reviewed recently [79]. Such complex models are currently not readily amenable to HTS. Not only has the production and handling of the 3-D structures to be automated, but also depending on the system being studied, miniaturization may be limited by the numbers of a given cell type actually being present in the model. Then, the automation of treatment application may be a challenge, as discussed previously, especially if the treatment needs to be spread across the surface of the model. Once the assay has been run, monitoring the desired phenotype of the skin 3-D model may then be limited depending on the disease-relevant markers being monitored, and finally, not only the methods to quantify the amount of an analyte present but also any change in its location within the 3-D structure need to be developed.

Thus, the complexity of 3-D models currently makes truly HTS impossible, and so, one common strategy is to use simpler 2-D models first to identify compounds of interest and to then characterize these compounds further in skin models [80]. Hand in hand with the complexity goes the heterogeneity of the individual organotypic skin models. In order to ensure unbiased conclusions based on the acquired data, all technical aspects in the assay protocols have to be carefully optimized to reduce well-to-well and plate-to-plate variation. One key issue is certainly the choice of the source of keratinocytes, fibroblasts, and other cell types to be incorporated, as mentioned earlier. Concerning image acquisition for traditional insert-grown skin models, the presence of a supportive membrane (polycarbonate) makes image acquisition by an inverted microscope unfeasible. Even in the case of transparent membranes, light scattering might remain an issue. The most striking aspect is the development of software packages that fulfill the technical requirements to analyze 3-D high-content images in respect to segmenting, tracking, visualizing, quantifying, and batch analyzing of the epithelial models. Although software and hardware systems exist for HCS, they have been optimized for 2-D cellular assays, while the transition to 3-D analysis is still ongoing. The field of skin tissue engineering is progressing dynamically in different directions (increase of model complexity, establishment of disease models, and standardization), but evaluation of equipment is lagging.

4.5 Emerging methods for HTS

A number of different mass spectrometry methods are being developed in differing fields that have the potential to be applied to models of skin. CyTOF has been developed primarily for immunologic applications, allowing the detailed characterization of subpopulations of cells and the response of such subpopulations to drugs [81,82]. This method though continues to be developed and has now been extended to allow the detection of antigens present in fixed cell samples [83,84].

In the attempt to analyze skin tissue models in a noninvasive manner, spectroscopic methods are having a greater impact. In order to gain deeper insight into the organization and composition of lipids in the *stratum corneum*, infrared spectroscopy (IR) can be used to deliver molecular information concerning the properties of the epidermal barrier [85]. Using this method, molecular conditions of psoriatic versus normal skin have been evaluated. The results from this study allowed the conclusion that an overall decrease of the structural organization occurs in psoriatic skin tissue, explaining the alteration of its barrier properties [86]. In the context of drug application and the study of ADME, concentration profiles of applied substances can be determined by IR spectroscopy. This method also allows the interaction of the compound with the different epidermal layers to be evaluated [87,88]. As IR spectroscopy exhibits a strong interference with water, it makes measurements in a liquid environment rather impractical [89]. A method that overcomes the limitations of IR spectroscopy is Raman spectroscopy. Raman spectroscopy represents a label-free technique based on the scattering of laser light that enables the identification of chemicals due to their different Raman spectra. This can be used for the identification of pathological alterations in the skin tissue [90], for example, in the case of cancer [91]. Moreover, it is feasible for Raman spectroscopy to identify the differences between in vivo and in vitro skin characteristics, revealing a significant difference concerning natural moisturizing factors [92]. In order to follow the distribution of an applied drug through the different layers of the epidermis, this noninvasive technique allows the quantification of the substance into and across the skin [93,94]. A drawback of this very promising technique persists in the rather long acquisition times and increases the risk of damaging the tissue by the laser [89].

Other methods that do not require novel labeling techniques are also being applied to gain further insight into the distribution of molecules over the surface of the skin [95]. One of the emerging methods in the context of drug delivery is electron paramagnetic resonance (EPR) spectroscopy. As most of the test systems of interest do not possess paramagnetic properties, reporter molecules have to be included [96]. Although EPR holds great promise in detecting the penetration pattern in a time- and cost-saving manner, it is restricted to the *stratum corneum* [96].

The advantage of such methods is that they may allow resolution of an analyst on a cell-by-cell basis, giving the potential for a detailed picture of different cell populations within an organ model noninvasively. However, this very detail offers challenges for the scientist in terms of representation and interpretation. Currently, very few methods have been reported for the analysis of populations of cells and their differing responses to environmental conditions, although efforts are being made in this regard [95].

4.6 Data analysis

Independent of the system and technique of image acquisition being used, the scientists will be confronted with tremendous amounts of data that they have to consolidate to draw a conclusion for further decisions. HTS analysis of skin models will produce confusing quantities of data that cannot be processed manually. In order to meet the demand of software-aided data integration, some open-source platforms were established, supporting 3-D image analysis (ImageJ, BioImageXD, Vaa3D, CellProfiler, and others). In addition to these image analysis packages, a number of open-source tools such as KNIME and Jenkins have been used to link such tools to allow their application to truly high-throughput applications. These tools have even been extended to allow tracking and monitoring of live cells [97]. Moreover, vendors of acquisition hardware also supply software packages that may offer more features compared with the open-source platforms. Referring to 3-D analysis of the tissue constructs, the software programs enable the measurement of shapes, volumes, and distances of the 3-D objects with precision. However, although much effort has been put in during the last decade, all these systems are not yet ready for high-throughput in combination with high-content analysis, which would be needed in the case of skin model evaluation in the context of drug development.

5. Conclusions

This chapter has briefly outlined the requirements for the development of skin equivalent models for research and safety assessment while also reviewing some of the developing assay methods that could be used in the future to further characterize these models.

The application of skin models for HTS holds great potential for the identification of new active compounds treating dermal dysfunctions. As these models are well established and mimic important properties of in vivo skin, they have already shown their relevance in helping to describe some of the basic biological features of the skin. These organotypic tissue models are suitable for ADME and toxicology studies, as multiple approaches have shown. Nevertheless, although such assays are practical and could be transferred to an HTS platform, there are still a few practical issues that have not yet been resolved.

When looking at the individual issues we have identified that need to be addressed in order to bring skin tissue models on to HTS platforms, they seem to be dealt with quite well when considered as a stand-alone solution. For example, there are some disease models of the human skin that are already well defined and characterized [63,98–100]. Some automated liquid-handling and robotic platforms exist that are able to handle the different assay formats used for the production of skin models. Even the issues concerning tissue analysis have recently begun to be addressed by the use of nondestructive methods to allow an unbiased picture of the cellular mechanisms and the temporal and spatial distribution of an applied substance across the epidermal layers.

However, there seems to be a break between the production and quality control of skin models and their application for the testing of the large numbers of samples. As the upstream process of commercially available in vitro skin models is performed by the incorporation of automated process steps, the downstream process of compound evaluation does not seem to have been transferred to robotic platforms in a comprehensive manner. During extensive investigation, two main bottlenecks in the process chain of drug evaluation have been identified, preventing the standardized application of in vitro skin models in HTS scenarios (Fig. 2). First, there is a necessity for community-wide

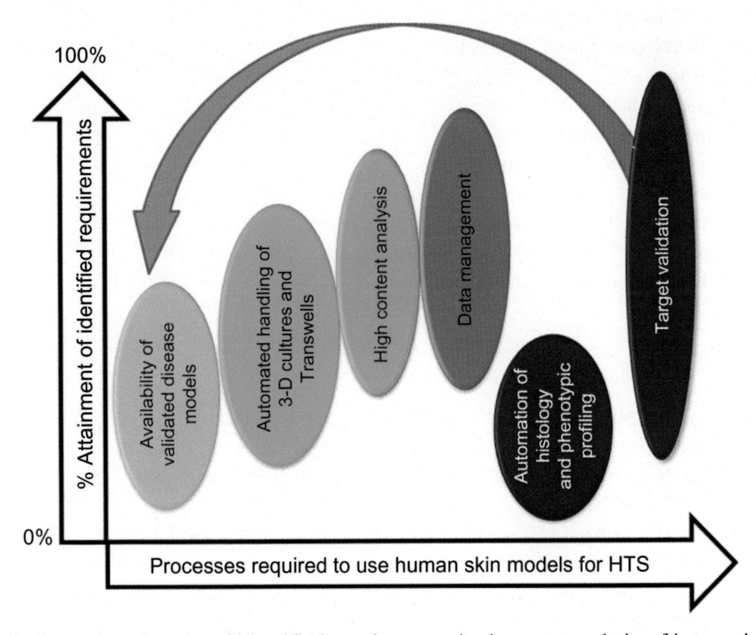

Fig. 2 Estimated attainment of identified requirements in the process chain of integration of in vitro skin tissue models to HTS.

acceptance and agreement concerning the phenotypes and targets to be monitored by such skin models. Community-wide acceptance of these criteria for model validation would then facilitate and enable interbatch comparisons and assay performance. Second, techniques and algorithms for characterization and trusted data analysis of the tissue models, as well as monitoring the interaction of these biological models with applied substances, still have to be adapted for HTS readouts. Developments to address these needs will be a dynamic process and will be driven by the demands of the cosmetic and pharmaceutical industries in order to meet the demands of the society.

Acknowledgments

We gratefully thank Simon Ströbel and Jens Kelm from the InSphero AG (Switzerland) for providing information and data concerning skin microtissues.

Abbreviations

ADME absorption, distribution, metabolism, and excretion
ANSI American National Standards Institute
ELISA enzyme-linked immunosorbent assay
EPR electron paramagnetic resonance
FBS fetal bovine serum
GC-MS gas chromatography-mass spectrometry
HPLC high-performance liquid chromatography
HTS high-throughput screening
IHC immunohistochemistry
iPS induced pluripotent stem cells
IR infrared
PK pharmacokinetics
PD pharmacodynamics
RHE reconstructed human epidermis
RNA ribonucleic acid
SAR structure-activity relationship
SLAS Society for Laboratory Automation and Screening
TERT telomerase reverse transcriptase

References

[1] Agabalyan NA, Borys BS, Sparks HD, Boon K, Raharjo EW, Abbasi S, et al. Enhanced expansion and sustained inductive function of skin-derived precursor cells in computer-controlled stirred suspension bioreactors. Stem Cells Transl Med 2016;6(2):434–43.

[2] Mathes SH, Ruffner H, Graf-Hausner U. The use of skin models in drug development. Adv Drug Deliv Rev 2014;69–70:81–102.

[3] Zhang Z, Michniak-Kohn BB. Tissue engineered human skin equivalents. Pharmaceutics 2012;4(1):26–41.

[4] Kalyanaraman B, Boyce S. Assessment of an automated bioreactor to propagate and harvest keratinocytes for fabrication of engineered skin substitutes. Tissue Eng 2007;13(5):983–93.
[5] Carlson MW, Alt-Holland A, Egles C, Garlick JA. Three-dimensional tissue models of normal and diseased skin. Curr Protoc Cell Biol 2008. [Chapter 19: Unit 19 9].
[6] Roy E, Neufeld Z, Cerone L, Wong HY, Hodgson S, Livet J, et al. Bimodal behaviour of interfollicular epidermal progenitors regulated by hair follicle position and cycling. EMBO J 2016;35(24):2658–70.
[7] Grindon C, Combes R, Cronin MT, Roberts DW, Garrod JF. Integrated decision-tree testing strategies for skin corrosion and irritation with respect to the requirements of the EU REACH legislation. Altern Lab Anim 2008;36(Suppl. 1):65–74.
[8] Fentem JH, Botham PA. Update on the validation and regulatory acceptance of alternative tests for skin corrosion and irritation. Altern Lab Anim 2004;32(Suppl. 1B):683–8.
[9] Macfarlane M, Jones P, Goebel C, Dufour E, Rowland J, Araki D, et al. A tiered approach to the use of alternatives to animal testing for the safety assessment of cosmetics: skin irritation. Regul Toxicol Pharmacol 2009;54(2):188–96.
[10] Kandarova H, Hayden P, Klausner M, Kubilus J, Kearney P, Sheasgreen J. In vitro skin irritation testing: improving the sensitivity of the EpiDerm skin irritation test protocol. Altern Lab Anim 2009;37(6):671–89.
[11] Schlotmann K, Kaeten M, Black AF, Damour O, Waldmann-Laue M, Forster T. Cosmetic efficacy claims in vitro using a three-dimensional human skin model. Int J Cosmet Sci 2001;23(5):309–18.
[12] Nelson K, Lyles JT, Li T, Saitta A, Addie-Noye E, Tyler P, et al. Anti-acne activity of Italian medicinal plants used for skin infection. Front Pharmacol 2016;7:425.
[13] Ascierto PA, Agarwala S, Botti G, Cesano A, Ciliberto G, Davies MA, et al. Future perspectives in melanoma research: meeting report from the “Melanoma Bridge”, Napoli, December 1st–4th 2015. J Transl Med 2016;14(1):313.
[14] Voiculescu V, Calenic B, Ghita M, Lupu M, Caruntu A, Moraru L, et al. From normal skin to squamous cell carcinoma: a quest for novel biomarkers. Dis Markers 2016;2016:4517492.
[15] Lin X, Huang T. Oxidative stress in psoriasis and potential therapeutic use of antioxidants. Free Radic Res 2016;50(6):585–95.
[16] Alepee N, Bahinski A, Daneshian M, De Wever B, Fritsche E, Goldberg A, et al. State-of-the-art of 3D cultures (organs-on-a-chip) in safety testing and pathophysiology. ALTEX 2014;31(4):441–77.
[17] Benam KH, Dauth S, Hassell B, Herland A, Jain A, Jang KJ, et al. Engineered in vitro disease models. Annu Rev Pathol 2015;10:195–262.
[18] Kandarova H, Letasiova S. Alternative methods in toxicology: pre-validated and validated methods. Interdiscip Toxicol 2011;4(3):107–13.
[19] Asbill C, Kim N, El-Kattan A, Creek K, Wertz P, Michniak B. Evaluation of a human bio-engineered skin equivalent for drug permeation studies. Pharm Res 2000;17(9):1092–7.
[20] Pappinen S, Pryazhnikov E, Khiroug L, Ericson MB, Yliperttula M, Urtti A. Organotypic cell cultures and two-photon imaging: tools for in vitro and in vivo assessment of percutaneous drug delivery and skin toxicity. J Control Release 2012;161(2):656–67.
[21] Jain P, Sonti S, Garruto J, Mehta R, Banga AK. Formulation optimization, skin irritation, and efficacy characterization of a novel skin-lightening agent. J Cosmet Dermatol 2012;11(2):101–10.

[22] Laco F, Kun M, Weber HJ, Ramakrishna S, Chan CK. The dose effect of human bone marrow-derived mesenchymal stem cells on epidermal development in organotypic co-culture. J Dermatol Sci 2009;55(3):150–60.
[23] van de Kamp J, Kramann R, Anraths J, Scholer HR, Ko K, Knuchel R, et al. Epithelial morphogenesis of germline-derived pluripotent stem cells on organotypic skin equivalents in vitro. Differentiation 2012;83(3):138–47.
[24] Ouwehand K, Spiekstra SW, Waaijman T, Scheper RJ, de Gruijl TD, Gibbs S. Technical advance: Langerhans cells derived from a human cell line in a full-thickness skin equivalent undergo allergen-induced maturation and migration. J Leukoc Biol 2011;90(5):1027–33.
[25] Tremblay PL, Berthod F, Germain L, Auger FA. In vitro evaluation of the angiostatic potential of drugs using an endothelialized tissue-engineered connective tissue. J Pharmacol Exp Ther 2005;315(2):510–6.
[26] Bellas E, Seiberg M, Garlick J, Kaplan DL. In vitro 3D full-thickness skin-equivalent tissue model using silk and collagen biomaterials. Macromol Biosci 2012;12(12):1627–36.
[27] Huber B, Link A, Linke K, Gehrke SA, Winnefeld M, Kluger PJ. Integration of mature adipocytes to build-up a functional three-layered full-skin equivalent. Tissue Eng Part C Methods 2016;22(8):756–64.
[28] Honardoust D, Kwan P, Momtazi M, Ding J, Tredget EE. Novel methods for the investigation of human hypertrophic scarring and other dermal fibrosis. Methods Mol Biol 2013;1037:203–31.
[29] Abhishek S, Palamadai KS. Epidermal differentiation complex: a review on its epigenetic regulation and potential drug targets. Cell J 2016;18(1):1–6.
[30] Lange J, Weil F, Riegler C, Groeber F, Rebhan S, Kurdyn S, et al. Interactions of donor sources and media influence the histo-morphological quality of full-thickness skin models. Biotechnol J 2016;7:.
[31] Reijnders CM, van Lier A, Roffel S, Kramer D, Scheper RJ, Gibbs S. Development of a full-thickness human skin equivalent in vitro model derived from TERT-immortalized keratinocytes and fibroblasts. Tissue Eng Part A 2015;21(17–18):2448–59.
[32] Iacovides D, Rizki G, Lapathitis G, Strati K. Direct conversion of mouse embryonic fibroblasts into functional keratinocytes through transient expression of pluripotency-related genes. Stem Cell Res Ther 2016;7(1):98.
[33] Gledhill K, Guo Z, Umegaki-Arao N, Higgins CA, Itoh M, Christiano AM. Melanin transfer in human 3D skin equivalents generated exclusively from induced pluripotent stem cells. PLoS ONE 2015;10(8):e0136713.
[34] Itoh M, Umegaki-Arao N, Guo Z, Liu L, Higgins CA, Christiano AM. Generation of 3D skin equivalents fully reconstituted from human induced pluripotent stem cells (iPSCs). PLoS ONE 2013;8(10):e77673.
[35] Petrova A, Celli A, Jacquet L, Dafou D, Crumrine D, Hupe M, et al. 3D in vitro model of a functional epidermal permeability barrier from human embryonic stem cells and induced pluripotent stem cells. Stem Cell Reports 2014;2(5):675–89.
[36] Dickson MA, Hahn WC, Ino Y, Ronfard V, Wu JY, Weinberg RA, et al. Human keratinocytes that express hTERT and also bypass a p16(INK4a)-enforced mechanism that limits life span become immortal yet retain normal growth and differentiation characteristics. Mol Cell Biol 2000;20(4):1436–47.
[37] Petrova A, Capalbo A, Jacquet L, Hazelwood-Smith S, Dafou D, Hobbs C, et al. Induced pluripotent stem cell differentiation and three-dimensional tissue formation attenuate clonal epigenetic differences in Trichohyalin. Stem Cells Dev 2016;25(18):1366–75.

[38] Michael S, Auld D, Klumpp C, Jadhav A, Zheng W, Thorne N, et al. A robotic platform for quantitative high-throughput screening. Assay Drug Dev Technol 2008;6(5):637–57.
[39] Booij TH, Klop MJ, Yan K, Szantai-Kis C, Szokol B, Orfi L, et al. Development of a 3D tissue culture-based high-content screening platform that uses phenotypic profiling to discriminate selective inhibitors of receptor tyrosine kinases. J Biomol Screen 2016;21(9):912–22.
[40] Ramaiahgari SC, den Braver MW, Herpers B, Terpstra V, Commandeur JN, van de Water B, et al. A 3D in vitro model of differentiated HepG2 cell spheroids with improved liver-like properties for repeated dose high-throughput toxicity studies. Arch Toxicol 2014;88(5):1083–95.
[41] Tung YC, Hsiao AY, Allen SG, Torisawa YS, Ho M, Takayama S. High-throughput 3D spheroid culture and drug testing using a 384 hanging drop array. Analyst 2011;136(3):473–8.
[42] Hamilton C. Automation of skin cell cultures. Available from: https://www.hamiltoncompany.com/applications/cell-biology/cell-culture/mammalian-culture-and-maintenance/automation-of-skin-cell-cultures; 2016.
[43] Kimlin L, Kassis J, Virador V. 3D in vitro tissue models and their potential for drug screening. Expert Opin Drug Discov 2013;8(12):1455–66.
[44] Kuratnik A, Giardina C. Intestinal organoids as tissue surrogates for toxicological and pharmacological studies. Biochem Pharmacol 2013;85(12):1721–6.
[45] Kooy AJ, Tank B, Vuzevski VD, van Joost T, Prens EP. Expression of interferon-gamma receptors and interferon-gamma-induced up-regulation of intercellular adhesion molecule-1 in basal cell carcinoma; decreased expression of IFN-gamma R and shedding of ICAM-1 as a means to escape immune surveillance. J Pathol 1998;184(2):169–76.
[46] Wei L, Debets R, Hegmans JJ, Benner R, Prens EP. IL-1 beta and IFN-gamma induce the regenerative epidermal phenotype of psoriasis in the transwell skin organ culture system. IFN-gamma up-regulates the expression of keratin 17 and keratinocyte transglutaminase via endogenous IL-1 production. J Pathol 1999;187(3):358–64.
[47] Corning. Excellent for ADMET screening, drug transport, and other assays. Available from: https://www.corning.com/media/worldwide/cls/documents/lp_HTS_Transwell_96_TissueCultureSystem_ss.pdf; 2004.
[48] Aufderheide M, Forster C, Beschay M, Branscheid D, Emura M. A new computer-controlled air-liquid interface cultivation system for the generation of differentiated cell cultures of the airway epithelium. Exp Toxicol Pathol 2016;68(1):77–87.
[49] Morales AR, Nassiri M. Automation of the histology laboratory. Lab Med 2007;38(7):405–10.
[50] Swinney DC, Anthony J. How were new medicines discovered? Nat Rev Drug Discov 2011;10(7):507–19.
[51] Swinney DC. The contribution of mechanistic understanding to phenotypic screening for first-in-class medicines. J Biomol Screen 2013;18(10):1186–92.
[52] Schirle M, Jenkins JL. Identifying compound efficacy targets in phenotypic drug discovery. Drug Discov Today 2016;21(1):82–9.
[53] Helliwell SB, Karkare S, Bergdoll M, Rahier A, Leighton-Davis JR, Fioretto C, et al. FR171456 is a specific inhibitor of mammalian NSDHL and yeast Erg26p. Nat Commun 2015;6:8613.
[54] Dell'anna ML, Picardo M. A review and a new hypothesis for non-immunological pathogenetic mechanisms in vitiligo. Pigment Cell Res 2006;19(5):406–11.

[55] Schallreuter KU, Bahadoran P, Picardo M, Slominski A, Elassiuty YE, Kemp EH, et al. Vitiligo pathogenesis: autoimmune disease, genetic defect, excessive reactive oxygen species, calcium imbalance, or what else? Exp Dermatol 2008;17(2):139–40 [discussion 41–60].

[56] Dell'anna ML, Cario-Andre M, Bellei B, Taieb A, Picardo M. In vitro research on vitiligo: strategies, principles, methodological options and common pitfalls. Exp Dermatol 2012;21(7):490–6.

[57] Macarron R, Banks MN, Bojanic D, Burns DJ, Cirovic DA, Garyantes T, et al. Impact of high-throughput screening in biomedical research. Nat Rev Drug Discov 2011;10(3):188–95.

[58] Vincent F, Loria P, Pregel M, Stanton R, Kitching L, Nocka K, et al. Developing predictive assays: the phenotypic screening “rule of 3”. Sci Transl Med 2015;7(293):293ps15.

[59] Carlson C, Koonce C, Aoyama N, Einhorn S, Fiene S, Thompson A, et al. Phenotypic screening with human iPS cell-derived cardiomyocytes: HTS-compatible assays for interrogating cardiac hypertrophy. J Biomol Screen 2013;18(10):1203–11.

[60] Sirenko O, Hancock MK, Hesley J, Hong D, Cohen A, Gentry J, et al. Phenotypic characterization of toxic compound effects on liver spheroids derived from iPSC using confocal imaging and three-dimensional image analysis. Assay Drug Dev Technol 2016;14(7):381–94.

[61] Horman SR, Hogan C, Delos Reyes K, Lo F, Antczak C. Challenges and opportunities toward enabling phenotypic screening of complex and 3D cell models. Future Med Chem 2015;7(4):513–25.

[62] Shamir ER, Ewald AJ. Three-dimensional organotypic culture: experimental models of mammalian biology and disease. Nat Rev Mol Cell Biol 2014;15(10):647–64.

[63] Weitz JE, Ritchlin CT. Mechanistic insights from animal models of psoriasis and psoriatic arthritis. Curr Rheumatol Rep 2013;15(11):377.

[64] Mulrane L, Rexhepaj E, Penney S, Callanan JJ, Gallagher WM. Automated image analysis in histopathology: a valuable tool in medical diagnostics. Expert Rev Mol Diagn 2008;8(6):707–25.

[65] Moutsatsos IK, Parker CN. Recent advances in quantitative high throughput and high content data analysis. Expert Opin Drug Discov 2016;11(4):415–23.

[66] Shinde V, Burke KE, Chakravarty A, Fleming M, McDonald AA, Berger A, et al. Applications of pathology-assisted image analysis of immunohistochemistry-based biomarkers in oncology. Vet Pathol 2014;51(1):292–303.

[67] Kummel A, Selzer P, Siebert D, Schmidt I, Reinhardt J, Gotte M, et al. Differentiation and visualization of diverse cellular phenotypic responses in primary high-content screening. J Biomol Screen 2012;17(6):843–9.

[68] Bickle M. The beautiful cell: high-content screening in drug discovery. Anal Bioanal Chem 2010;398(1):219–26.

[69] Swinney DC. Phenotypic vs. target-based drug discovery for first-in-class medicines. Clin Pharmacol Ther 2013;93(4):299–301.

[70] Feng Y, Mitchison TJ, Bender A, Young DW, Tallarico JA. Multi-parameter phenotypic profiling: using cellular effects to characterize small-molecule compounds. Nat Rev Drug Discov 2009;8(7):567–78.

[71] Westerink WM, Schirris TJ, Horbach GJ, Schoonen WG. Development and validation of a high-content screening in vitro micronucleus assay in CHO-k1 and HepG2 cells. Mutat Res 2011;724(1–2):7–21.

[72] Liang CC, Park AY, Guan JL. In vitro scratch assay: a convenient and inexpensive method for analysis of cell migration in vitro. Nat Protoc 2007;2(2):329–33.

[73] Johnston ST, Simpson MJ, McElwain DL. How much information can be obtained from tracking the position of the leading edge in a scratch assay? J R Soc Interface 2014;11(97):20140325.

[74] Gasparri F, Mariani M, Sola F, Galvani A. Quantification of the proliferation index of human dermal fibroblast cultures with the ArrayScan high-content screening reader. Drug Discov Today 2005;(Suppl):31–42.

[75] Lee JH, Chen H, Kolev V, Aull KH, Jung I, Wang J, et al. High-throughput, high-content screening for novel pigmentation regulators using a keratinocyte/melanocyte co-culture system. Exp Dermatol 2014;23(2):125–9.

[76] Li L, Zhou Q, Voss TC, Quick KL, LaBarbera DV. High-throughput imaging: focusing in on drug discovery in 3D. Methods 2016;96:97–102.

[77] Benninger RK, Piston DW. Two-photon excitation microscopy for the study of living cells and tissues. Curr Protoc Cell Biol 2013;1–24 [Chapter 4: Unit 4 11].

[78] Keller PJ, Ahrens MB. Visualizing whole-brain activity and development at the single-cell level using light-sheet microscopy. Neuron 2015;85(3):462–83.

[79] Kriston-Vizi J, Flotow H. Getting the whole picture: high content screening using three-dimensional cellular model systems and whole animal assays. Cytometry A 2016;12:.

[80] Naska S, Yuzwa SA, Johnston AP, Paul S, Smith KM, Paris M, et al. Identification of drugs that regulate dermal stem cells and enhance skin repair. Stem Cell Reports 2016;6(1):74–84.

[81] Chen G, Weng NP. Analyzing the phenotypic and functional complexity of lymphocytes using CyTOF (cytometry by time-of-flight). Cell Mol Immunol 2012;9(4):322–3.

[82] Bodenmiller B, Zunder ER, Finck R, Chen TJ, Savig ES, Bruggner RV, et al. Multiplexed mass cytometry profiling of cellular states perturbed by small-molecule regulators. Nat Biotechnol 2012;30(9):858–67.

[83] David BA, Rezende RM, Antunes MM, Santos MM, Freitas Lopes MA, Diniz AB, et al. Combination of mass cytometry and imaging analysis reveals origin, location, and functional repopulation of liver myeloid cells in mice. Gastroenterology 2016;151(6):1176–91.

[84] Leelatian N, Doxie DB, Greenplate AR, Mobley BC, Lehman JM, Sinnaeve J, et al. Single cell analysis of human tissues and solid tumors with mass cytometry. Cytom B: Clin Cytom 2017;92(1):68–78.

[85] Leroy M, Lafleur M, Auger M, Laroche G, Pouliot R. Characterization of the structure of human skin substitutes by infrared microspectroscopy. Anal Bioanal Chem 2013;405(27):8709–18.

[86] Leroy M, Lefevre T, Pouliot R, Auger M, Laroche G. Using infrared and Raman microspectroscopies to compare ex vivo involved psoriatic skin with normal human skin. J Biomed Opt 2015;20(6):067004.

[87] Mendelsohn R, Flach CR, Moore DJ. Determination of molecular conformation and permeation in skin via IR spectroscopy, microscopy, and imaging. Biochim Biophys Acta 2006;1758(7):923–33.

[88] Mendelsohn R, Chen HC, Rerek ME, Moore DJ. Infrared microspectroscopic imaging maps the spatial distribution of exogenous molecules in skin. J Biomed Opt 2003;8(2):185–90.

[89] Franzen L, Windbergs M. Applications of Raman spectroscopy in skin research-from skin physiology and diagnosis up to risk assessment and dermal drug delivery. Adv Drug Deliv Rev 2015;89:91–104.

[90] Tu Q, Chang C. Diagnostic applications of Raman spectroscopy. Nanomedicine 2012;8(5):545–58.

[91] Lieber CA, Majumder SK, Billheimer D, Ellis DL, Mahadevan-Jansen A. Raman microspectroscopy for skin cancer detection in vitro. J Biomed Opt 2008;13(2):024013.
[92] Fleischli FD, Morf F, Adlhart C. Skin concentrations of topically applied substances in reconstructed human epidermis (RHE) compared with human skin using in vivo confocal Raman microscopy. Chimia (Aarau) 2015;69(3):147–51.
[93] Franzen L, Selzer D, Fluhr JW, Schaefer UF, Windbergs M. Towards drug quantification in human skin with confocal Raman microscopy. Eur J Pharm Biopharm 2013;84(2):437–44.
[94] Pyatski Y, Zhang Q, Mendelsohn R, Flach CR. Effects of permeation enhancers on flufenamic acid delivery in ex vivo human skin by confocal Raman microscopy. Int J Pharm 2016;505(1–2):319–28.
[95] Bouslimani A, Porto C, Rath CM, Wang M, Guo Y, Gonzalez A, et al. Molecular cartography of the human skin surface in 3D. Proc Natl Acad Sci U S A 2015;112(17):E2120–9.
[96] Haag SF, Fleige E, Chen M, Fahr A, Teutloff C, Bittl R, et al. Skin penetration enhancement of core-multishell nanotransporters and invasomes measured by electron paramagnetic resonance spectroscopy. Int J Pharm 2011;416(1):223–8.
[97] Moutsatsos IK, Hossain I, Agarinis C, Harbinski F, Abraham Y, Dobler L, et al. Jenkins-CI, an open-source continuous integration system, as a scientific data and image-processing platform. SLAS Discov 2017;22(3):238–49.
[98] do Nascimento Pedrosa T, De Vuyst E, Mound A, Lambert de Rouvroit C, Maria-Engler SS, Poumay Y. Methyl-beta-cyclodextrin treatment combined to incubation with interleukin-4 reproduces major features of atopic dermatitis in a 3D-culture model. Arch Dermatol Res 2016;309(1):63–9.
[99] van Drongelen V, Haisma EM, Out-Luiting JJ, Nibbering PH, El Ghalbzouri A. Reduced filaggrin expression is accompanied by increased *Staphylococcus aureus* colonization of epidermal skin models. Clin Exp Allergy 2014;44(12):1515–24.
[100] Kumar R, Parsad D, Kanwar A, Kaul D. Development of melanocye-keratinocyte co-culture model for controls and vitiligo to assess regulators of pigmentation and melanocytes. Indian J Dermatol Venereol Leprol 2012;78(5):599–604.

Index

Note: Page numbers followed by *f* indicate figures and *t* indicate tables.

D

E

J

K

L

M

N

S

T

Printed in the United States
By Bookmasters